T0177088

Geophysical Monograph Series

Including

IUGG Volumes
Maurice Ewing Volumes
Mineral Physics Volumes

Geophysical Monograph 129

Environmental Mechanics
Water, Mass and Energy Transfer
in the Biosphere

The Philip Volume

Peter A. C. Raats
David Smiles
Arthur W. Warrick
Editors

Ⓢ American Geophysical Union
Washington, DC

Published in cooperation with CSIRO,
Australia

Published under the aegis of the AGU Books Board

John E. Costa, Chair; Gray E. Bebout, David Bercovici, Carl T. Friedrichs, James L. Horwitz, Lisa A. Levin, W. Berry Lyons, Kenneth R. Minschwaner, Darrell Strobel, and William R. Young, members.

Library of Congress Cataloging-in-Publication Data
Environmental mechanics : water, mass, and energy transfer in biosphere / Peter A. C. Raats, David Smiles, Arthur W. Warrick, editors.

     p. cm. - (Geophysical monograph ; 129)
     "The Philip Volume."
     Includes bibliographical references.
     ISBN 0-87590-988-4
1. Soil moisture. 2. Soil absorption and adsorption. 3. Groundwater flow. I. Raats, P. A. C. II. Smiles, David., 1936- III. Warrick, Arthur W. IV. Series.

S594 .E58 2002
631 .4'32-dc21   2002025428

ISSN 0065-8448
ISBN 0-87590-988-4

Copyright 2002 by the American Geophysical Union
2000 Florida Avenue, N.W.
Washington, DC 20009

Cover image: Thunderstorm over irrigated vineyards near Griffith, New South Wales, Australia. Photo by Greg Heath, CSIRO Land and Water.

Figures, tables, and short excerpts may be reprinted in scientific books and journals if the source is properly cited.

Authorization to photocopy items for internal or personal use, or the internal or personal use of specific clients, is granted by the American Geophysical Union for libraries and other users registered with the Copyright Clearance Center (CCC) Transactional Reporting Service, provided that the base fee of $1.50 per copy plus $0.35 per page is paid directly to CCC, 222 Rosewood Dr., Danvers, MA 01923. 0065-8448/02/$01.50+0.35.

This consent does not extend to other kinds of copying, such as copying for creating new collective works or for resale. The reproduction of multiple copies and the use of full articles or the use of extracts, including figures and tables, for commercial purposes requires permission from the American Geophysical Union.

Printed in the United States of America.

CONTENTS

CONTENTS

John Robert Philip AO FAA FRS

John Philip was Australia's most distinguished environmental physicist; his pioneering study of movement of water, energy, and gases in the natural environment is internationally acclaimed. On Saturday 26 June 1999, he was tragically struck by a car and killed in Amsterdam during a visit to the Centre for Mathematics and Information Science. The present volume is dedicated to John Philip and the scientific work he accomplished.

John Philip displayed prodigious mathematical talent at an early age and won an open scholarship to Scotch College in Melbourne. There his intellectual world expanded enormously, he developed his work ethic, and his English teacher encouraged him to write poetry, which remained a lifetime avocation. John matriculated at 13, and spent a further 2 years at school before he could enter Queens College in the University of Melbourne. He took a Bachelor of Civil Engineering degree at 19.

On graduation, he was appointed to the CSIRO Irrigation Research Station in Griffith to work on the hydraulics of furrow irrigation. With his acute mathematical and physical insights he quickly identified an array of scientific problems about water movement in the soil-plant-atmosphere environment. He remarked, "I blundered into a vocation that turned out, over the past 50 years, to be more fun than work."

He then joined the Queensland Water Supply Commission as a design engineer responsible for the Burdekin and Mareeba Irrigation Schemes. He also developed strong links with Brisbane bohemia, which gave him further entrée to the world of art and ideas and, throughout his life, he remained a catholic reader, a published poet and a connoisseur of architecture.

He was appointed to the Deniliquin laboratories of the CSIRO Division of Plant Industry in 1951. His boss, Sir Otto Frankel, strongly supported scientific freedom to get on with important tasks, but he was ill at ease with John's mathematical and physical approach to environmental problems. This approach owed much to insights and attitudes of L.A. Richards, which were exemplified in the Richards and Wadleigh paper in *Soil Physical Conditions and Plant Growth*. Philip's early paper on the use of mathematics as both a language and a logic reveals his need then to justify his position. Fortunately, Professors Pat Moran and John Jaeger at the Australian National University reassured Otto Frankel, and John Philip followed his scientific instincts.

John's Deniliquin work brought unity to the very complicated field of water and heat flow in unsaturated soils and the wonderful synergy of innovative mathematics and physical insights in the resulting papers culminated in a Doctorate in Science from the University of Melbourne.

Otto Frankel moved John to Canberra in 1959 to establish the Agricultural Physics Section, within Plant Industry and, broadly, to relate findings obtained in the laboratory to the "real" world and to develop mathematical descriptions of environmental processes. The Section represented each of the three components of the Soil-Plant-Atmosphere-Thermodynamic Continuum, a concept conceived by Gradmann and van der Honert, as the "seamless web" connecting these elements of the physical environment. It also included a fourth group, led by John himself. This group was named Applied Mechanics and it provided a framework in theoretical mechanics for the more experimental groups. The Agricultural Physics Section became the CSIRO Division of Environmental Mechanics in 1970.

John's rigorous pursuit of scientific quality ensured that Environmental Mechanics was recognised internationally, if not always within CSIRO, as a centre of excellence and, except for a 5-year period as Director of the CSIRO Institute of Physical Sciences, he was Chief until his retirement.

Distinctions between "basic" and "applied" research and the needs of "stakeholders" were cornerstones of John's philosophy before these ideas became fashionable. He argued that research should be "applicable", and his often-stated motivation was "to improve engineering practice" with mathematical models based on sound physics and a firm focus on Occam's Razor. He encouraged this philosophy within his Division, and sought to provide an environment conducive to creative research. The Pye laboratory, which John and his wife Frances conceived, is an elegant and functional building, cleverly planned to promote interaction between the occupants.

John Philip attached great importance to maintaining a preeminent personal role in research while he was a manager. His papers are skillfully crafted models of brevity and precision and remain among the most frequently cited in environmental science. At the same time he was personally difficult, mischievous and often outrageous. In science he was uncompromising, unforgiving, competitive and petulant. His acknowledgments were parsimonious. He disliked computers and for much of his career relied on a mechanical Monroe calculator he called Marilyn but later accepted an electronic one. He performed calculations lying on the floor, and drew graphs by hand. His "retirement" saw no diminution in the flood of ideas and original, incisive analysis, and he maintained close liaison with scientists round the world.

This complicated man was a Fellow of the Royal Society of London, a Fellow of the Australian Academy of Science, a Fellow of the American Geophysical Union, a Foreign Member of the All-Union (later Russian) Academy of Agricultural Sciences, and only the second Australian Foreign Associate of the U.S. National Academy of Engineering. He was the first non-American recipient of the Robert E Horton Medal, the highest award for hydrology of the American Geophysical Union. He was made Officer of the Order of Australia in 1998 for "services to the science of hydrology."

The Editors

PREFACE

Modern theories of mass and heat transfer in the biosphere, based on notions of a soil-plant-atmosphere thermodynamic continuum focused on water, were generally formulated by the mid-20th century. They tended to be reductionist and flow equations combined macroscopic laws of flow and of material and energy balance. They were difficult to solve because material transfer properties tend to be strongly related to the local concentration of an entity of concern, to the location, or to both. The architecture of the soil and the plant canopy also complicated their formulation, the scale of their application and their test.

Despite the complicated nature of the equations describing transfer processes, since the early 1950s solutions to a large number of problems have been developed. For the earliest solutions, Arnold Klute and John Philip used mixed analytical-numerical methods. Following this, analytical and numerical methods initially developed rather independently, but more recent studies often exploit the synergy of the two types. For nearly 50 years, John Philip played a dominant role among the small, worldwide group seeking analytical solutions. He pursued a wide variety of methods: similarity solutions, asymptotic time invariant travelling waves, linearization by appropriate specialization and/or transformation. As a result we now have good understanding of some key aspects of the hydrological cycle, a sound basis for design and interpretation of experiments seeking the physical properties of soils, and benchmarks against which numerical procedures may be tested.

John Philip was also very actively involved in extending the theory of soil water in numerous new directions, the most important being multiphase flow, simultaneous transport of water and heat, flow of water in soils subject to swelling and shrinkage, transport of solutes in unsaturated soils, and coping with flow and transport at a range of spatial and temporal scales. Furthermore, he made key contributions to two aspects of micrometeorology and physical ecology: two-dimensional transfer associated with advection and transfer of water in the soil-plant-atmosphere continuum.

This monograph assembles papers from across environmental fields influenced by John Philip and where integration of existing and essentially complementary approaches and their application are becoming key issues. A wide-ranging Introduction surveys the current state of water and solute transport in soils and other porous materials. It draws attention to the fundamental change in approach to flow equations and their solutions that has accompanied the development of, and wide access to, computers. It also briefly comments on some current issues in micrometeorology and physical ecology. Some areas where research is required are identified.

Subsequent sections of the monograph explore topical areas of current study and application. The focus of contributions is eclectic and authoritative. They include:

- a variety of analytical and numerical methods used to analyse flow of water in rigid soils: capillary rise and infiltration in uniform and heterogeneous soils, stability analysis of wetting fronts, and two- and three-dimensional flow associated with cavities in soils are illustrated;
- some aspects of equilibrium and flow of water in swelling systems and of solute transport during water flow in rigid and swelling, and saturated and unsaturated soils;
- stability bounds for a saline layer formed during upward flow induced by evaporation at a horizontal surface, analysis of velocity dependent two-dimensional dispersion, and immiscible liquid interactions in porous materials;
- general aspects of liquid flow and the effects of hysteresis and of temperature on equilibrium and flow, relations between behaviour at the Darcy scale and that at larger and smaller scales, and some aspects of water content measurement using heat pulse methods;
- current issues of concern relating to energy and water transfer in the lower atmosphere and in vegetation, and an example of application of theory and technology to water management in irrigated horticulture.

Most papers provide extensive literature reviews and most acknowledge John Philip's influence in their area of focus. Following the Introduction, the monograph leads with two sketches of John Philip's views on science and its application in the natural environment. The attached CD-ROM records his 1995 interview with Steve Burges on behalf of the American Geophysical Union. The CD-ROM also includes a recent address-in-print that reinforces the views expressed in the interview, and a bibliography of his work.

We gratefully acknowledge the encouragement of Rien van Genuchten, who initiated the project and inspired an AGU seminar in San Francisco in December 2000, where many of the papers of the monograph received their initial airing. We are also thankful for the enthusiasm of the authors and for the help of the reviewers who guided them and the Editors, thereby contributing to the quality of the monograph.

The Editors

CD-ROM INFORMATION

The CD-ROM that accompanies this volume includes papers by John Philip's colleagues, Philip's complete bibliography, and a 1995 video interview with Philip. Below, find a more detailed contents list.

A Convergence of Paths That Culminated in John Philip's 1995 Video Recorded History of Hydrology Interview
Stephen J. Burges

Simplification Plus Rigorous Analysis: The Modus Operandi of John Philip
James C. I. Dooge

Physics, Mathematics, and the Environment: The 1997 Priestley Lecture
J. R. Philip
(*Aust. Met. Mag. 47* [1998] 273-283. Copyright Commonwealth of Australia. Reproduced by permission.)

Full Publication List of John Robert Philip

History of Hydrology Film Interviews
John R. Philip interviewed by Stephen Burges, July 5, 1995

Contributions to Environmental Mechanics: Introduction

Peter A. C. Raats

Wageningen University and Research Centre, Wageningen, The Netherlands

David E. Smiles

CSIRO Land and Water, Canberra, Australia

Arthur W. Warrick

Department of Soil, Water and Environmental Science, University of Arizona, Tucson, Arizona

In the second half of the 20th century, environmental mechanics developed from a collection of loosely connected principles and techniques to a coherent quantitative treatment of flow and transport in the soil-plant-atmosphere continuum. John Philip was in many respects the life and soul of this adventure. He contributed foremost to the physics of water in unsaturated soils, but also to micrometeorology and physical ecology. In this introductory chapter we briefly review how his contributions influenced and are related to the activities of his colleagues and provide an overview of the present status of theory of soil water movement. We also indicate how the various contributions to this volume fit in this context. We start with a discussion of the nature, foundation, and application of the Richards equation, with emphasis on the dominant role of John Philip in finding analytical solutions of this equation. This is followed by a discussion of various developments beyond the Richards equation: multiphase flow, simultaneous transport of water and heat, flow of water in soils subject to swelling-shrinkage, transport of solutes in unsaturated soils, and flow and transport processes at various scales in space and time. The varied contributions of John Philip to micrometeorology and physical ecology are also reviewed briefly. In the concluding section, some challenges for environmental mechanics are indicated.

1. INTRODUCTION

Modern theories of heat and mass transfer in the biosphere, based on notions of a soil-plant-atmosphere thermodynamic continuum focused on water, were generally formulated by the mid-20th century. They tended to be reduc-

Environmental Mechanics: Water, Mass and Energy Transfer in the Biosphere
Geophysical Monograph 129
Copyright 2002 by the American Geophysical Union
10.1029/129GM01

tionist and flow equations combined macroscopic laws of material and energy balance and of flow. These equations were difficult to solve because material transfer properties tend to be strongly related to the local concentration of an entity of concern, to the location, or to both. The architecture of the soil and the plant canopy also complicated their formulation, the scale of their application, and their test.

John Philip and a relatively modest group of international colleagues were challenged to solve these equations for practically important conditions and, particularly, to integrate existing and essentially complementary approaches to permit a more holistic approach to vegetative production and management in the natural environment. Originally, the principal practical focus related to irrigation management, often in an environment where energy was advected from neighbouring deserts. The integrating principle was the notion of a thermodynamic continuum that specifically aimed to quantify the passage of water from its arrival at the surface of the soil as rain or irrigation, into the soil, and then back to the atmosphere. Return to the atmosphere might be direct as evaporation through the soil surface or as transpiration through the root, stem and leaves of plants. With computers in their infancy, these environmental physicists established a framework of analysis. The original vision of a comprehensive, integrated environmental model still evades us. Nevertheless, well focussed though more modest models of the biosphere and the advent of much more powerful and accessible computational procedures have greatly enhanced our ability to describe and more sustainably manage our resources.

This introductory chapter briefly examines the scientific context of John Philip's work and identifies how his early ideas have influenced developments in the second half of the 20th century. It also identifies areas where his approach might still resolve problems that prejudice management of land and water in the biosphere. Theory of soil water relations is central to his contributions and we deal with this in some detail. His somewhat lesser contributions in physical ecology and some of their consequences are also surveyed. Throughout, we indicate where the contributions to this volume fit in the wider context of the general physics and, to a lesser extent, the chemistry of the natural environment.

The organization of this introduction is as follows. Section 2 deals with the origin of the Richards equation, the identification of classes of soils based on different mathematical functions approximating the physical properties, and the measurement of soil physical properties and monitoring of processes in the field. Section 3 deals with solutions of the Richards equation, both analytical and numerical, with emphasis on the former. In Section 4 some developments

beyond the classical Richards equation for homogeneous, unstructured, isothermal, rigid soils are reviewed, including multiphase flow, simultaneous transport of water and heat, and flow of water in soils subject to swelling and shrinkage. Some issues related to transport of solutes in unsaturated soils and to flow and transport at various scales in space and time are also discussed. Section 5 deals with two aspects of micrometeorology and physical ecology, two dimensional transfer associated with advection and transfer of water in the soil-plant-atmosphere continuum. Finally, in Section 6 some challenges for environmental mechanics are identified.

2. THE RICHARDS EQUATION

2.1. Formulation of the Theory

About 70 years ago, Lorenzo A. Richards formulated a general, macroscopic theory for movement of water in unsaturated soils [Richards, 1928, 1931]. Richards' theory combines the simplest possible balances of mass, expressed in the equation of continuity, and of momentum, expressed in Darcy's law. Assuming the density ρ of the soil water to be constant, the balance of mass can be expressed as a volumetric balance equation:

$$\frac{\partial \theta}{\partial t} = -\nabla \cdot (\theta \mathbf{v}) - \lambda, \qquad (2.1)$$

where t denotes the time, θ the volumetric water content, $\mathbf{v}$ the velocity of the water, and λ the volumetric rate of uptake of water by plant roots. Further $\nabla = \mathbf{i}_x \partial/\partial x + \mathbf{i}_y \partial/\partial y + \mathbf{i}_z \partial/\partial z$ denotes the vector differential operator, where $\mathbf{i}_x, \mathbf{i}_y, \mathbf{i}_z$ are unit vectors in the orthogonal x, y, z directions. Given the constancy of the density ρ of the soil water, Darcy's law can be written as:

$$\theta \mathbf{v} = -k \nabla h + k \nabla z, \qquad (2.2)$$

where k is the hydraulic conductivity, z the vertical coordinate taken positive downward, and the capillary pressure head h is defined by:

$$h = \frac{p - p_a}{\rho g} = -\frac{p_c}{\rho g}. \qquad (2.3)$$

Here p and p_a are the pressures of the aqueous and gaseous phases, p_c is the capillary pressure, and g is the gravitational constant. The capillary pressure head and the hydraulic conductivity are nonlinear functions of the volumetric water content θ. Moreover, the relationship $h(\theta)$ is hysteretic.

Richards' theory consolidated the efforts of previous generations, notably from Franklin H. King, Charles S. Slichter, Lyman J. Briggs, Edgar Buckingham, Willard Gardner, and

W.B. Haines (see *Philip* [1974a]; *Gardner* [1986]). At the urging of King, *Slichter* [1899] used the mass balance and Darcy's law to describe flow of water in saturated soils. He also calculated the hydraulic conductivity for a packing of spheres based on a model of tubes with variable triangular cross section, thus hitting upon the quadratic dependence of the hydraulic conductivity upon the particle size. *Briggs* [1897] applied the Young-Laplace equation to unsaturated soils, thus discovering in essence the inverse dependence of the capillary pressure head upon the particle size. *Buckingham* [1907] described the hydrostatic vertical equilibrium, the principle of mass balance, and the extension of Darcy's law to unsaturated soils. *Gardner* [1919] was the first to actually write the one-dimensional form of the mass balance equation. *Gardner and Widtsoe* [1921] came particularly close to the modern interpretation of Darcy's law as a macroscopic force balance, when they wrote

> We may therefore say that for the chosen element [of liquid] there exists a force acting vertically downward proportional to the mass, a pressure on each of the six sides, and a frictional drag due to the relative slipping of the element, which may be zero for any or all of the six sides, depending upon the relative velocity at each side.

Haines [1930] applied the Young-Laplace equation to ideal soils, i.e., packings of monodisperse spheres, to calculate water retention, including the hysteresis effect, and the cohesion.

The foundation of the Richards equation on the basis of the principles of surface tension and viscous flow was treated comprehensively by *Miller and Miller* [1956]. They related the water retention and hydraulic conductivity characteristics of geometrically similar soils, each characterized by a length scale. They established that, for geometrically similar soils at the same volumetric water content, the Young-Laplace equation implies that the capillary pressure is inversely proportional to a characteristic length scale and the linearized Navier-Stokes equation implies that the hydraulic conductivity is proportional to its square.

In his studies of flow and transport in porous media, John Philip's primary interest was in the analysis of processes at the macroscopic (Darcian) scale. However, to establish a solid foundation for the macroscopic equations, he also made numerous studies of processes at the pore scale. Like the Miller brothers, he sought the foundation of the extension of Darcy's law to unsaturated soils in the linearity of the Navier-Stokes equation in the limit of zero Reynolds number, i.e. slow viscous flow sometimes referred to as Stokes or creeping flow, and sufficient homogeneity to allow averaging [*Philip*, 1970, 1973]. He early recognized the problem posed by the boundary condition at the fluid-fluid interfaces in pores [*Philip*, 1957c]. Later he explored justification for treating such interfaces as if they were rigid [*Philip*, 1972a, 1972b, 1973], thus putting speculations by *Buckingham* [1907], *Richards* [1931], and *Miller and Miller* [1956] (see also *Raats and Klute* [1968b]) on a firmer footing.

More recently, some further effort has gone into founding Richards' theory on the basis of the principles of surface tension and viscous flow at the pore scale, using the method of volume averaging [*Whitaker*, 1986]. The theory also fits in the framework of the modern continuum theory of mixtures, provided that one recognizes from the outset the existence of the separate solid, liquid, and gaseous phases (for review see *Raats* [1984a, 1998b]). The wide acceptance of the Richards equation is perhaps reflected by the fact that none of the contributions in this book seek to derive it. However, several of the papers deal with aspects such as swelling soils [*Smiles, Grant et al., White*, all this volume], general implications of the Richards equation [*Raats*, this volume], extension to non-isothermal systems [*Grant and Bachmann*, this volume], hysteretic water retention characteristics [*Haverkamp et al.*, this volume], and complications arising from soil structure and spatial variability [*Hopmans et al., Kutilek et al., Cislerova et al*, all this volume].

2.2. Classes of Soils

In the context of Richards equation, the relationships among the volumetric water content θ, pressure head h, and hydraulic conductivity k define the hydraulic properties of a soil. John Philip advocated that the soil physical characteristics preferably be measured, as they generally cannot be predicted. Nevertheless, many solutions of Richards equation hinge upon mathematical functions that match the physical characteristics in a reasonable way. Different classes of soils have been identified with different functions approximating the physical properties. We distinguish two groups of parametric expressions describing the hydraulic properties:

1. A group yielding flow equations that can be solved analytically, in most cases as a result of linearization following one or more transformations;

2. A group that is favored in numerical studies and to a large extent shares flexibility with a rather sound basis in Poiseuillean flow in networks of capillaries.

To the first group belong the classes of: i) linear soils; ii) Green and Ampt or delta function soils; iii) Brooks and Corey or power function soils; iv) Gardner soils; and v) versatile nonlinear soils, including the subclass of Knight soils. For these particular classes of soils, definitions as well as solutions and applications of the corresponding special forms of the Richards equation are discussed in Subsections 3.1, 3.2, 3.5, 3.6, and 3.7.

In the second group of relationships among θ, h and k, the hydraulic conductivity is calculated from the water retention characteristic and certain assumptions concerning the geometry of the pore system (see *Raats* [1992] for a review). The procedure in essence links physico-mathematical models at the Darcy and Navier-Stokes scales. By introducing a very general equation for the water retention characteristic, various well-known classes of soils can be regarded as subclasses of one and the same superclass, thus summarizing a widely scattered literature [*Raats*, 1990a, 1990b, 1992]. The water retention characteristic of this superclass of soils is given by:

$$\lambda h = \left(\frac{1 - S^{b_1}}{S^{b_2}} \right)^{b_3}, \qquad (2.4)$$

where λ is the inverse characteristic length of the soil, S is the reduced water content, and b_1, b_2, and b_3 are empirical parameters. The reduced water content is defined by:

$$S = \frac{\theta - \theta_r}{\theta_s - \theta_r}, \qquad (2.5)$$

where θ_r is the residual water content and θ_s is the water content at saturation. The following table shows values of the indices b_1, b_2, b_3 in equation (2.4) for four important subclasses of this superclass of soils:

Equation (2.4)	b_1	b_2	b_3
Van Genuchten [1980]	b_1	b_1	b_3
Visser [1968], *Su and Brooks* [1975]	1	b_2	b_3
Brooks and Corey [1964, 1966]	∞	b_2	b_3
Brutsaert [1966, 1967], and			
Ahuja and Swartzendruber [1972]	1	1	b_3

Note that the popular subclass of *Van Genuchten* [1980] is included as a special case, as is the *Brooks and Corey* [1964, 1966] or power function class of the first group.

A rather general predictive model for the relative hydraulic conductivity is given by:

$$k_{rel} = S^{c_1} \left(\frac{\int_0^S (\lambda h)^{-c_2}}{\int_0^1 (\lambda h)^{-c_2}} \right)^{c_3}, \qquad (2.6)$$

where c_1, c_2 and c_3 are constants. Introducing the retention curve (2.4) for the superclass of soils in (2.6) gives:

$$k_{rel} = S^{c_1} \left(I_s^{b_1} [(c_2 b_2 b_3 + 1)/b_1, (1 - c_2 b_3)] \right)^{c_3}, \qquad (2.7)$$

where $I_s^{b_1}[\cdot, \cdot]$ is the incomplete beta function. Important special cases of (2.6) and (2.7) are the *Burdine* [1953] model with

$$c_1 = n + m_1 - 1, \qquad c_2 = 2 + m_2, \qquad c_3 = 1, \quad (2.8)$$

and the *Mualem* [1976] model with

$$c_1 = n + m_1 - 2, \qquad c_2 = 1 + m_2, \qquad c_3 = 2, \quad (2.9)$$

where n is the connectivity parameter with $1 < n < 2$, and m_1 and m_2 are tortuosity parameters. Numerical solutions using models implied by (2.4) and (2.6) are discussed in Subsection (3.8).

The description of hysteretic water retention characteristics on the basis of the independent domain model was pioneered by *Collis-George* [1955], *Poulovassilis* [1962] and *Philip* [1964]. Hysteretic water retention characteristics can now be described by various models, such as the modified dependent-domain model [*Mualem*, 1984] and a generalization of the Van Genuchten model [*Dirksen et al.*, 1993]. An alternative generalization of the Van Genuchten model, based on the extrapolation theory of *Parlange* [1976] and supported by a detailed comparison with experimental data, is described by *Haverkamp et al.* [this volume].

The U.S. Salinity Laboratory and the University of California at Riverside organized two major conferences on physical characterization in 1989 and 1997 [*Van Genuchten et al.*, 1992, 1999a]. The models representing the physical properties are now routinely used to handle field data. For example, *Vereecken* [1995] used 11 prediction models to analyze 44 measured curves, arriving at some specific conclusions regarding the importance of particular parameters. With increasing experience, soil survey data can be used to infer physical characteristics using so-called pedo-transfer functions.

2.3. Characterization of Soils and Monitoring of Processes

Experimental methods in soil physics are treated in several books [*Klute*, 1986; *Dirksen*, 2000; *Smith and Mullins*, 2000]. In soil physics research and application, key material properties are measurable in the field and in the laboratory, but generally highly specialized instrumentation is needed. Besides papers, conferences and books, companies initiated and sometimes run by soil physicists, are playing a key role, especially in the USA. The close connections between the research groups and scientific instrument companies have stimulated rapid commercial development and widespread use of new methods. Improvement of already existing methods is of course also strongly stimulated by advances in other sectors of physics and engineering, particularly in electronics and data collection systems.

Thirty-five years ago methods for measuring the water content and the water potential and its components were already available (see e.g. *Rijtema and Wassink* [1969]). The water content was measured not only gravimetrically, but also methods based on scattering of neutrons and adsorption of gamma rays were widely used. The major instrumental development of the last 25 years has been in the realm of dielectric methods, which allow one to infer not only the water content, but also the electrolyte concentration. Tensiometry has changed from mainly water or mercury filled U-tubes or vacuum gauges to electrical transducers, allowing also the development of rapid response micro-tensiometers.

Thus far no routine methods are available for the measurement of the water status in the drier range and of the flux of water. An osmotic tensiometer, intended to cover a wider range than the conventional tensiometer was conceived more than thirty years ago [*Peck and Rabbidge*, 1966, 1969], but a recent paper [*Biesheuvel*, 1999] shows that a practical design is still to be delivered. Direct measurement of the flux of water was pioneered by Richards in the 30s, and by Cary and Dirksen in the late 60s and early 70s (see *Dirksen* [1972] for a brief review). An improved version was developed by *Van Grinsven et al.* [1988] in the context of a study of water balance and chemical budgets of the unsaturated zone. Recently *Brye et al.* [1999, 2000] successfully used equilibrium tension lysimeters to measure and compare drainage from prairie, fertilized no-tillage, and fertilized chisel plow systems.

Inverse methods to determine the physical properties of soils on the basis of equilibrium and steady flow date back to the beginnings of modern soil physics by Buckingham and Richards. The solutions of the Richards equation in the 1950s became the basis for the next generation of methods. An early example is the Boltzmann-Matano method for determining the diffusivity from horizontal absorption experiments [*Bruce and Klute*, 1956; see Subsection 3.3]. This method has been widely used not only for rigid soils, but also for swelling soils and in studies of hydrodynamic dispersion (*Smiles*, this volume; see also Subsections 4.3 and 4.4). Another early example is the outflow method introduced by *Gardner* [1956]. This method was originally based on the linearized theory (see Subsection 3.1), but data are now generally analyzed using numerical solutions of the nonlinear Richards equation for a particular class of soils, such as the Van Genuchten - Mualem class of soils, combined with a procedure for parameter optimization. The evaluation of the parameters in the Green and Ampt model from the solution of infiltration problems and corresponding experiments by

Elrick et al. [this volume] nicely illustrates the tradition of using solutions of a flow equation inversely. In the last quarter century there was much progress with experimental design, both for use in the laboratory and in the field. The combination of analytical or numerical solutions with optimization algorithms led to a wide variety of new methods, while the shortcomings with regard to non-uniqueness were identified and repaired. Several examples of such applications of analytical and numerical solutions will be indicated later. The two Riverside conferences on physical characterization mentioned earlier provide much further details [*Van Genuchten et al.*, 1992, 1999a].

3. SOLUTIONS OF THE RICHARDS EQUATION AND THEIR APPLICATION

3.1. Linear Soils

Both for rigid soils and non-rigid soils (see Subsection 4.3), the earliest solutions were for linearized equations (*Childs* [1936]; *Terzaghi* [1923]). The linear diffusion equation was also used initially by *Gardner* [1956] for his outflow method for estimating D (see Subsection 2.3). For the class of linear soils, the diffusivity $D = kdh/d\theta$ is constant and the hydraulic conductivity k is linear in θ. This leads to a linear Fokker-Planck equation, which can be solved relatively easily. At a 1966 landmark symposium on water in the unsaturated zone [*Rijtema and Wassink*, 1969], three of the seven papers on solutions of the Richards equation were presented by *Philip* [1969a, 1969b, 1969d]. The central idea in these papers is a linearization technique based on matching sorptivities associated with the one-dimensional linear and nonlinear diffusion equations. The linearization was boldly generalized to one-dimensional infiltration and capillary rise and to adsorption and infiltration in two- and three-dimensional systems, based on the philosophy that in all these problems the short time cumulative uptake results from one-dimensional diffusion [*Philip*, 1969c]. The integral mass balance is exact for the sorption problem and quite accurate for the other problems. However, the resulting distribution of the water content is very inaccurate. The paper by *Swartzendruber* [this volume] appears to be the first analytical solution for capillary rise based on a nonlinear form of the Richards equation.

3.2. Green and Ampt or Delta Function Soils

The early work of *Green and Ampt* [1911] was put on a modern footing by *Philip* [1954]. Later *Philip* [1957i]

showed that this work is consistent with Richards equation for the class of Green and Ampt or delta function soils with the diffusivity given by $D = 1/2S^2 (\theta_1 - \theta_0)^{-1} \delta (\theta_1 - \theta)$. This class implies discontinuities of the water content at wetting fronts, but gives reasonable results for one-dimensional, vertical infiltration and absorption [*Philip*, 1969e, 1990]. It also yields very useful results for integral aspects of the water balance, but fails to give any details of the distribution of the water content.

Four contributions to this volume show that the Green and Ampt model remains remarkably vital. *Elrick et al.* [this volume] derive infiltration equations based on only sorption as well as based on sorption and gravity, in both cases for several boundary conditions: constant head, falling head, sequential constant head 1 / constant head 2, and sequential constant head / falling head. *Youngs* [this volume] analyzes flow around tunnel cavities in the capillary fringe on the basis of the Laplace equation for the total head. He notes that this corresponds to using the Green and Ampt model. Several of the studies on stability of soil water flow reviewed by *Parlange et al.* [this volume] use the Green and Ampt model. Examples are the somewhat simplistic analysis by Raats and a subsequent more rigorous analysis by Philip, both using the hydraulic conductivity as a known function of depth. In a generalization of John Philip's analysis, *Tartakovsky et al.* [this volume] consider three-dimensional randomly heterogeneous soils. They derive analytical results for wetting front evolution and complement these with numerical Monte Carlo simulations.

3.3. Early Solutions

It was early realized that the scope of the linear and Green and Ampt models was limited and soon the need to tackle the nonlinear equations was faced up to, for rigid soils starting in the early fifties and for non-rigid soils starting in the late sixties. Most early solutions of the nonlinear Richards equation are of the form [*Raats*, 1988]:

$$z = \zeta_\theta (\theta, t), \qquad z = \zeta_h (h, t). \qquad (3.1)$$

Examples of this are (i) solutions for steady upward and downward flows, (ii) the Boltzmann solution with $t^{1/2}$ proportionality for horizontal absorption, (iii) Philip's series expansion in $t^{1/2}$ for vertical infiltration, and (iv) solutions in the form of traveling, time-invariant waves.

Steady, one-dimensional upward or downward flows have received a lot of attention in locations with shallow water tables in humid regions and irrigation agriculture (see *Raats and Gardner* [1974] for a review).

The Boltzmann solution was first applied by *Klute* [1952; see also *Philip*, 1955a, 1957b, f], following the formulation of the gravity-free form of the Richards equation as a nonlinear diffusion equation by *Childs and George* [1948]. It is well-suited to validate the theory and to establish the physical characteristics of a particular soil [*Bruce and Klute*, 1956]. In his calculations, *Klute* [1952] used the data of *Moore* [1939] for Yolo Light Clay. These same data became the favorite of John Philip, starting with the first paper in his series of papers on infiltration [*Philip*, 1957f] and ending with his paper on redistribution and air diffusion [*Philip and van Duijn*, 1999]. *Constantz* [1987] gives interesting historical background on R.E. Moore and the benchmark soil that resulted from his PhD thesis.

The series expansion in $t^{1/2}$ for vertical infiltration provides a good description for short term infiltration (*Philip* [1957f]; see the reviews by *Philip* [1969e, 1974b, 1988]. Traveling, time-invariant waves are often good approximations of long term water content distributions near wetting fronts [*Philip*, 1957g, h] and regions above both rising and falling water tables [*Childs and Poulovassilis*, 1962; *Raats and Gardner*, 1974]. *Dooge* [this volume] evaluates John Philip's research strategy on the basis of the theory of infiltration. *Burges* [this volume] describes the impact of the two term infiltration equation of *Philip* [1957i] on watershed modelling.

3.4. Approximate Methods Involving Integral Constraints

The first of the two functional relationships expressed in (3.1) is consistent with the water content θ and the time t being the independent variables, while the position z and the flux θv are the dependent variables. With these choices of variables, the volumetric mass balance is expressed as

$$\left(\frac{\partial z}{\partial t}\right)_\theta \doteq \left(\frac{\partial \theta v}{\partial \theta}\right)_t. \qquad (3.2)$$

Integration of (3.2) between θ_∞ and θ gives

$$\theta v - \theta_\infty v_\infty = \frac{\partial}{\partial t} \int_{\theta_\infty}^{\theta} z \, d\theta. \qquad (3.3)$$

This θ-integrated mass balance equation is the common starting point of two approximate, iterative methods that were first introduced in the early 1970s, respectively, by Parlange and by Philip and Knight. The two methods use different constraints in the iterative procedures. Integration of (3.3) with respect to position z gives

$$\int_0^\infty (\theta v - \theta_\infty v_\infty) \, dz = \frac{\partial}{\partial t} \int_{\theta_0}^{\theta_\infty} \frac{1}{2} z^2 \, d\theta =$$

$$\frac{\partial}{\partial t} \int_0^\infty z \left(\theta - \theta_\infty \right) dz. \quad (3.4)$$

Considering a plot of $z^2/2$ versus θ, it is clear that the second and third integrals appearing in (3.4) are equal. This integral moment balance is the constraint used by Parlange in an initial series of 10 papers published in Soil Science in the period 1971-1975, and in numerous later papers (for a review see *Parlange*, [1980]). Alternatively, integration of (3.3) with respect to time t gives

$$\int_0^t \left(\theta_0 v_0[t] - \theta_\infty v_\infty \right) dt = \int_{\theta_\infty}^{\theta_0[t]} z[\theta, t] d\theta, \quad (3.5)$$

where the subscript 0 denotes values at the soil surface. This integral mass balance is the constraint used in the flux-concentration method developed by Philip and Knight in response to the work by Parlange (for a review see *Philip*, [1988]).

The approximate, iterative methods using either one of these two integral constraints have yielded solutions of horizontal absorption and vertical infiltration with a given water content or flux at the soil surface and without or with a surface crust, including soils subject to swelling and shrinkage (see Subsection 4.3). Some of these solutions involve interesting scaling rules. For constant flux, horizontal adsorption and short time constant flux vertical infiltration, the surface flux $\theta_0 v_0$ enters the solution only via the reduced position $Z_{\text{fluxb.c.}}$ and time $T_{\text{fluxb.c.}}$ defined by [*Smiles*, 1978; *White et al.*, 1979; *Perroux et al.*, 1981]:

$$Z_{\text{fluxb.c.}} = (\theta_0 v_0) z, \qquad T_{\text{fluxb.c.}} = (\theta_0 v_0)^2 t. \quad (3.6)$$

For horizontal adsorption and small time vertical infiltration with constant potential boundary conditions via a crust, a constant crust conductance enters the solution only via the reduced position $Z_{\text{crustxb.c.}}$ and time $T_{\text{crustb.c.}}$ defined by *Smiles et al.* [1982]:

$$Z_{\text{crustb.c.}} = \gamma z, \qquad T_{\text{crustb.c.}} = \gamma^2 t. \quad (3.7)$$

Prior to this, flux and crust boundary conditions were mainly dealt with using numerical models. Flux boundary conditions are appropriate in dealing with non-ponding infiltration from rainfall or sprinkling, especially also for establishing the ponding time, i.e. the instant at which ponding starts. Surface crusts have a strong influence on the ponding time and on the subsequent infiltration under ponded conditions. Excellent reviews of physical and chemical aspects of surface crusts are given in the Proceedings of The First International Symposium on Soil Crusting [*Sumner and Stewart*, 1992). Crust formation plays an important role in soil struc-

ture deterioration. Perforation of crusts by soil biota may alleviate the problem.

3.5. *Solutions for the Class of Brooks and Corey or Power Function Soils*

This class has turned up in recent years regularly as resulting from fractal models of pore structure (*Crawford et al.* [1999]; the Preface to this Special Issue of Geoderma gives an eloquent vision of the fractal approach to problems in soil science).

The class of Brooks-Corey power function soils plays a key role in the mathematical literature on the so-called porous medium equation, i.e. the nonlinear diffusion equation with the diffusivity a power function of the volumetric water content, and on the corresponding special case of the Richards equation. The main interest has been in so-called similarity solutions. The roots lie in the 1950's in the work of two mathematicians, the Russian Barenblatt and the Englishman Pattle. Anyone interested in some of the contributions by mathematicians in this area should read the review paper by *Gilding* [1991] on qualitative mathematical analysis of the Richards equation. There one can learn about aspects like generalized solutions, existence, uniqueness, regularity, boundedness in the sense of distributions, wetting fronts / free surfaces / interfaces, comparison theorems for generalized supersolutions and subsolutions. Interestingly, the two integral constraints playing key roles in approximate solutions of the Richards equation, namely the integral moment balance constraint in the work of Parlange and the integral mass balance constraint in the work of Philip and Knight (see Subsection 3.4), also figure prominently in the work of the mathematicians. Further attempts to bridge the gap between the quantitative results sought by soil physicists and the qualitative results of mathematicians appears worthwhile.

An important feature of some of these similarity solutions is the appearance of free surfaces. These were considered initially by *Philip* [1957h] in the third paper of his series of seven papers on the theory of infiltration, where he discussed the special case with the diffusivity corresponding to the initial water content vanishing. He showed that for this case there is a sharp boundary between the region in which flow and wetting occurs and the region in which the initial condition prevails. Nevertheless, *Philip* [1990] felt that in many cases the consideration of free surfaces should and could be avoided.

Ignoring the role of hysteresis, *Philip* [1992a] and *Philip and Knight* [1991] developed similarity solutions for redistribution of finite slugs of soil water applied from instantaneous plane, line, and point sources near the soil surface, including illuminating pictures and examples of practical calculations. Although the results apply to the restricted class of Brooks-Corey power function soils, they could serve as a benchmark for related numerical studies.

Lessoff et al. [this volume] consider purely gravitational infiltration and redistribution in a Brooks and Corey soil. Purely gravitational flows are governed by the kinematic wave equation [*Raats*, 1983]. It can be shown that for purely gravitational drainage of deep profiles in Brooks and Corey soils and in Gardner soils (see subsection 3.6) the average water content above a particular depth is linearly related to the water content at that depth. Such a linear relationship was actually observed in the University of California (Davis) field data that generated the intense interest ever since from soil physicists in spatial variability [*Simmons et al.*, 1979; *Libardi et al.*, 1980].

3.6. Solutions for the Class of Gardner Soils: Quasilinear Analysis of Multi-dimensional Steady Flows

The class of Gardner soils is defined by $k = k_0 \exp \alpha (h - h_0)$, where α is an inverse characteristic length of the soil [*Gardner*, 1958]. For this class, the steady flow equation reduces to the linear equation:

$$\nabla^2 \phi = \alpha \partial \phi / \partial z, \qquad (3.8)$$

where the matric flux potential ϕ is defined by

$$\phi - \phi_0 = \int_{h_0}^{h} k \mathrm{d}h = \int_{\theta_0}^{\theta} D \mathrm{d}\theta. \qquad (3.9)$$

The transformation from h and θ to ϕ is often referred to as the Kirchhoff transformation. This linearization was first noted by *Gardner* [1958] and the first solutions were given by *Philip* [1968b] and *Wooding* [1968] . *Philip* [1988, 1989; see also *Raats* [1988]] and *Pullan* [1990] gave comprehensive reviews, including the scattering analog solutions of Waechter and Philip and the boundary element solutions of Pullan. Numerous further problems have been solved in the last decade. *Basha* [1994, 1999] used the Green's function method to solve problems with rather complicated boundary conditions and root uptake forcing functions. Quasilinear analysis of steady flows can be used for a wide variety of problems: flow from surface and subsurface drip irrigation sources of various geometries; flows to sinks, of interest in connection with the operation of porous cup samplers and suction lysimeters; flows involving specified extraction patterns by plant root systems [*Raats*, 1974,

1982b]; flows around obstructions in the form of solid objects or air filled cavities, especially the unique properties of parabolic barriers (see e.g. *Philip et al.* [1989, 1998d]; *Warrick and Fennemore* [1995]); flows from surface disc and bore-hole permeameters (see next paragraph); steady infiltration and seepage flows with sloping boundaries; analysis of steady air-sparging [*Philip*, 1998a].

Infiltration from a shallow circular pond is both truly challenging and of great practical importance. Using a Hankel transform, *Wooding* [1968] solved this problem with the mixed boundary condition of an equipotential surface representing the pond and a stream surface representing the remainder of the soil surface. For the steady, integral flux into the soil, Wooding found the simple, approximate expression:

$$Q = \left(\pi R^2 + 4R/\alpha \right) k_0 =$$
$$\pi R^2 k_0 + 2\pi R \left(2/\pi \right) \left(k_0/\alpha \right), \quad (3.10)$$

where R is the radius of the pond. On the right hand side of (3.10), the first term represents the contribution from the gravitational force and the second term represents the contribution from capillarity. Later this solution was improved by *Weir* [1987] and recently an alternative derivation was given by *Basha* [1994]. The expression for Q given by (3.10) is the original basis for disc-permeametry [*Scotter et al.*, 1982], a method particularly suited for determining hydraulic properties in the field. *Simunek et al.* [1999] compared three estimates of the hydraulic properties of a crusted soil based, respectively, on a numerical solution, Wooding's analytical solution, and a neural network model requiring textural input information. All three methods gave in essence the same result. *White and Sully* [1987] recast (3.10) in the form

$$Q = \pi R^2 \left(k_0 - k_n \right) + 4 R b S^2 \left(2/\pi \right) \left(\theta_0 - \theta_n \right), \quad (3.11)$$

where the subscripts 0 and n refer to values at the soil surface and initially in the soil, S is the sorptivity, and the constant $b = 0.55$ for field soils. Small time disc permeameter data can be used to estimate the sorptivity [*Minasny and McBratney*, 2000].

Warrick and Or [this volume] review the contributions of John Philip to steady flow from spheroidal sources, without and with the influence of gravity. They also compare the analytical solutions for Gardner soils with numerical solutions for Van Genuchten-Mualem soils. To further explore the limitations of the linearization implied in the Gardner soil model, more such comparisons should be made.

3.7. The Class of Versatile Nonlinear Soils, Including the Subclass of Knight Soils

Philip [1973] noted that assuming the diffusivity to be constant and the hydraulic conductivity to be a quadratic function of the water content leads to the Burgers equation. This equation can be linearized with the Hopf-Cole transformation. John Knight obtained solutions of this Burgers equation for infiltration with constant water content and with constant flux at the soil surface (see *Philip* [1974b]) and for a periodic surface boundary condition (personal communication in 1974). *Clothier et al.* [1981] used the solution for constant flux vertical infiltration to interpret a set of field data. *Philip* [1987b] used the solution for infiltration with constant potential boundary condition to fill the gap between the series expansion in $t^{1/2}$ for small/intermediate time and the traveling, time-invariant wave for large time.

Swartzendruber [this volume] uses the Burgers equation to analyze capillary rise. Motivated by the results of *Philip* [1987b] for infiltration, he hoped the solution would cover the whole range from small time sorption proportional to $t^{1/2}$ to the large time equilibrium distribution. However the approach to the static equilibrium solution is not clear-cut: the cumulative water uptake is found to be proportional to the natural logarithm of the square root of time and thus increases without limit. Further study of this problem for other classes of soils seems appropriate.

Knight [1973] also showed that for gravity free flows and a diffusivity of the form $D = a_1 (\theta^* - \theta)^{-2}$, where a_1 and θ^* are constants, the flow equation can be linearized by the so-called Storm transformation. Based on this, *Knight and Philip* [1974] gave solutions for an instantaneous source and for redistribution in a finite region. More recently this work and work in related fields [*Fokas and Yortsos*, 1982; *Rosen*, 1982; *Rogers et al.*, 1983] became the foundation for several studies of the class of versatile nonlinear soils defined by Knight's $D = a_1 (\theta^* - \theta)^{-2}$ and by $k = a_2 + a_3 (\theta^* - \theta) + a_4/(\theta^* - \theta)$, where a_2, a_3, and a_4 are constants [*Broadbridge and White*, 1988; *White and Broadbridge*, 1988; see also *Sander et al.*, 1988]. For these hydraulic properties the Richards equation can be transformed into the Burgers equation by successively applying the Kirchhoff transformation and the Storm transformation. The Burgers equation can in turn be transformed into the linear diffusion equation by applying the Hopf-Cole transformation. Broadbridge and White give physical arguments to reduce the dependence upon the five parameters a_1, θ^*, a_2, a_3, and a_4 to dependence upon the single parameter C and some easily determined physical properties. It turns out that the parameter C ranges from unity for Green and Ampt soils to infinity for so-called Knight soils.

3.8. Numerical Solutions of Flow Problems

In the 1950s it was already clear that most flow problems in the unsaturated zone require numerical solutions [*Klute*, 1952; *Philip*, 1955a, 1957b]. Yet at the 1966 IASH/AIHS-UNESCO symposium on "Water in the unsaturated zone", Philip made a plea to seek analytical solutions of problems less amenable to analysis than the one-dimensional infiltration problem [*Philip*, 1969a]:

> High-speed computer techniques can, of course, be used to secure solutions to these more intractable problems. One drawback of computer solutions is that, although a large body of numerical results may be obtained, it still remains for the investigator to organize the results and discern a pattern in them. It, therefore, seems at least of equal importance to explore such problems, so far as possible, by analytical means.

These days the interdependence of analytical and numerical techniques often arises when partial and/or approximate analytical results require a complementary numerical analysis. Examples in this volume are i) numerical Monte Carlo simulations complementing the analytical results for wetting front evolution in randomly heterogeneous soils by *Tartakovsky et al.* [this volume]; ii) numerical experiments verifying theoretical stability bounds for a saline boundary layer formed by evaporation induced upward throughflow at a horizontal surface by *Van Duijn et al.* [this volume]; and iii) use of the program DYMSYM in the symmetry analysis of the two-dimensional solute transport equation with velocity-dependent dispersion by *Broadbridge et al.* [this volume].

In the detailed review by [*Braester et al.*, 1971] most studies concerned one-dimensional flow, including complications arising from hysteresis, ponding and moving water tables. *Raats* [this volume] briefly reviews the development of numerical solutions of multi-dimensional problems, starting in 1968 with the alternating direction implicit method and rapidly advancing in the 1970s with the introduction of finite element and control volume methods. These methods, and the fast computers that are now available, have made it possible to efficiently solve transient flow problems involving complications such as (i) multi-dimensional regions that are partly and variably saturated; (ii) spatial variation of soil physical properties; (iii) hysteresis of water retention; and (iv) uptake by plant roots. The popular HYDRUS-2D software package [*Simunek et al.*, 1996] exemplifies the state of the art. *Warrick and Or* [this volume] use it in an analysis of steady infiltration from spheroidal sources. In a review of the usefulness of small-scale hydraulic property measurements for large scale vadose zone modeling, *Hopmans et al.* [this volume] mention the use of HYDRUS-2D, both in laboratory and field studies.

The numerical methods for solution of the Richards equation now seem to have evolved to a satisfactory state and attention of model builders has shifted to post-Richards factors such as multi-phase flow [*White et al.*, 1995; *Lenhard et al.*, 1995], simultaneous transport of water and heat [*Pruess*], 1991], flow in structured media involving local non- equilibrium [*Selim and Ma*, 1998; *Vogel et al.*, 2000], flow in soils subject to swelling and shrinkage [*Garnier et al.*, 1997a], and water uptake by plant roots [*Feddes and Van Dam*, 1999; *Heinen*, 1997; *Heinen and de Willigen*, 1998], and to linkage with other processes such as solute transport, aeration, chemical and biochemical reactions, and activity of plant roots. The next section reviews some of the concepts involved in the post-Richards factors.

4. DEVELOPMENTS BEYOND THE CLASSICAL RICHARDS EQUATION

4.1. Multiphase Flow

Traditionally, multiphase flow has been studied intensely in petroleum engineering. *Tang and Morrow* [this volume] present data related to the effect of brine composition upon the recovery of crude oil from sandstones containing clay. Currently, multiphase models are used often to study the flow of liquid contaminants in aquifers and of air injected below the groundwater table for remediation purposes, so-called air-sparging (see e.g., *Philip* [1998a]). In fact, drawn by funding opportunities, recently some mutually beneficial cross-fertilization between soil physicists and petroleum engineers studying these problems has occurred (see several papers in *Van Genuchten et al.*, [1992, 1999a]).

Following Richards' lead, in soil physics the air pressure is traditionally assumed to be atmospheric everywhere and at any time. Yet, field and laboratory evidence that the gaseous phase is not always at atmospheric pressure date at least from the 1920s. Within the framework of the continuum theory of mixtures it is therefore natural to relax the assumption of atmospheric air pressure and derive expressions for the fluxes of both the water and the air in an unsaturated soil (cf., *Raats*, [1984a, 1998b]):

$$\begin{pmatrix} \theta_w \, \mathbf{v}_w \\ \theta_a \, \mathbf{v}_a \end{pmatrix} = - \begin{pmatrix} k_{ww} & k_{a\,w} \\ k_{wa} & k_{a\,a} \end{pmatrix} \begin{pmatrix} \nabla p_w - \gamma_w \, \mathbf{g} \\ \nabla p_a - \gamma_a \, \mathbf{g} \end{pmatrix}. \tag{4.1}$$

In this equation each of the fluxes depends on both pressure gradients. *Whitaker* [1986] derived this equation on the basis of volume averaging the force balance equations in two fluid phases at the pore scale. In its full generality, equation (4.1) has rarely been applied. However, since the 1930s related expressions, with the off-diagonal terms in the conductivity matrix set equal to zero, have been used in petroleum engineering. Naturally, in soil physics a full-fledged two-phase approach to unsaturated soils has been advocated primarily by people influenced by experience in petroleum engineering (for reviews see *Morel-Seytoux* [1983]; *McWhorter and Marinelli* [2000]). Effects arising from restricted access of air have often been observed in the field and the laboratory. These effects include retarded infiltration and unstable wetting fronts. Particularly interesting effects occur if locally one of the two phases completely fills the pores [*Philip and van Duijn*, 1999]. This was the subject of John Philip's last lecture, on June 24, 1999 at the Centre for Mathematics and Information Science at Amsterdam. It concluded a series of high spirited debates between John and his host Hans van Duijn about the boundary condition for two-phase flow at the interface between two different soils.

4.2. Simultaneous Transport of Water and Heat

Simultaneous flow of water in the liquid phase and diffusion of water vapor in the gaseous phase attracted the early attention of *Philip* [1955b, 1957a]. He estimated the components of the soil water diffusivity function in the liquid and vapor phases for isothermal conditions, using soil hydraulic properties of *Moore* [1939], which he extrapolated to low water contents, and the method of *Childs and George* [1950] to calculate the hydraulic conductivity. The resulting total diffusivity shows a minimum in the region of transition from predominant movement in the vapor phase to predominant movement in the liquid phase. This characteristic feature has since then been verified experimentally for a wide variety of soils and other porous media [*Jackson*, 1964a, b, c; *Scotter*, 1976]. Combining these experiences with those of Dan de Vries in heat conduction and gas diffusion, eventually resulted in their celebrated theory for simultaneous transport of heat and moisture (*Philip and de Vries* [1957]; see also *Philip* [1998c] and references giving there). This theory included the role of liquid islands at the pore scale in enhancing transport of heat and moisture. Water condenses at the upstream side of such islands and evaporates at the downstream side. *Grant and Bachmann* [this volume] critically examine the effect of temperature on capillary pressure.

An alternative approach, based on the theoretical framework of thermodynamics of irreversible processes, was formulated a few years later by Taylor and Cary in the USA and by Mikov and Mikhailov in Russia. In the 1970s these two theories were reconciled as follows (for details see *Raats* [1975] and references given there, especially the PhD thesis by Jury and the papers by Groenevelt and Kay):

1. Balances of mass for the water in *n* distinct phases and a balance of heat for the medium as a whole are formulated.

2. Following Philip and de Vries, it is assumed that the flux of water in each of the phases is proportional to the gradient of the pressure in that phase and that the diffusive component of the flux of heat is proportional to the gradient of the temperature.

3. Clapeyron equations are used to express the gradient of the pressure in any phase in terms of the gradient of the pressure in some reference state and at the same temperature. The reference state may be the water in one of the phases or the water in some measuring device such as a tensiometer or a psychrometer. It then turns out that the resulting expressions, for the total flux of water and for the diffusive flux of heat plus the convective flux of heat associated with the conversion from any phase to the reference state, satisfy the Onsager reciprocal relations.

4. A theorem due to Meixner can be used to delineate the class of transformed fluxes and forces that preserves these relations. Using this theorem, it can be shown in particular that if one follows Philip and de Vries and chooses the gradients of water content and temperature as the driving forces, the Onsager relations are no longer satisfied. But this is no cause for alarm: the expressions for the fluxes, with the gradients of water content and temperature as the driving forces as formulated by Philip and de Vries, remain valid just the same. These expressions merely do not fit in the framework of thermodynamics of irreversible processes.

Using essentially the same approach, *Ten Berge and Bolt* [1988] generalised the theory to include interaction of the liquid phase with the surface of the solid phase. To analyze temperature gradient induced transport in a lyophilic matrix, they considered an idealized capillary model. With a thermal gradient along such a capillary, the surface tension governed flow is from the warm side to the cool side and the thermo-osmosis is in the opposite direction. An uptake of heat is associated with the desorption of water at the warm side. The heat is carried as sensible heat to the cool side, where with the adsorption of water is associated a release of heat. Ten Berge and Bolt show that, for particular choices of the

fluxes and driving forces, the Onsager reciprocal relations turn out again to be satisfied.

For further discussion of the alternative mechanistic and irreversible thermodynamic theories, interested readers are referred to the papers of *Jury and Miller* [1974], *Jury and Letey* [1979], *Milly* [1982], and *Chu et al.* [1983].

4.3. Flow of Water in Soils Subject to Swelling-Shrinkage

Volume change associated with water flow characterizes very large areas of low lying, often organic rich, soils as well as soils with high contents of clay, especially montmorillonite clay. In agriculture, these soils are generally chemically fertile but physically very difficult to deal with because they tend to be sticky when they are wet and very hard and strong when they dry. In addition, many low lying soils with high water and organic matter contents close to the sea have acid properties that are activated when the soil is drained. Understanding of water flow and accompanying volume change is central to management of these soils. Volume change has also been of great historical concern to civil engineers and the first coherent approach seems to have been that of *Terzaghi* [1923], who perceived that in saturated swelling soils consolidation associated with loading could be considered to result from the escape of water. Terzaghi then formulated a linear, one-dimensional theory of consolidation based on material continuity and Darcy's law. There has been a good deal of controversy in engineering literature about what his formulation really entailed. It seems that he assumed the deformation to be so small that it produced no significant geometrical change of the region occupied by the solid phase of the soil and his material coordinates then reduced to a coordinate system attached to the solid phase but moving rigidly in space (see *Raats and Klute* [1968a]).

The adaptation of the theory of Richards for rigid soils to non-rigid soils was not accomplished till the 1960s and early 1970s. It involved almost simultaneous contributions from civil engineers (e.g.*Gibson et al.* [1967]), chemical engineers (e.g. *Atsumi et al.* [1973]), and soil scientists (for reviews see *Philip* [1992b, 1995]; *Philip and Smiles* [1982]; *Raats* [1984a, 1987a, b, 1998b, 2001, 2002]; *Smiles* [1986, 2000a]; *Smiles* [this volume]). This adaptation involves the following ideas:

1. The composition is described in terms of the void ratio and the liquid phase ratio.

2. The solid phase is used as reference continuum: the spatial coordinates are transformed to the material coordinates of the solid phase.

3. Darcy's law is written for the flux of water relative to the solid phase, making it independent of the movement of the observer.

4. The dependence of the capillary pressure head, defined by (2.3), upon the volumetric water content is generalized to a dependence upon the liquid phase ratio and the load, leading to a decomposition of the capillary pressure in a matric component representing the capillary pressure of the unloaded soil and an overburden component representing the effect of the external load.

Compared with the theory of Richards for rigid soils, the additional physical characteristic is the shrinkage curve, i.e., the relationship between the void ratio and the liquid ratio with the load as a parameter. Generally empirical relationships are used to represent the shrinkage characteristics of particular soils. *Grant et al.* [this volume] discuss the shrinkage curves in detail, including the application of the empirical relationship proposed by Groenevelt and Bolt to represent such curves. Interestingly, the energetic discussions surrounding the concept of overburden potential around 1970 still reverberate in their paper. Most available experimental data are restricted to the special case of zero-load. For shrinkage curves at various loads the paper by *Talsma* [1977] still remains the most important source [*Grant et al.*, this volume].

For 1-dimensional flows, the volumetric flux of the water relative to the solid phase is given by (for a detailed derivation see e.g. *Raats* [2002]):

$$\theta \left(v - v_s \right) = -k \left(\frac{\partial h}{\partial z} - 1 \right) =$$
$$-k \left(\frac{\partial h_u}{\partial z} - 1 + \zeta \gamma \right), \quad (4.2)$$

in which h is the pressure head defined by (2.3) and measured by a tensiometer, h_u is the unloaded capillary pressure head, γ is the wet specific gravity of the soil, and ζ is a factor that reflects the way the overburden load is communicated to the soil water. The second equality illustrates how the pressure head h measured by a tensiometer is related to the unloaded capillary pressure head h_u and the overburden. It also reveals the basic analogy with flow in a non-swelling soil. In a saturated soil $\zeta = 1$, in a unsaturated swelling soil $\zeta < 1$, and in a non-swelling soil $\zeta = 0$. Initially, John Philip equated ζ with the slope of the shrinkage curve. Bolt and Groenevelt formally corrected that approximation, but their amended formulation introduces numerical and experimental problems that have yet to be resolved and, operationally

the Philip approximation persists. The issue is revisited by *Grant et al.* [this volume].

Gravity components of the pressure head are reflected in the $(1 - \zeta \gamma)$ term of (4.2), with the second term representing the effort required to raise the wet soil in the gravitational field. Explicitly,

$$(1 - \zeta \gamma) = \left(1 - \zeta \frac{\gamma_s + \vartheta}{e + 1} \right), \quad (4.3)$$

where γ_s is the specific gravity of the solid phase, the void ratio e is the ratio of the porosity and the volume fraction of the solid phase, and the water ratio ϑ is the ratio of the volume fractions of the aqueous and solid phases. For mineral soils, where the solid specific gravity $\gamma_s = 2.6$, this term reduces the net effect of gravity. So, for a saturated potentially acid sulfate soil [*White*, this volume], where $\theta = 0.8 \, (e = \vartheta = 4)$ and $\zeta = 1$, $(1 - \zeta \gamma) = -0.32$ and the overburden effect reverses the effect of gravity. When the soil is unsaturated with $\zeta = 0.3$, and γ_s, then $(1 - \zeta \gamma) = 0.7$ and the effect of gravity is much reduced compared with a non-swelling soil. *John Philip* [1995] was much taken with what he called this "bouleversement" and he explores it in detail. Thus in non-rigid soils, infiltration may be analogous to capillary rise in rigid soils, and long time, traveling waves in non-rigid soils arise when the water moves upward relative to the solid phase. Sedimentation of suspensions is an example of the latter.

The nonlinear, one-dimensional theory has been widely applied. The combination of solutions of boundary value problems with experiments has been used by Smiles et al. for slurries, transferring ideas from soil physics to technological applications (see *Smiles et al.* [1982]; *Philip and Smiles* [1982]; *Smiles* [1986, 2000a]; *Smiles* [this volume]). In many cases the flux-concentration method of Philip and Knight was used (see Subsection 3.4), using analogs of either equations (3.4) or (3.5) as integral constraints. The resulting data amply validate the approach for these materials. *Smiles and Harvey* [1973] derived an expression for the diffusivity in terms of the sorptivity. *Kirby and Smiles* [1988] used the same method to determine the influence of solution salt concentration upon the physical properties of bentonite suspensions. They found that both the dependencies of the unloaded capillary pressure head and the hydraulic conductivity upon the liquid ratio are affected by the solution salt concentration, but that the capillary pressure/conductivity relationship is not. In a study of consolidation of soft sulfidic coastal clay soils, *White* [this volume] shows that the chemical composition has significant practical consequences.

John Philip made key contributions to flow of water in non-rigid soils, mostly in cooperation with David Smiles. It is important to note that these developments have been almost entirely focussed on one-dimensional flow and material characteristics are defined per unit area of cross section of the system. On several occasions John Philip expressed his surprise that experts in mechanics, soil mechanics in particular, were paying little attention to unsaturated soils *Philip* [1973, 1974a]. In a discussion with him at Amsterdam on June 24, 1999, two days before his fatal accident, he strongly expressed his opinion that the area of multidimensional, nonlinear problems for soils subject to swelling and shrinkage was one of the challenges some of us should take up next. Clearly, John still planned to participate in that venture. The limited progress with multi-dimensional problems is reviewed briefly below (see *Raats* [2001, 2002] for more details).

Drying of a non-rigid soil may start as a combination of purely one-dimensional flow of the aqueous phase and deformation of the solid phase. This changes when cracks appear. Most attempts to cope with such systems describe them in terms of one-dimensional flow of the aqueous phase and transversely isotropic deformation of the solid phase. The deformation gradient tensor of the solid phase

$$\mathbf{F}_s = \begin{pmatrix} \frac{\partial x}{\partial X_s} & 0 & 0 \\ 0 & \frac{\partial y}{\partial Y_s} & 0 \\ 0 & 0 & \frac{\partial z}{\partial Z_s} \end{pmatrix}, \qquad (4.4)$$

where $\mathbf{X}_s = (X_s, Y_s, Z_s)$ are the material coordinates of parcels of the solid phase, is then given by:

$$\mathbf{F}_s = \begin{pmatrix} \left(\frac{\partial z}{\partial Z_s}\right)^{\frac{1-n}{2n}} & 0 & 0 \\ 0 & \left(\frac{\partial z}{\partial Z_s}\right)^{\frac{1-n}{2n}} & 0 \\ 0 & 0 & \left(\frac{\partial z}{\partial Z_s}\right) \end{pmatrix}, \qquad (4.5)$$

where

- $n = 1$ for purely axial deformation;

- $n = 1/2$ for balanced axial and lateral deformation;

- $n = 1/3$ for isotropic deformation;

- $n = 0$ for purely lateral deformation.

From (4.4) and (4.5) it follows that for transversely isotropic deformation:

$$J_s = \det \mathbf{F}_s = \frac{\partial x}{\partial X_s}\frac{\partial y}{\partial Y_s}\frac{\partial z}{\partial Z_s} = \left(\frac{\partial z}{\partial Z_s}\right)^{\frac{1}{n}}. \qquad (4.6)$$

From (4.6) and from Euler's equation for the mass density and the definition of the void ratio as the ratio of the porosity and the volume fraction of the solid phase, it follows that:

$$\frac{\partial z}{\partial Z_s} = \left(\frac{\rho_s}{\rho_{s\kappa}}\right)^{-n} = \left(\frac{1+e}{1+e_\kappa}\right)^n. \qquad (4.7)$$

Introducing (4.7) in (4.5) gives:

$$\mathbf{F}_s = \begin{pmatrix} \left(\frac{\rho_s}{\rho_{s\kappa}}\right)^{\frac{n-1}{2}} & 0 & 0 \\ 0 & \left(\frac{\rho_s}{\rho_{s\kappa}}\right)^{\frac{n-1}{2}} & 0 \\ 0 & 0 & \left(\frac{\rho_s}{\rho_{s\kappa}}\right)^{-n} \end{pmatrix}, \qquad (4.8)$$

and

$$\mathbf{F}_s = \begin{pmatrix} \left(\frac{1+e}{1+e_\kappa}\right)^{\frac{1-n}{2}} & 0 & 0 \\ 0 & \left(\frac{1+e}{1+e_\kappa}\right)^{\frac{1-n}{2}} & 0 \\ 0 & 0 & \left(\frac{1+e}{1+e_\kappa}\right)^n \end{pmatrix}. \qquad (4.9)$$

The expression (4.8) for the deformation gradient for the class of transversely isotropic deformations in terms of the mass density was introduced in 1969 in an analysis of axial fluid flow in swelling and shrinking rods (see *Raats* [1969, 1984b]). The equivalent expression (4.9) for the deformation gradient for the class of transversely isotropic deformations in terms of the void ratio is due to *Garnier et al.* [1997a, b]. The r_s-factor, introduced by *Rijniersce* [1983, 1984] in a study of physical changes in the soils of the IJsselmeerpolders following reclamation, is the reciprocal of the parameter n introduced above. More recently, this $r_s = n^{-1}$-factor has been used extensively to characterize the kinematics of swelling and shrinkage in clay soils (see e.g. *Bronswijk* [1990]; *Bronswijk and Evers-Vermeer* [1990]; *Garnier et al.* [1997a, b]; *Kim et al.* [1999]). According to (4.6), the $r_s = n^{-1}$-factor relates one-dimensional vertical deformation measured by $\partial z/\partial Z_s$ to the three-dimensional volume change measured by the Jacobian J_s of the deformation gradient tensor, with $r_s = n^{-1} = 3$ for isotropic shrinkage with cracking and $r_s = n^{-1} = 1$ for one-dimensional subsidence without cracking.

Drainage of mudflats and peatlands generally leads to subsidence. If the subsidence is the result of compaction of sediments then, for a given drainage system, the situation may stabilize eventually. On the other hand, if the subsidence is in part the result of decomposition of organic material, then the intended land use may lead to successive lowerings of the water table, until eventually most organic material has disappeared. The material coordinate approach can be extended to such situations, by using either markers or the non-reactive mineral fraction as the frame of reference [*Raats*, 1998a, 2001]. Subsidence resulting from decomposition of peat is a major factor in coastal areas in the Netherlands, the Fens in England, and the Everglades in the USA.

4.4. Transport of Solutes in Unsaturated Soils

Stimulated by the success of using material coordinates of the solid phase in dealing with swelling soils, around 1975 interest arose in using material coordinates of the soil water as a framework to analyze the transport and reactivity of solutes (see *Raats* [this volume]; *Smiles* [this volume]). John Philip was involved in the earliest studies of hydrodynamic dispersion in unsaturated soil at the Pye Laboratory. These studies showed, theoretically as well as experimentally, that during absorption of a water and a non-reactive solute the water content and the solute concentration both preserve similarity in terms of the Boltzmann variable, i.e., distance divided by the square root of time (*Smiles et al.* [1978]; *Smiles and Philip* [1978]).

Smiles et al. [1981] showed that, for one-dimensional movement of soil water and solute, the solute concentration c satisfies a diffusion equation of the form:

$$\left(\frac{\partial c}{\partial t}\right)_{X_w} = \left(\frac{\partial}{\partial X_w} \theta D \frac{\partial c}{\partial X_w}\right)_t, \qquad (4.10)$$

where X_w is the material coordinate of the water defined by the distribution of the water, θ is the local water content, and D the diffusion/dispersion coefficient of the solute in water. Convective transport and dispersion during constant rate absorption was analyzed in detail by *Smiles et al.* [1981], using the flux-concentration method described in subsection 3.4 to solve the Richards equation for horizontal flow and (4.10) to describe the solute transport.

Some details of the early history of the use of material coordinates of the water to describe transport and dispersion of solutes in unsaturated soils are given by *Raats* [1982a]. Equation (4.10) can be generalized to multi-dimensional flow and transport, possibly including the influence of adsorption and exchange and uptake of water and solute by plant roots [*Raats*, 1987a, b, 2001].

The early papers were the basis for an active program on solute transport and reactivity at Canberra throughout the 1980s, first at the Pye Laboratory and later at the CSIRO Division of Soils (for a summary, see e.g. *Bond et al.* [1990]). Recently the approach has been extended by *Smiles* [2000b, this volume] to hydrodynamic dispersion and chemical reaction in swelling systems where the solid, the water and the solute are all in motion relative to an external observer.

In addition to the paper by (*Smiles*, this volume), five more papers in this volume deal with some aspects of solute transport. *White* [this volume] and *Tang and Morrow* [this volume] consider the influence of the chemical composition upon the hydraulic properties. *Lessoff et al.* [this volume] analyze solute transport in infiltration-redistribution cycles in heterogeneous soils on the basis of three mechanisms: i) advection by gravitational flow of water in a Brooks and Corey soil, regarding the hydraulic conductivity at saturation a random value; ii) linear-equilibrium sorption; and iii) linear decay. This paper is part of a vast body of theory dealing with flow and transport in heterogeneous soils waiting to be applied in the field. *Van Duijn et al.* [this volume] analyze gravitational stability of a saline boundary layer formed by evaporation induced by upward flow at the horizontal surface of a porous medium. Stability criteria are derived, using both the energy method and the method of linearized stability. The results are relevant for evaporating salt lakes (*Wooding et al.*, 1997a, b). This paper deserves attention from soil physicists, since thus far they have rarely considered the possibility of soil solution density driven instability. *Broadbridge et al.* [this volume] discuss analytical solutions for two-dimensional solute transport with velocity-dependent dispersion. Using symmetry analysis, they find new solutions for non-radial solute transport on a background of radial water flow. This Lie group approach is akin to earlier studies for the Richards equation leading to solutions for the class of versatile nonlinear soils, including the subclass of Knight soils, as discussed in sub-section (3.7) (cf., *Sposito*, 1990).

4.5. Flow and Transport Processes at Various Scales in Space and Time

We need information about water movement and solute transport in soils at the *local*, *field* and *regional* scales. For each scale appropriate models are needed, with ample attention for the determination of the parameters in the models. Studies of processes at a particular scale should preferably also pay attention to the related processes at adjoining smaller and larger scales. Researchers increasingly try to come to terms with the dilemma that processes are understood best at small spatial and short temporal scales and that answers are needed to societal questions at large spatial and long temporal scales (see e.g., *Hopmans et al.* [this volume]). However, *Philip* [1975, 1991b] warned that the problems posed may be trans-scientific and he challenged us to provide the best possible judgments based on deep understanding at the small scales. It is perhaps not surprising that the contributions to this book, in line with John Philip's work, concentrate on the scientifically safe, small scales and have, by and large, stayed away from the potentially trans-scientific large scales. Following is a brief review of progress at the various scales in space and time.

4.5.1 The local scale. The *local scale* refers to several scales: the *micro-scale* of solid particles and pores, the *meso-scale* of soil structural elements and of individual roots and associated volumes of soil, and the *macro-scale* of the soil profile and of individual plants. Uptake of water by plant roots is discussed in Section 5 on micrometeorology and physical ecology.

In the older literature, two mechanisms were already invoked to account for water retention at the pore scale: the surface tension at air-water interfaces and the diffuse double layer at solid-water interfaces [*Bolt and Miller*, 1958]. Yet it remained difficult to deal in a quantitative fashion with soils in which both mechanisms operate side by side, until *Philip* [1977a, b, 1978, 1979] initiated a new approach to the interplay of adsorption and capillary condensation. This approach has been pursued vigorously in recent years by Or et al. (cf., *Or and Tuller* [1999]; *Tuller et al.* [1999]; *Tuller and Or* [2000]), yielding not only water retention characteristics, but also hydraulic conductivity characteristics incorporating physico-chemical influences of the solid phase upon the water.

Meso-scale related information is needed whenever representative elementary volumes at the macro-scale lack internal equilibrium. John Philip made an early, detailed analysis of absorption of water in aggregated media [*Philip*, 1968a, c]. For the meso-scale geometries he considered, the characteristic times associated with local equilibration turned out to be small and he concluded that lack of local equilibrium was not a threat to the validity of analyses based on the Richards equation.

Despite this early conclusion, lack of internal equilibrium was suggested regularly by laboratory experiments in the 1960s and early 1970s (see e.g. *Smiles et al.* [1971]) and it received considerable attention in the following decades. The current consensus is that layering, aggregates, cracks, and channels left behind by penetrating roots and burrowing animals often have a large influence on movement of water and transport of solutes in soils. Many models for flow and transport in such soils distinguish a mobile and a stag-

nant phase, roughly corresponding to networks of large and small pores. Important further ingredients in these models are the mechanisms of transport in the mobile phase and the nature of the storage capacities of the phases and the associated exchange between the phases. Such models are often referred to as dual-porosity models. For linear systems these models lead to linear partial integro-differential equations [*Raats*, 1981b; *Van Genuchten and Dalton*, 1986]. For slow motion these equations can be converted into partial differential equations in which appear derivatives of all orders with respect to time. The derivation involves the use of Laplace transforms and is similar to a procedure long familiar in visco-elasticity. Commonly used models, involving an equivalent film resistance between the phases, or an effective dispersion coefficient, or equilibrium between the phases, correspond to successive approximations of these partial differential equations. The results can also be cast in the form of linear systems dynamics, yielding expressions for the transfer functions and the associated moments of the residence time distributions. The theory can be worked out explicitly for various geometries of the stagnant phase, e.g. the slabs considered by *Skopp and Warrick* [1974].

In a wide ranging review paper, *Van Genuchten et al.* [1999b] show that now root water uptake, multicomponent transport, and preferential flow can be dealt with. More complete overviews of flow and transport in structured soils and related image analysis are given in two special issues of the journal Geoderma [*Van Genuchten et al.*, 1990; *Mermut and Norton*, 1992] and a recent book [*Selim and Ma*, 1998]. Unfortunately, effects arising from genuine lack of local equilibrium are difficult to distinguish from effects arising from unstable flows described in the review paper by *Parlange et al.* [this volume].

For the micro- and meso-scales nondestructive techniques, typically adapted from medical technology, are now used to map the pore space and the fluids filling it (see e.g., *Cislerova* [this volume]). There is also an increased interest in flow and transport at the pore scale, using network modelling and percolation theory. Available funding for such studies these days often serves technological interests, such as multiphase flow in petroleum and environmental engineering, and moisture and heat transfer in building materials. With few exceptions, most studies related to agriculture or the natural environment take the macroscopic scale as their starting point, although often using some guidance from the underlying micro- and meso-scales and aiming at extrapolation to the field and regional scales.

4.5.2. The field scale. The *field-scale* is the scale of farm management, especially of soil structure, water, nutrients, pesticides, and herbicides. There is an inherent potential conflict of interest between the farming community and society at large. On the one hand, the farmers' first priority is to economically optimise the yield and the quality of a crop by satisfying the demand of the crop for water, nutrients, proper aeration and trafficability conditions, and minimum interference from plant diseases and weeds. On the other hand, with regard to the use of fertilizers and plant protection chemicals, society increasingly insists on production methods that minimize input, and avoid accumulation in the soil and emission to groundwater, surface water and the atmosphere. With respect to plant nutrients, traditional field plot research served the farmers interest simply by relating the amount of nutrients supplied at the beginning, and perhaps during the growing season, to crop yield and quality, without paying much attention to emissions. But the current wider interests are better served by in-depth experimental studies of soil structure and of accumulation, uptake, transformation, accumulation, and leaching processes in the soil-plant-atmosphere system. Field-scale implementations of local-scale models, such as the two-dimensional HYDRUS-2D are the proper basis for the design and interpretation of detailed field experiments and for extrapolation from the limited number of experiments one can afford to carry out. *Jarvis* [1999] shows that preferential flow models also have reached the stage of being applicable for management purposes. Such models are particularly needed for the evaluation of the impact of temporal variations in soil structure by farm management practices and natural processes. Field data and models are also becoming key elements in site specific farming technology.

The main bottleneck in field scale studies is determining the parameters in the models. Sometimes systems approaches can be used, directly linking inputs and outputs of the system by means of transfer functions. This is generally possible if all transport and reaction processes are linear. Linearization has also been dominant in stochastic approaches to flow in heterogeneous media. Going beyond this, recent papers by *Severino and Santini* [1999] on nonlinearly reactive transport by means of temporal moments and by *Attinger and Kinzelbach* [1999] deriving effective transport parameters for non-linear transport in heterogeneous porous media represent important advances. These advances critically depend on the large time structure of solutions of nonlinear transport equations worked out by mathematicians [*Dawson et al.*, 1996; *Van Duijn et al.*, 1997]. These advances also illustrate the synergy of mathematical and computational efforts.

4.5.3. The regional scale. The *regional scale* is the scale of greatest interest to policy makers. It is the scale at which pollution of ground and surface waters and damage to biotopes become most evident. Developing and implementing models at the regional scale remains a big challenge. It appears that at the regional scale the concept of a single model with effective parameters generally becomes untenable and that regional models by necessity are aggregates of representative elements (fields). *Carrera and Medina* [1999] give an excellent review of calibration of regional groundwater models. They stress the proper selection of processes consistent with the grid size, the importance of time variability, the role of geology in defining spatial variability, the treatment of areal recharge and river-aquifer interactions, and the use and role of concentration and temperature data. The overall impression is a severe shortage of sufficiently detailed data, resulting in correspondingly severe nonuniqueness in calibration exercises.

5. MICROMETEOROLOGY AND PHYSICAL ECOLOGY

John Philip is best known for his research in soil physics and hydrology, but he also made benchmark contributions in micrometeorology, in plant water relations, and in physical aspects of ecology. Much of this work originated in the semi-arid irrigation area of Deniliquin, where John was located at the Regional Pastoral Laboratory of the CSIRO Division of Plant Industry in the period 1951-1958, during the last three years of this period together with Dan de Vries. That association led both independently to explore advection [*De Vries* 1959, *Philip* 1959], the classic example of which is an irrigation area in the middle of a desert, and the way in which soil and atmosphere interact to control evaporation from soils and wilting in plants.

John Philip's influence in the physics of the biosphere was different in style from that in soil water physics, where he contributed aggressively to the science for more than 40 years. In micrometeorology his role, after some incisive applications of his considerable knowledge of diffusion theory, was that of influential critic and patron of studies that he perceived to be central to the hydrological cycle. Under his aegis systematic measurement of heat and mass transfer and theoretical development were enthusiastically supported and he encouraged his people to see beyond what he called "flat-earth micrometeorology" with the focus on vertical flux-gradient relationships in a homogeneous, semi-infinite plain. Rather, he encouraged study of the complexities arising from surface heterogeneity that are characteristic of practical applications of micrometeorology. As he put it in his 1959 paper on local advection, "In this real world, irrigated fields adjoin deserts, reservoirs are of finite extent, dry lands exist beside seas, and cornfields beside close-grazed pasture".

Environmental physics is a wide field that includes soil physics and we do not attempt its survey here. Rather we focus on two related areas that John Philip influenced and which are the subject of papers here. The first dealt with the physical nature of heat and matter transfer in the biosphere and lower atmosphere and here he focussed on two-dimensional transfer associated with advection. The second major area of Philip's concern was the soil-plant-atmosphere continuum (SPAC) of Gradmann and van den Honert (see *Kirkham* [this volume]), where he saw an energetically unifying structure within which, particularly, water transfer might be considered. In both areas he made important theoretical contributions where existing methods of measurement prejudiced critical test of theory.

John Philip's unique contribution to basic micrometeorology was to derive quasi-analytical solutions of 2-dimensional, steady convection-diffusion equations of the form

$$u \frac{\partial \psi}{\partial x} = \frac{\partial}{\partial z} K \frac{\partial \psi}{\partial z}, \qquad (5.1)$$

which were believed to describe transfer of heat and matter in the air flow above surfaces representing sources and sinks for the entity of concern [*Philip*, 1959]. In this equation, x and z are the horizontal and vertical directions, u is the mean wind speed, K is the vertical component of the eddy diffusivity, and ψ is the concentration of the entity, such as heat or water vapor, that concerns us; windward diffusion is neglected. The form of (5.1) is the same as that of the one-dimensional diffusion equation with time replaced by the horizontal coordinate x. He sought simultaneous solutions for humidity and heat that also satisfied the surface energy balance. As he later [*Philip*, 1987a] pointed out, the use of an eddy diffusivity to represent turbulent transport is demonstrably inexact, but he sought justification in its merit of providing a simple analysis of a complicated process. *Novak* [this volume] also argues this view. Both *Philip* [1959] and *De Vries* [1959], who attacked the same problem at the same time, used power law profiles for the vertical profiles of u. Only Philip, however, apparently following *Timofeev* [1954], took advantage of a power law for the K profile to solve (5.1). His solutions, for the first time, provided concentration fields downwind of a change in surface type for radiation, concentration, and flux boundary conditions. Companion papers covered direct experimental tests of the theory in landmark experiments conducted by colleagues, downwind of the junction between tarmac and grass at Canberra airport (see for example, *Rider et al.* [1963]). Some difficulties in Philip's original analysis were examined

in later studies, where he showed, for example, that effects of surface resistance associated with soils, water bodies and leaves could strongly reduce advection close to the leading edge, but less so at larger distances downwind. In recent years, he also explored development of the boundary layer and blending heights for checkerboard patterns, where the wind blows across many alternating surfaces with different properties [*Philip*, 1996a, b, 1997a, c].

Three related papers in this volume exemplify aspects of the scientific issues that remain of concern. *Raupach* [this volume] explores the transfer of finely divided and non-buoyant solids in a turbulent air flow. The problem is central to wind erosion and transport of soil and the approach also applies to the movement and fate of spray and other droplets in the air. *Finnigan* [this volume] discusses the effect of topography on the wind flow and momentum transfer to the terrain and offers an entrée to air movement in complex terrain. *Novak* [this volume] follows the Philip reasoning that one might use gradient based eddy facilitated diffusion cautiously, knowing it to be incorrect, when the scale of turbulence approximates that of the diffusive mean flow. The approach is justified by its working reasonably well.

John Philip was also concerned with the effect of the structure of plant stands and included interactions between canopy geometry and source-sink distributions for heat, water vapor and CO_2 within the canopy [*Philip*, 1966]. He also raised concerns about John Monteith's pragmatic proposal that an overall "crop resistance" might deal with these complexities. Philip argued that the neglect of canopy geometry and the simplification of two-dimensional canopy exchange processes inherent in Monteith's one-dimensional, "big leaf" model were unrealistic and possibly misleading and that the zero-plane could not have the physiological significance claimed for it. Monteith later still maintains [*Monteith and Unsworth* 1990] that effects of crop resistance can be inferred from micrometeorological measurements of the heat and water vapor fluxes above the canopy and profiles of temperature and humidity extrapolated to a zero-plane endowed with the physiological significance of a leaf. The model forms a firm foundation for the approach of *Green et al.* [this volume] to water management in olives as well as to other horticultural crops.

As first enunciated, the SPAC concept recognized that water in the soil, the plant and the atmosphere forms a continuum on a thermodynamic basis. Water passes from one domain to the next along gradients of water potential. To get some insight in the acquisition of water from the soil, both *Philip* [1957d] and *Gardner* [1960] considered uptake of water by a plant root with radius r_0 surrounded by a hollow cylinder of soil with radius r_1. They assumed the plant roots to be relatively thin and the spatial distribution of the roots to be relatively sparse. Then $r_0 << r_1$, so that the geometrical number $\rho_0 = r_0/r_1 \to 0$ and the volume fraction of the soil occupied by the roots is negligible. The solution for a line sink embedded in a soil of infinite extent is an appropriate approximation [*Carslaw and Jeager*, 1959]:

$$\theta_i - \theta = \frac{r_1^2 \lambda}{4D} \int_{\frac{r^2}{4Dt}}^{\infty} \frac{e^{-x}}{x} dx = \frac{r_1^2 \lambda}{4D}(-\text{Ei}(-\frac{r^2}{4Dt})), \quad (5.2)$$

where $\text{Ei}(\ldots)$ is the exponential integral. For small values of $r_1^2/(4Dt)$, and thus for large values of t, this solution can be approximated by:

$$\theta_i - \theta = \frac{r_1^2 \lambda}{4D}(\ln\frac{4Dt}{r^2} - \gamma + \frac{4Dt}{r^2} + O\frac{4Dt}{r^2}^2) \approx$$
$$\frac{r_1^2 \lambda}{4D}(\ln\frac{4Dt}{r^2} - \gamma). \quad (5.3)$$

This two-term approximation was used by *Gardner* [1960] to calculate the water depletion pattern around an individual root. Earlier *Philip* [1957d] used this solution to evaluate the time dependence of the water content at the soil-root interface. Philip was particularly interested in the instant $t = t_w$ at which the water content $\theta = \theta_0$ at the soil-root interface reaches the value θ_{0w} at which the plant wilts. From equation (5.3) at that instant:

$$\theta_i - \theta_{0w} \approx \frac{r_1^2 \lambda}{4D}(\ln\frac{4Dt_w}{r_0^2} - \gamma). \quad (5.4)$$

The overall water balance implies that at time $t = t_w$ the total volume λt_w taken up is equal to the average depletion $(\theta_i - \bar{\theta_w})$, so that

$$t_w = \frac{(\theta_i - \bar{\theta_w})}{\lambda}. \quad (5.5)$$

Introducing (5.5) in (5.4) and solving for the average water content $\bar{\theta}$ at wilting gives:

$$\bar{\theta}_w = \theta_i - \frac{e^\gamma r_0^2 \lambda}{4D}\exp(\frac{4D(\theta_i - \theta_{0w})}{r_1^2 \lambda}). \quad (5.6)$$

The older literature of plant physiology and soil science regarded the water content at which a plant wilted to be a soil property, called the wilting point of a soil. *Philip* [1957d] used (5.6) to demonstrate that the average water content at wilting does not only depend on the soil properties θ_{0w} and D, but also on the geometrical parameters of the plant root system, i.e. the radii r_0 and r_1, and the water demand λ. He concluded that 'uncritical use of the "wilting point" as an

invariant index of the lower limit of the availability of soil moisture to plants can be very misleading'. Despite this pioneering study, Philip in later years became rather skeptical about models at the scale of individual roots. In fact, when he returned to the problem of uptake of water by plants in recent years [*Philip*, 1991a, 1997b], he introduced the root extraction rate directly as a sink term in the flow equation, recognizing, of course, that it depends on the water status of the soil, the characteristics of the soil and the root system, and the meteorological conditions.

Gardner [1960] used the two-term approximation (5.3) has a point of departure for a simpler model in which the uptake is treated as a series of steady state flows in cylindrical shells of soil surrounding the roots with the soil-root interface at the inner edge and the water coming from the outer edge. This model has been a benchmark for more complicated models involving geometrical complications and upscaling (see *Raats*, 1990a).

The analysis by Philip and Gardner predicted that water was not equally available to plants in the range between field capacity and the permanent wilting point. It also anticipated that transpiration could be restricted and plants could wilt over a wide range of soil moisture contents, depending on root density, the soil hydraulic properties and the evaporative demand of the atmosphere. *Denmead and Shaw* [1962] verified these predictions. These early studies provided a point of departure for many current models of evaporation and soil water balance and set the scene for a dynamic approach to plant water relations, which prevailed for many years. More recent research indicates that other factors such as the rate of CO_2 assimilation, atmospheric humidity, soil water status and soil strength also influence stomatal closure. Uptake of water is complicated by factors such as poor contact between roots and soil and influences from solutes (for discussion of such aspects see *Gardner* [1991] and *Raats* [1990a]). *Feddes and Van Dam* [1999] evaluate in detail the influence of soil water and salinity status on water uptake. The challenge now is to marry both physical and physiological influences into a coherent framework of plant water relations. At the same time, the paper by *Green et al.* [this volume] shows how application of the current relatively simple concepts, together with careful measurement of plant physiological processes, can already be used to better manage water in an orange plantation.

Finally, John Philip offered analyses of processes and measurements required by a thoroughgoing theory of heat and mass transfer in the biosphere. Examples include his treatment of evaporation from bare soil as a constant rate phase in which the evaporation rate is that from a saturated surface and is determined only by atmospheric conditions, and a falling rate phase in which the evaporation rate depends only on the hydraulic properties of the soil [*Philip*, 1957d]. This simple, but elegant, either/or description of the evaporation behavior of drying soils allows relatively easy parameterization of the time course of soil evaporation from field experiments and, as in the case of the SPAC, the subsequent development of models of soil evaporation and soil water balance. He also developed a theory of heat flux measurement needed to close the surface energy balance based on eddy correlation measurements of sensible and latent heat fluxes. In support of experimental techniques, he analyzed the damping of turbulent fluctuations in atmospheric scalar concentrations imposed by drawing air through a tube for remote measurement [*Philip*, 1963]. The work illustrates his extraordinary ability to perceive the essential physics in many measurement problems. His theoretical analysis is applied in modern closed-path gas analyzers for purposes of eddy correlation to reconstruct from the measured damped concentration fluctuation of CO_2 and water vapor the sampled turbulent fluctuations [*Leuning and King*, 1992].

6. CHALLENGES FOR ENVIRONMENTAL MECHANICS

John Philip frequently asserted that the point of his science was its practical application. In general, soil physics is a relatively mature discipline that is becoming the basis for engineering technology that can be routinely applied across a number of fields. Examples are found in areas such as intensive horticulture, and the use of land to ameliorate poor quality water and to use nutrient rich effluents of urban or agricultural origins. In the case of intensive horticulture, relatively uniform porous media and an engineered environment permit highly efficient and productive use of scarce resources. These developments presuppose that we understand the whole-plant physiology in the context of such a managed environment. The challenge is attaining a product of desired quality by matching the physiological needs of the plant to the engineered properties of the environment. For example, partial root zone drying, where the root system of grapes is divided so that the different halves are alternately stressed and irrigated out of phase, is an interesting technique of irrigation and water stress manipulation to optimize water use and enhance product quality.

In the natural environment, issues are not so clear and discriminating application of existing knowledge is as uncertain as is the application of good science to influence land managers and people who set public policy. It remains important however to attempt to optimise both biological and environmental outcomes. Field measurement is costly but important. Models are often pretty persuasive, but it remains important that models are systematically tested in the field.

This challenge to ensure that conventional and well-tried science is used to advantage remains restricted by some significant conceptual, as opposed to implementation, problems. They tend to focus on water, which, increasingly is becoming a critically important commodity to be used advisedly and carefully. The problems include:

6.1. Issues of scale and heterogeneity. We still have difficulties dealing with the issues of scale and heterogeneity. They range from problems of preferred pathway flow in an otherwise homogeneous or deterministically variable profile to problems that limit extension of deterministic models derived at the Darcy scale to wider areas of the landscape, such as a catchment or a region. Approaches to these problems are explored in recent workshop proceedings edited by *Feyen and Wiyo* [1999] and *Van Genuchten et al.* [1999a]. An approach based on so-called pedo-transfer functions, that purport to relate hydraulic characteristics to soil types, remains most promising where geomorphological variation is limited. Empirical models may remain the only recourse at a regional scale.

This issue extends to transfer of water and heat in plant canopies. It relates to the consequences of the distribution and orientation of leaves and assemblages of leaves and to the physiological variation across plants and populations of plants that effect, for example, stomatal transfer of gases and water vapor. Tensions between the empirical usefulness of Monteith's extended leaf model as a practical way to deal with small-scale heterogeneity within the canopy and the uncertain physical and physiological bases remain to be resolved. This also applies to tensions exemplified by *Novak* [this volume] between a diffusion model based on eddy diffusivity, that is demonstrably inexact physically but practically expedient, and more 'correct' but much more complicated statistical approaches.

6.2. Swelling soils. Soils that change volume with water content also present a significant challenge. They represent some of the most productive lands, with large areas in Eastern Europe, North America, India, and Australia. Water management lies at the core of their sustainable and productive use but the consequences of volume change including cracking present challenges that have yet to be dealt with. Philip and Smiles presumed that generally flow in these soils is best treated as 1-dimensional vertical. This sensible presumption brings with it problems of 3-dimensional aggregate volume change within the profile, the appearance of cracks as the profile dries out, and local disequilibrium of water content and pressure head within and between aggregates. Systematic measurement of water content is a problem, as is the scale at which the key profile properties of water content and bulk density should be expressed. Practical models of flow in these systems have been formulated but basic issues of theory remain to be accepted. In particular, the use of mass based coordinates, that ensure that material continuity is respected, are rarely used, although the consequences in terms of the water balance are significant. There is a need for comprehensive field studies of swelling soils. In such studies the focus on the overburden and unloaded water potentials should be reconsidered in favor of Ed Miller's suggestion that we focus on things we can measure like the manometric pressure [*Miller*, 1975]. And finally, John Philip's challenge that we study multi-dimensional problems more fully should be faced up to.

6.3. Water extraction by plant roots. These problems are analogous to those in cracked soils. Plant root systems impose meso-scale distributions of water content and pressure head, determined by evaporative demand, root distribution, and local hydraulic properties. Upscaling from the meso-scale to the more practical macro-scale remains difficult. Thus far the consequences of the root distribution seem not to offer much better estimates than Wilford Gardner's model of 40 years ago. *Gardner* [1991] is probably correct in asserting that further progress requires incorporation of the influence of solutes on water uptake.

6.4. Processes in the rhizosphere. Although this volume does not deal with it, the nature and function of the rhizosphere seems an area where the laws of transfer at a Darcy scale have uncertain application and where the dynamics of interacting populations in a managed or natural physical environment are important for land management. The rehabilitation of denuded sites, such as mined areas, demands the re-establishment of an ecosystem rather than the mere application of seed and fertilizer.

Much of environmental science touched on in this monograph emanated from the application of reductionist scientific principles to environmental processes. The origins of this work lie early in the last century and developed greatly during John Philip's most creative decades from 1955 to 1985. Integration of these basic notions in well-defined situations has also been successful so that in many areas, environmental science now forms a sound basis for environmental engineering. At the same time issues of scale and heterogeneity, swelling and shrinkage of soils, and uptake of water by plant roots present conceptual problems and recourse to quite basic but well founded theory often offers useful if only semi-quantitative guidance as well as test of black box models, which may be the only other approach. It is therefore important not to forget where we have been and the bases of the science. Application is also affected by the social prejudice against science [*Smiles et al.*, 2000], that *Medawar* [1985] calls 'postural anti-scientism', which

interferes with application of important insights, particularly in the area of public policy.

Acknowledgements. We wish to acknowledge very useful advice from John Philip's micromet colleagues at Canberra, including Frank Bradley, Tom Denmead, John Finnigan and Mike Raupach.

REFERENCES

Ahuja, L.R., and D. Swartzendruber, An improved form of the soil-water diffusivity function, *Soil Sci. Soc. Amer. Proc.*, *36*, 9-14, 1972.

Atsumi, K., T. Akiyama, and S. Miyagawa, Expression of thick slurry of titanium dioxide in water, *J. Chem. Eng. of Japan*, *6*, 236-240, 1973.

Attinger, S., and W. Kinzelbach, Effective transport parameters for nonlinear transport in heterogeneous porous media, in *Modelling of Transport Processes in Soils at Various Scales in Time and Space*, edited by J. Feyen and K. Wiyo, Proc. Int. Workshop of EurAgEng's Field of Interest on Soil and Water, held 24-26 November 1999 at Leuven, Belgium, pp. 501-508, Wageningen Pers, Wageningen, The Netherlands, 1999.

Basha, H.A., Multidimensional steady infiltration with prescribed boundary conditions at the soil surface, *Water Resour. Res.*, *30*, 2105-2118, 1994.

Basha, H.A., Multidimensional linearized nonsteady infiltration with prescribed boundary conditions at the soil surface, *Water Resour. Res.*, *35*, 75-83, 1999.

Biesheuvel, P.M., R. Raangs, and H. Verweij, Response of the osmotic tensiometer to varying temperatures: modeling and experimental validation, *Soil Sci. Soc. Am. J.*, *63*, 1571-1579, 1999.

Bolt, G.H., and R.D. Miller, Calculation of total and component potentials of water in soil, *Trans. Amer. Geophys. Union*, *39*, 917-928, 1958.

Bond, W.J., and I.R. Philips, Ion transport during unsteady water flow in an unsaturated clay soil, *Soil Sci. Soc. Am. J.*, *54*, 636-645, 1990.

Braester, C., G. Dagan, S. Neuman, and D. Zaslavsky, *A Survey of the Equations and Solutions of Unsaturated Flow in Porous Media*, First Annual Report (Part 1), Project No. A10-SWC-77, Grant No. FG-Is-287 made by USDA under PL480, Technion Israel Institute of Technology and Research and Development Foundation LTD, Hydrodynamics and Hydraulic Eng. Lab., 1971.

Briggs, L.J., *The Mechanics of Soil Moisture*, USDA Division of Soils, Bull. 10, 24 pp., Government Printing Office, Washington, D.C., 1897.

Broadbridge, P., and I. White, Constant rate rainfall infiltration: a versatile nonlinear model I. Analytic solution, *Water Resour. Res.*, *24*, 145-154, 1988.

Bronswijk, J.J.B., Shrinkage geometry of a heavy clay soil at various stresses, *Soil Sci. Soc. Am. J.*, *54*, 1500-1502, 1990.

Bronswijk, J.J.B., and J.J. Evers-Vermeer, Shrinkage of Dutch clay soil aggregates, *Neth. J. Agric. Sci.*, *38*, 175-194, 1990.

Brooks, R.H., and A.T. Corey, *Hydraulic Properties of Porous Media*, Hydrology Paper 3, 27 pp., Colorado State University, Fort Collins, Colorado, 1964.

Brooks, R.H., and A.T. Corey, Properties of porous media affecting fluid flow *J. Irrigation and Drainage Div., Proc. Amer. Soc. Civil Engs.*, *92(IR2), 61-88*, 1966.

Bruce, R.R., and A. Klute, The measurement of soil water diffusivity, *Soil Sci. Soc. Am. Proc.*, *20*, 458-462, 1956.

Brye, K.R., J.M. Norman, L.G. Bundy, and S.T. Gower, An equilibrium tension lysimeter for measuring drainage through soil, *Soil Sci. Soc. Amer. J.*, *63*, 536-543, 1999.

Brye, K.R., J.M. Norman, L.G. Bundy, and S.T. Gower, Water-budget evaluation of prairie and maize ecosystems, *Soil Sci. Soc. Amer. J.*, *64*, 715-724, 2000.

Brutsaert, W., Probability laws for pore size distributions, *Soil Sci.*, *101*, 85-92, 1966.

Brutsaert, W., Some methods of calculating unsaturated permeability, *Trans. Amer. Soc. Agr. Eng.*, *10*, 400-404, 1967.

Buckingham, E., *Studies on the movement of soil moisture*, USDA Bureau of Soils, Bull. 38, 61 pp., Government Printing Office, Washington, D.C., 1907.

Burdine, N.T., Relative permeability calculations from pore size distribution data, *Petroleum Trans., Amer. Inst. Mining Eng.*, *198*, 71-77, 1953.

Carrera, J., and A. Medina, A discussion on the calibration of regional groundwater models, in *Modelling of Transport Processes in Soils at Various Scales in Time and Space*, edited by J. Feyen and K. Wiyo, Proc. Int. Workshop of EurAgEng's Field of Interest on Soil and Water, held 24-26 November 1999 at Leuven, Belgium, pp. 629-640, Wageningen Pers, Wageningen, The Netherlands, 1999.

Carslaw, H.S., and J.C. Jaeger, *Conduction of Heat in Solids, 2nd. Ed.*, Oxford University Press, London, 1959.

Childs, E.C., The transport of water through heavy clay soils: I, *J. Agr. Sci.*, *26*, 114-127, 1936.

Childs, E.C., and A. Poulovassilis, The moisture profile above a moving water table, *J. Soil Sci.*, *13*, 272-285, 1962.

Childs, E.C., and N. C. George, Soil geometry and soil-water equilibria, *Faraday Soc. Disc.*, *3*, 78-85, 1948.

Childs, E.C., and N. Collis-George, The permeability of porous materials, *Proc. Roy. Soc.*, *201A*, 392-405, 1950.

Chu, S. Y., G. Sposito, and W.A. Jury, The cross-coupling transport coefficient for the steady flow of heat in soil under a gradient of water content, *Soil Sci. Soc. Am. J.*, *47*, 21-25, 1983.

Clothier, B.E., J.H. Knight, and I. White, Burgers equation: application to field constant-flux infiltration, *Soil Sci.*, *132*, 252-261, 1981.

Collis-George, N., Hysteresis in moisture content-suction relationships in soils, *Proc. Natl. Acad. Sci. India*, *24A*, 80-85, 1955.

Constantz, J., R.E. Moore and Yolo Light Clay, in *The History of Hydrology*, edited by E.R. Landa and S. Ince, *History of Geophysics*, *3*, 99-101, American Geophysical Union, Washington DC, 1987.

Crawford, J.W., Ya.A. Pachepsky, and W.J. Rawls (Eds.), Special Issue: Integrating Processes in Soils Using Fractal Models, *Geoderma*, *88*, 109-362, 1999.

Dawson, C.N., C.J. van Duijn, and R.E. Grundy, Large time asymptotics in contaminant transport in porous media, *SIAM J. Appl. Math.*, *56*, 965-993, 1996.

Denmead, O.T., and R.H. Shaw, Availability of soil water to plants as affected by soil moisture content and meteorological conditions, *Agron. J.*, *54*, 385-390, 1962.

De Vries, D.A., The influence of irrigation on the energy balance and the climate near the ground, *J. Meteorol.*, *16*, 256-270, 1959.

Dirksen, C., A versatile soil water flux meter, in *Proc. Second IAHR-ISSS Symposium on Fundamentals of Transport Phenomena in Porous Media*, edited by D.E. Elrick, Vol. 2, pp. 425-442, University Guelph, Canada, 1972.

Dirksen, C., *Soil Physics Measurements*, Catena Verlag, Reiskirchen, Germany, 2000.

Dirksen, C., J.B. Kool, P. Koorevaar, and M.Th. Van Genuchten, HYSWASOR-Simulation model of hysteretic water and solute transport in the root zone, in *Water Flow and Solute Transport in Soils*, edited by D. Russo and G. Dagan, pp. 99-122, Springer-Verlag, New York, N.Y, 1993.

Feddes, R.A., and J.C. Van Dam, Effects of plants on the upper boundary condition, in *Modelling of Transport Processes in Soils a Various Scales in Space and Time*, edited by J. Feyen, and K. Wiyo, Proc. Int. Workshop of Eur-AgEng's Field of Interest on Soil and Water, held 24-26 November 1999 at Leuven, Belgium, pp. 391-405, Wageningen Pers, Wageningen, The Netherlands, 1999.

Feyen, J., and K. Wiyo (Eds.), *Modelling of Transport Processes in Soils a Various Scales in Space and Time*, Proc. Int. Workshop of Eur-AgEng's Field of Interest on Soil and Water, held 24-26 November 1999 at Leuven, Belgium, Wageningen Pers, Wageningen, The Netherlands, 1999.

Fokas, A.S., and Y.C. Yortsos, On the exactly solvable equation occurring in two phase flow in porous media, *Siam J. Appl. Math.*, *42*, 318-332, 1982.

Gardner, W., The movement of moisture in soil by capillarity, *Soil Sci.*, *7*, 313-317, 1919.

Gardner, W., and J.A. Widtsoe, The movement of soil moisture, *Soil Sci.*, *11*, 215-232, 1921.

Gardner, W.H., 1986. Early soil physics into the mid-20th century, *Advances in Soil Sci.*, *4*, 1-101, 1986.

Gardner, W.R., Calculation of capillary conductivity from pressure plate outflow data, *Soil Sci. Soc. Am. Proc.*, *20*, 317-320, 1956.

Gardner, W.R., Some steady-state solutions of the unsaturated moisture flow equation with application to evaporation from a water table, *Soil Sci.*, *85*, 228-232, 1958.

Gardner, W.R. Dynamic aspects of water availability to plants, *Soil Sci.*, *89*, 63-73, 1960.

Gardner, W.R., Modeling water uptake by roots, *Irrig. Sci.*, *12*, 109-114, 1991.

Garnier, P., E. Perrier, R. Angulo Jaramillo, and P. Baveye, Numerical model of 3-dimensional anisotropic deformation and 1-dimensional water flow in swelling soils, *Soil Sci.*, *162*, 410-420, 1997a.

Garnier, P., M. Rieu, P. Boivin, M. Vauclin, and P. Baveye, Determining the hydraulic properties of a swelling soil from a transient evaporation experiment, *Soil Sci. Soc. Am. J.*, *61*, 1555-1563, 1997b.

Gibson, R.E., G.L. England, and M.J.L. Hussey, The theory of one-dimensional consolidation of saturated clays, *Geotechnique*, *17*, 261-273, 1967.

Gilding, B.H., Qualitative mathematical analysis of the Richards equation, *Transport in Porous Media*, *5*, 561-566, 1991.

Green, W.H., and G.A. Ampt, Studies in soil physics, 1. The flow of air and water through soils, *J. Agr. Sci.*, *4*, 1-24, 1911.

Haines, W.B., Studies in the physical properties of soils, V. The hysteresis effect in the capillary properties and the modes of moisture distribution associated therewith, *J. Agric. Sci. (Cambridge)*, *20*, 97-116, 1930.

Heinen, M., *Dynamics of Water and Nutrients in Closed, Recirculating Cropping Systems in Glasshouse Horticulture, with Special Attention to Lettuce Grown in Irrigated Sand Beds*, PhD Thesis, Wageningen Agricultural University, Wageningen, The Netherlands, 1997.

Heinen, M., and P. De Willigen, FUSSIM2: A two-dimensional simulation model for water flow, solute transport, and root uptake of water and nutrients in partly unsaturated porous media, *Quantitative Approaches in Systems Analysis*, No 20, 140 pp., DLO Research Institute for Agrobiology and Soil Fertility and the C.T. de Wit Graduate School for Production Ecology, Wageningen, The Netherlands, 1998.

Jackson, R.D., Water vapor diffusion in relatively dry soil: I. Theoretical considerations and sorption experiments, *Soil Sci. Soc. Am. Proc.*, *28*, 172-176, 1964a.

Jackson, R.D., Water vapor diffusion in relatively dry soil: II. Desorption experiments, *Soil Sci. Soc. Am. Proc.*, *28*, 464-466, 1964b.

Jackson, R.D., Water vapor diffusion in relatively dry soil: III. Steady state experiments, *Soil Sci. Soc. Am. Proc.*, *28*, 467-470, 1964c.

Jarvis, N.J., Using preferential flow models for management purposes, in *Modelling of Transport Processes in Soils at Various Scales in Time and Space*, edited by J. Feyen and K. Wiyo, Proc. Int. Workshop of EurAgEng's Field of Interest on Soil and Water, held 24-26 November 1999 at Leuven, Belgium, pp. 521-535, Wageningen Pers, Wageningen, The Netherlands, 1999.

Jury, W.A. and E.E. Miller, Measurement of the transport coefficients for coupled flow of heat and moisture in a medium sand, *Soil Sci. Soc. Am. Proc.*, *38*, 551-557, 1974.

Jury, W.A. and J. Letey, Water vapor movement in soil: reconciliation of theory and experiment, *Soil Sci. Soc. Am. J.*, *43*, 823-827, 1979.

Kim, D.J., R. Angulo-Jaramillo, M. Vauclin, J. Feyen, and S.I. Choi, Modelling of soil deformation and water flow in swelling soil, *Geoderma*, *92*, 217-238, 1999.

Kirby, J.M., and D.E. Smiles, Hydraulic conductivity of aqueous bentonite suspensions, *Aust. J. Soil Res.*, *26*, 561-574, 1988.

Klute, A., A numerical method for solving the flow equation for water in unsaturated materials, *Soil Sci.*, *73*, 105-116, 1952.

Klute, A. (Ed.), *Methods of Soil Analysis, Part 1, Physical and Mineralogical Methods*, Agronomy Monograph 9, 2nd ed., American Society of Agronomy, Madison, Wisconsin, USA, 1986.

Knight, J.H., *Solutions of the Nonlinear Diffusion Equation: Existence, Uniqueness, and Estimation*, PhD Thesis, Australian National University, Canberra, ATC, Australia, 1973.

Knight, J.H., and J.R. Philip, Exact solutions in nonlinear diffusion, *J. Eng. Math.*, *8*, 219-227, 1974.

Lenhard, R.J., M. Oostrom, and M.D. White, Modeling fluid flow and transport in variably saturated porous media with the STOMP simulator: 2. Verification and validation exercise, *Adv. Water Resour.*, *18*, 365-373, 1995.

Leuning, R., and K.M. King, Comparison of eddy-covariance measurements of CO_2 fluxes by open- and closed-path CO_2 analyzers, *Boundary-Layer Meteorol.*, *59*, 297-311, 1992.

Libardi, P.L., K. Reichardt, D.R. Nielsen, and J.W. Biggar, Simple field methods for estimating hydraulic conductivity, *Soil Sci. Soc. Am . J.*, *44*, 3-7, 1980.

McWhorter, D.B., and F. Marinelli, Theory of soil-water flow, in *Agricultural Drainage*, edited by R.W. Skaggs and J. van Schilfgaarde, *Agronomy Monograph*, *38*, American Society of Agronomy, Crop Science Society of America, and Soil Science Society of America, Madison, Wisconsin, USA, 2000.

Medawar, P.B., *The Limits of Science*, Oxford University Press, Oxford, UK, 1985.

Mermut, A.R., and L.D. Norton (Eds.), Special issue on "Digitization, processing and quantitative interpretation of image analysis in soil science and related areas", *Geoderma, 53*,(3-4), 1992.

Miller, E.E., Physics of swelling and cracking soils *J. Colloid Interface Sci., 52*, 434-443, 1975.

Miller, E.E., and R.D. Miller, Physical theory of capillary flow phenomena, *J.Appl. Phys., 27*, 324-332, 1956.

Milly, P.C.D., Moisture and heat transport in hysteretic, inhomogeneous porous media: a matric head-based formulation and a numerical model, *Water Resour. Res., 18*, 489-498, 1982.

Minasny, A.R., and A. McBratney, Estimation of sorptivity from disc permeameter measurements, *Geoderma, 95*, 305-324, 2000.

Monteith J.L., and M.H. Unsworth, *Principles of Environmental Physics (2nd Edn)*, 291 pp., Edward Arnold, London, 1990.

Moore, R.E., Water conduction from shallow water tables, *Hilgardia, 12*, 383-426, 1939.

Morel-Seytoux, H.J., Multiphase flows in porous media, in *Developments in Hydraulic Engineering, 4*, edited by P. Novak, pp. 103-174, Elsevier, London, 1983.

Mualem, Y., A new model for predicting the hydraulic conductivity of unsaturated porous media, *Water Resour. Res., 12*, 513-522, 1976.

Mualem, Y., A modified dependent-domain theory of hysteresis, *Soil Sci., 137*, 283-291, 1984.

Or, D., and M. Tuller, Liquid retention and interfacial area in variably saturated porous media: Upscaling from single-pore to sample scale model, *Water Resour. Res., 35*, 3591-3606, 1999.

Parlange, J.-Y., Capillary hysteresis and the relationship between drying and wetting curves, *Water Resour. Res., 12*, 224-228, 1976.

Parlange, J.-Y., Water transport in soils, *Ann. Rev. Fluid Mech., 12*, 77-102, 1980.

Peck A.J., and R.M. Rabbidge, Soil-water potential: Direct measurement by a new technique, *Science, 151*, 1385-1386, 1966.

Peck A.J., and R.M. Rabbidge, Design and performance of an osmotic tensiometer for measuring capillary potential, *Soil Sci. Am. Proc., 33*, 196-202, 1969

Perroux, K.M., D.E. Smiles, and I. White, Water movement in uniform soils during constant-flux infiltration, *Soil Sci. Soc. Amer. J., 45*, 237-240, 1981.

Philip, J.R., An infiltration equation with physical significance, *Soil Sci., 77*, 153-157, 1954.

Philip, J.R., Numerical solution of equations of the diffusion type with diffusivity concentration dependent, *Trans. Faraday Soc., 51*, 885-892, 1955a.

Philip, J.R., The concept of diffusion applied to soil water, *Proc. Nat. Acad. Sci. India, Sect. A, 24*, 93-104, 1955b.

Philip, J.R., Evaporation, and moisture and heat fields in the soil, *J. Meteorol., 14*, 354-366, 1957a.

Philip, J.R., Numerical solution of equations of the diffusion type with diffusivity concentration dependent: 2. *Australian J. Phys., 10*, 29-42, 1957b.

Philip, J.R., Remarks on the analytical derivation of the Darcy equation, *Trans. Am. Geophys. Union, 38*, 782-784, 1957c.

Philip, J.R., The physical principles of soil water movement during the irrigation cycle, *Proc. Congr. Int. Comm. Irrig. Drain. 3rd, San Francisco*, 8.125-153, 1957d.

Philip, J.R., The role of mathematics in soil physics, *J. Aust. Inst. Agric. Sci., 23*, 293-301, 1957e.

Philip, J.R., The theory of infiltration: 1. The infiltration equation and its solution, *Soil Sci., 83*, 345-357, 1957f.

Philip, J.R., The theory of infiltration: 2. The profile at infinity, *Soil Sci., 83*, 435-448, 1957g.

Philip, J.R., The theory of infiltration: 3. Moisture profiles and relation to experiment, *Soil Sci., 84*, 163-178, 1957h.

Philip, J.R., The theory of infiltration: 4. Sorptivity and algebraic infiltration equations, *Soil Sci., 84*, 257-264, 1957i.

Philip, J.R., The theory of local advection: 1, *J. Meteorology, 16*, 535-547, 1959.

Philip, J.R., The damping of a fluctuating concentration by continuous sampling through a tube, *Aust. J. Phys., 16*, 454-463, 1963.

Philip, J.R., Similarity hypothesis for capillary hysteresis in porous meterials, *J. Geophys. Res., 69*, 1553-1562, 1964.

Philip, J.R., Plant water relations: Some physical aspects, *Ann. Rev. Plant Physiol., 17*, 245-268, 1966.

Philip, J.R., Diffusion, dead-end pores, and linearized absorption in aggregated media, *Aust. J. Soil Res., 6*, 21-30, 1968a.

Philip, J.R., Steady infiltration from buried point sources and spherical cavities, *Water Resour. Res., 4*, 1039-1047, 1968b.

Philip, J.R., The theory of absorption in aggregated media, *Aust. J. Soil Res., 6*, 119, 1968c.

Philip, J.R., A linearization technique for the study of infiltration, in *Water in the Unsaturated Zone*, edited by P.E. Rijtema and H. Wassink, Proc. 1966 IASH/UNESCO Symposium at Wageningen, Vol. 1, 471-478, UNESCO, Paris, 1969a.

Philip, J.R., Absorption and infiltration in two- and three-dimensional systems, in *Water in the Unsaturated Zone*, edited by P.E. Rijtema and H. Wassink, 1966

IASH/UNESCO Symposium at Wageningen, Vol. 1, 503-516, UNESCO, Paris, 1969b.

Philip, J.R., Early stages of infiltration in two- and three-dimensional systems, *Aust. J. Soil Res.*, *7*, 213-221, 1969c.

Philip, J.R., The dynamics of capillary rise, in *Water in the Unsaturated Zone*, edited by P.E. Rijtema and H. Wassink, Proc. 1966 IASH/UNESCO Symposium at Wageningen, Vol. 2, 559-564, UNESCO, Paris, 1969d.

Philip, J.R., Theory of infiltration, *Adv. Hydrosci.*, *5*, 215-296, 1969e.

Philip, J.R., Flow in porous media, *Ann. Rev. Fluid Mech.*, *2*, 177-204, 1970.

Philip, J.R., Flows satisfying mixed no-slip and no-shear conditions, *J. Appl. Math. Phys. (ZAMP)*, *23*, 353-372, 1972a.

Philip, J.R., Integral properties of flows satisfying mixed no-slip and no-shear conditions. *J. Appl. Math. Phys. (ZAMP)*, *23*, 960-968, 1972b.

Philip, J.R., Flow in porous media, *Proceedings of the 13th International Congress of Theoretical and Applied Mechanics, Moscow*, Springer-Verlag, Heidelberg, pp. 279-294, 1973.

Philip, J.R., Fifty years progress in soil physics, *Geoderma*, *12*, 265-280, 1974a.

Philip, J.R., Recent progress in the solution of nonlinear diffusion equations, *Soil Sci.*, *117*, 257-264, 1974b.

Philip, J.R., Some remarks on science and catchment prediction, in *Prediction in Catchment Hydrology*, edited by T.G. Chapman and F.X. Dunin, Proc. Symp. Aust. Acad. Science, Canberra, pp. 2330, Aust. Acad. Science, Canberra, 1975.

Philip, J.R., Unitary approach to capillary condensation and adsorption, *J. Chem. Phys.*, *66*, 5069-5075, 1977a.

Philip, J.R., Adsorption and geometry: the boundary layer approximation, *J. Chem. Phys.*, *67*, 1732-1741, 1977b.

Philip, J.R., Adsorption and capillary condensation on rough surfaces, *J. Phys. Chem.*, *82*, 1379-1385, 1978.

Philip, J.R., Remarks on Comment by B.V. Derjaguin and N.V. Churaev, *J. Chem. Phys.*, *70*, 598, 1979.

Philip, J.R., Advection, evaporation, and surface resistance, *Irrig. Sci.*, *8*, 101-114, 1987a.

Philip, J.R., The infiltration joining problem, *Water Resour. Res.*, *23*, 2239-2245, 1987b.

Philip, J.R., Quasianalytic and analytic approaches to unsaturated flow, in *Flow and Transport in the Natural Environment: Advances and Applications*, edited by W.L. Steffen, and O.T. Denmead, Proceedings of the International Symposium on Flow and Transport in the Natural Environment, held in September 1987 at Canberra, Australia, pp. 30-47, Springer-Verlag, Berlin, 1988.

Philip, J.R., The scattering analog for infiltration in porous media, *Rev. Geophys.*, *27*, 431-448, 1989.

Philip, J.R., How to avoid free boundary problems, in *Free Boundary Problems: Theory and Applications", Research Notes in Mathematics, 185*, edited by K.H. Hoffman, and J. Sprekels, pp. 193-207, Longman, London, 1990.

Philip, J.R., Effects of root and subirrigation depth on evaporation and percolation losses, *Soil Sci. Soc. Amer. J.*, *55*, 1520-1523, 1991a.

Philip, J.R., Soils, natural science, and models, *Soil Sci.*, *151*, 91-98, 1991b.

Philip, J.R., Exact solutions for redistribution by nonlinear convection-diffusion, *J. Aust. Math. Soc., Ser. B*, *33*, 363-383, 1992a.

Philip, J.R., Flow and volume change in soils and other porous media, in *Mechanics of Swelling*, edited by T.K. Karalis, pp. 3-32, Springer-Verlag, Berlin, 1992b.

Philip, J.R., Phenomenological approach to flow and volume change in soils and other media, *Appl. Mech. Rev.*, *48*, 650-658, 1995.

Philip, J.R., One-dimensional checkerboards and blending heights, *Boundary-Layer Meteorol.*, *77*, 135-151, 1996a.

Philip, J.R., Two-dimensional checkerboards and blending heights, *Boundary-Layer Meteorol.*, *80*, 1-18, 1996b.

Philip, J.R., Blending heights for winds oblique to checkerboards, *Boundary-Layer Meteorol.*, *82*, 263-281, 1997a.

Philip, J.R., Effect of root water extraction on wetted regions from continuous irrigation sources *Irrigation Sci.*, *17*, 127-135, 1997b.

Philip, J.R., One-dimensional checkerboards and blending heights, *Boundary-Layer Meteorol.*, *84*, 85-98, 1997c.

Philip, J.R., Full and boundary-layer solutions of steady air sparging problem, *J. Contam. Hydrol.*, *33*, 337-345, 1998a.

Philip, J.R., Infiltration, in *Encyclopedia of Hydrology and Water Resources*, edited by R. Herschy, pp. 418-426, Chapman and Hall, London, 1998b.

Philip, J.R., Physics, mathematics, and the environment: the 1997 Priestley lecture, *Aust. Meteorol. Mag.*, *47*, 273-283, 1998c.

Philip, J.R., Seepage shedding by parabolic capillary barriers and cavities, *Water Resour. Res.*, *34*, 2827-2835, 1998d.

Philip, J.R., and C.J. van Duijn, Redistribution with air diffusion, *Water Resour. Res.*, *35*, 2295-2300, 1999.

Philip, J.R., and D.A. de Vries, Moisture movement in porous materials under temperature gradients, *Trans. Am. Geophys. Union*, *38*, 222-232, 1957.

Philip, J.R., and J. H. Knight, Redistribution from plane, line, and point sources, *Irrigation Sci.*, *12*, 169-180, 1991.

Philip, J.R., and D.E. Smiles, Macroscopic analysis of the behavior of colloidal suspensions, *Adv. Colloid. Interface Sci.*, *17*, 83-103, 1982.

Poulovassilis, A., Hysteresis of pore water, an application of the concept of independent domains, *Soil Sci.*, *93*, 405-412, 1962.

Pruess, K., *TOUGH2 - A General-Purpose Numerical Simulator for Multiphase Fluid and Heat Flow, LBL-29400, UC-251*, 103pp., Lawrence Berkeley Laboratory, University of California Earth Science Division, 1991.

Pullan, A.J., The quasilinear approximation for unsaturated porous media flow, *Water Resour. Res.*, 26, 1219-1234, 1990.

Raats, P.A.C., Axial fluid flow in swelling and shrinking porous rods, *Abstracts 40th Annual Meeting of the Society of Rheology*, p. 13, 1969.

Raats, P.A.C., Steady flows of water and salt in uniform soil profiles with plant roots, *Soil. Sci. Soc. Am. Proc.*, 38, 717-722, 1974.

Raats, P.A.C., Transformations of fluxes and forces describing the simultaneous transport of water and heat in unsaturated porous media, *Water Resour. Res.*, 11, 938-942, 1975.

Raats, P.A.C., Transport in structured porous media, in *Flow and Transport in Porous Media*, edited A. Verruijt, and F.B.J. Barends, Proc. Euromechanics Colloquium 143, held at Delft, The Netherlands, pp. 221-226, Balkema, Rotterdam, The Netherlands, 1981.

Raats, P.A.C., Convective transport of ideal tracers in unsaturated soils, *Proc. Symposium on Unsaturated Flow and Transport Modelling*, edited by E.M. Arnold, G.W. Gee, and R.W. Nelson, NUREG/CP-0030, PNL-SA-10325, pp. 249-265, Pacific Northwest Laboratory, Richland, Washington, USA, 1982a.

Raats, P.A.C., The distribution of the uptake of water by plants: inference from hydraulic and salinity data, in *Séminaires sur l'irrigation localisée. 1. Mouvement de l'eau et des sels en function des charactéristiques des sols soumis l'irrigation localisée*, pp.35-46, Institut d'Agronomie de l'Université de Bologne, Bologna, Italy, 1982b.

Raats, P.A.C., Implications of some analytical solutions for drainage of soil water, *Agric. Water Man.*, 6, 161-175, 1983.

Raats, P.A.C., Applications of the theory of mixtures in soil science, Appendix 5D, p. 326-343, in C. Truesdell, *Rational Thermodynamics, with an appendix by C.-C. Wang. Second Edition, corrected and enlarged, to which are adjoined appendices by 23 authors*, pp. 326-343, Springer Verlag, New York, 1984a.

Raats, P.A.C., Mechanics of cracking soils, in *Proc. Symp. on Water and Solute Movement in Heavy Clay Soils, ILRI publication 37*, edited by J. Bouma and P.A.C. Raats, pp. 23-38, International Institute for Land Reclamation and Improvement, Wageningen, The Netherlands, 1984b.

Raats, P.A.C., Applications of material coordinates in the soil and plant sciences, *Neth. J. Agric. Sci.*, 35, 361-370, 1987a.

Raats, P.A.C., Applications of the theory of mixtures in soil science, *Math. Modelling*, 9, 849-856, 1987b.

Raats, P.A.C., Quasianalytic and analytic approaches to unsaturated flow: commentary, in *Flow and Transport in the Natural Environment: Advances and Applications*, edited by W.L. Steffen and O.T. Denmead, Proceedings of the International Symposium on Flow and Transport in the Natural Environment, held in September 1987 at Canberra, Australia, Springer-Verlag, Berlin, pp. 48-58, 1988.

Raats, P.A.C., Characteristic lengths and times associated with processes in the root zone, in *Scaling in Soil Physics: Principles and Applications*, edited by D. Hillel, and D.E. Elrick, chapter 6, pp. 59-72, Soil Science Society of America, Madison, Wisconsin, USA, 1990a.

Raats, P.A.C., On the roles of characteristic lengths and times in soil physical processes, in *Vol. 1 of Proc. 14th Int. Congr. Soil Sci., held 12-18 August 1990, at Kyoto, Japan*, pp. 202-207, 1990b

Raats, P.A.C., A superclass of soils, *Indirect methods for estimating the hydraulic properties of unsaturated soils*, edited by M.Th. Van Genuchten, F.J. Leij, and L.J. Lund, Proceedings of an International Workshop organized by the U.S. Salinity Laboratory, USDA-ARS, and the Department of Soil and Environmental Sciences of the University of California, both at Riverside, CA, USA, and held 11-13 Oct. 1989 at Riverside, Calif., USA, pp. 45-51, Univ. of Calif., Riverside, 1992.

Raats, P.A.C., Kinematics of subsidence of soils with a non-conservative solid phase, in *Proceedings 16th World Congre of Soil Science* [CD-ROM] Cirad, Montpellier, France, 1998a. o

Raats, P.A.C., Spatial and material description of some processes in rigid and non-rigid saturated and unsaturated soils, in *Poromechanics: A Tribute to Maurice A. Biot*, edited by J.-F. Thimus, Y. Abousleiman, A.H.-D. Cheng, O. Coussy, E. Detournay, Proceedings of the Biot Conference on Poromechanics, held September 14-16, 1998 at Louvain-la-Neuve, Belgium, pp. 135-140, Balkema, Rotterdam, The Netherlands, 1998b.

Raats, P.A.C., Developments in soil-water physics since the mid 1960s, *Geoderma*, 100, 355-387, 2001

Raats, P.A.C., Flow of water in rigid and non-rigid, saturated and unsaturated soils, in *Modeling and Mechanics of Granular and Porous Materials*, edited by G. Capriz, V.N. Ghionna, and P. Giovine, in the series *Modeling and Simulation in Science, Engineering and Technology*, edited by N. Bellomo, pp.181-211, Birkhäuser, Boston, 2002.

Raats, P.A.C. and A. Klute, Transport in soils: The balance of mass, *Soil Sci. Soc. Amer. Proc.*, 32, 161-166, 1968a.

Raats, P.A.C. and A. Klute, Transport in soils: The balance of momentum, *Soil Sci. Soc. Amer. Proc.*, 32, t 452-456, 1968b.

Raats, P.A.C. and W.R. Gardner, Movement of water in the unsaturated zone near a water table, in *Drainage for Agriculture*, edited by J. van Schilfgaarde, *Agronomy Monograph 17*, 311-357, American Society of Agronomy, Madison, Wisconsin, USA, 1974.

Richards, L.A., The usefulness of capillary potential to soil moisture and plant investigators, *J. Agric. Res.*, *37*, 719-742, 1928.

Richards, L.A., Capillary conduction of liquids through porous mediums, *Physics*, *1*, 318-333, 1931.

Richards, L.A., and C.H. Wadleigh, Soil water and plant growth, in *Soil Physical Conditions and Plant Growth*, edited by B.T. Shaw, *Agronomy*, *2*, 73-251, Academic Press, New York, 1952.

Rider, N.E., J.R. Philip, and E.F. Bradley, The horizontal transfer of heat and moisture - a micrometeorological study, *Quart. J. Roy. Meteorol. Soc.*, *89*, 507-531, 1963.

Rijniersce, K., *A Simulation Model for Physical Soil Ripening in the IJsselmeerpolders*, Lelystad, The Netherlands, 216pp, 1983.

Rijniersce, K., Crack formation in newly reclaimed sediments in the IJsselmeerpolders, in *Proc. Symp. on Water and Solute Movement in Heavy Clay Soils, ILRI publication 37*, edited by J. Bouma and P.A.C. Raats, pp. 59-62, International Institute for Land Reclamation and Improvement, Wageningen, The Netherlands, 1984.

Rijtema, P.E. and H. Wassink (Eds.), *Water in the Unsaturated Zone*, Proc. 1966 IASH/UNESCO Symposium held at Wageningen, 2 Vols., UNESCO, Paris, 1969.

Rogers, C., M.P. Stallybrass and D.L. Clements, On two-phase filtration under gravity and with boundary infiltration: application of a Bäcklund transformation, *J. Nonlinear Anal.*, *7*, 785-799, 1983.

Rosen, G., Method of the exact solution of a nonlinear diffusion-convection equation, *Phys. Rev. Lett.*, *49*, 1844-1846, 1982.

Sander, G.C., J.-Y. Parlange, V. Kühnel, W.L. Hogarth, D. Lockington, and J.P.J. OKane, Exact nonlinear solution for constant flux infiltration, *J. Hydrol.*, *97*, 341-346, 1988.

Scotter, D.R., Liquid and vapor phase transport in soil, *Aust. J. Soil Res.*, *14*, 33-41, 1976.

Scotter, D.R., B.E. Clothier, E.R. Harper, Measuring saturated hydraulic conductivity and sorptivity using twin rings, *Aust. J. Soil Res.*, 20, 295-304, 1982.

Selim, H.M., and L. Ma (Eds.), *Physical Nonequilibrium in Soils, Modelling and Application*, Ann Arbor Press, Chelsia, Michigan, 1998.

Severino, G. and A. Santini, Analysis of nonlinearly reactive transport by means of temporal moments, in *Modelling of Transport Processes in Soils at Various Scales in Time and Space*, edited by J. Feyen and K. Wiyo, Proc. Int. Workshop of EurAgEng's Field of Interest on Soil and Water, held 24-26 November 1999 at Leuven, Belgium, pp. 323-330, Wageningen Pers, Wageningen, The Netherlands, 1999.

Simmons, C.S., D.R. Nielsen, and J.W. Biggar, Scaling of field-measured soil water properties, I. Methodology, II. Hydraulic conductivity and flux, *Hilgardia*, *47*, 73-173, 1979.

Simunek, J., M. Senja, and M.Th. Van Genuchten, The HYDRUS-2D software package for simulating water flow and solute transport in two-dimensional variably saturated media, Version 1.0, *Research Report*, U.S. Salinity Laboratory, Riverside, CA., 1996.

Simunek, J., R. Angulo-Jaramillo, M.G. Schaap, J.P. Vandervaere, and M.Th. Van Genuchten, Using an inverse method to estimate the hydraulic properties of crusted soils from tension-disc infiltrometer data, *Geoderma*, *86*, 61-81, 1999.

Skopp, J., and A.W. Warrick, A two-phase model for the miscible displacement of reactive solutes in soils, Soil Sci. Soc. Amer. Proc., 38, 545-550, 1974.

Slichter, C.S., *Theoretical investigation of the motion of ground water*, U.S. Geol. Survey, 19th Annu. Rep., Part 2, 295-384, 1899.

Smiles, D.E., Constant flux filtration of bentonite, *Chem Engng. Sci.*, *33*, 1355-1361, 1978.

Smiles, D.E., Principles of constant pressure filtration, in N.P. Cheremisinoff (Ed.), *Encyclopedia of Fluid Mechanics, vol. 5 Slurry Flow Technology*, Gulf Publ. Co., Houston, Texas, USA, pp. 791-824, 1986.

Smiles, D.E., Hydrology of swelling soils: a review, *Aust. J. Soil Res.*, *38*, 501-521, 2000a.

Smiles, D.E., Material coordinates and solute movement in consolidating clay, *Chem. Eng. Sci.*, *55*, 773-781, 2000b.

Smiles, D.E.. and A.G. Harvey, Measurement of moisture diffusivity of wet swelling materials, *Soil Sci.*, *116*, 391-399, 1973.

Smiles, D.E., and J.R. Philip, Solute transport during absorption of water by soil: laboratory studies and their practical implications, *Soil Sci. Soc. Am. J.*, *42*, 537-544, 1978.

Smiles, D.E., G. Vachaud, and M. Vauclin, A test of the uniqueness of the soil moisture characteristic during transient non-hysteric, horizontal flow of water in a rigid soil, *Soil Sci. Soc. Am. Proc.*, *35*, 534-539, 1971.

Smiles, D.E., J.R. Philip, J.H. Knight, and D.E. Elrick, Hydrodynamic dispersion during absorption of water by soil, *Soil Sci. Soc. Am. J.*, *42*, 229-234, 1978.

Smiles, D.E., K.M. Perroux, S.J. Zegelin, and P.A.C. Raats, Hydrodynamic dispersion during constant rate absorption of water by soil, *Soil Sci. Soc. Am. J.*, *45*, 453-458, 1981.

Smiles, D.E., P.A.C. Raats, and J.H. Knight, Constant pressure filtration: the effect of a filter membrane, *Chem. Eng. Sci.*, *37*, 707-714, 1982.

Smiles, D.E., I. White, and C.J. Smith, Soil science education and society, *Soil Sci.*, *165*, 87-97, 2000.

Smith, K.A. and C.E. Mullins, (Eds.), *Soil and Environmental Analysis: Physical Methods, Second Edition, Revised and Expanded*, Marcel Dekker, New York, 2000.

Sposito, G., Lie group invariance of the Richards equation, in *Dynamics of Fluids in Hierarchical Porous Media*, edited by J.H. Cushman, pp. 327-347, Academic Press, London, 1990.

Su, C., and R.H. Brooks, Soil hydraulic properties from infiltration tests, in *Proc. Watershed Management*, pp. 516-542, Irrigation and Drainage Division, American Society of Civil Engineering, 1975.

Sumner, M.E., and B.A. Stewart (Eds.), *Soil Crusting: Chemical and Physical Processes*, Lewis Publishers, Boca Raton, Fla,, USA, 1992.

Talsma, T., A note on shrinkage behavior of a clay paste under various loads, *Aust. J. Soil Res.*, *15*, 275-277, 1977.

Ten Berge, H.F.M., and G.H. Bolt, Coupling between liquid flow and heat flow in porous media: a connection between two classical approaches, *Transport in Porous Media*, *3*, 35-49, 1988.

Terzaghi, K.T., Die Berechnung der Durchlässigkeitsziffer des Tones aus dem Verlauf der hydrodynamischen Spannungserscheinungen, *Akademie der Wissenschaften in Wien, Sitzungsberichte, Mathematisch-naturwissenschaftliche Klasse, Part IIa, 132*, 125-128, 1923.

Timofeev, M.P., Change in the meteorological regime on irrigation (in Russian), *Izv. Akad. Nauk. SSSR, Ser. Geograf.*, *2*, 108-113, 1954.

Tuller, M. and D. Or, Hydraulic conductivity of variably saturated porous media: Film and corner flow in angular pore space, *Water Resour. Res.*, *36*, 2000.

Tuller, M., D. Or, and L.M. Dudley, Adsorption and capillary condensation in porous media: Liquid retention and interfacial configurations in angular pores, *Water Resour. Res.*, *35*, 1949-1964, 1999.

Van Duijn, C.J., R.E. Grundy, and C.N. Dawson, Large time profiles in reactive solute transport, *Transport in Porous Media*, *27*, 57-84, 1997.

Van Genuchten, M.Th., A closed form equation for predicting the hydraulic conductivity of unsaturated soils, *Soil Sci. Soc. Am. J.*, *44*, 892-898, 1980.

Van Genuchten, M.Th., D.E. Rolston, and P.F. Germann (Eds), Special Issue on Transport of Water and Solutes in Macropores, *Geoderma*, *46*, 1-297, 1990.

Van Genuchten, M.Th., and F.N. Dalton, Models for simulating salt movement in aggregated field soils, *Geoderma*, *38*, 165-183, 1986.

Van Genuchten, M.Th., F.J. Leij, and L.J. Lund (Eds.), *Indirect methods for estimating the hydraulic properties of unsaturated soils*, Proceedings of an International Workshop organized by the U.S. Salinity Laboratory, USDA-ARS, and the Department of Soil and Environmental Sciences of the University of California, both at Riverside, CA, USA, and held 11-13 Oct. 1989 at Riverside, CA, USA, University of California, Riverside, 1992.

Van Genuchten, M.Th., F.J. Leij, and L. Wu (Eds.), *Characterization and Measurement of the Hydraulic Properties of Unsaturated Porous Media*, Proceedings of an International Workshop organized by the U.S. Salinity Laboratory, USDA-ARS, and the Department of Soil and Environmental Sciences of the University of California, both at Riverside, CA, USA, and held 22-24 October, 1997 at Riverside, CA, USA, University of California, Riverside, 2 Vols, 1999a.

Van Genuchten, M.Th., M.G. Schaap, B.P. Mohanty, J. Simunek, and F.J. Leij, Modeling flow and transport processes at the local scale, in *Modelling of Transport Processes in Soils at Various Scales in Time and Space*, edited by J. Feyen and K. Wiyo, Proc. Int. Workshop of EurAgEng's Field of Interest on Soil and Water, held 24-26 November 1999 at Leuven, Belgium, pp. 23-45, Wageningen Pers, Wageningen, The Netherlands, 1999b.

Van Grinsven, J.J.M., H.W.G. Booltink, C. Dirksen, N. van Breemen, N. Bongers, and N. Waringa, Automated in situ measurement of unsaturated soil water flux, *Soil Sci. Soc. Am. J.*, *52*, 1215-1218, 1988.

Vereecken, H., Estimating the unsaturated hydraulic conductivity from theoretical models using simple soil properties, *Geoderma*, *65*, 81-92, 1995.

Visser, W.C., An empirical expression for the desorption curve, in *Water in the Unsaturated Zone*, edited by P.E. Rijtema and H. Wassink, Proc. 1966 IASH/UNESCO Symposium at Wageningen, Vol. 1, 329-335, UNESCO, Paris, 1969.

Vogel, T, H.H. Gerke, R. Zhang, and M.Th. van Genuchten, Modeling flow and transport in a two-dimensional dual-permeability system with spatially variable hydraulic properties, *J. of Hydrology*, *238*, 78-89, 2000.

Warrick, A.W., and G.G. Fennemore, Unsaturated water flow around obstructions simulated by two-dimensional Rankine bodies, *Adv. Water Resour.*, *18*, 375-382, 1995.

Weir, G.J., Steady infiltration from small shallow circular ponds, *Water Resour. Res.*, *23*, 733-736, 1987.

Whitaker, S., Flow in porous media II: The governing equations for immiscible, two-phase flow, *Transport in Porous Media*, *1*, 105-125, 1986.

White, I., and M.J. Sully, Macroscopic and microscopic capillary length and time scales for field infiltration, *Water Resour. Res.*, *23*, 1514-1522, 1987.

White, I., and P. Broadbridge, Constant rate rainfall infiltration: a versatile nonlinear model II. Applications of solutions, *Water Resour. Res.*, *24*, 155-162, 1988.

White, I., D.E. Smiles, and K.M. Perroux, Absorption of water by soil: the constant flux boundary condition, *Soil Sci. Soc. Amer. J.*, *43*, 659-664, 1979.

White, M.D., M. Oostrom, and R.J. Lenhard, Modeling fluid flow and transport in variably saturated porous media with the STOMP simulator: 1. Nonvolatile three-phase model description, *Adv. Water Resour.*, *18*, 353-364, 1995.

Wooding, R.A., Steady infiltration from a shallow circular pond, *Water Resour. Res.*, *4*, 1259-1273, 1968.

Wooding, R.A., S.W. Tyler, and I. White, Convection in groundwater below an evaporating salt lake: 1. Onset of instability, *Water Resour. Res.*, *33*, 1199-1217, 1997a.

Wooding, R.A., S.W. Tyler, and I. White, Convection in groundwater below an evaporating salt lake: 2. Evolution of fingers or plumes, *Water Resour. Res.*, *33*, 1219-1228, 1997b.

Peter A.C. Raats, Paaskamp 16, 9301 KL Roden, The Netherlands (email: pac.raats@home.nl).

David E. Smiles, CSIRO Land and Water, PO Box 1666, Canberra ACT 2601, Australia (email: david.smiles@cbr.clw.csiro.au).

Arthur W. Warrick, Soil, Water and Environmental Science Department, PO box 210038, The University of Arizona, Tucson, AZ 85721-0038, USA (email: aww@Ag.Arizona.Edu).

A Convergence of Paths That Culminated in John Philip's 1995 Video Recorded History of Hydrology Interview

Stephen J. Burges

Department of Civil and Environmental Engineering, University of Washington, Seattle, Washington

INTRODUCTION

John Philip influenced me in many ways. His influence started when I first read some of his work on infiltration and later after I read some of his ideas about hydrologic science. His greatest impact, however, was made in the mid 1980s when I had the good fortune to have my first long talk with him. I detail here a series of events that led ultimately to my interviewing him in July 1995 for the American Geophysical Union video recorded History of Hydrology series.

I first became aware of John's work when I was a Graduate Student at Stanford University (autumn 1967 to summer 1970). I was working with Ray Linsley on storage reservoir water supply reliability, but was trying to read as much as I could in the field of hydrology and water resources. Professor Ven te Chow's annual review series "Advances in Hydroscience", published by Academic Press, appeared to me to be an effective way for a novice to be introduced to developments in diverse fields of research in hydroscience, and John's review paper in Advances in Hydroscience [*Philip*, 1969] was my introduction to his scholarship. This excellent paper provided me with the background to his justly famous series solution for infiltration rate and cumulative infiltration into the soil, which I had only seen previously in summary. I am not sure that I appreciated John's commitment to precise scholarship, but that sharpness of scholarship influenced how I prepared for our July 1995 interview. I have chosen material from Philip (1969) to illustrate his commitment to the precise conduct of and reporting of science.

Environmental Mechanics: Water, Mass and Energy Transfer in the Biosphere
Geophysical Monograph 129
Copyright 2002 by the American Geophysical Union
10.1029/129GM02

GETTING IT RIGHT

John used the English language precisely and was an absolute stickler for getting things right. On page 275 of Philip (1969) he wrote:

"Philip [33, 84] suggested the two-parameter infiltration equations

$$i = St^{1/2} + At$$

$$v_0 = 1/2 St^{-1/2} + A \qquad (195)$$

for use in applied hydrology when t is not too large. Clearly the relation

$$v_0 = K_1 \ (196) \ \text{also Eq. (109)}]$$

holds in the limit as $t \rightarrow \infty$.

It has been suggested recently [19] that this implies that A must be K_1. However, the coefficients of corresponding terms in a series expansion with limited radius of convergence and in an asymptotic (large argument) expansion of the same function are not necessarily equal. We have, in fact, from Eqs. (193) that $A = K_1/3$, $2 \ K_1/3$, and 0.38 K_1 for the linearized model, the delta-function model, and for our nonlinear example."

There are no hints in this paper of John's working environment at the time he developed his solutions to Richards' equation. During his interview with me in July 1995 he explains why he brought quantitative tools to bear on what he considered the dominant problems of hydrology. He also sought to solve Richards' equation numerically, but soon realized the impossibility of that with the primitive computers of that era. He was forced to obtain analytical solutions. A slight hint for his lifelong interest in the vadose zone is given in the opening paragraph [Philip, 1969, p 216].

"A very large fraction of water falling as rain on the land surfaces of the earth moves through unsaturated soil during the subsequent processes of infiltration, evaporation, and the absorption of soil-water by plant roots. Hydrologists and their text books and handbooks, have tended, nevertheless, to pay relatively little attention to the phenomenon of water movement in unsaturated soil. Most research on this topic has been done by soil physicists, concerned ultimately with agronomic or ecological aspects of hydrology; but their colleagues in engineering hydrology have exhibited an increasing interest in this field in recent years."

JOHN PHILIP'S LITTLE KNOWN INFLUENCE ON HSPF

There were many reasons for my paying more attention to John's theory of infiltration rather than to that of any others working in that field when I was a Graduate Student. Norman Crawford and Ray Linsley started to work on a new version of the Stanford Watershed Model [*Crawford and Linsley*, 1966] about the time I arrived at Stanford in the autumn of 1967. All of us who were working with Ray were influenced by what he and Norm were doing with the Stanford Model, no matter what we were working on ourselves. The model was being rewritten from Burroughs Algol -- BALGOL -- (both the Burroughs Corporation and the ALGOL language are long since defunct) and with some different algorithms into the new IBM supported language, PL1, a language that "was destined to be the language of the future". PL1 has disappeared as well.

There are two brilliant conceptual representations of hydrologic process spatial variation in this "lumped model". One of them treats infiltration as a uniform spatial probability distribution across the catchment or sub-catchment that is being modeled. The other treats evaporation as a different spatial probability distribution that is uncorrelated with infiltration. These two representations capture, relatively simply, much of what later model builders have attempted to do using far more complicated spatial model representations.

One component of the model required coupling the rate of infiltration to the accumulated amount of infiltration. Ray and Norm took a bit of liberty with John's series solution - eq. 195 in [*Philip*, 1969] - by assuming that A was small, yielding $iv_0 = S$ or a constant. Ray and Norm were well aware that John's solution was for ponded infiltration. They used his work largely as an index to infiltration. The model is based on continuous accounting of moisture, so this approach provided a means for estimating the infiltration rate as a function of current conceptual modeled soil water storage. This rate was set equal to the basin median infiltration rate, with infiltration rate varying uniformly from zero to a maximum. The

new form for handling infiltration was first included in the "Hydrocomp Simulation Program" (HSP). That program was later rewritten and coded efficiently in Fortran (Fortran is still with us!) and made part of a major software package, now widely available from the US Environmental Protection Agency, known as HSPF. Details are given in Johanson et al (1984). John's work certainly influenced Ray Linsley. Ray Linsley had a profound influence on me. Ray was a brilliant man with many interests. Little did I know how much I would be influenced in later years by John Philip. Ray and John are two of few I have been privileged to know who were close to being, if not, Renaissance men.

A YOUNG EDITOR MEETS A GIANT

I first met John in Philadelphia at the reception that followed the Honors ceremony at the AGU Spring Annual Meeting on Tuesday, June 1, 1982. The Honors ceremony was held in the Ballroom of the Philadelphia Centre Hotel. John was resplendent in a tuxedo and had just received the Horton Medal, the highest recognition for a hydrologist within AGU. John was the third recipient of the medal and the first of only two from outside North America. In his acceptance speech, John reminded us of the excellent work published in the early issues of the Transactions of AGU and singled out Horton's (1931) paper, the first published in the Transactions by the newly formed Hydrology Section, for our attention. I chatted with him briefly at the reception and he commented that he thought that the beverage that he was holding was not of the highest class. My sense was that he was a shy man who was not overly comfortable in such social settings. I was at the time serving my second year as editor (for physical sciences) of WRR. Soon after our meeting I received the first of what became a steady stream of his papers. We corresponded frequently about his papers and it appears that John was comfortable with the way the WRR community handled them.

In December 1983, at the end of my third year as editor of WRR, I consulted a group of senior colleagues about a special issue of WRR. My objective was to provide an opportunity for some of the leaders of the profession to write about directions of our science and practice. It was with considerable trepidation that I sent out invitation letters early in July 1984. The response was more than heartening. Almost all agreed to participate, but all wanted some time to think about what subject matter they would cover. The scope of what we attempted is given in the introduction to the August 1986 supplemental issue of WRR (Burges, 1986). I had invited John to participate, hoping that he would be comfortable writing a deeply thoughtful paper along similar lines to his paper in the 1975 book "Prediction in Catchment Hydrology", edited by Tom Chapman and Frank Dunin [*Philip*, 1975]. He pondered my request for a long time, but had to decline. He did not feel he had something fresh to contribute. My failure to recruit him for that issue was a major loss.

VISITS TO SEATTLE

John wrote me on April 4, 1985 and arranged to call in to visit me at the University of Washington. He insisted on "singing for his supper" -- translation: he wished to present a seminar or two and talk with colleagues while he was here in Seattle. John arrived on June 15 and gave a seminar on June 17 titled "The quasilinear analysis of soil-water movement; basic theory and application". This was the first of a series of visits on an approximately two-year schedule. John and I worked in completely different areas, but had many long and, for me at least, enormously beneficial discussions about the whole gamut of hydrologic science and science and engineering in general as well as the humanities. My wife, Sylvia, and I looked forward to our dinner time chats with him at our home. We covered a large range of subject matter with him and the evenings always ended too soon. We were fortunate that his wife, Frances (Fay to all who know her), could accompany him on his visit in September 1989. During this visit John presented a seminar "Pollution plumes from hydrocarbon leaks beneath the water table". He also explored informally with our group some of his ideas that ultimately were incorporated into a UNESCO/IAHS committee report on education in hydrology [Nash et al., 1990].

John presented seminars each time he visited. Three that were particularly notable reflect his wide interests. He presented what he referred to as a travelogue on November 20, 1992. This was his "Desperately seeking Darcy in Dijon", Philip (1995). He absolutely captivated the audience with this presentation and reminded us that "fame is fleeting". John finished with:

"Let me conclude with this sad reflection: 134 years after the whole town of Dijon mourned the death of Darcy, nobody in Dijon knows who he was, and nobody cares. So much for the 'imperishable right to the remembrance of his native town' that the Municipal Council believed was his due."

John presented his work on "blending heights" [Philip, 1996a, b] on October 4 1996. This was timely because several of our doctoral student colleagues were in the early stages of planning major field measurement programs. John was particularly impressed with our younger colleagues on that occasion. He came back from a chat with three of them and announced quite cheerfully that they had "healthy scepticism". That, indeed was praise, and he was genuinely delighted to have spent time with them. John's last visit was in 1998. On July 10 he

presented his 1997 "Priestley Lecture", reproduced in this volume. It was in this lecture that most of us had a chance to track the many facets of his work. It is as close as we will get to a written personal summary of how he saw his work.

SCIENCE AND TRANS-SCIENCE

John's 1975 paper "Some remarks on science and catchment prediction" introduced me to Alvin Weinberg's concept of trans-science, "questions which can be asked of science and yet cannot be answered by science". I had read some of Weinberg's work previously, but that was largely in the context of the prospects for a nuclear energy based world economy. John challenged us in his paper to think about how could we prepare rigorous tests of hydrologic conceptualizations. John's final paragraph provided much food for though and influenced the thinking that led me to propose a scheme for testing hydrologic models [Burges, 1985]. The ideas that I first outlined in 1984 [Burges, 1985] were developed by my doctoral colleague, Thian Yew Gan and our findings were published by Gan and Burges, (1990a, b). I presented some early thoughts on the planned work at a seminar at the Pye Laboratory, CSIRO, Canberra, in May 1986. John presided over the seminar and was a most gracious host. I think he was genuinely pleased to know that we were attempting less than ideal, but rigorous tests for models. I repeat here the final paragraph from Philip (1975):

"All this may seem pessimistic, defeatist even. What, you may ask, is the point of scientific hydrology if the problems it seeks to solve are ultimately trans-scientific? In my opinion, the answer is that it remains our obligation to ensure that our methods are as scientific and objective as possible. Let us at least work towards a situation where the trans-scientific judgements which practical hydrologists are forced to make are informed and sustained by a truly scientific hydrology: a sceptical science with a coherent intellectual content firmly based on the real phenomena. Finally, let me repeat the text of my sermon, 'The most science can do is to inject some intellectual discipline into the republic of trans-science.'"

The last sentence is "vintage John Philip."

1995 AGU HISTORY OF HYDROLOGY INTERVIEW

I was president of the Hydrology Section of AGU between July 1994 and June 1996 and I worked closely with David Dawdy, chairman of or committee on the "History of Hydrology", to make video taped interviews with the senior leaders of our field. We made our first major effort in 1995. David and Marshall Moss (a former

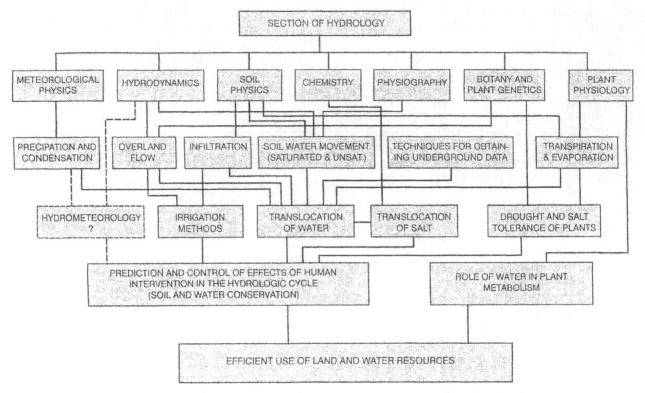

Figure 1. A Proposed Section of Hydrology. Exact From Report to CSIRO Executive, 1953 (Figure 2 from "Physics, Mathematics and the Environment: the 1997 Priestley Lecture," by J. R. Philip).

hydrology section president and then general secretary of AGU) recommended that we should invite John to participate. Marshall approached him and asked him to indicate with whom he would prefer to do the interview. John called me late one evening at home and mentioned that Marshall had approached him. He asked me to be the interviewer. There was only one answer I could give him. We planned to conduct the interview at the University of Colorado at Boulder where we would all be assembling in early July 1995 for the XXI Assembly of the International Union of Geodesy and Geophysics. What most concerned me when I was preparing for the interview was how I could capture something of the breadth and depth of interests of this remarkable man. I had seen first hand the architectural success of the Pye laboratory and his home, knew of his deep interest in art and architecture, and had read some of his poetry. We have included a CD version of the interview [*Burges*, 1995] with this publication because in it we learn much about the time and place of his work. This interview sheds light on aspects of his work that his immediate colleagues, John Knight, Ian White, and David Smiles, did not know until they viewed it in June 2000.

During the interview, John comments on his early schooling and the influence of an exceptional mathemat-

ics teacher at Scotch College. John completed high school at age thirteen, but had to wait until age sixteen before he could enter university. He spent the intervening years in detailed additional studies at Scotch, the best private school in Victoria at the time.

John developed a passion for research during a one-year appointment as a research assistant at the CSIR Irrigation research Station, Griffith starting in February 1947. He rejoined the then renamed CSIRO Division of Plant Industry at Deniliquin, New South Wales in 1951. It was at Deniliquin that he taught himself the needed physics and mathematics to tackle his famous solution to Richards' equation for infiltration. John mentions that it was a publication by Richards and Wadleigh (1952) that gave him a basis for his vision for how hydrologic research ought to be done. Richards and Wadleigh were working at the USDA Salinity Laboratory at Riverside, California at the time. (I had not read the particular work until December 2000. It is an extraordinarily thoughtful book chapter and I regret that I had not read it decades ago). I did not know how extensive was John's vision and plan until I received a copy of John's 1997 Priestley Lecture, presented at CSIRO. Figure 1, which is Figure 2 from his lecture, is provided here because it details his 1953 vision for a CSIRO "Section of Hydrology". This figure helps explain John's approach to his science and his acute awareness of how that science should be directed to the needs of society.

John comments on the influence of colleagues from his days at Deniliquin on his work and how they shaped some of his activities for decades into the future. Few, if any, could match his grasp of the interdisciplinary nature and needs of hydrologic research. Towards the latter part of the interview, John comments on serendipity in science and the multiple benefits of analogies. His work on the "scattering analog" drew heavily on the extensive body of work that had been developed for magneto-hydrodynamics. Some would view this serendipitous occasion as the result of a chance encounter. It was, however, John's trained mind that recognized opportunity. There is considerable food for thought for all in this interview.

SUMMARY OBSERVATIONS

John Philip was exceptional in many ways. He cared deeply about science and the nurturing of young scientists. He wrote about scientific needs and challenges at various times during his career. His paper "Future Problems of Soil Water Research" [Philip, 1972] remains fresh and relevant today. On one of his visits to Seattle, I asked if he had considered writing a book that would cover his range of scientific interests. He replied that he did not anticipate writing a book but had attempted over the years to write appropriate summary and review papers to achieve comparable ends. Philip (1969) definitely met the bill. His short 1988 paper "Infiltration of Water into Soil" is on the recommended reading list for my Graduate Student colleagues as is Philip (1975). John had a deep interest in the history of science. His paper "Fifty Years of Progress in Soil Physics" [Philip, 1974] demonstrates skill at placing work into historical context and assessing contributions to and impediments to progress. His "Desperately Seeking Darcy in Dijon" tells us much about John as well as Darcy's "fleeting fame". The research community can learn much from his published work, particularly the gems that record parts of his vision and scientific philosophy. His work is replete with rich insights that are often only gleaned after multiple readings. (His complete bibliography is included in the enclosed CD). We learn much more about this polymath from the enclosed 1995 interview.

REFERENCES

Burges, S. J., Rainfall-runoff model validation: the need for unambiguous tests", Proc, 9th World Congress of the Int. Fed. of Automatic Control, Budapest, Hungary, 1984, in A Bridge Between Control Science & Technology, edited by J. Gertler and L. Keviczky, Vol. 6 Pergamon Press, 1985.

Burges, S. J., Trends and directions in hydrology, Water Resources Research, 22(9), 1s-5s, 1986.

Burges, S. J., Interview with John R. Philip FRS, NAE, Chief Emeritus, Division of Environmental Mechanics, CSIRO, Canberra, Australia (University of Colorado, Boulder, Colorado), Video Taped History of Hydrology Interviews -- American Geophysical Union, July 5, 1995.

Crawford, N. H. and Linsley R. K. (Jr.), Digital simulation in hydrology: Stanford watershed model IV, Technical report No. 39, Department of Civil Engineering, Stanford University, p210, 1966.

Gan T. Y., and Burges, S. J., An assessment of a conceptual rainfall-runoff model's ability to represent the dynamics of small hypothetical catchments", 1: models, model properties, and experimental design, Water Resources Research, 26(7), 1595-1604, 1990.

Gan T. Y., and Burges, S. J., An assessment of a conceptual rainfall-runoff model's ability to represent the dynamics of small hypothetical catchments", 2: hydrologic responses for normal and extreme rainfall, Water Resources Research, 26(7), 1605-1619, 1990.

Horton, Robert, E., The field, scope, and status of the science of hydrology", Transactions, American Geophysical Union, 189-202, 1931.

Johanson, R. C., Imhoff, J. C., Kittle, J. L. (Jr.), and Donigian, A. S. (Jr.), Hydrologic simulation program -- fortran (HSPF): Users manual for release 8.0, US Environmental Protection Agency, Athens Georgia, p767, 1984.

Nash, J. E., Eagleson, P. S., Philip, J. R., and van der Molen, W. H., The education of hydrologists, Hydrologic sciences Journal, 35(6), 597-607, 1990.

Philip, J. R., Theory of infiltration, Advances in Hydroscience, 5, 215-296, Academic Press, New York, 1969.

Philip, J. R., Future problems in soil water research, Soil Science, 113(4), 294-300, 1972.

Philip, J. R., Some remarks on science and catchment prediction, in Prediction in Catchment Hydrology, edited by T. G. Chapman and F. X. Dunin, pp. 23-30 Australian Academy of Science, 1975.

Philip, J.R. (1988) "Infiltration of Water into Soil", Animal and Plant Sciences, Vol. 1, pp. 231-235.

Philip, J. R., Desperately seeking Darcy in Dijon, Soil Science Society of America Journal, 59(2), 319-324, March-April, 1995.

Philip, J. R., One-dimensional checkerboards and blending heights, Boundary-Layer Meteorology, 77, 135-151, 1996.

Philip, J. R., Two-dimensional checkerboards and blending heights, Boundary-Layer Meteorology, 80, 1-18, 1996.

Richards, L.A and Wadleigh, C.H., Soil water and plant growth, in Soil physical conditions and plant growth, Agronomy, edited by B.T. Shaw (ed.), (2 ed.): pp. 73-251, 1952.

Stephen J. Burges, Department of Civil and Environmental Engineering, University of Washington, Seattle, Washington 98195-2700, USA, sburges@u.washington.edu

Simplification Plus Rigorous Analysis: The *Modus Operandi* of John Philip

James C. I. Dooge

Centre for Water Resources Research, University College Dublin

The legacy of John Philip to scientific research consists of (a) a large volume of research output in relation to the soil-plant-atmosphere continuum, (b) a wide impact on subsequent research in a number of areas in environmental physics, and (c) the example of a consistent application of a reliable research strategy. This research strategy is identified and outlined in relation to the various problems which he tackled. It consisted essentially of a three stage process: (1) careful formulation of the appropriate physical equations and their possible simplifications, (2) rigorous analysis of these simplified equations and (3) thorough review of the analysis. In order to exemplify it in more detail, his application of this research strategy to the problem of infiltration is presented in more detail. This more detailed description includes both the solutions for the initial simplified cases and an outline of the gradual extension of these solutions to more complex cases.

1. INTRODUCTION

The legacy of John Philip to environmental physics and to the natural sciences generally is characterized by three particular features. First, one notices the volume of published work which is available to us in the text and publication list in Smiles(2001) which lists 307 papers of high quality in a wide variety of journals. Secondly, there is the huge impact of many of these publications on developments in the study of diverse water-related problems in environmental physics. Thirdly, there is his remarkable consistency and success in following a single research strategy based on: (a) a physically sound formulation of the problem of interest, (b) an intuitive simplification of this complex problem, and (c) the rigorous analysis of the resulting simplified equations. The third component of this legacy is the main concern of the following contribution which deals both with his general research strategy and its application to the process of infiltration.

In order to concentrate on the approaches, concepts and techniques, rather than on the resulting formulae, an attempt is made to avoid the use of formulae. For most of those already acquainted with the work of John Philip, this should create no real difficulty. Any reader who does find some difficulty in this treatment is advised to refer to the fuller treatment in John's review chapter in the 1969 volume of Advances in Hydroscience (Philip 1969e) or to his last review paper on infiltration published as a contribution to the Encyclopaedia of Hydrology and Water Resources (Philip 1998a). Equally the number of references cited has been curtailed so as not to obscure the account of John Philip's research strategy.

2. THE TRIPLE LEGACY OF JOHN PHILIP

2.1. The Scope of his Research

The scope of the research work of John Philip is not merely a matter of his large volume of output in the field of soil physics and allied topics. It is also characterized by a broad field of interest in environmental physics generally which is evident from his earliest works and throughout his career. As early as 1957 he was emphasizing separately to both irrigation engineers and to plant experts the importance of considering water in relation to the whole soil-plant-atmosphere continuum. In an invited contribution to the

Environmental Mechanics: Water, Mass and Energy Transfer in the Biosphere
Geophysical Monograph 129
Copyright 2002 by the American Geophysical Union
10.1029/129GM03

Annals of Plant Physiology he stressed (Philip 1966, page 246):

> *"Because water is generally free to move across the plant-soil, soil-atmosphere and plant-atmosphere interfaces, it is necessary and desirable to view the water transfer systems in the three domains of soil, plant and atmosphere as a whole".*

He addressed the same message in greater detail to hydrologists three years later in a contribution wholly devoted to this broad continuum concept (Philip 1969d).

His interest in the atmospheric part of the continuum also began early and continued throughout his career. A paper on local advection published in 1959 broke new ground in providing a quasi–analytical solution of the two dimensional equation for atmospheric diffusion. In this area also, he continued to develop his theories incorporating the effect of surface resistance and later on introduced the concept of blending heights in an important series of three papers in the Journal of Boundary Layer Meteorology.

2.2. The Impact of his Research

There is little need to emphasize to readers of this monograph the immense impact of the work of John Philip in the field of environmental physics. However, it is interesting to note in passing that a search under the citation index for the period 1974 to 2000 reveals a total of over five thousand citations including twelve publications each with over a hundred citations. The three highest figures were 237 citations for the first paper of the ground-breaking series of seven papers in 1957 on "The infiltration equation and its solution" (Philip 1957b), 301 citations for the fourth paper in the same series on "Sorptivity and algebraic infiltration equations" (Philip 1957d), and 594 citations for the comprehensive article on the Theory of Infiltration published in the 1969 volume of Advances in Hydroscience (Philip 1969e).

John's legacy of revolutionizing the approach to a number of problems in soil physics was widely appreciated throughout his career. The importance of his related insights into key problems in the remaining parts of the soil-plant-atmosphere continuum have recently been acknowledged after some decades of relative neglect.

The third part of his legacy in the form of his research strategy has received much less attention and consequently has had less influence. His manner of reaching his important results, as well as the results themselves can, provide a vital input into the continuing endeavour of research in environmental physics.

2.3. His Approach to Research

Attention is concentrated in this contribution on the approach of John Philip to a number of research problems in environmental physics. It is based on the careful reading of a large number of his published papers. When reading papers in the field of mathematical physics, we tend to concentrate on the succession of equations rather than on the explanatory prose that links them. Concentration on the equations in this way enables us to connect with our existing knowledge and to evaluate the advance that has been made in this particular publication. If we want to follow the thought processes of an author, as is the case here, it is necessary to concentrate on the textual material in which these thought processes are revealed. Accordingly, in what follows the use of equations will be avoided in order to concentrate on the research strategies, and the concepts and assumptions that lie behind them. As indicated by the title of this contribution, the key features of his strategy were simplification followed by rigorous analysis.

3. THE STRATEGY OF SIMPLIFICATION

3.1. The Simplification Tradition

The search for progress by initial simplification, followed by subsequent study of the solution obtained for this simplified case as a preliminary to extending it to more realistic cases, has an honoured place in the history of human thought and of scientific research. Though used before his time in philosophical writings, the principle of parsimony was first used systematically in the writings of William of Ockham (1285-1349). Typical of his statements was:

> *"Pluralitas non est ponenda sine necessita"*

which may be translated as "plurality should not be assumed without necessity". John Philip followed this principle in his work and would have sympathized with the modern paraphrase which applies this principle to hydrologic modeling by saying that the number of parameters requiring calibration should not be increased beyond what is absolutely necessary for the problem in hand.

In mathematics, this approach has been well described by George Polya both in his eminently readable work entitled "How to solve it" (Polya 1945) which is a perennial best seller and in his more formal work on "Mathematics and Plausible Reasoning" (Polya 1954). In the former he writes:

> *"If you cannot solve the proposed problem do not let that failure afflict you too much but try to find consolation with some easier success, try to solve first some related problem; then you may find courage to attack your original problem again."*

In geophysics, Joseph Pedlosky (1987) has characterized this approach as follows:

> *"One of the key features of geophysical fluid dynamics is the need to combine approximate forms of the basic*

fluid-dynamical equations of motion with careful and precise analysis. The approximations are required to make any progress possible, while precision is demanded to make the progress meaningful."

This, in the opinion of the present writer, is a succinct description of the *modus operandi* of John Philip.

There are many types of simplification from which to choose. The most common approaches are: (a) to simplify the equation itself by concentrating on the most significant physical forces involved; (b) to simplify the solution space by reducing the number of independent variables; (c) to reduce the state space by reducing the number of dependent variables; and (d) to simplify the characterization of the physical parameters involved in the problem. All of these and others were used by John Philip. The process of simplification is far from trivial. Physical knowledge is necessary to formulate the complex problem adequately and physical intuition is equally necessary in order to choose the most appropriate type of simplification.

3.2. John Philip's use of Simplification

A reading of John's papers throughout his career reveals a constant adherence to the strategy of simplification as a starting point for deeper analysis. This attitude is clear in all of his landmark papers. It is interesting that in his later papers he does not omit the preliminary steps of simplification in his exposition. Even when reviewing the development of the more advanced forms of his analysis, he repeats the simplified problem and its solution as the original starting point which leads to the subsequent development being described in more detail.

In his first paper on the topic of infiltration (Philip 1954, p156), he wrote:

"It is emphasized that the analysis developed here is for a homogeneous soil of stable structure. This simple case must, of course, be studied before progress is likely on those of greater complexity."

This approach to the infiltration problem and its solution is used later in this presentation as an illustration of his research strategy.

Twelve years later in his review paper to the Annals of Plant Physiology in 1966, John Philip tackles problems of plant physiology in the same fashion by first presenting the problem in the daunting complexity of its physical formulation and then seeking its simplification as a way forward. In that paper he says (Philip 1966, page 253):

"After presenting this alarmingly complicated formalism and going on to point out some of it limitations, I am under some obligation to examine the possibility of simplifying our model of soil-plant- atmosphere continuum to the point where it might be tractable."

This strategy is also employed in his comprehensive treatment of the theory of infiltration in (Philip 1969e, page 230), where he writes in relation to the absorption equation in which gravity is neglected:

"This then is the equation describing absorption, i.e. infiltration into horizontal systems, or into fine textured soils in which the influence of the moisture gradients is much more important than that of gravity. As we shall see, absorption solutions have the additional theoretical importance they yield the limiting small-time behaviour of transient infiltration processes even when gravity cannot be neglected, and so provide a basic point of departure for the solution of the (more complicated) infiltration equations."

The extension of the basic solutions to more complex problems in infiltration is also the subject of a later section of this contribution.

John Philip was prepared to use numerical methods to derive solutions based on soil moisture characteristics from the real world of field measurements, but was keenly aware of their dangers. He criticized the approach based on the use of parameter fitting to complicated models based on an inadequate physical foundation. John was also aware of the dangers of over-reliance on analysis. He states this explicitly in one of his papers which explores the relationship between science, transcience and society (Philip 1991, page 93):

"On the other hand, preoccupation with soluble problems must not become an excuse for timidity on the part of scientists."

In his own work he always followed up the solution of the simplified problem by a broader attack on the original complex problem based on the physical insight gained by that preliminary solution. In his last review paper on the topic of infiltration he is explicit along the same lines (Philip 1998a, page 419):

"The foregoing is the simplest transient solution of the unsaturated flow equation. As we shall see it plays a central role in the study of more difficult problems involving the effect of gravity and/or geometrical considerations."

However, he was also a great exemplar of the use of rigorous analysis of these simplified equations which is equally important.

3.3. Rigorous Analysis and Review

It is significant that the title of one of his early papers (Philip 1957e) is "The role of mathematics in soil physics". The opening sentences of the summary of that paper are well exemplified by his work throughout his career:

"Mathematics performs a dual role of language and logic, and in a concentrated form which makes possible manipulations which would be otherwise unmanageable. Mathematical operations can be applied to the phenomenon of the real world only by means of the process of abstraction. This involves conserving in the problem only those factors with appropriate entities which have a bearing on the answer we seek, and then identifying these factors with appropriate entities which are amenable to mathematical treatment. Obviously, successful use of mathematics in science requires insight into the phenomena as well as mathematical competence."

His own use of mathematics was exemplary. John Philip also adopted the second half of the precept quoted above from Pedlosky to the effect that for meaningful progress the analysis of the simplified equations must be rigorous. Thus in his comprehensive treatment of the theory of infiltration, having simplified from the total potential to the capillary potential, he stresses (Philip 1969e, pp 223-224), both (a) the need to proceed to the form of the equation which allows for gravity potential as well as capillary potential in the terms of the basic dynamic equation and (b) the need to allow for the inherent non-linearity of the soil-moisture characteristics typical of real soils. In that paper, there follows a section of six pages on the question of "limits to applicability of approach" (Philip 1969e, pages 224-229). It is abundantly clear that John used simplification of the basic equation not as a device to avoid difficult problems but rather as an efficient first step towards their eventual solution.

At the end of his career he shows the same concern with these problems of the successive relaxation of simplifying constraints. In the Encyclopaedia of Hydrology and Water Resources, his articles on "Infiltration" (Philip 1998a), and on "Water movement in unsaturated soils" (Philip 1998b), devote what many would consider a disproportionate amount of space to the complications of physical behaviour not allowed for even in the extended analysis.

4. THE SIMPLIFIED INFILTRATION PROBLEM

4.1. *The Position Before 1955*

In order to appreciate the revolution in thinking on the infiltration problem brought about by the work of John Philip in the 1950s, it is important to reflect for a moment on the restricted body of knowledge readily available to hydrologists and engineers prior to 1950. This is best done by consulting such standard works as the Hydrology Handbook of the American Society of Civil Engineers (ASCE 1949) or the comprehensive pioneer text on Applied Hydrology by Linsley, Kohler and Paulhus (1949).

John's early papers in 1954 and 1955 reflect two important additions to that background in the form of (a) the two

papers by Klute (1952a, 1952b) on the application of the non-linear diffusion equation to the problem of ponded infiltration under capillary action alone and (b) the substantial review of "Soil physical conditions and plant growth" by Richards and Wadleigh (1952). Klute had arrived at the solution whereby the depth of penetration for any particular moisture content varied with the moisture content but was always proportional to the square root of the elapsed time.

In regard to the latter review by Richards and Wadleigh, John Philip twenty years later wrote (Philip 1972: p 295):

"The 1952 paper by Richards and Wadleigh came as a breath of fresh air. It was a definitive review of the relations between soil water and plant growth. But it was no routine review. The authors brought to the task their personal insights in soil physics and plant physiology, and they creatively explored and explained the interactions between the two."

He went on to comment how timely this encounter was for his own development:

"I was, at that time, struggling to find my way in this field and the Richards and Wadleigh article became my Bible and my bedtime reading. The directions of most research on the soil-plant-atmosphere continuum over the next twenty years lie implicit in 'Soil water and plant growth'."

A sabbatical with E. C. Childs at Cambridge in 1954 brought John Philip into still closer contact with the whole tradition of soil physics. The progress made in soil physics up to that time has been well reviewed by Gardner (1986).

By 1955, John Philip had the mastery of mathematical techniques and the physical insights which enabled him to formulate the basic infiltration problem in a satisfactory physical form and to control the simplification of the equation. This led rapidly to the derivation of his classical series solution for one-dimensional infiltration under ponding conditions as proposed in 1957 (Philip 1957b), which was refined and extended in his later expositions (Philip 1969e; Philip 1998a). In these and other papers, John at all times acknowledged the contribution of the early pioneers in the analytical formulation of flow in porous media. The introductory section of his papers were models of how to combine a literature review, an acknowledgement of key contributions in the past, and an appropriate context for the work being described in the current paper.

4.2. *Simplifying the Infiltration Problem*

The first step in the physical formulation of the problem was the use of Darcy's Law promulgated in 1856 for saturated flow, assumed as applicable to unsaturated flow by Richards in 1931, and subsequently confirmed experimentally for such flows by Childs and Collis-George in 1950. The first two steps in the simplification by John Philip of Richard's equation were (a) reducing the original general 4-dimensional solution space to the 2-dimensional solution space representing unsteady vertical flow in a soil column; and (b) restricting the total potential to the two elements of capillary potential and gravity potential. These two restrictions reduce the general Richard's equation to

$$\frac{\partial \theta}{\partial t} = \frac{\partial}{\partial z}\left(K\frac{\partial \Psi}{\partial z}\right) - \frac{\partial K}{\partial z} \tag{1}$$

where t is the elapsed time and z the depth below the surface, $\theta(z,t)$ is the moisture content, $\Psi(\theta)$ is the moisture potential and $K(\theta)$ is the unsaturated conductivity. If hysteresis is neglected, the 2-dimensional state space (θ, Ψ) can be reduced to a 1-dimensional solution space in terms of either moisture content (θ) or moisture potential Ψ by means of the concept of moisture diffusivity (Buckingham 1907, Childs 1936, Childs and Collis-George 1950, Philip 1955) which is defined by

$$D = K\frac{d\Psi}{d\theta} \tag{2}$$

In the former case equation (1) reduces to

$$\frac{\partial \theta}{\partial t} = \frac{\partial}{\partial z}\left(D\frac{\partial \theta}{\partial z}\right) - \frac{dK}{d\theta}\frac{\partial \theta}{\partial z} \tag{3}$$

which is the basic equation for vertical infiltration into a stable homogeneous soil in the absence of hysteresis.

The next step in the simplification procedure was to neglect the second term on the right hand side due to gravity and draw on existing solutions from other fields in physics of the resulting equation of non-linear diffusion. The use of the Boltzmann similarity transformation

$$\phi_1 = zt^{-0.5} \tag{4}$$

then reduces the latter partial differential to an ordinary differential equation in the single variable $\phi_1(\theta)$. For the case of infiltration into a semi-infinite column at constant initial moisture content, the boundary conditions are compatible with such a transformation and the solution of the transformed equation takes the form

$$\phi_1(\theta) = z(\theta,t).t^{-\frac{1}{2}} \tag{5a}$$

$$\text{or } z(\theta,t) = \phi_1(\theta).t^{\frac{1}{2}} \tag{5b}$$

which gives the shape of the infiltrating moisture profile at any elapsed time in terms of the coefficient $\phi_1(\theta)$ which is the solution of that ordinary differential equation. Integration of equation (5b) over the range from the initial moisture content to the surface moisture content gives the excess of the cumulative surface infiltration over the bottom drainage at any given time t so that the cumulative infiltration can be written as:

$$F(t) = S.t^{\frac{1}{2}} + K_o.t \tag{6}$$

where the coefficient S is known as the sorptivity and K_o is the conductivity at the initial constant moisture content.

4.3. General Asymptotic Solutions

The next step taken by John Philip led to a remarkable breakthrough. Taking the solution of the simplified equation, which treated gravity as negligible compared with capillarity, he solved for the perturbation from this result when the full form of equation (3) allowing for gravity is used. Applying a Boltzmann-like transformation

$$\phi_2 = z.t^{-1} \tag{7}$$

to this perturbation, and again solving the resulting ordinary differential equation numerically, a first order correction is obtained. The procedure is then repeated by substituting the resulting two term solution in equation (3) to obtain a second order perturbation. This process is repeated thus obtaining a solution for the depth of penetration z in terms of a power series in $t^{\frac{1}{2}}$ whose coefficients are function only of θ (Philip 1957b, p 351):

$$z(\theta, t) = S.t^{\frac{1}{2}} + a(\theta).t + b(\theta).t^{3/2} \quad + \tag{8}$$

By applying the basic continuity equation to the soil profile, the cumulative inflow at the upper surface F(t) can be equated to the outflow at the bottom of the semi-infinite soil column $(K_o.t)$ and the increase in the volume of the soil moisture in the column which can be obtained by integrating each term in equation (8) from the initial moisture content (θ_o) to the surface moisture content (θ_1). This results in the expression:

$$F(t) = S.t^{\frac{1}{2}} + (A + K_o).t + B.t^{3/2} + C.t^2 + \ldots \tag{9}$$

where A, B, C etc. are the result of such integration from θ_o to θ_1 of $a(\theta)$, $b(\theta)$, $c(\theta)$ etc. respectively.

The basic series solution of equation (9) for the infiltration problem allowing for both capillarity and gravity can be solved numerically for any set of empirical data describing the variation of the diffusivity D and unsaturated conductivity K as functions of the moisture content θ. Variations in these soil moisture characteristics will inevitably produce differences in the shape and movement of the soil moisture profile and in the resulting values for cumulative infiltration F(t) and the corresponding rate of infiltration f(t).

Checking on the convergence properties of the series represented by equation (9), revealed that the series, while not convergent for very long values of the elapsed time, did converge rapidly for smaller and intermediate values of time in the case of real soils (Philip 1957b, Philip 1969e). The limit of applicability of the above series solution was taken as equal to the elapsed time t_g at which the effect of gravity becomes equal to the effect of capillarity. For the case of Yolo light clay (Moore 1939) the range of convergence was estimated as 10^6 seconds (i.e. about 12 days). This soil with its distinctive non-linear properties became the one most commonly used in the soil physics literature as an example because of its use by Klute, Philip, and de Vries (Constanz 1987).

For longer time periods, the profile proved to be asymptotic to a stable profile (Philip 1957c). In a number of subsequent papers John Philip presented successively more rigorous and more insightful results for the joining of the series solution with the large t solution of the stable profile, either using 2 terms or 4 terms in equation (8).

Forty years after John's initial presentation of the series solution, Salvucci (1996) achieved sufficient improvement in convergence to obviate need for a joining technique by replacing the series in $t^{1/2}$ by a similar series in terms of the transformed variable t′ defined by

$$t' = \frac{t}{t+a} \qquad (10)$$

which has the effect of replacing the original infinite range from zero to infinity by a finite range from zero to one. In his paper, Salvucci used a value of the parameter a equal to one half of the characteristic time (t_g) defined by John Philip as the time when the effect of the gravity and capillary forces became equal and used by him as an estimate of the limit of applicability of the series solution (Philip 1969c pp250-251)

4.4. Special Analytical Solutions

In order to gain insight into the range of variation in profile shapes and infiltration rates, it is desirable to seek analytical solutions which will reveal the extent to which the differences in soil characteristics materially effect the soil moisture movement in the unsaturated zone. The first attempt at an analytical solution was that by Green and Ampt (1911), based on a physical analogy with the capillary rise in narrow cylindrical tubes. John Philip in an early paper

pointed out that the abrupt wetting front of the Green and Ampt model, with its rectangular moisture profile, could be generalized to deal with the case of similar moisture profiles through a small modification of the original equation (Philip 1954, pp 155-156). He pointed out later (Philip 1957d, p.260), that the physical feature of an abrupt wetting front was equivalent to the mathematical discontinuity of a delta-function form in the diffusivity function D(θ). Thus, throughout his subsequent work one will find references to the similarity solution and to the delta-function solution, both of which correspond essentially to the modified Green and Ampt approach.

In 1966 John presented to the Wagingenen Conference on Water in the Unsaturated Zone a paper entitled "The linearization technique for the study of infiltration" (Philip 1969a). This was concerned with a mathematical simplification of the basic infiltration equation accounting for both capillarity and gravity through the assumption of a constant diffusivity D and linear variation between unsaturated conductivity K and moisture content θ. This particular model gives an analytical solution involving the complementary error function which became a matter of fascination for him for some years. In this Wagingenen paper (Philip 1969a), John Philip plotted in dimensionless form the decline in infiltration for three particular cases (a) the non-linear series solution for Yolo light clay (Moore 1939) with four terms of the series used; (b) the equivalent similarity or delta-function solution derived from Green Ampt; and (c) the equivalent solution of the linear equation obtained by taking diffusivity D as a constant and the unsaturated conductivity K as a linear function of the moisture content. In spite of the markedly different shapes for the moisture profiles, the dimensionless curves for the variation of infiltration with time did not differ greatly. Even though the dimensionless curves of declining infiltration rate plotted closely together, the non-linear series solution plotted outside and below the region bounded by the two linear solutions with their contrasting assumptions of isolated delta function diffusivity and uniform constant diffusivity. (Philip 1969a p476, Philip 1969e p274)

In a paper published a few years later, John Philip presented another interesting comparison of the limiting cases of delta–function diffusivity and constant diffusivity, combined with linear conductivity (Philip 1973). The comparison was made by plotting a dimensionless flux against a dimensionless moisture content (Philip 1973 p333). The delta-function model produced a lower limit and the linear case an upper limit with a small sub-area of the possible variation in between. Real soils have a D(θ) relationship between these two models and this carries over into the flux-concentration relationship. This was exemplified by including the result of the numerical solution for Yolo light clay which (Moore 1939) plots approximately half-way between the two limiting analytical solutions.

Later, John Philip used another special solution based on the Fujita-Knight soil which can be solved analytically in the form of Burgers equation (Fujita 1952; Philip 1974; Philip and Knight 1974; Knight and Philip 1974). This solution proved useful in the joining of the two asymptotic solutions because it provides the best available estimate for the approach to the traveling wave solution which is the asymptotic solution for large values of elapsed time t (Philip 1987, Philip 1990).

5. EXTENDING THE BASIC SOLUTION

5.1. Thermal Effects

At the same time that John Philip was studying the basic problem of one-dimensional downward infiltration in a homogeneous stable medium, he was also working on problems arising in the extension of this basic solution to problems such as thermal transport and capillary rise, the extension beyond one dimensional solutions, and extension to non-homogeneous and to unstable soils.

John Philip's work in relation to thermal transport was largely carried out as a result of his close research partnership with Daniel de Vries in the three year period from 1956 to 1958. This work was stimulated by the discrepancy between the measured and predicted values for vapour diffusion in porous media. The results of this collaboration were published in a notable series of papers which made a landmark breakthrough in the study of this problem (Philip 1957a, Philip and de Vries 1957, de Vries 1958). The discrepancy between theory and measurement was resolved by a complex analysis which revealed that the water isolated around the points of contact of the soil particles were regions of enhanced transport rather than barriers to water transport. In 1960, John Philip wrote two further papers giving theoretical reasons why the thermal effects on unsaturated flow due to heat of wetting were quite unimportant. This conclusion was later verified by experimental studies.

During the thirty years following the publication of these papers much research both experimental and theoretical was carried out by others on this particular topic but the revolutionary nature of the breakthrough by Philip and de Vries was confirmed rather than replaced by this work (de Vries 1987).

It was typical of John Philip's interest in the whole environmental water continuum that this analysis was applied to the problem of evaporation from bare soil across the boundary of the soil-atmosphere continuum. John returned to the question of thermal effects in water transport some 40 years later when he studied the effect of heterogeneiety on this problem (Philip and Kluitenberg 1999).

5.2. Capillary Rise

The basic infiltration problem, as discussed in section 4 above, was concerned with the semi-infinite profile of homogeneous stable soil with an initial moisture content less than saturation, combined with the imposition of an upper boundary condition of a higher moisture content at the initial time t = 0. The corresponding problem of a uniform semi-infinite column of homogeneous stable soil with an initial moisture content less than saturation and a boundary condition at the lower boundary of moisture content greater than this initial condition was also tackled by John (Philip 1969c). His discussion of this capillary-rise problem paralleled the development of the theory of infiltration in his series of seven papers on the theory of infiltration in 1957 but differences and difficulties soon appeared.

The solution for the case of absorption (i.e. for very small values of elapsed time t) paralleled the solution for the basic infiltration case (Philip 1957b, page 349), and thus provided a similar starting point (Philip 1969c, p 560). For the capillary rise case at intermediate values of elapsed times, the terms in the series solution were alternating in sign but with the same absolute value of the co-efficients as in the basic case for downward infiltration. The limit for an intermediate limiting value of t can be calculated for the capillary rise case in the same way as for the downward infiltration case and an expression can also be found for the equilibrium profile in terms of the diffusivity function $D(\theta)$ and the conductivity function $K(\theta)$.

However, the solution which is represented in the downward infiltration case by the large t asymptotic solution (Philip 1957c) is not readily applicable and the gap is more serious between (a) the profile given for small values of elapsed time t by the series solution in its convergent range and (b) the solution at infinity. This was illustrated for the case of Yolo light clay with the following result (Philip 1969c, p. 562). The infiltration rate for the case of downward flow reaches over 90% of its final quasi-equilibrium value after an elapsed time $t_9 = 10^6$ seconds (i.e. about 12 days). For the case of capillary rise after the same lapse of time the total moisture content of the column is only about 20% of the value at final equilibrium. John filled this gap by use of linearisation using the special soil of constant diffusivity (D) and linear conductivity $K = k.\theta$ (Philip 1969a), discussed in the section 4.4.

5.3. Two- and Three-Dimensional Problems

Extending the basic problem of one-dimensional infiltration to 2 and 3 dimensions naturally created difficulties in analysis. The strategy of simplifying from the infiltration problem (involving both capillarity and gravity), to the absorption problem (involving only capillarity), again proved to be a useful starting point. Such a solution would be applicable to a very fine-grained soil and to other soils for small values of elapsed time. The analysis (Philip 1969b;Philip 1969e), was first conducted for the two-dimensional case of a semi-circular furrow and the three dimensional case of a hemispherical cavity, both of which are of practical significance being related respectively to furrow infiltration and infiltration from a shallow ring infiltrometer. In each case, there is now an added parameter of significance in the shape of the radius r_0 of the semi-circle or of the hemisphere. For the case of capillarity alone (and the allied case of the solution for small values of the elapsed time), the single term solution for one-dimensional absorption is replaced by a series solution in terms of the square root of the elapsed time and the reciprocal of the radius r_0. The solution for large values of the elapsed time t indicates that for the case of three-dimensional absorption, there exists a steady phase of infiltration even in the absence of gravity.

For the full infiltration equation, including both gravity and capillarity, the situation is necessarily still more complex and turns out to be even more difficult to resolve. The steady state form of the equation can be analysed by taking the unsaturated conductivity as an exponential function of the moisture potential and thus reducing the steady state equation to linear form. For the full equation in its unsteady form the situation is more complicated still. In this case, progress can only be made by retaining the exponential relationship between unsaturated conductivity K and soil moisture potential Ψ and adding the further simplification that (a) the diffusivity is constant or (b) the unsaturated conductivity K varies linearly with the moisture content θ. These are equivalent since each of which implies the other in the case of the K-Ψ relationship being exponential.

5.4. Non-Homogenous Stable Soils

The assumption of a stable homogenous soil in the basic solution described in section 4.3 above was relaxed in further studies dealing with heterogeneous soils, aggregated soils, and layered soils. Because the profiles of capillary potential preserve similarity under geometrical scaling, whereas the moisture profiles do not, the analysis is performed with the potential Ψ rather than the moisture content θ as the single dependent variable. The solution for the infiltration case consists of a series in powers of $t^{1/2}$ where the leading term refers to the absorption case which applies strictly only at very small values of t. In moving beyond these gravity free solutions, John Philip chose three special forms of homogeneity for further study. These were (a) scale heterogeneous media (Philip 1967), (b) aggregated media (Philip 1968), and (c) crusted soils (Philip 1998).

In his first paper on the subject (Philip 1967), John described his simplification of the type of heterogeneity as follows:

"We propose to study a limited class of heterogeneiety that we shall call scale-heterogeneiety. *We shall mean by a* scale-heterogeneous *medium one in which the internal geometry is everywhere geometrically similar but in which the characteristic internal length scale is free to vary spatially."*

In order to advance the analysis beyond the simple absorption case, John made the simplifying assumption that, when scaled according to the characteristic lengtrh scale, the capillary potential (ψ) decreases exponentially with the moisture content (θ) and the unsaturated conductivity (K) is inversely proportional to the square of the capillary potential (ψ). These combine according to equation (2) in section 4.2 above to give diffusivity (D) as an exponential function of the moisture content. The subsequent analysis shows that (a) the history of the profile of capillary potential is independent of the spatially varying length scale (λ), (b) the potential profiles preserve similarity at all times, and (c) the cumulative infiltration is given by an equation identical to equation (6) in section 4.2. John draws attention to the consequence of (c) that the observation of a sorption rate inversely proportional to the square root of the elapsed time does not necessarily indicate homogeneiety of the porous medium.

In his second paper dealing with non-homogeneous media, John Philip deals with what he calls an aggregated medium. He writes (Philip 1968 p.2):

"We here regard an aggregated medium as one in which the pore space is made up of two distinct types of porosity, macroporosity *and* microporosity. *The* macropore space *consists of a continuous, but multiply connected, part of the pore space characterised by large pore dimensions. The* micropore space *consists of the remaining pore space, characterized by small pore dimensions and the tendency to occur in a large number of small and isolated regions of the total pore space".*

The analysis makes the following basic assumptions (Philip 1968 p.3):

"(1) Water transfer on the Darcy scale (i.e. over distances large compared with a characteristic macropore dimension) is assumed to be via the macroporosity only, and the process of transfer is taken to be similar to that in a classical porous medium. (2) The exchange of water between the macroporosity and the microporosity is describable as a distributed sink (or source), the strength of which is a function of ψ_1 and ψ_2, the local capillary

potentials in the macroporosity and the microporosity respectively."

Absorption in the region was first studied for the two limiting cases of (a) the delta-function solution and (b) the linear case of constant D and linear conductivity K. The region bounded by these two cases was found to be small and the further analysis was based on the former approach which is the easier one.

A particular form of heterogeneity occurs in natural soils which are frequently layered and sometimes crusted. Previous attempts to deal with layered soils were based on the Green-Ampt approach or on the zero gravity simplification or a combination of both. Philip (1998c) wrote two separate equations for the crust and for the main soil profile and derived solutions for both small values and for large values of the elapsed time t. For very small values of t, the infiltration was confined to the crust and the absorption similarity solution with a single term in $t^{\frac{1}{2}}$ can be applied. For large values of t, the solution for a homogeneous soil carries over into the crusted soil case with only minor modifications. The solution for the infiltration rate at intermediate times is based on the flux-concentration relationship (Philip 1973).

5.5. Unstable Homogeneous Soils

In the same year that John Philip published his 82-page comprehensive review of the theory of infiltration (Philip 1969e), there appeared two important papers from his group on the problems arising in extending the classical solution of the basic problem to the case of soils subject to colloidal swelling (Philip and Smiles 1969; Philip 1969f). These papers were concerned with analyzing the equilibrium conditions of such soils in which the total potential includes an overburden potential as a third element in addition to the moisture potential and the gravity potential dealt with in the basic analysis. In order to allow for the overburden potential in the total potential it was necessary to add to the two original soil properties of moisture potential (Ψ) and conductivity (K) the new property of the void ratio (e) and to express it as a function either of the moisture content (θ) or of the moisture potential (Ψ). It is also necessary to express the equation in terms of material co-ordinates relative to the soil particles rather than to the original fixed natural co-ordinates.

In this important extension of his classical work on unsaturated vertical flow, John Philip followed his customary pattern of proceeding from a known result for a simple case to the analysis of the more complex problem. This is made clear in one of the earlier papers on the topic (Philip and Smiles 1969 p.3) which states:

"We limit consideration here to systems in which gravity is unimportant i.e. to horizontal systems and other systems in which differences in gravitational potential is unimportant i.e. to horizontal systems and other systems in which differences in gravitational potential are negligible in comparison with differences in moisture potential (e.g. the early stages of sorption in vertical columns)."

In another paper published in the same year (Philip 1969h p.1071), John writes:

"Mathematical methods developed in connection with the classic 'diffusion analysis' (Philip 1969e) thus either apply directly or provide a useful point of departure."

When this approach was applied to the problem of absorption in the absence of gravity, the derived equation for the profile in terms of material coordinates was of the same form as the original equation for a stable soil, i.e. the depth of penetration was proportional to the square root of the elapsed time. However, on conversion of this result back into natural co-ordinates, the coefficient function $\phi(\theta)$ which varied with the moisture content (θ) was not of such form that would allow for a solution through inversion as in the basic case.

Subsequent papers published in the next three years (Philip 1969g, Philip 1970, Philip 1971)include analyses and illustrated examples revealing a number of surprising results in regard to the moisture profiles. The effect of including the overburden potential and the void ratio is to reduce the effect of gravity. Under certain conditions this can result in an upward flow against the moisture gradient. This analysis synthesizing aspects of both classical infiltration theory and classical soil mechanics comes as a shock to the hydrologist who has also to come to terms with a new nomenclature involving such items as the pycnotatic point which separates the hydric range of reverse flow from the xeric range of classical infiltration or capillary rise flow. The above anomalous behaviour for a downward moisture gradient has an interesting analogy with the classical solution for capillary rise. John Philip continued to work on this problem and still published papers in this connection in the 1990s (Philip 1992, Philip 1995).

6. CONCLUDING REMARKS

The above attempt to describe John Philip's research strategy is of necessity broad and limited in detail. The emphasis is on the nature of the strategy and on the continuity of his approach over a period of 50 years of research activity. Though the single topic of infiltration was chosen to illustrate the general points put forward, the same combination of insightful simplification, rigorous analysis, and thorough post-hoc review is evident in John Philip's approach to other problems. His success in a number of research areas should encourage the rising generation of young researchers to appreciate the place of analysis in science and to realise that key advancements in knowledge do not depend on replacing the pursuit of insight through analysis by still more complex models, extra parameters to be calibrated, and larger computers to crunch out numbers.

REFERENCES

ASCE, *Hydrology Handbook,* American Society of Civil Engineers, New York.,1949.

Buckingham, E., Studies on the movement of soil moisture, *Bur. of Soils Bull. 38,* U.S. Department of Agriculture, Washington D.C., 1907.

Childs, E.C., The transport of water through heavy clay soils, 1, *Journal of Agricultural Science,* 26, 114-127, 1936.

Childs, E.C., and N. Collis-George, The permeability of porous materials, *Proceedings of the Royal Society of London,* A201, 395-405, 1950

Constanz, R., R.E. Moore and Yolo Light Clay, in *History of Geophysics Volume 3,* pp 99-101. American Geophysical Union, Washington. 1987.

de Vries, D.A., Simultaneous transfer of heat and moisture in porous media, *Trans AGU,* 39 900-919, 1958.

de Vries, D.A., The theory of heat and moisture transfer in porous media revisited, *International Journal of Heat and Mass Transfer,* 30, 1343-1350, 1987.

Fujita, H., The exact pattern of concentration dependent diffusion in a semi-infinite medium. Part II, *Textile Res. J.,* 22, 823-827, 1952.

Gardner, W.H., Early soil physics into the mid-20th century, *Adv. Soil Science,* 4, 1-101, 1986.

Green, and Ampt, Studies on Soil Physics: 1-Flow of Air and Water Through Soils, *Jour. Agr. Res.,* 4, 1-24, 1911.

Klute, A., A numerical method for solving the flow equation for water in unsaturated soils, *Soil Science,* 73, 105-16, 1952a.

Klute, A., Some theoretical aspects of the flow of water in unsaturated soils, *Soil Science of America Proc.,* 16, 144-48, 1952b.

Knight, J.H., and J.R. Philip, Exact solutions in non-linear diffusion, *J. Eng. Math.,* 8, 219-227, 1974.

Linsley, R.K., Kohler, M.A., and J.L.H. Paulhus, *Applied Hydrology,* McGraw-Hill, New York, 689 pp. 1949.

Moore, R.E., Water conduction from shallow water tables, *Hilgardia,* 12, 383-426, 1939.

Pedlosky, J., *Geophysical Fluid Dynamics,* Springer-Verlag, New York, 1987.

Philip, J.R., An infiltration equation with physical significance, *Soil Sci.,* 77,153-157, 1954.

Philip, J.R., The concept of diffusion applied to soil water, *Proc. Natl. Acad. Sci. India,* 24(A),93-104, 1955.

Philip, J.R., Transient fluid motions in saturated porous media, *Aust. J. Phys.,* 10, 43-53, 1957a.

Philip, J.R., The theory of infiltration: 1. The infiltration equation and its solution, *Soil Sci.,* 83, 345-357, 1957b.

Philip, J.R., The theory of infiltration: 2. The profile at infinity, *Soil Sci.,* 83, 435-448, 1957c.

Philip, J.R., The theory of infiltration: 4. Sorptivity and algebraic infiltration equations, *Soil Sci.,* 84,257-264, 1957d.

Philip, J.R., The role of mathematics in soil physics, *Aust. Inst. Agric.Sci.,* 23, 293-301, 1957e.

Philip, J.R., Plant water relations: some physical aspects, *Annu. Rev. Plant Physiol.,* 17, 245-268, 1966.

Philip, J.R., The theory of absorption in aggregated media, *Aust. J. Soil Res.,* 6, 1-19, 1968.

Philip, J.R., A linearization technique for the study of infiltration, in *Water in the Unsaturated Zone,* edited by R.E. Rijtema and H. Wassink, Proc. IASH/UNESCO Symp. Wageningen, 1966, 1, pp. 471-478, UNESCO: Paris, 1969a.

Philip, J.R., Absorption and infiltration in two-and three-dimensional systems, in *Water in the Unsaturated Zone, edited by* R.E. Rijtema and H. Wassink, Proc. IASH/UNESCO Symp. Wageningen, 1966, 1, pp. 503-525, UNESCO Paris. 1969b.

Philip, J.R., The dynamics of capillary rise, in *Water in the Unsaturated Zone,* edited by R.E. Rijtema and H. Wassink, Proc. IASH/UNESCO Symp.Wageningen. 1966. Vol.2, pp. 559-564, UNESCO Paris. 1969c.

Philip, J.R., The soil-plant-atmosphere continuum in the hydrological cycle, *Hydrological Forecasting,* Tech. Note No. 92(1969), W.M.O. No.228.TP No.122, pp. 5-12. 1969d.

Philip, J.R., Theory of infiltration, *Adv.Hydrosci.,* 5, 215-296 1969e.

Philip, J.R., Moisture equilibrium in the vertical in swelling soils. 1. Basic theory, *Aust J. Soil Res.,* 7, 99-120, 1969f.

Philip, J.R., Moisture equilibrium in the vertical in swelling soils. II. Applications, *Aust J.Soil Res.,* 7, 121-141, 1969g.

Philip, J.R., Hydrostatics and hydrodynamics in swelling soils, *Water Resour. Res.,* 5, 1070-1077, 1969h.

Philip, J.R., Hydrostatics in swelling soils and soil suspensions: unification of concepts, *Soil Sci.,* 109, 294-298, 1970a.

Philip, J.R., Flow in porous media, *Ann.Rev.Fluid Mech.,* 2, 177-204, 1970b.

Philip, J.R., Hydrology of swelling soils, in *Salinity And Water Use,* edited by T. Talsma and J.R. Philip, Proc. Symp. Aust. Academy of Science, Canbara, 1971, pp. 95-107, Macmillan, London, 1971.

Philip, J.R., Future problems of soil research, *Soil Sc.,* 113, 294-300, 1972.

Philip, J.R., On solving the unsaturated flow equation. 1. The Flux-concentration relation, *Soil Sci.,* 116, 328-335, 1973.

Philip, J.R., Recent progress in the solution of non-linear diffusion equations, *Soil Sci.,* 117, 257-264, 1974.

Philip, J.R., The infiltration joining problem, *Water Resour. Res.,* 23, 2239-2245, 1987.

Philip, J.R., Inverse solution for one-dimensional Infiltration and the ratio A/K_1, *Water Resour. Res.,* 26(9), 2023-2027, 1990.

Philip, J.R., Soils, natural science and models, *Soil Sci.,* 151, 91-98, 1991.

Philip, J.R., Flow and volume change in soils and other porous media, and in tissues, in *Mechanics of Swelling: From Clays to Living Cells and Tissue,* edited by T.K. Karalis, pp. 3-31, Springer-Verlag, Berlin, 1992.

Philip, J.R., Phenomenological approach to flow and volume change in soils and other media, *Applied Mechanics Review,* 48(10), 650-658, 1995.

Philip, J.R., Infiltration, in *Encyclopedia of Hydrology and Water*

Resources, edited by R. Herschy and R. W. Fairbridge, Kluwer,Dordrecht, pp. 418-426, 1998a.

Philip, J.R., Water movement in unsaturated soils, in Encyclopedia of Hydrology and Water Resources, edited by R. W. Herschy and R. W. Fairbridge, Kluwer, Dordrecht, pp. 699-706, 1998b.

Philip, J.R., Infiltration into crusted soils, Water Resour. Res., 34, 1919-1927, 1998c.

Philip, J.R., Physics, mathematics and the environment: The 1997 Priestley lecture, Aust. Meteorological Mag., 47, 273-283, 1998d.

Philip, J.R., and D.A, de Vries, Moisture movement in porous materials under temperature gradients, Trans Am. Geophys. Union, 38, 222-232, 1957.

Philip, J.R., and C.J. Kluitenberg, Errors of the dual thermal probes due to soil heterogeneity across a plane interface, Soil Sci. Amer. J., 63, 1579-1585, 1999.

Philip, J.R., and J.H. Knight, On solving the unsaturated flow equation. 3. New quasi-analytical technique, Soil Sci., 117, 1-13, 1974.

Philip, J.R., and D.E. Smiles, Kinetics of sorption and volume change in three-component systems, Aust. J. Soil Res., 7, 1-19, 1969.

Polya, G., How to solve it, Princeton University Press, 1945.

Polya, G., Mathematics and Plausible Reasoning, Vol 1.Induction and Analogy in Mathematics 280 pp. Vol II. Patterns of Plausible Inference 190 pages, Princeton University Press, 1954.

Richards, L.A., Capillary conduction of liquids through porous mediums, Physics I., 318-33, 1931.

Richards, L.A., and C.H. Wadleigh, Soil water and Plant growth, in Soil Physical Conditions and Plant Growth, edited by B.T. Shaw, pp. 73-251, 1952.

Salvucci, G., Series solution for Richards equation under concentration boundary conditions and uniform initial conditions, Water Resour. Res., 32(8), 2401-2407, 1996.

Smiles, D., The Environmental Mechanic, Aust. J. Soil Sc., 39(4), 649-681, 2001

James C. I. Dooge, Centre for Water Resources Research, University College Dublin, Earlsfort Terrace, Dublin 2, Ireland.

Infiltration Under Constant Head and Falling Head Conditions

D.E Elrick[1], R. Angulo-Jaramillo[2], D.J. Fallow[3], W.D. Reynolds[4], and G.W. Parkin[5]

Prediction of the infiltration of water into field soils requires knowledge of the field-saturated hydraulic conductivity and a second parameter, such as the matric flux potential (corresponding to field saturation), or the Green and Ampt wetting-front pressure head, or the alpha parameter. Analytical solutions of 1-D infiltration under both constant and falling-head conditions are reviewed and several new solutions are developed based on the Green and Ampt assumptions. A laboratory experiment using the falling head technique is analyzed using several approximate analytical solutions.

1. MEMORIES OF JOHN PHILIP

John Philip was the keynote speaker at the "Conference on Advances in Infiltration", sponsored by the American Society of Agricultural Engineers, Chicago, 1983. Here are some selected quotes [Philip, 1983]. "I am uncomfortably aware that it is now 30 years since I did the basic work on solving the infiltration equation [present authors- now more than 45 years] ... " and "Over the last decade much attention has been given to what is commonly called the Green-Ampt model of infiltration. I wonder how many of you have gone back to the original paper of Green and Ampt [1911]. If you have you may have noticed that W.

Heber Green was "Lecturer and Demonstrator" in Chemistry at the University of Melbourne, and that G.A. Ampt was a graduate "Research Scholar". Thirty two years later when I did freshman chemistry at that same University of Melbourne, Gussy Ampt (as he was known) still lurked in the Chemistry school". John relished his historical digressions. Perhaps his best known digression from an announced talk was at a Retirement Symposium for him in 1992, sponsored by the Soil Science Society of America in Minneapolis. Rather than "Capillarity, Gravity and Geometry interact to shape Unsaturated Flow", his talk was entitled "Desperately Seeking Darcy in Dijon". It was a remarkable performance, based on his recent trip to Dijon, and a welcome relief from his seminar performances, where he studiously read from his notes and then terrified the audience by not tolerating uninformed questions. Indeed he was, at times, a "curmudgeon", as he often referred to himself.

For many of us John Philip was best known for his work on infiltration. Here we take another look at infiltration under both constant head and falling-head conditions, areas where John Philip published extensively, and we offer some new insights using the Green-Ampt model.

2. INTRODUCTION

Measurements of both field-saturated hydraulic conductivity, K_{fs} (L/T) and a second parameter characterizing the unsaturated flow properties, such as the field-saturated matric flux potential, ϕ_m (L^2/T), the Green and Ampt wetting-front pressure head ψ_f (L) or the alpha parameter $\alpha*$ (L^{-1}) are required for the description and

[1] Department of Land Resource Science, University of Guelph, Guelph, ON, Canada
[2] Laboratoire d'édute des Transferts en Hydrologie et Environnement (UMR 5564 CNRS, UJF, INPG, IRD), Grenoble, France
[3] Department of Land Resource Science, University of Guelph, Guelph, ON, Canada
[4] Greenhouse and Processing Crops Research Centre, Agriculture and Agri-Food Canada, Harrow, ON, Canada
[5] Department of Land Resource Science, University of Guelph, Guelph, ON, Canada

Environmental Mechanics: Water, Mass and Energy Transfer in the Biosphere
Geophysical Monograph 129
Copyright 2002 by the American Geophysical Union
10.1029/129GM04

prediction of ponded infiltration into unsaturated soil. Most field procedures for measuring K_{fs} and ϕ_m, ψ_f or $\alpha*$ use rings (single or double) and measure the flow under constant head conditions until the flow has reached steady state. *Elrick et al.* [1995)] pointed out that a difficulty with this approach is that insufficient information is obtained from the measurement of the steady flow rate under a single constant head to evaluate both K_{fs} and ϕ_m, ψ_f or $\alpha*$. Furthermore, waiting for steady flow using either single or double ring infiltrometers may take very long periods of time, particularly in slowly permeable soils. However, measurements under falling head conditions, or a combination of a constant head followed by a falling head, can provide sufficient information from which both K_{fs} and ϕ_m, ψ_f or $\alpha*$ can be obtained with reasonable accuracy. Early time transient flow measurements in slowly permeable soils can reduce the measurement times from days to hours, relative to steady flow rate measurements. The expressions presented by *Elrick and Reynolds, [1992] and Elrick et al.* [1995] and *Fallow et al.* [1994] are re-examined and alternative, and possibly more accurate solutions are presented.

3. INFILTRATION EQUATIONS: BASED ONLY ON SORPTION

Sorption is the term used when gravitational effects during infiltration into unsaturated soil can be considered to be negligible. Sorption applies directly to horizontal infiltration, as well as to the early time period of vertical infiltration. We also assume that the early time period of vertical infiltration is restricted to one-dimensional flow. The three-dimensional aspects of ring infiltration in the field can be ignored during this early time.

3.1 Constant Head

Philip [1957,1958] showed that early time one-dimensional infiltration can be described by:

$$I(t) = S_H t^{1/2} \qquad (1)$$

where S_H ($L/T^{1/2}$) is the soil sorptivity at the ponded head H (L) and I (L) is the cumulative infiltration. *White and Sully* [1987] developed an approximate equation for S_H:

$$S_H = \left\{ \frac{(\Delta\theta)\phi_m}{b} + 2(\Delta\theta)K_{fs}H \right\}^{1/2} \qquad (2)$$

where $(\Delta\theta)$ is the difference between the field-saturated water content, θ_{fs} (L^3/L^3), and the initial water content, θ_i (L^3/L^3). Setting b = 0.55 gives an error of less than 10% in

S_H. The first term in (2) gives the sorptivity, S_0, for H = 0 and the second term gives the increase in sorptivity due to the positive (ponded) head H.

In the *Green and Ampt* [1911] (G&A) approach the soil is assumed to be saturated down to a presumed sharp wetting front, x_f [L]. We use x_f for distance when infiltration is based only on sorption. In section 4, we use z for distance when infiltration is based on both sorption and gravity. The wetting front forms a boundary between the field-saturated and the initially unsaturated soil where the soil water pressure head at the wetting front is ψ_f (L), with ψ_f negative. The hydraulic conductivity in the field-saturated zone is given by K_{fs}. The water flux density or infiltration rate, q (LT^{-1}), is thus given by the following form of Darcy's law:

$$q = -\frac{K_{fs}(\psi_f - H)}{x_f} \qquad (3)$$

From continuity q is also given by:

$$q = (\Delta\theta)\frac{dx_f}{dt} \qquad (4)$$

Equating (3) and (4), carrying out the integration and noting that the cumulative infiltration is given by

$$I = (\Delta\theta)x_f \qquad (5)$$

gives (1) with S_H given by [*Philip*, 1958]:

$$S_H = \left[2K_{fs}(\Delta\theta)(H - \psi_f)\right]^{1/2} \qquad (6)$$

Eq. (2) and (6) represent two different approximations of S_H. Note that ϕ_m and ψ_f are related by:

$$\phi_m = -K_{fs}\psi_f \qquad (7)$$

Substituting (7) into (6) shows that the only difference between (2) and (6) is in the estimation of b. In theory, b = 0.5 for G&A soils and b = 0.785 for Linear soils [*White and Sully*, 1987]. In addition, *White and Sully* [1987] have shown that b = 0.55 is a good approximation for ponded infiltration into most soils.

3.2 Falling Head

For the falling head condition, replace H by H(t) in (3). If $H(0) = H_1$, the cumulative infiltration I can then be obtained experimentally from:

$$I(t) = R[H_1 - H(t)] \qquad (8)$$

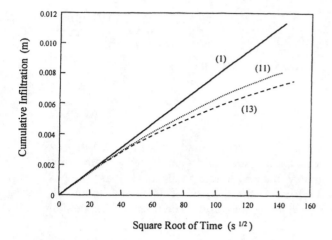

Figure 1. Cumulative infiltration under constant head [Eq. (1)] and falling head conditions [Eq. (11) and (13)]. Equation (1) uses (6) for S_H. Gravity effects were assumed to be negligible relative to sorption.

where $R = a/A$ is the ratio of the cross sectional area of the falling head reservoir, a, to the cross-sectional area of the infiltrating surface, A. For the falling head with H replaced by H(t) in (3), use of (4), (5) and (8) and rearrangement gives:

$$\frac{x_f dx_f}{x_f - B} = -\frac{K_{fs} dt}{R} \qquad (9)$$

where the constant B is given by:

$$B = R\left(H_1 - \psi_f\right)/\Delta\theta \qquad (10)$$

Integrating (9) assuming that $x_f = 0$ at $t = 0$ gives the following implicit equation in I:

$$t = -\frac{R}{K_{fs}(\Delta\theta)}\left[I(t) + (\Delta\theta)B \ln\left(1 - \frac{I(t)}{(\Delta\theta)B}\right)\right] \qquad (11)$$

Equation (11) is a newly developed expression and is valid until the falling head drops to zero, given by:

$$0 \le I \le RH_1 \qquad (12)$$

Figure 1 is a plot of the cumulative Infiltration, I, vs. square root of time, $t^{\frac{1}{2}}$, for a typical slowly permeable soil such as clay liner material. Here $K_{fs} = 10^{-8} ms^{-1}$, $\psi_f = -0.25m$, $H_1 = 1m$ and $\Delta\theta = 0.25$. The constant head expression [(1) with (6)] is linear with $t^{\frac{1}{2}}$, whereas the

G&A falling head expression, (11), falls off with time as expected with a falling head. We have also plotted an approximate solution based on substituting H(t) from (8) into (1) and using (6), as suggested by *Fallow et. al.* (1994), to obtain:

$$t = \frac{I^2(t)}{2(\Delta\theta)K_{fs}\left[H_1 - \frac{I(t)}{R} - \psi_f\right]} \qquad (13)$$

Fallow et. al. (1994) expressed I(t) in (13) in terms of H(t) [from (8)] when fitting for the hydraulic parameters:

$$t = \frac{R^2\left[H_1 - H(t)\right]^2}{2(\Delta\theta)K_{fs}\left[H(t) - \psi_f\right]} \qquad (14)$$

Note that (11) and (13) in Figure 1 are similar at early times but deviate as time increases. Both expressions are approximations. However, (11) is a true solution under G&A falling head conditions, whereas (14) is an approximation based on a succession of constant head solutions. Due to the use of the sorptivity expressions which assume negligible gravity effects, it is expected that more accurate estimates of K_{fs} and ψ_f will be obtained when (11) or (13) are fitted to only early-time I vs. t or H vs. t data.

3.3 Sequential Constant Head 1/Constant Head 2

Here we assume a constant head, H_1, for $0 \le t \le t_c$ and a different constant head, H_2, for $t \ge t_c$. The solution can be obtained using a similar approach to that used in *Section 3.1*, with the exception that the solution uses the initial condition that for $t = t_c$, $x_f = S_{H1}t_c^{\frac{1}{2}}/\Delta\theta = I_c/\Delta\theta$:

$$I = \left[2(\Delta\theta)K_{fs}(H_1 - \psi_f)t_c + 2(\Delta\theta)K_{fs}(H_2 - \psi_f)(t - t_c)\right]^{\frac{1}{2}}$$

$$t \ge t_c \qquad (15)$$

Note that I is not linear with $t^{\frac{1}{2}}$ in (15). Equation (15), a newly developed expression, is not linear because the initial condition that θ be constant with depth no longer holds as a result of the initial infiltration period up to t_c. Thus sorptivity can only be measured using $I \propto t^{\frac{1}{2}}$ during the initial constant head period

Figure 2 (using the same parameter values as in Figure 1) illustrates (15) for $H_1 > H_2$. Note that (15) starts at $t = t_c$, falls off with time, and asymptotically approaches the straight line for the second constant head H_2.

3.4 Sequential Constant Head/Falling Head

Here we assume a constant head for $0 \le t \le t_c$ and a falling head for $t \ge t_c$. The G&A solution can be obtained using a similar approach to that used in *Section 3.2* with the exception that the solution uses the initial condition that for $t = t_c$, $x_f = S_{H1}t_c^{1/2}/\Delta\theta$:

$$t = t_c - \frac{R}{K_{fs}(\Delta\theta)}\left[I(t) - I_c + (\Delta\theta)B\ln\left(\frac{B}{B - \dfrac{I_c}{(\Delta\theta)}} - \frac{I(t)}{(\Delta\theta)\left(B - \dfrac{I_c}{(\Delta\theta)}\right)}\right)\right]$$

(16)

where $t > t_c$. In (16), a newly developed expression, $I_c = S_{H1}t_c^{1/2}$ and t_c is set arbitrarily or determined experimentally. An experimental advantage of using (16) where t_c is known is that H_1 is also known and does not have to be fitted in a non-linear least squares program. B is given by (10) where $H_1 = H(t_c)$. Note that (16) reduces to (11) for $t_c = I_c = 0$.

An alternative approximate solution based on the assumption that H_2 is a function of time can be obtained by substituting $H(t)$ from (8) into H_2 in (15) giving:

$$t = t_c + \frac{I^2(t) - I_c^2}{2(\Delta\theta)K_{fs}\left[H_1 - \dfrac{I(t)}{R} - \psi_f\right]}$$

(17)

Elrick et al. [1995] expressed (17) in terms of $H(t)$ when fitting for the hydraulic parameters. However, (13) in *Elrick et al.* (1995) is not correct as written and should be replaced by:

$$t = t_c + \frac{R^2[H_1 - H(t)]^2}{2(\Delta\theta)K_{fs}[H(t) - \psi_f]}$$

(18)

where $H_1 = H(t_c)$ and $H(t)$ is defined by (8).

Note that (16) and (17) are similar at early times in Figure 2, but like the comparable curves in Figure 1, deviate as time increases. As in section 3.2 both expressions are approximations; however, (16) is a true solution for the G&A assumptions whereas (17) is an approximation based on a succession of constant head solutions. As with (11) and (13), it is expected that more accurate estimates of K_{fs} and ψ_f would be obtained if (16) and (17) were fitted to only early-time I vs. t data, which are less affected by gravity.

4. INFILTRATION EQUATIONS: BASED ON SORPTION AND GRAVITY

For G&A infiltration (3) is replaced by:

$$q = -\frac{K_{fs}(\psi_f - H)}{z_f} + K_{fs}$$

(19)

where z_f is distance and defined as positive vertically downwards. Note that (19) includes both sorption and gravity in the hydraulic gradient, whereas (3) includes only sorption. With the use of (19) and following similar procedures as for sorption the following expressions can be obtained.

4.1 Constant Head

$$t = \frac{(\Delta\theta)}{CK_{fs}}\left[\frac{I(t)}{\Delta\theta} - \frac{(H - \psi_f)}{C}\ln\left(1 + \frac{CI(t)}{(\Delta\theta)(H - \psi_f)}\right)\right]$$

(20)

where

$$C = 1.$$

(21)

An equation, similar in form to (20), was derived *by Green and Ampt* (1911).

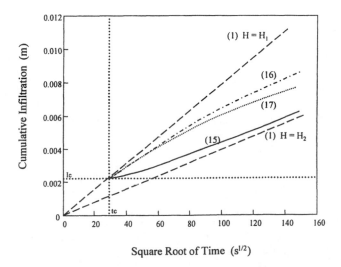

Figure 2. Cumulative infiltration under sequential constant head 1 / constant head2 [Eq. (15)] and constant head / falling head [Eq. (16) and (17)] conditions. Also shown are cumulative infiltration under continuous constant head conditions H_1 [Eq. (1)] and constant head conditions H_2 [Eq. (1)] where $H_2 < H_1$. Note that Eq. (1) plots as a straight line vs. $t^{1/2}$. Gravity effects were assumed to be negligible relative to sorption.

4.2 Falling Head

The G&A solution for a falling head was derived by *Philip* [1992] and is identical to (20) but with

$$C = 1-(\Delta\theta)/R \qquad 0 \le I \le RH_0 \qquad (22)$$

4.3 Sequential Constant Head 1/Constant Head 2

Here we assume a constant head, H_1, for $0 \le t \le t_c$ and a different constant head, H_2, for $t \ge t_c$. The solution can be obtained using a similar approach to that used in *Section 4.1* with the exception that the solution uses the initial condition that for $t = t_c$, $z_f = [S_{H1}t_c^{1/2}+(K_{fs}t_c)/3]/\Delta\theta = I_c/\Delta\theta$:

$$t = t_c + \frac{(\Delta\theta)}{CK_{fs}}\left[\frac{I-I_c}{\Delta\theta} - \frac{(H_2-\psi_f)}{C}\ln\left(\frac{(H_2-\psi_f)+CI/\Delta\theta}{(H_2-\psi_f)+CI_c/\Delta\theta}\right)\right] \quad (23)$$

and $C = 1$. In (23), which is a newly developed expression, $I_c = S_{H1}t_c^{1/2}+(K_{fs}t_c)/3$, where the second term takes into account the effect of gravity. If t_c is small, as it generally is, the contribution of the gravity term in the above Ic expression is negligible. The Philip two-term equation (a truncation of Philip's series solution and therefore an approximation (Philip,1957)) was used to calculate I_c rather than (20), as (20) is not explicit in I (or z_f).

4.4 Sequential Constant Head/Falling Head

The G&A solution for a falling head ($t \ge t_c$) is identical to (23) but with C given by (22) and $H_2 = H_1$.

4.5 Other Approximate Falling Head Solutions

In Section (3) we approximated the falling head solution by inserting H(t) into the constant head solution. We can use the same approach for solutions based on both sorption and gravity by using the Philip two-term infiltration equation:

$$I(t) = S_H t^{1/2} + At \qquad (24)$$

where A can be approximated by a constant with a value given approximately by $A = K_{fs}/3$. More exact infiltration equations have been derived and reviewed by *Parlange et al.* (1999) but it is likely that only the crude approximation given by (24) is necessary, especially given errors of observation in field data.

Plots of the infiltration equations are similar in form to those for sorption (Figures 1 and 2) with the exception that the gravity effect during infiltration becomes more dominant with time and increases the amount of infiltration.

5. LABORATORY TESTS

Fallow et al. [1994] report on a laboratory experiment using a falling head technique on a compacted clay soil. They compacted air-dried, sieved clay in a Proctor Density Apparatus to a bulk density of approximately 1.6 Mg/m³. A 1.5m long vertical tube was connected to the soil surface, filled with water, and falling head readings, H(t), collected. The first reading was at 30s and readings continued for 7 min at which time the water level in the vertical tube was approximately at the soil surface. The difference in water content, $\Delta\theta$, was determined to be 0.32. The data were plotted as H vs. $t^{1/2}$.

Equations (11) and (14) with (8) and (10) inserted, and (20) with (8) and (22) inserted, were fitted to the H vs. $t^{1/2}$ data using nonlinear least squares procedures to obtain estimates of K_{fs}, ϕ_m, and H at t = 0 (H_1). The number of data points used to obtain the fits was limited to 20.

Note that the plots of (11) and (20) in Figure 3 are indistinguishable, indicating the negligible influence of gravity at this very early stage of infiltration. The Fallow et al. [1994] approximation, based on (14), differs only at early times and in the approximation of H_1.

From Fig. 1 it appears that (14) [which is equivalent to (13)] and (11) are very similar at early times but they diverge progressively as time increases. This suggests that the fitting of the K_{fs} and ϕ_m parameters should give comparable results at early times, but progressively

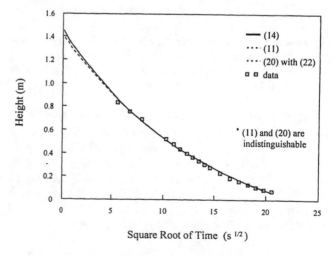

Figure 3. H vs. $t^{1/2}$ data from *Fallow et al.* (1994) plus fits of Eq. (11), (14) and (20) as described in the text. Note that all these equations produce similar fits to the data, and (11) and (20) are indistinguishable.

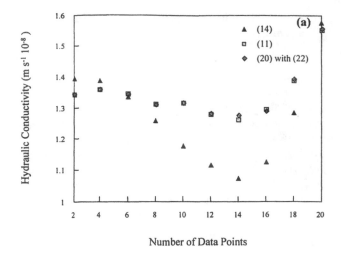

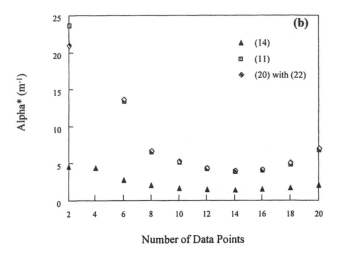

Figure 4. (a) Calculated hydraulic conductivity values (K_{fs}) and (b) calculated alpha* (α^*) values, both as a function of the number of data points used in the least squares analysis. Note that the smaller the number of data points, the earlier the time.

divergent results as time increases. To test this possibility, we conducted a series of 10 fits by progressively restricting the data to earlier times through sequential elimination of the two largest time data points. Consequently, the first fit used all 20 data points, the second fit used only the first 18 data points, and so on until the tenth fit was reached which only used the first two data points. The results for K_{fs} and α^* are shown in Fig. (4a) and (4b), respectively. Note that

$$\alpha^* = K_{fs}/\phi_m = -\psi_f^{-1}. \qquad (25)$$

We chose to plot α^* as it is the most sensitive parameter to the fitting procedure, being the ratio of the two hydraulic

parameters. The α^* parameter can also be interpreted as the slope parameter in [Gardner, 1958]:

$$K = K_{fs} e^{\alpha^* \psi} \qquad (26)$$

Fig. 4a shows that the calculated K_{fs} values are essentially identical for (11) and (20) using both early times (small number of data points) and later times (large number of data points). Equation (14) gives K_{fs} values that are very close to the above calculated values between n= 2 and n = 8 (early times), a maximum difference at n = 14 (intermediate time) and then values that converge again as n increases to 20 (later time). Unfortunately, this experiment does not give a good test of the early time period because of a combination of the small value of R (R = 1.1×10^{-3}) and the large K_{fs} value. R is the ratio of the cross-sectional area of the falling head reservoir to the cross-sectional area of the infiltrating surface. At n = 2 (early time), the head had already fallen by approximately 40%. We have no explanation of why the results tend to converge for n > 14, other than the sensitivity of the equations and the errors in reading the falling head. However, from a practical point-of-view, the maximum change in K_{fs} of approximately 50% is not important given that K_{fs} ranges over a factor of about 10^5 from sandy/structured material to tightly packed clayey material.

Figure 4b shows α^* vs. number of data points and echoes the comments made above for K_{fs} values. A negative value was obtained for α^* at n = 4, which indicates an invalid result and is not plotted. An invalid or negative result for α^* can be a result of inhomogeneous soil properties and/or observational errors. Note that α^* values obtained using (11) or (20) are consistently larger than those calculated using (14) by a factor of about 3. Although a factor of 3 is large, an α^* value of $4 \pm 2 m^{-1}$ encompasses most of the values for n > 8.

6. CONCLUDING REMARKS

Analytical solutions of infiltration under both constant head and falling head conditions have been reviewed and several new solutions have been developed based on the Green and Ampt assumptions. An analysis of some laboratory data indicated that if only early-time data were collected, it made little difference in the calculation of K_{fs} and ϕ_m (or ψ_f or α^*) if gravity was, or was not, included in the infiltration equation.

Acknowledgments. The authors acknowledge support from the Natural Sciences and Engineering Research Council of Canada and the Ontario Ministry of Agriculture, Food and Rural Affairs.

REFERENCES

Elrick, D.E., G.W. Parkin, W.D. Reynolds and D.J. Fallow, Analysis of early-time and steady state single-ring infiltration under falling head conditions, *Water Resour. Res.*, 31, 1883-1893, 1995.

Elrick, D.E. and W.D. Reynolds, Infiltration from constant-head well permeameters and infiltrometers, in *Advances in Measurement of Soil Physical Properties: Bringing theory into practice*, Soil Sci. Soc. Am. Special Pub. 30, edited by G.C. Topp, W.D. Reynolds, and R.E. Green, pp. 1-24, Soil Sci. Soc. Am., Madison, WI, 1992.

Fallow, D.J., D.E. Elrick, W.D. Reynolds, N. Baumgartner, and G.W. Parkin, Field measurement of hydraulic conductivity in slowly permeable materials using early-time infiltration measurements in unsaturated media, in *Hydraulic conductivity and waste contaminant transport in soils*, ASTM STP 1142, edited by D.E. Daniel and S.J. Treautwein, American Soc. for Testing and Materials, Philadelphia, 1994.

Gardner, W.R., Some steady-state solutions of the unsaturated moisture flow equation with application to evaporation from a water table, Soil Sci., 85, 228-232, 1958.

Green, W.H. and G.A. Ampt, Studies in soil physics. I. The flow of air and water through soils, *J. Agr. Sci.* 4, 1-24, 1911.

Parlange, J.-Y., T.S. Steenhuis, R. Havercamp, D.A. Barry, P.J. Culligan, W.L. Hogarth, M.B. Parlange, P. Ross and F. Stagnitti, Soil properties and water movement, in *Vadose Zone Hydrology*, edited by M.B. Parlange and J.W. Hopmans, pp. 99-129, Oxford University Press, new York, 1999.

Philip, J.R., The theory of infiltration: 4, Sorptivity and the algebraic infiltration equation., *Soil Sci.*, 84, 329-339, 1957.

Philip, J.R., The theory of infiltration: 6, Effect of water depth over soil, *Soil Sci.*, 84, 329-339, 1958.

Philip, J.R., Infiltration in one, two and three dimensions, in *Advances in Infiltration*, Proc. National Conference on Advances in Infiltration, Dec. 12-13, 1983, Chicago, IL, pp. 1-13, Amer. Soc. Agr. Eng., St. Joseph, MI, 1983.

Philip, J.R., Falling head ponded infiltration, *Water Resour. Res.*, 28, 2147-2148, 1992.

White, I. and M.J. Sully, Macroscopic and microscopic capillary length and time scales from field infiltration, *Water Resour. Res.*, 23, 1514-1522, 1987.

R. Angulo-Jaramillo, LTHE, Domaine Universitaire, BP 53, 38041 Grenoble, Cedex 9, France.

D.E. Elrick and D.J. Fallow and G.W. Parkin, Department of Land Resource Science, University of Guelph, Guelph, ON Canada N1G 1W8.

W.D. Reynolds, Greenhouse and Processing Crops Res. Ctr., Agriculture and Agri-Food Canada, 2585 County Road 20, Harrow, ON, Canada N0R 1G0.

Capillary Rise of Water Into Soil as Described by a Solution of Burgers' Equation

D. Swartzendruber

Department of Agronomy and Horticulture, University of Nebraska-Lincoln,
Lincoln, Nebraska

An exact mathematical solution for vertically upward capillary rise of water, into initially unsaturated, uniform soil or porous medium, was found by solving the applicable form of Burgers' equation subject to the appropriate initial and boundary conditions. From the classic Richards equation for unsaturated flow of soil water, Burgers' equation was obtained by imposing a constant soil water diffusivity and a parabolic concave-upward hydraulic conductivity as a function of the volumetric water content. J. R. Philip was the first to point out the possibility of using Burgers' equation for the opposite case of vertically downward infiltration of water into soil. One impetus for seeking the capillary-rise solution was to obtain an improved quantitative description of the time course by which the quantity of water taken up by the soil would approach a finite, fixed value as time increased without limit. This exact solution, however, did not produce such a finite value. Instead, for large values of time, the quantity of water uptake was found to be proportional to the natural logarithm of the square root of time, and was thus predicted to increase without limit as time increased without limit. Therefore, in this respect, the new Burgers-equation solution was less satisfactory physically than an earlier solution of the linearized Richards equation. Remaining to be assessed is whether early time ranges of the new solution would have utility in describing experimental capillary-rise data.

INTRODUCTION

The one-dimensional upward movement of water into an initially unsaturated soil or porous medium is commonly known as capillary rise. This phenomenon has not received nearly as much research attention as its counterparts in the horizontal and vertically-downward directions, termed absorption and infiltration, respectively, by *Philip* [1958, pp. 279, 280; 1969a, pp. 230, 231]. The absorption case is mathematically the simpler, and also serves as the starting point for either

infiltration or capillary rise. The infiltration case has been studied extensively because of its relevance for the process by which the water present in the soil is renewed and augmented, whether by rainfall or surface irrigation. Capillary rise, of course, could have relevance for some types of subsurface irrigation, for example, as in the so-called flooded bench fertigation system [*Otten et al.*, 1999].

In the classic works of *Klute* [1952] and *Philip* [1955], solution of the absorption problem (gravity effect absent) disclosed a dependence on the square root of time. As extended by *Philip* [1957a] for infiltration, the solution comprised an infinite series in the square root of time, but the commonly used four-term truncated series held only for small to moderate times. *Philip* [1957c] also developed an asymptotic infinite-time infiltration solution, but joining it with the truncated series solution was not without some difficulty of its

Environmental Mechanics: Water, Mass and Energy Transfer in the Biosphere
Geophysical Monograph 129
Copyright 2002 by the American Geophysical Union
10.1029/129GM05

own, particularly if a single-form infiltration equation were desired. Alternative relief was sought, first by linearizing the governing differential equation before solving it [*Philip*, 1969b], and then by employing the minimally nonlinear governing differential equation of *Burgers* [1948] as solved by Knight [*Philip*, 1974, p. 261] in follow-up of the earliest suggestion [*Philip*, 1973] for such use of Burgers' equation. Further efforts with Burgers-equation solutions have been carried out [*Clothier et al.*, 1981; *Knight*, 1983; *Hills and Warrick*, 1993], but with focus mainly on constant surface-flux infiltration.

Applying his series-solution approach to the case of capillary rise, *Philip* [1969c] found the solution to be a square-root-of-time series counterpart of the infiltration case, but with the series being alternating in sign. A counterpart asymptotic infinite-time solution, being the equilibrium water-content profile, was also obtained, but with the expressly stated caution that the problem of bridging the gap between the square-root-of-time series solution and the infinite-time solution would appear to be more serious than the corresponding problem for infiltration. Once again, the governing differential equation was linearized and a solution found.

For the case of infiltration, *Philip* [1974] viewed the Knight solution of Burgers' equation to be more informative and more accurate than simple linearization, apparently because the Burgers-equation solution gives appropriate limits at infinite time, whereas the linearized solution does not. It would therefore appear worthwhile to investigate whether this might also be the case for capillary rise. If so, further and more accurate description might be provided on how the final equilibrium of water intake during capillary rise is approached at large and infinite times, for which only approximations are now available–the Green-Ampt approach [*Green and Ampt*, 1911], and the simple linearization of *Philip* [1969c]. Therefore, we shall here seek a solution to Burgers' equation for capillary rise.

MATHEMATICAL ANALYSIS

The Problem

Consider one-dimensional upward rise of water into an infinitely long, uniform, rigid soil column of constant bulk cross-sectional area. We shall begin with the governing partial differential equation of *Richards* [1931] written in the form [*Philip*, 1957b, p. 349; *Swartzendruber*, 1969, p. 222]

$$\frac{\partial \theta}{\partial t} = \frac{\partial}{\partial z}\left[D\frac{\partial \theta}{\partial z}\right] + \frac{\partial K}{\partial z}, \tag{1}$$

where $\theta = \theta(z,t)$ is the volumetric water content at time t and

positive-upward position coordinate z. $D = D(\theta)$ and $K = K(\theta)$ are the soil water diffusivity and hydraulic conductivity functions, respectively. Because $K = K(\theta)$, then

$$\partial K/\partial z = (dK/d\theta)(\partial \theta/\partial z). \tag{2}$$

To facilitate solution of (1), we next introduce two assumptions, the first being that $dK/d\theta$ is linear in θ, or

$$dK/d\theta = A\theta + B, \tag{3}$$

where A and B are constants. The second assumption is to take D as constant, notwithstanding its general variation over at least three or four orders of magnitude as θ changes from near zero to its water-saturated value. We thus follow not only Knight [*Philip*, 1974], whose use of constant D for infiltration seemed to work reasonably well [*Philip*, 1987] in several respects, but *Philip* [1969b,c] as well. Combining (3), (2) and (1), along with D constant (from here on), gives

$$\frac{\partial \theta}{\partial t} = D\frac{\partial^2 \theta}{\partial z^2} + (A\theta + B)\frac{\partial \theta}{\partial z}, \tag{4}$$

which is the minimally nonlinear Burgers equation already mentioned in the introduction. Note that (4) embodies improvement over the earlier and more drastic linearizations [*Philip*, 1969b] of also taking A = 0, so that the integration of (3) led to $K(\theta)$ being linear with θ. For A > 0, the integration of (3) thus yields a parabolic dependence of K on θ, which would seem distinctly better than only a linear dependence. Nonetheless, we cannot evade or gainsay the somewhat ruthless-appearing assumption of taking D to be constant. Admittedly, $K(\theta)$ now has a much stronger dependence on θ than does the constant D, but it also would have been even more satisfying if D could somehow have been invested with more than a zero dependence on θ. Finally, to complete the statement of the problem, we take (4) subject to

$$\theta = \theta_n, z > 0, t = 0, \tag{5}$$

$$\theta = \theta_o, z = 0, t > 0, \tag{6}$$

where (5) and (6) are the initial and boundary conditions, respectively, θ_n is the initial constant volumetric water content throughout the uniform soil column, and θ_0 is the constant higher ($\theta_0 > \theta_n$) water content applied instantaneously at the bottom end of the soil column ($z = 0$) and maintained for all time ($t > 0$).

The Solution

The governing differential equation solved for infiltration by Knight [*Philip*, 1974] is of the same form as (4) except that a minus sign appears in front of $(A\theta + B)$. Therefore, we here write the Hopf-Cole transformation [*Hopf*, 1950; *Cole*, 1951] without a minus sign also, namely

$$A\theta + B = 2D\frac{\partial(\ln V)}{\partial z} = \frac{2D}{V}\frac{\partial V}{\partial z}, \qquad (7)$$

with $V = V(z,t)$ considered to be a solution of the heat equation

$$\frac{\partial V}{\partial t} = D\frac{\partial^2 V}{\partial z^2}, \qquad (8)$$

with D constant, of course, as already noted. We examine whether the specification of θ by (7) and (8) is consistent with (4). Solving (7) for θ, we find the partial derivatives of θ with respect to z and t to be

$$\frac{\partial\theta}{\partial t} = \frac{2D}{A}\left[\frac{1}{V}\frac{\partial^2 V}{\partial t\partial z} - \frac{1}{V^2}\left(\frac{\partial V}{\partial z}\right)\frac{\partial V}{\partial t}\right], \qquad (9)$$

$$\frac{\partial\theta}{\partial z} = \frac{2D}{A}\left[\frac{1}{V}\frac{\partial^2 V}{\partial z^2} - \frac{1}{V^2}\left(\frac{\partial V}{\partial z}\right)^2\right], \qquad (10)$$

$$\frac{\partial^2\theta}{\partial z^2} =$$

$$\frac{2D}{A}\left[\frac{1}{V}\frac{\partial^3 V}{\partial z^3} - \frac{3}{V^2}\left(\frac{\partial^2 V}{\partial z^2}\right)\frac{\partial V}{\partial z} + \frac{2}{V^3}\left(\frac{\partial V}{\partial z}\right)^3\right]. \qquad (11)$$

If the order of differentiation is immaterial, the mixed second derivative on the right-hand side of (9) becomes

$$\frac{\partial^2 V}{\partial t\partial z} = \frac{\partial^2 V}{\partial z\partial t}. \qquad (12)$$

Substituting (7), (9), (10), (11), and (12) into (4), canceling and combining terms, and rearranging, yields ultimately

$$\frac{\partial}{\partial z}\left[\frac{\partial V}{\partial t} - D\frac{\partial^2 V}{\partial z^2}\right] = \frac{1}{V}\frac{\partial V}{\partial z}\left[\frac{\partial V}{\partial t} - D\frac{\partial^2 V}{\partial z^2}\right]. \qquad (13)$$

Because the bracket terms of (13) are zero by virtue of (8), (13) reduces to $0 = 0$, and thus the θ of (7) is indeed the solution of (4) given that $V(z,t)$ is a solution of (8). Hence, it now remains to find a $V(z,t)$ that satisfies (8) such that the resulting $\theta(z,t)$ will also meet conditions (5) and (6).

After extensive trial-and-error efforts of seeking insights from the Knight solution [*Philip*, 1974] for infiltration, the following trial form was devised,

$$V = \alpha\,\text{erf}\,v + \alpha(\exp x)\text{erfc}\,w + \beta, \qquad (14)$$

where α and β are constants, erf v is the error function of v defined by

$$\text{erf}\,v = \frac{2}{\pi^{1/2}}\int_0^v e^{-\omega^2}d\omega,$$

and ω is the dummy variable of integration. Also, erfc w $= 1 - \text{erf}\,w$ in (14) is the complementary error function of w, while v, w, and x are defined by

$$v = z/2D^{1/2}t^{1/2}, \qquad (15)$$

$$w = (z/2D^{1/2}t^{1/2}) + kt^{1/2}/D^{1/2}, \qquad (16)$$

$$x = k(z + kt)/D, \qquad (17)$$

where k is a constant that will be examined in more detail later. Since v, w, and x are all functions of z and t, then V of (14) is also ultimately a function of z and t. Note also that (15), (16), and (17) can be combined to give

$$w^2 = v^2 + x .\tag{18}$$

Since the term α erf v is recognizable as a solution of (8), it suffices to determine whether the two remaining terms of (14) will satisfy (8). If so, then we know that (14) is also a solution, because a sum of solutions of the heat equation is likewise a solution. The partial time derivative of $[\alpha(\exp x)\,\mathrm{erfc}\ w + \beta]$ is, after obtaining the necessary partial time derivatives of v, w, and x,

$$\frac{\alpha z \exp(x - w^2)}{2(\pi D)^{1/2} t^{3/2}} - \frac{\alpha k \exp(x - w^2)}{(\pi D)^{1/2} t^{1/2}}$$

$$\tag{19}$$

$$+ \frac{\alpha k^2 (\exp x)(\mathrm{erfc}\ w)}{D} .$$

Similarly, the partial derivative of $[\alpha(\exp x)\,\mathrm{erfc}\ w + \beta]$ with respect to z is

$$\frac{\alpha[\exp(x - w^2)]}{(\pi Dt)^{1/2}} + \frac{\alpha k(\exp x)(\mathrm{erfc}\ w)}{D} .\tag{20}$$

Taking the partial derivative with respect to z of the expression in (20) gives the second partial derivative of $[\alpha(\exp x)\,\mathrm{erfc}\ w + \beta]$ with respect to z, and multiplying this second derivative by D gives exactly the expression in (19). This means that $[\alpha(\exp x)\,\mathrm{erfc}\ w + \beta]$ is indeed a solution of (8), and therefore that the V of (14) satisfies (8) as well. We hence use the V of (14) to evaluate the far right-hand side of (7), where then

$$\frac{\partial V}{\partial z} = \frac{\alpha[\exp(-v^2) - \exp(x - w^2)]}{(\pi Dt)^{1/2}}$$

$$\tag{21}$$

$$+ \frac{\alpha k(\exp x)(\mathrm{erfc}\ w)}{D} ,$$

but from (18) we observe that $x - w^2 = -v^2$ so that the first term on the right-hand side of (21) vanishes, leaving $\partial V/\partial z$ to be given solely by the second term. Using this result in (7), along with the V of (14), yields

$$A\theta + B = \frac{2\alpha k(\exp x)(\mathrm{erfc}\ w)}{\alpha[\mathrm{erf}\ v + (\exp x)(\mathrm{erfc}\ w)] + \beta}\tag{22}$$

as the expression for θ as a solution of the Burgers equation (4). This solution still must be specialized to satisfy the initial and boundary conditions.

To examine whether (22) will accommodate the initial condition, first employ (5) in (22). For $z > 0$, $t \to 0$ gives: $v \to \infty$ so erf $v \to 1$; $w \to \infty$ so erf $w \to 1$ and erfc $w \to 0$; and $x = kz/D$ so $\exp x = \exp(kz/D)$ remains finite. Putting these quantities into (22) along with $\theta = \theta_n$ yields

$$A\theta_n + B =$$

$$\tag{23}$$

$$\frac{2\alpha k[\exp(kz/D)](0)}{\alpha\{1 + [\exp(kz/D)](0)\} + \beta} = 0 ,$$

from which $B = -A\theta_n$.

To determine whether (22) will accommodate the boundary condition, next employ (6) in (22). For $t > 0$, $z = 0$ gives: $v = 0$ so erf $v = 0$; $w = k/(Dt)^{1/2}$ so erfc$[k/(Dt)^{1/2}]$ remains finite; and $x = k^2 t / D$ so $\exp(k^2 t / D)$ remains finite. Putting these quantities into (22) along with $\theta = \theta_0$ and $B = -A\theta_n$ yields

$$A(\theta_0 - \theta_n) =$$

$$\tag{24}$$

$$\frac{2\alpha k[\exp(k^2 t/D)]\{\mathrm{erfc}[k/(Dt)^{1/2}]\}}{0 + \alpha[\exp(k^2 t/D)]\{\mathrm{erfc}[k/(Dt)^{1/2}]\} + \beta} .$$

If finite $\beta > 0$, then the right-hand side of (24) would vary with time t, in contradiction of $A(\theta_0 - \theta_n)$ required to be constant. Selecting $\beta = 0$ will produce cancellations in (24) to yield acceptably that

$$A(\theta_0 - \theta_n) = 2k .\tag{25}$$

Setting $\beta = 0$, $B = -A\theta_n$, and substituting 2k from (25) into (22) will cause α and A to cancel out so that (22) after rearrangement becomes

$$\theta - \theta_n = \frac{\theta_0 - \theta_n}{1 + g} \qquad (26)$$

as the solution of the problem, where

$$g = [\exp(-x)](\text{erf } v)/\text{erfc } w . \qquad (27)$$

Whereas the form of (26) is essentially that of Knight [*Philip*, 1974], the function g of (27) and its arguments v of (15), w of (16) and x of (17) differ from their respective Knight counterparts. Lastly, we note that if the constant coefficient α of (exp x)erfc w in (14) were replaced by a different constant, say, γ, then the resulting modification of (14) would still satisfy (8), but to meet initial condition (5) would require taking γ = α.

Further evaluation of the constant k can be made by separating the variables in (3) and integrating to obtain

$$K = A(\theta^2/2) + B\theta + c , \qquad (28)$$

where c is the constant of integration. Using the conditions K = K_n for θ = θ_n and K = K_0 for θ = θ_0 successively in (28), along with B = -Aθ_n and A evaluated from (25), form the difference K_0 - K_n and rearrange to obtain

$$k = (K_0 - K_n)/(\theta_0 - \theta_n) . \qquad (29)$$

In the infiltration setting, k is the constant downward velocity of the asymptotic large-time, fixed-shape water-content profile [*Philip*, 1957c].

The behavior of the function g of (27) at extreme values of z is examined first for z = 0 at any positive fixed t. So, for z = 0 at fixed t: v = 0 from (15) so erf v = 0; w = $kt^{1/2}/D^{1/2}$ from (16) so erfc w = 1 - erf($kt^{1/2}/D^{1/2}$); and x = k^2t/D from (17) so exp(-x) = exp(-k^2t/d). Putting these quantities into (27) yields

$$g_{z=0} = g_0 = \frac{[\exp(-k^2t/D)](0)}{[1 - \text{erf}(kt^{1/2}/D^{1/2})]} = 0 , \qquad (30)$$

thus verifying that the solution (26) satisfies the boundary condition (6). As z → ∞ for fixed t, then v → ∞ from (15), w → ∞ from (16), and x → ∞ from (17); putting these quantities into (27), the resulting indeterminate form requires use of l'Hospital's rule to find ultimately

$$g_{z \to \infty} = g_\infty \to \infty , \qquad (31)$$

thus verifying that the solution (26) implies for x → ∞ that the water content retains its initial value θ_n for all t.

As a final check on the validity of the function g in the solution θ of (26), we examine a condition sometimes stated in problems of this kind. That is, at any given fixed t > 0, there should be no change in θ with respect to z at great distances z at and beyond the point at which the initial water content θ_n has not yet been altered by the application of free water at the inlet end (z = 0) of the soil column. Analytically,

$$\partial\theta/\partial z = 0, \quad z \to \infty, \quad t > 0 . \qquad (32)$$

Partial differentiation of θ of (26) with respect to z, and making use of (27) both directly and in the modified form [exp(-x)](erf v) = g(erfc w), enables writing the still general form

$$\frac{\partial\theta}{\partial z} = -\frac{(\theta_0 - \theta_n)}{(1 + g)^2}\left[\frac{(1 + g)[\exp(-w^2)]}{(\pi Dt)^{1/2}(\text{erfc } w)} - \frac{kg}{D}\right] . \qquad (33)$$

Next, substitute for (erfc w) in (33) its equivalent erfc w = [exp(-x)](erf v)/g from (27), along with the exponential argument x - w^2 = -v^2 as obtained from (18), and multiply the denominator $(1 + g)^2$ through the large-bracket terms in (33), to provide

$$\frac{\partial\theta}{\partial z} =$$

$$-(\theta_0 - \theta_n)\left[\frac{[\exp(-v^2)]g}{(\pi Dt)^{1/2}(\text{erf } v)(1 + g)} - \frac{kg}{D(1 + g)^2}\right] . \qquad (34)$$

For fixed t > 0, letting z → ∞ in (15) and (31) will produce v → ∞, erf v = erf ∞ = 1, exp(-v^2) = exp(-∞) = 0, and g → ∞ in

(34) so that both terms in the large bracket become zero, and $\partial\theta/\partial z$ vanishes, so condition (32) is indeed satisfied.

Quantity of Water Intake

The Buckingham-Darcy flux equation [*Swartzendruber,* 1969, p. 219], written for only the vertically upward direction, is applied at the bottom end of the soil column, $z = 0$, in the form of the upward flux q_0,

$$q_0 = -D\left(\frac{\partial\theta}{\partial z}\right)_{z=0} - K_0. \tag{35}$$

The first term on the right-hand side of (35) is the upward water-flux tendency, and is countered by the second term, $-K_0$, which expresses the downward influence of gravity-induced unit hydraulic gradient at the hydraulic conductivity existing at the inlet-end water content θ_0. To evaluate $\partial\theta/\partial z$ of (33) at $z = 0$, we set $z = 0$, which from (16) produces $w^2 = k^2 t / D$ along with $g = g_0 = 0$ from (30) which we substitute into (33), and then multiply by D to get from (35) that

$$q_0 = \frac{dy}{dt} = \frac{D(\theta_0 - \theta_n)\exp(-k^2 t/D)}{(\pi D t)^{1/2}[1 - \mathrm{erf}(k t^{1/2}/D^{1/2})]} - K_0. \tag{36}$$

Here y is the total volume of water, per unit bulk cross-sectional area of soil column, that has moved upward past the inlet end of the column at $z = 0$. Make the substitution

$$W = 1 - \mathrm{erf}(k t^{1/2}/D^{1/2}), \tag{37}$$

from which we find

$$dW = -\{k[\exp(-k^2 t/D)]dt\}/(\pi D t)^{1/2}. \tag{38}$$

After separating variables in (36) and combining (37) and (38) with (36) the result is

$$dy = -[D(\theta_0 - \theta_n)/k](dW/W) - K_0 dt. \tag{39}$$

This integrates to provide, after back substitution of W from (37),

$$y = -[D(\theta_0 - \theta_n)/k]\ln[1 - \mathrm{erf}(k t^{1/2}/D^{1/2})] - K_0 t, \tag{40}$$

after also having employed the condition $y = 0$ at $t = 0$ that eliminates the constant of integration.

DISCUSSION AND APPLICATION OF SOLUTION

To simplify the argument of the error function in (40), we set $C = k/D^{1/2}$, from which

$$D = k^2/C^2. \tag{41}$$

Putting (41) into the $D(\theta_0 - \theta_n)/k$ of (40), with k expressed by (29), yields

$$D(\theta_0 - \theta_n)/k =$$

$$(k^2/C^2)(\theta_0 - \theta_n)/k = (K_0 - K_n)/C^2, \tag{42}$$

and using (42) and (41) in (40) then gives the three-parameter form (in K_0, K_n, and C)

$$y = -\frac{(K_0 - K_n)}{C^2}\ln[1 - \mathrm{erf}(C t^{1/2})] - K_0 t. \tag{43}$$

Also, using standard series expansions in (40) for very small t gives

$$y = 2(\theta_0 - \theta_n)(D/\pi)^{1/2} t^{1/2}, \tag{44}$$

which is the square-root-of-time expression for the case of absorption as mentioned in the introduction. This enables the coefficient of $t^{1/2}$ in (44) to be identified as the constant sorptivity S of *Philip* [1957d], namely $2(\theta_0 - \theta_n)(D/\pi)^{1/2} = S$, which when solved for D yields

$$D = \pi S^2/4(\theta_0 - \theta_n)^2. \tag{45}$$

This expresses D in terms of S and matches the small-time

behavior of (40) with horizontal absorption. The relationship between D and S of (45) for infiltration has also been found by *Philip* [1969b] and by Knight [*Philip*, 1974], and for *Philip's* [1969c] case of linearized capillary rise. That (45) also appears in the present capillary-rise solution therefore fits in well with previous work, and reflects again the sense in which the solution for absorption is a starting point for vertical flow at very small times, whether for infiltration or capillary rise.

In considering capillary-rise behavior at large times t → ∞, we first follow *Philip* [1974] by neglecting K_n in comparison with K_0 but not necessarily requiring $\theta_n = 0$. Although doing this will restrict our scope to soils initially in the drier ranges, it is nonetheless such conditions that are most likely to be subjected to experimental scrutiny and assessment. Moreover, once K_n is sensibly above zero, it will become more difficult experimentally to sustain a vertical soil column at a satisfactorily constant θ_n under the influence of gravity. Setting $K_n = 0$ in (43) gives the two-parameter form (in K_0 and C)

$$y = -(K_0/C^2)\ln[1 - erf(Ct^{1/2})] - K_0t. \qquad (46)$$

For t → ∞, the right-hand side of (46) would require resolution of the indeterminate form ∞ - ∞, but, as an easier and more instructive exercise, we shall employ the first two terms of the infinite series expansion of the error function for large t to write

$$erf(Ct^{1/2}) = 1 - [\exp(-C^2t)]/C(\pi t)^{1/2}. \qquad (47)$$

Substituting (47) into (46) yields

$$y = -(K_0/C^2)\{\ln[\exp(-C^2t)] \\ - \ln[C(\pi t)^{1/2}]\} - K_0t, \qquad (48)$$

which after further cancellations and manipulation gives

$$y = (K_0/C^2)\ln[C(\pi t)^{1/2}] \qquad (49)$$

as the large-time asymptotic relationship of y versus t.

From (49) it is clear that, as t → ∞, y also increases without limit rather than approaching a fixed, finite value. It is there-

fore pertinent to make sure of how the time rate of change in y behaves as t → ∞. Taking the ordinary time derivative of y in (46), and then letting t → ∞ in the resulting time function, we will encounter an indeterminate form (0/0) that multiplies the coefficient $(-K_0/C^2)$. Applying l'Hospital's rule, and again taking t → ∞, will multiply $(-K_0/C^2)$ by $(-C^2)$ to yield $(dy/dt)_{t\to\infty} = K_0 - K_0 = 0$. This same result is also found by differentiating (49) with respect to t and letting t → ∞. Thus, even though t → ∞ will cause dy/dt to approach zero, its manner of approach seems not strong enough to cause y to approach a fixed, finite value. It is, of course, physically very unrealistic for (49) to predict that the quantity of water drawn up by the soil column against gravity would increase indefinitely. Surely it is required that the upward attraction of the soil for water would ultimately somehow be balanced by the downward attraction of gravity all along the column at every water content between θ_n and θ_0. In the sense of a flux balance, (35) and its outworking via (36) through (49) admittedly does achieve dy/dt → 0 for t → ∞ as just shown, but this has not been sufficient to ensure a finite y at infinite t.

In terms of original objectives of the present work, we have, of course, succeeded in the major thrust of finding the exact solution to Burgers' equation for the problem of capillary rise– the counterpart to Knight's [*Philip*, 1974] solution for infiltration. Notwithstanding this success, however, our secondary, corollary expectation has not been realized. That is, our exact solution has not provided a physically acceptable description of how the quantity of water intake into the soil during capillary rise would attain equilibrium with time. Whereas for infiltration the Burgers-equation solution of Knight provided a distinct improvement over both simple linearization [*Philip*, 1969b, 1974] and the Green-Ampt [*Green and Ampt*, 1911] step profile, a corresponding improvement has not accrued from our present Burgers-equation solution. In contrast with *Philip's* [1969c] reporting of a qualitatively reasonable time course of capillary rise toward an ultimate (t → ∞) finite quantity of water intake, for both simple linearization and the Green-Ampt analysis, the hoped-for quantitative improvement from our present solution is belied by the less-than-satisfactory nature of (49) as t → ∞. This might well be underscoring *Philip's* [1969c] express caution that bridging the gap between his moderate- and infinite-time solutions would appear more difficult for capillary rise than for infiltration. He later [*Philip*, 1987] strongly advocated Knight's solution for a more accurate bridging of the gap in the case of infiltration.

Even so, the present author finds it disappointing that the Burgers-equation solution for capillary rise has not provided the improvement originally sought and expected. Admittedly, in the case of infiltration, the improvement over simple linearization stems from the Knight solution in its provision of the so-called traveling-wave feature [*Philip*, 1987], which qualitatively accommodates the fixed-shape profile as it translates linearly downward at infinite time. Such a feature is not present or needed for capillary rise, and may well be the underlying reason that the parabolic $K(\theta)$, while more realistic than a linear function, nonetheless offers no improvement for capillary rise, and, in the foregoing light, might even possibly be a drawback.

In an effort to utilize some possible benefit from the present solution for capillary rise, we note that there are classic experimental data [*Loughridge*, 1892-1894] which suggest that the time approach to capillary-rise equilibrium is extremely slow. It might therefore be of interest to examine (43), (46), and possibly even (49) in relation to these and any other available data, to determine whether any of these equations would be applicable for descriptive or fitting purposes. In doing this, special care would be needed to determine a suitable upper limit of time, above which no experimental data points would be allowed in the fitting process. This would not, however, be an attempt to gainsay or ignore the very real limitations in (49), (46), and (43) that have already been pointed out and discussed.

Further understanding, and possible insight regarding modification and improvement, might be afforded by a graphical examination of the θ solution embodied in (26), (27), (15), (16), and (17). Dimensionless curves of water-content profiles at a progression of dimensionless times could be prepared, and then compared with similar curves from *Philip's* [1969c] solutions, linearized and otherwise, and possibly also with any experimental data if such can be found. Moreover, it may be of interest to determine whether (43) or (46) would have any useful ranges of applicability for fitting to experimental data points, provided that time t is not too large. Assuming an ample supply of data points, such efforts could proceed by using nonlinear least-squares fitting, by first including all points. The fitting exercises could then be repeated by successive deletion of data points, starting with the point at the largest time value, and continuing until some measure such as the root-mean-square error would no longer decrease. If such error remained essentially constant over the whole time range, it would be concluded that the fitted equation was applicable over the complete time range of the experimental points. Analyses of the types here mentioned are currently under consideration and study.

CLOSING REMARKS

In reflecting upon the monumental and prolific legacy of accomplishment in soil hydrology and soil physics bequeathed us by John R. Philip, let us direct particular attention to the now-classic seven-paper series entitled "The theory of infiltration" published in *Soil Science* in 1957-58. The reader can observe that four of these papers [*Philip*, 1957 b,c,d; 1958] have been cited in the present work. It can also be noted that the manuscripts of the first four installments of the series were received during a four-month period in 1956 (three of the manuscripts on the same day!). On the premise that editorial and reviewer anonymity serve no useful purpose after a span of 45 years, it therefore seems permissible to disclose that Editor-in-Chief Firman E. Bear assigned the processing of these manuscripts to Consulting Editor Don Kirkham, who in turn requested my involvement as a referee. We found, of course, that the manuscripts were very worthy of publication, although Dr. Bear did express a bit of reticence about accepting what he considered to be so many contributions in one fell swoop from a single author. It is probably doubtful, however, that any of us foresaw the lofty place eventually to be attained by this seven-paper series. It is perhaps fortunate that Dr. Philip completed this crucial work before the onset of pervasive computer technology and software. He was therefore motivated to pursue analytical mathematical solutions to the utmost; then, when he did turn to numerical evaluation, he developed simple and rapidly converging procedures capable of use even on an ordinary desk calculator. This persistence in seeking analytical solutions became almost a leitmotif of his career, and he reiterated it explicitly in his writing. His involvement with Burgers' equation can thus be viewed as an expression of his predilection and preference for closed-form analytical solutions.

Acknowledgments. Contribution from the Department of Agronomy and Horticulture, Agricultural Research Division, University of Nebraska-Lincoln. Special appreciation is expressed to Bette L. Schernikau, Project Assistant, for processing and printing of the manuscript.

REFERENCES

Burgers, J.M., A mathematical model illustrating the theory of turbulence, *Adv. Appl. Mech., 1*, 171-199, 1948.

Clothier, B.E., J.H. Knight, and I. White, Burgers' equation: Application to field constant-flux infiltration, *Soil Sci., 132*, 255-261, 1981.

Cole, J.D., On a quasi-linear parabolic equation occurring in aerodynamics, *Q. Appl. Math, 9*, 225-236, 1951.

Green, W.H., and G.A. Ampt, Studies on soil physics, I, The flow of air and water through soils, *J. Agric. Sci., 4*, 1-24, 1911.

Hills, R.G., and A.W. Warrick, Burgers' equation: A solution for soil water flow in a finite length, *Water Resour. Res. 29*, 1179-1184, 1993.

Hopf, E., The partial differential equation $u_t + uu_x = u_{xx}$, *Commun. Pure Appl. Math., 3*, 201-230, 1950.

Klute, A., A numerical method for solving the flow equation for water in unsaturated materials, *Soil Sci., 73*, 105-116, 1952.

Knight, J.H., Infiltration functions from exact and approximate solutions of Richards' equations, in *Advances in Infiltration, Proc. Nat. Conf. Advan. Infiltration*, pp. 24-33, American Society of Agricultural Engineers, St. Joseph, Michigan, 1983.

Loughridge, R.H., Investigations in soil physics; the capillary rise of water in soils, *Calif. Agric. Exp. Sta. Ann. Rept.*, pp. 91-100, 1892-1894.

Otten, W., P.A.C. Raats, R. Baas, H. Calla, and P. Kabat, Spatial and temporal dynamics of water in the root environment of potted plants on a flooded bench fertigation system, *Netherlands J. Agric. Sci.*, 47, 51-65, 1999.

Philip, J.R., Numerical solution of equations of the diffusion type with diffusivity concentration-dependent, *Trans. Faraday Soc., 51*, 885-892, 1955.

Philip, J.R., Numerical solution of equations of the diffusion type with diffusivity concentration-dependent, II, *Austr. J. Phys., 10*, 29-42, 1957a.

Philip, J.R., The theory of infiltration, 1, The infiltration equation and its solution, *Soil Sci., 83*, 345-357, 1957b.

Philip, J.R., The theory of infiltration, 2, The profile of infinity, *Soil Sci., 83*, 435-448, 1957c.

Philip, J.R., The theory of infiltration, 4, Sorptivity and algebraic infiltration equations, *Soil Sci., 84*, 257-264, 1957d.

Philip, J.R., The theory of infiltration, 6, Effect of water depth over soil, *Soil Sci., 85*, 278-286, 1958.

Philip, J.R., Theory of infiltration, *Adv. Hydrosci., 5*, 215-296, 1969a.

Philip, J.R., A linearization technique for the study of infiltration, in *Proc. Symp. Water Unsat. Zone*, vol. 1, pp. 471-478, Int. Assoc. Hydrol. Sci. and UNESCO, Gentbrugge, Belgium, 1969b.

Philip, J.R., The dynamics of capillary rise, in *Proc. Symp. Water Unsat. Zone*, vol. 2, pp. 559-564, Int. Assoc. Hydrol. Sci. and UNESCO, Gentbrugge, Belgium, 1969c.

Philip, J.R., Flow in porous media, in *Theoretical and Applied Mechanics, Proc. 13th Internat. Congr. Theoret. Appl. Mech.*, edited by E. Becker and G.E. Mikhailov, pp. 279-294, Springer, New York, 1973.

Philip, J.R., Recent progress in the solution of nonlinear diffusion equations, *Soil Sci., 117*, 257-264, 1974.

Philip, J.R., The infiltration joining problem, *Water Resour. Res., 23*, 2239-2245, 1987.

Richards, L.A., Capillary conduction of liquids through porous mediums, *Physics, 1*, 318-333, 1931.

Swartzendruber, D., The flow of water in unsaturated soils, in *Flow Through Porous Media*, edited by R.J.M. DeWiest, pp. 215-292, Academic, San Diego, Calif., 1969.

Dale Swartzendruber, Department of Agronomy and Horticulture, University of Nebraska-Lincoln, 246 Keim Hall, East Campus, Lincoln, NE 68583-0915.

Effect of Gravity and Model Characteristics on Steady Infiltration From Spheroids

A. W. Warrick

Department of Soil, Water & Environmental Science, University of Arizona, Tucson, Arizona

Dani Or

Department of Plants, Soils and Biometerology, Utah State University, Logan, Utah

J. R. Philip contributions to flow analysis from spheroidal sources include steady-state solutions for absorption, small-time solutions for spheres with and without gravity, and steady-state solutions for infiltration. We review some of Philip's contributions in these areas. Three types of calculations and comparisons are then performed. In the first example, the pressure distributions are computed for absorption from spheres using a Gardner hydraulic conductivity function (K exponentially related to pressure head h) and using van Genuchten functions. In the second example, Philip's analytical expressions are used to show the effects of gravity and spheroidal shape on infiltration rate. Comparisons of spheroids with different shapes were made relative to spheres which have either equal absorption rates or which have an equal source area. Finally, infiltration results using a numerical solution with van Genuchten functions are compared to the analytical Gardner calculations to gain information on whether the simpler analytical results can be used to predict infiltration rates for other functions. The infiltration comparisons were compared by matching the capillary lengths. Analytical results are compared to numerical computations and agree within about 5 percent. The ability to use the simple algebraic forms of Philip to make predictions within 5 percent is deemed a positive result and suggestive that application to a wider range of soils and hydraulic functions exists.

1. INTRODUCTION

J. R. Philip (1985, 1986) developed solutions describing the potential and steady-flow rates from spheroids. Spheroids are ellipsoids formed by rotating an ellipse around one of the axes. A prolate spheroid is formed by rotation about the major axis and results in an elongated object (Figure 1A). With ω the ratio of the length of the axis used for rotation to the other, the prolate case is defined with $\omega > 1$. The opposite case is the oblate spheroid (Figure 1C) for which ω is less that 1. For the sphere, the value of ω is one

Environmental Mechanics: Water, Mass and Energy Transfer in the Biosphere
Geophysical Monograph 129
Copyright 2002 by the American Geophysical Union
10.1029/129GM06

(Figure 1A). The sorption rate (without gravity) Q_{sorp} (L^3T^{-1}) is

$$Q_{sorp} = 4\pi(K_{wet} - K_{dry}\tau\lambda_c r_0) \qquad (1)$$

with r_0 defined in Figure 1.

The conductivities K_{wet} and K_{dry} correspond to the supply surface and the background pressure potentials, respectively. The shape factor τ is 1 for a sphere. For the prolate case ($\omega > 1$), τ is

$$\tau = \frac{(\omega^2 - 1)^{1/2}}{\ln[\omega + (\omega^2 - 1)^{1/2}]} \qquad (2)$$

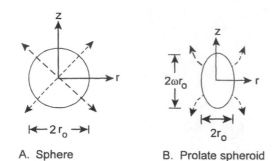

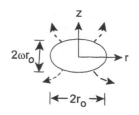

A. Sphere   B. Prolate spheroid

C. Oblate spheroid

Figure 1. Sphere, prolate spheroid ($\omega > 1$) and oblate spheroid ($\omega < 1$).

and for the oblate case ($\omega < 1$), it is

$$\tau = \frac{(1 - \omega^2)^{1/2}}{\cos^{-1}\omega} \tag{3}$$

The capillary length λ_c is

$$\lambda_c = (K_{wet} - K_{dry})^{-1} \int_{h_{dry}}^{h_{wet}} K(h)dh \tag{4}$$

with the integral of the hydraulic conductivity $K(h)$ evaluated between the background pressure head ($h_{dry} < 0$) and the pressure at the spheroid surface.

The solution (1) is equivalent to earlier work for flow from auger holes and wells. For example, *Maasland* (1957) gives solutions for flow from spheroids based on earlier work by *Smythe* (1939), and similar to results of *Kirkham* (1945) and *Hvorslev* (1951).

When gravity is included, the further assumption is made of a Gardner soil (*Gardner*, 1958):

$$K(h) = K_s \exp(\alpha h) \tag{5}$$

Philip (1986) derived relationships for Q with gravity for the Gardner soil. The results can be expressed in terms of the ratio of Q to Q_{sorp}:

$$\frac{Q}{Q_{sorp}} = 1 + \tau s + \frac{A(\omega)s^2}{\tau} \tag{6}$$

with $A(\omega)$ from

$$A(\omega) = \frac{-2\tau}{7}\left(2.5\tau^2 - 0.99615319(\omega\tau)^{2/3} - 1\right) \tag{7}$$

where $s = 0.5\ \alpha r_o$ is assumed less than τ^{-1} (an alternative form is provided for larger s).

2. CALCULATIONS AND EXAMPLES

2.1. Pressure Head Distributions for Absorption

Absorption rates from spheroids are independent of the hydraulic function provided capillary length λ_c and hydraulic conductivities do not change. The pressure head distributions away from the source, however, will necessarily be different. Here we will compare the effects of using the Gardner function for hydraulic conductivity (Eq. 5) vs. the *van Genuchten* functions (1980)

$$K = K_s S_e^{0.5}[1 - (1 - S_e^{1/m})^m]^2 \tag{8}$$

$$S_e = (1 + |\alpha_{vg}h|^n)^{-m} = \frac{\theta - \theta_r}{\theta_s - \theta_r} \tag{9}$$

In (8) and (9), K_s is the saturated hydraulic conductivity, S_e the reduced water content, θ_s the saturated water content and θ_r the residual water content.

For the comparison, values of the Vinton fine sand will be chosen with $K_S = 152$ cm d^{-1}, m = 0.522, n = 1/(1-m), θ_s = 0.419, θ_r = 0.031 and α_{vg} = 0.0331 cm^{-1} for which λ_c = 13.0 cm for h_{wet} = 0 and h_{dry} = -∞ (*Or et al.*, 2000). Gardner's α is taken as 1/13.0 cm^{-1} which will give an identical sorption rate. A sphere with radius of 5 cm is considered.

For all forms of K, the matric potential ϕ is defined by

$$\phi = \int_{-\infty}^{h} K(h)dh \tag{10}$$

and is related inversely to r:

$$\phi = \frac{\lambda_c K_s r_o}{r} \tag{11}$$

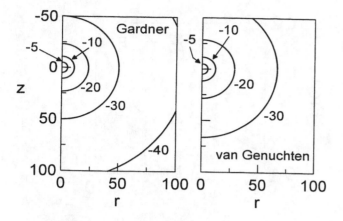

Figure 2. Pressure head comparison for absorption in Vinton fine sand. Contour values are given in cm.

For any particular value of r, ϕ can be found and then h can be found from (10). The comparison between pressure head distributions for the two functions is given by Figure 2. At the spherical source ($r_o = 5$ cm), both distributions correspond to h = 0. At larger distance, Gardner's function gives larger suctions than the van Genuchten function. Before making general conclusions, however, note Figure 3. For h of about -60 cm, the pressure heads are approximately equal, beyond which the earlier trend reverses.

2.2. Shape Effects on Infiltration Using the Gardner Function

The effect of shapes can be compared by looking at the "error" introduced compared to a sphere. Two cases are considered by comparing flow from spheroids to that of a sphere with an equal sorption rate and to that of a sphere of equal area. The "error" is defined by

$$\text{Error} = \frac{(Q_{sphere} - Q_{spheroid})}{Q_{sphere}}$$

For the first case Q_{sorp} from (1) is matched by setting $\tau\, r_o$ equal for the sphere and spheroid. The shape factor τ gives a corresponding value of ω through (2) or (3). Results are compared in Figure 4A for two values of $s = 0.5\ \alpha\, r_o = 0.1$ and 1. For the smaller value of $s = 0.1$, the error is very small and the effects of gravity are nil. For the larger value of $s = 1$, the effects of gravity are underestimated for the oblate case ($\omega < 1$) and overestimated for the prolate case ($\omega > 1$) as expected intuitively. The overestimation for the prolate case is about 15 percent at $\omega = 10$.

For absorption based on equal areas, Table 1 is used and the results are compared for $s = 0.5\ \alpha\, r_o = 0.01$, 0.1 and 1. The errors are reasonably small, except for the smallest value of s and larger values of ω for which the error approaches 20 percent. Differences are less than 10 percent, however, if ω is within the range of 0.25 to 4.

2.3. Comparison of Infiltration With Two Hydraulic Functions

For comparisons of infiltration with alternate hydraulic functions, *van Genuchten* (1980) and *Gardner* (1958) functions are again chosen. For results with the van Genuchten functions it is necessary to use a numerical solution. This was done using "HYDRUS-2D" (*Simunek et al.*, 1999). A cylindrical domain of radius 100 cm and length 200 cm was chosen with boundaries z = -100 cm above and z = 100 cm below the source at z = 0. No-flow boundaries were chosen at the top and for r = 100 cm. A unit hydraulic gradient was taken at the lower boundary.

Results are compared for two soils, the Vinton fine sand and the Millville silt loam, and two spherical radii $r_o = 2.5$ and 5 cm.. The parameter values for the Vinton are the same as used above for comparing absorptions. The Millville values for the van Genuchten parameters are $K_s = 50.4$ cm d^{-1}, m = 0.300, n = 1/(1-m), $\theta_s = 0.434$, $\theta_r = 0.045$ and $\alpha_{vg} = 0.0166$ cm^{-1} for which $\lambda_c = 11.6$ cm for $\theta_{wet} = \theta_{sat}$ and $\theta_{dry} = \theta_r$ (*Or et al.*, 2000). Gardner's K_s is taken the same and α is taken as 1/11.6 which will give an identical sorption rate.

Values for absorption, infiltration (with gravity) with the Gardner function and infiltration with the van Genuchten function are given in Table 2. The results with gravity are larger than without gravity as expected with a smaller difference for the 2.5 radius than for the larger 5.0 radius. In the case of the Vinton the calculated infiltration rates are larger than for the Gardner function, but by less than 5 percent. The rates for the Millville showed similar behavior, except that the calculated van Genuchten rates were smaller than for infiltration using the Gardner function. Times to reach a limiting value were about 2 days for Vinton and 100 d for the Millville. The steady-state pressure head distributions are compared

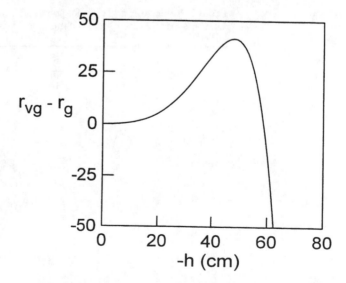

Figure 3. Pressure head related to difference of radii for chosen van Genuchten and Gardner functions.

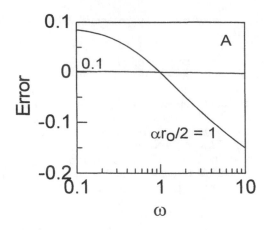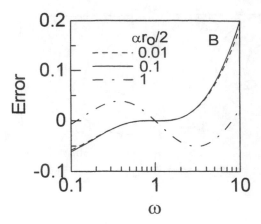

Figure 4. Fractional error when assume flow from spheroid is the same as a sphere of equal sorption (A) or of equal area (B). In both cases, r_o is the radius of the sphere.

in Figure 5 for the Vinton fine sand. The results from the numerical model (Figure 5A) show some effects of the finite radial boundary in that the pressure heads are orthogonal to the boundary. The results for the Gardner function (Figure 5B) are based on the point solution, which is a reasonable approximation in that the pressure head near the source is very nearly spherical. In comparing Parts A and B of Figure 5, the pressure heads show nearly the same behavior, except for the boundary effects in Part A.

The effects of shape were compared with $\omega = 0.25$ (oblate), 1 (sphere) and 4 (prolate) and all of an area of a sphere of radius 2.5 cm. Steady values of flow rate are given in Table 2. There is an increase in the rates in agreement with Figure 4B for increasing values of ω. However, the differences between the Gardner and van

Genuchten functions remain on the order of 5 percent for any given shape. The pressure head distributions are given in Figure 6 for the Millville soil. Very little effect of the shape persists beyond a distance equal to a few radii; the most noticeable effect is that the equal areas gave a wetter condition as the value of ω increased (from 0.25 to 1 to 4). A comparison with $s = 0.5 \, \alpha \, r_o$ in Figure 4B reveals the maximum "error" for $0.25 < \omega < 4$ should not exceed 4 per cent. This is verified numerically in Table 2.

3. SUMMARY AND CONCLUSIONS

For absorption, pressure head distributions have been compared for the *Gardner* (1958) and *van Genuchten* (1980) hydraulic functions. The rates are the same in both

Table 1. Values of τ and r_{eq}/r_o for spherical, prolate and oblate spheroids of equal volumes. Axes of spheroids are r_{eq} and e is the eccentricity of the corresponding ellipse.

	τ	r_{eq}/r_o
1. Spherical $\omega = 1$ ($e = 0$)	1	1
2. Prolate $\omega > 1$ ($e^2 = 1 - \omega^{-2}$)	$\dfrac{\omega e}{\ln[\omega(1 + e)]}$	$2^{0.5}\left(1 + \dfrac{\omega\sin^{-1}e}{e}\right)^{-0.5}$
3. Oblate $\omega < 1$ ($e^2 = 1 - \omega^2$)	$\dfrac{e}{\cos^{-1}(\omega)}$	$2^{0.5}\left(1 + \dfrac{\omega^2[(1+e)/(1-e)]}{2e}\right)^{-0.5}$

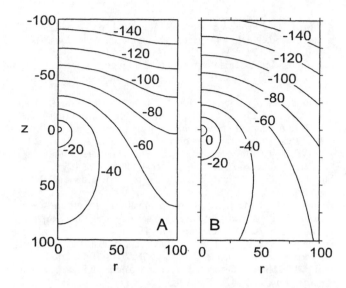

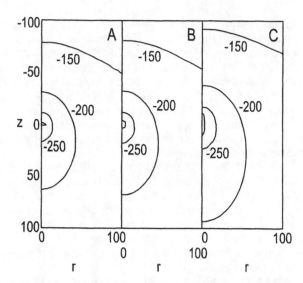

Figure 5. Pressure head contours for van Genuchten (A) and Gardner (B) functions using a sphere of radius 5 cm. Contour values are given in cm.

Figure 6. Pressure contours using the Millville silt loam for oblate (A), spherical (B) and prolate (C) spheroids, all of source area 4π $(2.5)^2$ cm^2 Contour values are given in cm.

cases (and for all other functions) by matching values of capillary length λ_c and hydraulic conductivity range of $K_{wet} - K_{dry}$. The pressure head was found to drop off faster for the van Genuchten function than for Gardner's close to the source and then reverse at larger distances (beyond about 60 cm for the chosen example). The methods used apply to all hydraulic functions.

For infiltration, the effects of shape were considered based on two alternative criteria. The first criterion

compared flow from spheroids to spheres which had equal absorption values. This worked well for $s = 0.5$ αr_o of 0.1 or smaller for all values of ω; however, for $s = 1$ differences approached 20 percent for $\omega = 1$. The second criterion used was for spheres which had equal surface areas. In this case the values agreed within about 10 percent for all $0.1 < \omega < 10$.

Infiltration rates as affected by hydraulic function were compared as well as the resulting pressure

Table 2. Comparison of spheroidal flow rates (r_o is given in cm and all rates are divided by 10^4 and are cm^{3}d^{-1}).

Spheres	Vinton		Millville	
	$r_o = 2.5$	$r_o = 5.0$	$r_o = 2.5$	$r_o = 5.0$
No gravity	6.23	12.5	1.84	3.67
Gravity (Gardner)	6.82	14.8	2.03	4.41
Gravity (van Genuchten)	7.19	15.4	2.00	4.05

Spheroids [Area = $4\pi(2.5)^2$ cm^2]	Vinton			Millville		
	Oblate	Sphere	Prolate	Oblate	Sphere	Prolate
No gravity	6.08	6.23	6.51	1.79	1.84	1.92
Gravity (Gardner)	6.65	6.82	7.15	1.98	2.03	2.13
Gravity (van Genuchten)	6.97	7.19	9.09	1.91	2.00	2.48

distribution around the sources. This required a numerical solution in the case of the van Genuchten functions which were compared to the results for the Gardner functions. Philip's predictions for the infiltration rate from spheres were within about 5% for van Genuchten parameters with matching λ_c for the two soils considered (Vinton fine sand and Millville silt loam). Pressure distributions had the same general pattern, although there was a numerical artifact for the radial boundary for the van Genuchten results as a consequence of choosing a finite-sized domain.

Effects of the shapes of the spheroids on the pressure-head distribution were compared for three different shapes ($\omega = 0.25$, 1 and 4) for the van Genuchten function only. As expected, the shapes were similar beyond distances of a few radii from the source. The results were based on equal areas of the sources and there was a larger effect due to differences in flow rates than the shapes for points remote from the sources.

It is encouraging that the infiltration rates using *Philip's* (1986) algebraic results are reasonable approximations to those for the *van Genuchten* (1980) functions which were much more difficult to evaluate. The methods and results are believed to be generally applicable for all soils and hydraulic functions.

Acknowledgment. Contribution from Western Regional Project W-188.

REFERENCES

Gardner, W.R., Some steady state solutions of unsaturated moisture flow equations with application to evaporation from a water table, *Soil Sci.,* 85, 228-232, 1958.

Kirkham, D., Proposed methods for field measurement of permeability below the water table. *Soil Sci. Soc. Am. Proc.,* 10, 58-68, 1945.

Maasland, M., Soil anisotropy and load drainage, in Drainage of Agricultural Lands, *Am. Soc. Agronomy, Monogr.* 7, 215-285, 1957.

Simunek, J., M. Sejna, and M.Th. van Genuchten, The HYDRUS-2D software package for simulating the two-dimensional movement of water, heat, and multiple solutes in variably-saturated media. Version 2.0, GWMC-TPS-56. International Ground Water Modeling Center, Colorado School of Mines. Golden, CO, 1999.

Smythe, W.R., *Static and dynamic electricity,* McGraw-Hill, New York, 1939.

Hvorslev, J., Time lag and soil permeability in groundwater observations, *Bull. 36,* 50 pp., Waterw. Exp. Stn., Corps of Eng., U.S. Army, Vicksburg, MS, 1951.

Or, D., U. Shani, and A. W. Warrick, Subsurface tension permeametry, *Water Resour. Res.,* 36, 2043-2053, 2000.

Philip, J. R., Steady absorption from spheroidal cavities, *Soil Sci. Soc. Am. J.,* 49, 828-830, 1985.

Philip, J. R., Steady infiltration from spheroidal cavities in isotropic and anisotropic soils, *Water Resour. Res.,* 22, 1874-1880, 1986.

van Genuchten, M.Th., A closed-form equation for predicting the hydraulic conductivity of unsaturated soils, *Soil Sci. Soc. Am. J.,* 44, 892-898, 1980.

Dr. Arthur W. Warrick, Soil, Water and Environmental Science Dept., PO Box 210038, The University of Arizona, Tùcson, AZ 85721-0038 USA

Dr. Dani Or, Plants, Soils and Biometeorology Dept., Utah State University, Logan, Utah 84322-4821 USA

The Seepage Exclusion Problem for Tunnel Cavities in the Saturated Capillary Fringe

E.G. Youngs

Institute of Water and Environment, Cranfield University, Silsoe, Bedfordshire, England

Vertical downward flow of water through the tension-saturated capillary fringe above a water table is perturbed by the presence of cavities that may exclude or allow water entry over different parts of their surface, with some critical shape that just excludes water entry over their whole surface. The analysis by Philip and his colleagues of the seepage exclusion problem was for a soil with an exponential hydraulic conductivity function that does not show a tension-saturated capillary fringe and was for an infinite flow field. The analysis given here is for the critical condition of water exclusion from two-dimensional tunnel cavities located above a water table in the saturated capillary fringe with a horizontal upper boundary where the soil-water pressure is the air-entry value for the soil. Conformal mapping, used to solve Laplace's equation with these boundary conditions, gives the critical shape of the tunnel cavity and also the flux density distribution through the upper boundary of given height. The shape of the cavity near its top can be approximated by a parabola that over-estimates the width of the cavity at depth below the apex. This is in contrast to the exact parabolic shape obtained for soils with an exponential hydraulic conductivity. Upper and lower bounds for a uniform flux density that raises the top of the saturated region in the vicinity above the cavity are obtained for the calculated critical shape.

1. INTRODUCTION

Philip et al. [1989a] recall Charles and Philippe de la Hire's unsuccessful attempt in 1690 to intercept downward percolation through unsaturated soil by means of a buried vessel. They provide a clear account of the physics of the perturbation of downward flows due to the presence of cavities that explains the exclusion of water from the de la Hire's buried vessel. The physical basis for water entry into cavities connected to the atmosphere is that the soil-

water pressure at any point on the cavity wall must be at least atmospheric. The analysis given in a series of papers by Philip and his colleagues [*Philip,* 1989a, b; 1990; *Philip et al.,* 1989a, b; *Knight et al.,* 1989] considers the effect of cavity shape, hydraulic soil properties and flow velocity on water entry and exclusion. It illuminates the role of macropores in unsaturated flow, as well as having engineering applications concerning the design of underground repositories for nuclear waste and the design and performance of interceptor drains for unsaturated seepage, while giving an insight into the emergence and growth of stalactites in caves.

The analysis of the seepage exclusion problem given by Philip and his colleagues [*loc. cit.*] assumes an exponential relationship between the hydraulic conductivity and the

Environmental Mechanics: Water, Mass and Energy Transfer in the Biosphere
Geophysical Monograph 129
Copyright 2002 by the American Geophysical Union
10.1029/129GM07

soil-water pressure. This relationship was suggested by *Gardner* [1958] and allows analytical solutions of Richards' equation to many two- and three-dimensional soil-water problems. The progress that Philip made into the understanding of two- and three-dimensional situations generally was for soils showing this relationship, soils that have been termed "Gardner" soils by *Youngs et al.* [1993]. A comprehensive insight into the way cavities of various shapes affect the soil-water regime was provided through the use of this relationship. It gave the critical shape of cavities when the soil-water pressure at any point on the cavity wall is just less than atmospheric thus excluding water entry, and showed the occurrence of retarded regions of flow and shadow regions of reduced water content.

The critical shape is shown to be parabolic cylindrical for two-dimensional tunnel cavities and paraboloidal for three-dimensional cavities [*Philip et al.*, 1989b] for Gardner soils, assumed to be infinite in extent, and depends on the downward flow velocity. An increase in the flow velocity towards a cavity of critical shape will initiate leakage over its whole surface. Cavities with an apically blunter shape (for example, a circular cylinder or sphere) will start leaking at their apices, whereas apically sharper ones (for example, a wedge, cone, hyperbolic cylinder, hyperboloid) will start leaking at some point below the apices at distance of order of the sorptive length above the cavity floor.

Many soils show a finite air-entry value of soil-water pressure above which the soil is tension-saturated with a constant value of hydraulic conductivity of the saturated soil. "Green and Ampt" soils [*Youngs at al.,* 1993] are a special case where the hydraulic conductivity at soil-water pressures less than the air-entry value is zero. No discussion of this class of soils was given by *Philip* (1989a, b; 1990), *Philip et al.* [1989a, b] and *Knight et al.* [1989] concerning the seepage around cavities. However, they argued that "multidimensional flows and critical and supercritical seepage flows are determined primarily by the (hydraulic conductivity) function for small values of (the soil-water pressure)". The implication is that the seepage behavior and the critical shape of the cavities in soils having a tension-saturated region of soil-water pressure would be different from that shown by Gardner soils.

In a tension-saturated region (not shown by a Gardner soil) Richards' equation for soil-water flow reduces to Laplace's equation Thus we can use methods of analysis to investigate problems of seepage in this zone that are employed in the analysis of groundwater problems, such as electric analogue methods and well established finite difference and finite element numerical methods of solution. For two-dimensional flows we can apply conformal mapping techniques that have also been used by *Kacimov and*

Obnosov [2000] in their study of water flow around parabolic cavities and through parabolic inclusions. This method of analysis is used here to determine the critical shape of tunnel cavities located above a drained gravel substratum in the tension-saturated capillary fringe of such soils. Unlike the analysis with a Gardner soil that considers infinite flow fields, the present analysis is for finite depths of saturated soil between the top of the capillary fringe and the gravel substratum.

2. VERTICAL FLOW TO A WATER TABLE

With steady infiltration over a large area to a horizontal freely draining gravel substratum that acts as a water table, the moisture profile can be calculated from [*Youngs*, 1957]

$$z = -\int_0^p \frac{dp'}{1 - q/K} \qquad (1)$$

where p is the soil-water pressure head (that is negative in the flow region being considered) and K the hydraulic conductivity, generally a function of p, at height z above the water table at the interface of the gravel substratum and soil, and q is the steady vertical downward flow through the soil profile. In soils showing a distinct value of p for air entry when $p = P$, above which the soil is tension-saturated, (1) gives the height of the capillary fringe H as

$$H = \frac{-P}{1 - q/K} \qquad (2)$$

where K is now the hydraulic conductivity of the saturated soil. A cavity introduced into the saturated capillary fringe has the effect, not only of disturbing the vertical flow occurring without the presence of such a cavity, but also of increasing the height of the capillary fringe in the vicinity above. This is illustrated in Figure 1 for a uniform flux density through the surface forming the top of the capillary fringe. This corresponds to the case for precipitation on to the surface of a Green and Ampt soil. It is also a very good approximation for real soils that are tension-saturated over a finite soil-water pressure range, since, as argued by *Childs* [1945] concerning the flow to land drains, the decrease of water content with decreasing soil-water pressure less than the air entry value ensures almost vertical flow with uniform precipitation on the soil surface until it reaches the saturated soil where it then diverges towards the lower sinks.

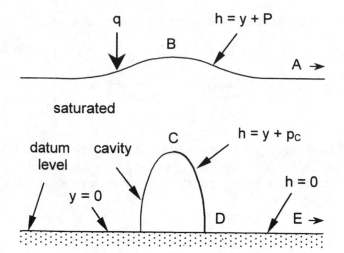

Figure 1. The saturated capillary fringe above a well-drained gravel substratum.

The soil-water pressure head p_c over the surface of the cavity is determined by its shape, its height above the water table and the flow velocity. No water will enter a cavity connected to the atmosphere so long as the soil-water pressure on its walls is less than atmospheric. With a cavity of critical shape, the soil-water pressure is just less than zero ($p_c = -\varepsilon, \varepsilon \to 0$) everywhere on its surface so that the hydraulic head h is just less than the height above the given datum where $h = 0$. Also, the walls of the cavity are a stream surface. In the tension-saturated soil between the top of the capillary fringe and the gravel substratum, the hydraulic head is given by Laplace's equation subject to the imposed boundary conditions. Two-dimensional flows can then be analyzed by conformal mapping [see, for example, *Aravin and Numerov*, 1965; *Harr*, 1962: *Polubarinova-Kochina*, 1962] which is used here to calculate the critical shape of tunnel cavities located above the gravel substratum.

3. FORMULATION OF THE PROBLEM

With a uniform downward flow to a water table, the position of the top of the capillary fringe is raised in the vicinity above any cavity as illustrated in Figure 1. The extent of the saturated soil emerges as part of the solution to the problem. The conformal mapping of the problem is simplified if the top of the capillary fringe is assumed horizontal. The flux density distribution along the top of the capillary fringe to allow this emerges in the solution as well as the critical shape of the two-dimensional cavity for water exclusion when the soil-water pressure head p_c on it is zero so that the hydraulic head on it is $h = p_c + y = y$.

Conclusions can then be reached concerning the shape for water exclusion with a uniform flux density.

4. CONFORMAL MAPPING SOLUTION

Assuming the capillary fringe is bounded by an upper horizontal boundary, the complex position plane $z = x + iy$ is shown by ABCDE in Figure 2a. With the gravel boundary as the datum level, the soil-water pressure head there is zero and the hydraulic head h is also zero. On the cavity wall, in the critical condition $p = -\varepsilon$ ($\varepsilon \to 0$) so that

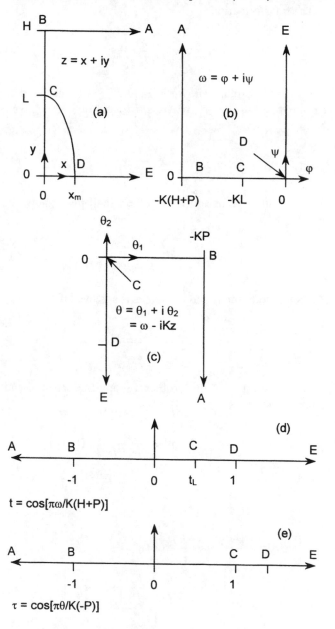

Figure 2. Conformal mapping of the flow problem.

$h = y - \varepsilon \approx y$. The cavity wall is also a streamline. At the top of the capillary fringe $y = H$ and $h = H + P$ where P is the air-entry value of the soil-water pressure head. The complex potential plane $\omega = \varphi + i\psi$ where φ is the seepage velocity potential equal to $-Kh$ and ψ is the stream function, is shown in Figure 2b.

Introducing the Zhukovsky function

$$\theta = \theta_1 + i\theta_2 = \omega - iKz = (\varphi + Ky) + i(\psi - Kx)$$

we obtain the θ-plane shown in Figure 2c.

The transformation

$$t = \cos\left(\frac{\pi\omega}{K(H+P)}\right) \tag{3}$$

gives the t-half-plane shown in Figure 2d with the value of t at the top of the cavity, where $z = iL$ and $\omega = -KL$, given by

$$t_L = \cos\left(\frac{\pi(-L)}{H+P}\right) \tag{4}$$

The θ-plane is transformed into the τ-half-plane shown in Figure 2e by

$$\tau = \cos\left(\frac{\pi\theta}{K(-P)}\right) \tag{5}$$

Coincidence of the t- and τ-planes is achieved if

$$\tau = \frac{2t + (1 - t_L)}{1 + t_L} \tag{6}$$

Thus the solution of the problem giving the potential and stream function is

$$\cos\left(\frac{\pi\theta}{K(-P)}\right) = \frac{2\cos\left(\dfrac{\pi\omega}{K(H+P)}\right) + (1 - t_L)}{1 + t_L} \tag{7}$$

or

$$y - ix = -\frac{\varphi + iy}{K} + \frac{(-P)}{\pi} \cdot$$
$$\cos^{-1}\left[\frac{2\cos\left(\dfrac{\pi(\varphi + i\psi)}{K(H+P)}\right) + (1 - t_L)}{1 + t_L}\right] \tag{8}$$

Along the cavity wall where $\omega = -Ky$ and $\theta = -iKx$

$$x = \frac{(-P)}{\pi}\cosh^{-1}\left[\frac{2\cos\left(\dfrac{-\pi y}{H+P}\right) + (1 - t_L)}{1 + t_L}\right] \tag{9}$$

Equation (9) gives the critical shape of the cavity that just excludes water entry. The position D is where $\omega = 0$ so that the half-width x_m of the cavity at its base where it is widest, is given by

$$x_m = \frac{(-P)}{\pi}\cosh^{-1}\left[\frac{3 - t_L}{1 + t_L}\right] \tag{10}$$

Along the top of the capillary fringe where $y = H$ so that $\omega = -K(H+P) + i\psi$, $\theta = K(-P) + i(\psi - x)$, which gives

$$\cosh\left(\frac{\pi(\psi - Kx)}{K(-P)}\right) = 1 + 2\left[\cosh\left(\frac{\pi\psi}{K(H+P)}\right) - 1\right]\Big/(1 + t_L) \tag{11}$$

Differentiating this we obtain the flux density q along this boundary:

$$q = \left|\frac{\partial\psi}{\partial x}\right|_{x,y=H} =$$
$$\frac{K\sinh\left(\dfrac{\pi(\psi - Kx)}{K(-P)}\right)}{\sinh\left(\dfrac{\pi(\psi - Kx)}{K(-P)}\right) - \dfrac{2(-P)}{(H+P)(1+t_L)}\cdot\sinh\left(\dfrac{\pi\psi}{K(H+P)}\right)} \tag{12}$$

The flux density q_0 on the upper boundary at $x = 0$ is obtained by taking limits as $x \to 0$ and $\psi \to 0$ that leads to

$$q_0 = \left|\frac{\partial\psi}{\partial x}\right|_{0,y=H} = \frac{K}{1 + \dfrac{(-P)}{(H+P)}\sqrt{\dfrac{2}{(1+t_L)}}} \tag{13}$$

For large x ($x \to \infty$) the flux density q_∞ at $y = H$ is from (2)

$$q_\infty = \frac{K(H+P)}{H} \tag{14}$$

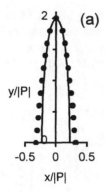
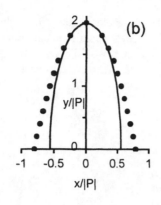
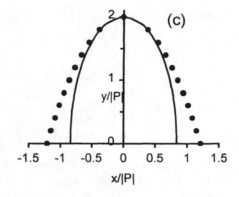

Figure 3. The critical shape of tunnel cavities for $L/|P| = 2$ for (a) $H/|P| = 10$, $q_0 = 0.89K$, $q_\infty = 0.9K$; (b) $H/|P| = 5$, $q_0 = 0.74K$, $q_\infty = 0.8K$; and (c) $H/|P| = 4$, $q_0 = 0.60K$, $q_\infty = 0.75K$. The dots are values given by (15).

5. CRITICAL SHAPES OF CAVITIES

The critical shapes of cavities with a height $L = 2|P|$ for a range of values of the height of the top boundary H calculated from (9) are shown in Figure 3. These have the general appearance of the parabolic cylindrical shapes obtained in the studies of *Philip et al.* [1989b], approximating near the apex to the parabolas

$$\left(\frac{x}{P}\right)^2 = \frac{4(L-y)}{\pi(H+P)}\sqrt{\left(\frac{1-t_L}{1+t_L}\right)} \qquad (15)$$

which are also shown in Figure 3. However, it is seen that (9) predicts that the cavity width increases with depth at a slower rate than that predicted by (15). The non-parabolic form of (9) compared to the parabolic cavity shape found by *Philip et al.* [1989 b] results from assuming a Green and Ampt soil instead of a Gardner soil and from considering a finite, instead of an infinite, flow region.

6. STREAMLINE PATTERN

The streamline pattern calculated from (8) is shown in Figure 4 for the case of $L/|P| = 2$ and $H/|P| = 4$. It is seen that the streamlines become almost vertical at a distance $x/|P| = 3$. If we identify the exponential exponent in the Gardner relationship for hydraulic conductivity with the reciprocal of the air entry value $|P|$ of Green and Ampt soils, then this agrees with the conclusions reached in the analysis used by Philip and his colleagues concerning the extent of the region of influence of a cavity.

7. FLUX DENSITY ON UPPER BOUNDARY

In Figure 5 is shown the flux density across the upper boundary $y = H$ of the capillary fringe, calculated from (12) with ψ given by (11), for $L/|P| = 2$ and for a range of values of $H/|P|$. These show that the flux density variation is smallest when $H/|P|$ is large, corresponding to high flow rates. It is also noted that the value of q approaches q_∞ rapidly as x increases. For example, for $L/|P| = 2$ and $H/|P| = 4$ for which the lowest value of q at $x = 0$ is $q_0 = 0.6K$ and its maximum value $q_\infty = 0.75K$, the value of q is $0.972q_\infty$ at $x/|P| = 2$ and $0.994q_\infty$ at $x/|P| = 3$. It can be argued that if the flux density were uniformly q_0, then the cavity would not allow water entry over any part of its surface if constructed with the calculated shape for that value. However, it would drip over some parts of its surface if the flux density were uniformly at the greater value q_∞.

8. MAXIMUM WIDTH OF CAVITY

The height of the cavity L for a flux density q_0 at $x = 0$ for a maximum cavity half-width x_m is obtained from (4), (10) and (13):

$$L = \frac{(-P)}{\pi} \frac{q_0/K}{1 - q_0/K} \cdot$$
$$\cos^{-1}\left[\frac{3 - \cosh\left(\dfrac{\pi x_m}{(-P)}\right)}{1 + \cosh\left(\dfrac{\pi x_m}{(-P)}\right)}\right]\cosh\left(\dfrac{\pi x_m}{2(-P)}\right) \qquad (16)$$

L for a flux density q_∞ at $x \to \infty$ is obtained from (4), (10) and (14):

$$L = \frac{(-P)}{\pi} \frac{q_\infty/K}{1-q_\infty/K} \cos^{-1}\left[\frac{3-\cosh\left(\frac{\pi x_m}{(-P)}\right)}{1+\cosh\left(\frac{\pi x_m}{(-P)}\right)}\right] \quad (17)$$

The uniform flux density q for a given cavity height L and a given cavity width $2x_m$ that would just exclude water everywhere, lies between q_0 and q_∞. Thus (16) and (17) provide bounds for L. These are shown in Fig.6 in which L, expressed as a fraction of $|P|$, is plotted against the flux density q, expressed as a fraction of K, for values of $2x_m = 2.0$, 1.0 and 0.2. Alternatively bounds for the dimensionless maximum width of cavity $2x_m/|P|$ at a given value of $L/|P|$ can be plotted against q/K, as shown in Fig.7. It is seen that the bounds become closer the smaller the cavity width, practically coalescing for $x_m < 0.2$.

For small values of $L/|P|$ and large values of $H/|P|$ (corresponding to large values of q/K) for which $x_m/|P|$ is small, we can write

$$\cosh\left(\frac{\pi x_m}{(-P)}\right) \approx 1 + \frac{1}{2}\left(\frac{\pi x_m}{(-P)}\right)^2 \quad (18)$$

and

$$\frac{3-t_L}{1+t_L} = 1 + 2\tan^2\left(\frac{\pi(-L)}{2(H+P)}\right) \approx 1 + 2\left(\frac{\pi L}{2(H+P)}\right)^2 \quad (19)$$

Since for these conditions

$$q_0 \to q_\infty = \frac{K(H+P)}{H}$$

using (18) and (19) in (10) we obtain

$$\frac{x_m}{L} = \frac{1-q/K}{q/K} \quad (20)$$

In Figure 8, $2x_m/L$ given by (20) is plotted against q/K and compared with the calculated values of the bounds for different values of $L/|P|$. The agreement is good for all values of $L/|P|$ near $q/K = 1$. The range of values of q/K for which (20) is a good approximation is better for small values of $L/|P|$ when the maximum and minimum bounds

are reasonably close. It is interesting to note that (20) lies between all the calculated maximum and minimum bounds, so it would appear that this simple equation is a good guide to the maximum width of the cavity for water exclusion.

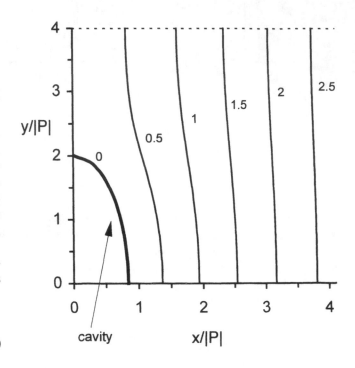

Figure 4. The streamline pattern for flow around a critical tunnel cavity for $L/|P| = 2$ and $H/|P| = 4$. The numbers by the lines are values of ψ/K.

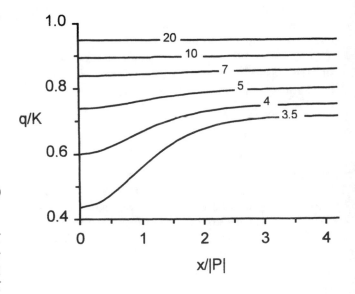

Figure 5. The flux density across $y = H$ for $L/|P| = 2$. Numbers by the lines are values of $H/|P|$.

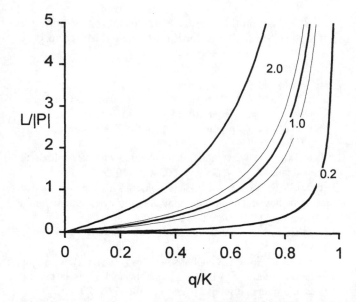

Figure 6. The bounds given by (16) (thick upper lines) and (17) (thin lower lines) for the relationship between the dimensionless cavity height $L/|P|$ and dimensionless flux density q/K for values of the dimensionless critical cavity widths $2x_m/|P|$ shown by the curves.

9. DISCUSSION

Elegant analytical solutions of Richards' equation to two- and three-dimensional soil-water flow problems are possible when soils are assumed to behave as Gardner soils with hydraulic conductivity assumed to be an exponential function of the soil-water pressure. A concern of analytical work based on this assumption that Philip accepted so readily since it could produce such elegant analytical solutions, is that real soils often show a tension-saturated region. While analyses of Gardner soils cannot take this into account, it is the essence of analyses of Green and Ampt soils that Philip often dismissed as being too unlike real soils. By considering results obtained for both types of soil, a more complete understanding of soil-water behavior is obtained. Generally, solutions applicable to real soils are possible with numerical methods employing hydraulic conductivity functions given in numerical form that *Philip* [1956, 1957] used to make his early important advances in infiltration theory but thereafter abandoned in favor of analytical studies that he considered gave a better insight into the physical mechanisms although they used hypothetical soil hydraulic properties. It is surprising that today, when computers can more easily handle such data, analytical hydraulic functions seem universally accepted (often without validation) even in numerical studies of Richards' equation.

This paper considers the seepage exclusion from tunnel cavities of critical shape located in the saturated capillary fringe, employing the method of conformal mapping to solve the simpler Laplace's equation, made possible by the uniform hydraulic conductivity, instead of Richards' equation. The critical shape of tunnel cavities calculated for this situation is approximately parabolic, similar to that obtained for a Gardner soil by *Philip et al.* [1989b]. Furthermore, a similar conclusion can be reached concerning the extent of the region of influence of a cavity. Thus, this is another example where results from the two analyses are in fair agreement, as is the case for two- and three-dimensional infiltration discussed by *Youngs* [1988].

Philip et al. [1989a, b] discuss the application of their work on seepage exclusion from underground cavities in infinite Gardner soils, particularly to the design of repositories for nuclear waste. The analysis given here considers the critical shape of cavities for water exclusion when they are located in a finite saturated capillary fringe (not produced in Gardner soils). Besides the application to the design of underground repositories, the analyses also identify the size and shape of tunnels that can be constructed in soils to maintain soil-water pressures below atmospheric and thus minimize collapse in such circumstances. They are also relevant to activities connected with the collection and sampling of water flowing through industrial waste heaps as well as to agricultural drains.

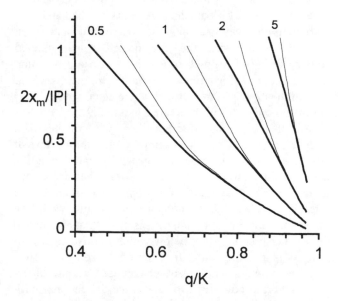

Figure 7. The bounds given by the analysis (shown by the thick and thin lines) for the relationship between the dimensionless critical cavity width $2x_m/|P|$ and dimensionless flux density q/K for values of the dimensionless cavity height $L/|P|$ shown by the curves.

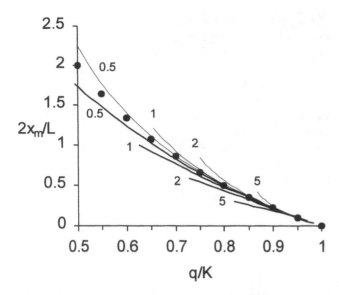

Figure 8. The relationship between the dimensionless critical cavity width $2x_m/L$ and dimensionless precipitation q/K. The thick lower lines are values for the bounds given by (16) and the thin upper lines are values for the bounds given by (17). Numbers by the curves are values of $L/|P|$. The dots are values given by (20).

Philip et.al. [1989b] give design criteria for water exclusion calculated for cavities in Yolo light clay soil whose hydraulic conductivity function does not exhibit a tension-saturated region. The results shown in Figures 6, 7 and 8 and approximated by equation (20) provide the maximum width of tunnel cavities for water exclusion for a given vertical flow rate when they are located in the saturated capillary fringe of a given soil. Thus, with a percolation rate equal to $0.5K$ for a tunnel height of 5m, the maximum width of the base is 10m, but with a rate equal to $0.75K$ it is 3.33 m and only 1.11 m for a rate equal to $0.9K$.

While this paper has discussed the two extremes of a Gardner and of a Green and Ampt soil in the context of water exclusion from cavities, generally more progress needs to be done to investigate soil-water behavior for soils having other soil hydraulic functions. For example, the hydraulic conductivity function of many soils would be better represented by a region of tension-saturation down to a soil-water pressure of P below which the hydraulic conductivity is an exponential function of the soil-water pressure. Recent work by this author has used a finite difference method to consider two- and three-dimensional infiltration into soils showing such a hydraulic conductivity relationship, adding to the confidence in adopting results from more idealized soil hydraulic functions that John Philip's prolific publications provided.

Acknowledgement. The author gratefully acknowledges discussions with the late John Philip and especially with Dr. J.H. Knight on the analysis given in this paper during a visit to the CSIRO Centre for Environmental Mechanics.

REFERENCES

Aravin, V.I., and S.N. Numerov, *Theory of Fluid Flow in Undeformable Porous Media* (translated from the Russian by A, Moscana), Israel Program for Scientific Translations, Jerusalem, 1965.

Childs, E.C., The water table, equipotentials and streamlines in drained land, III, *Soil Sci.*, 59, 405-415 , 1945.

Gardner, W.R., Some steady-state solutions of the unsaturated moisture flow equation with application to evaporation from a water table, *Soil Sci.*, 85, 228-232, 1958.

Harr, M.E., *Groundwater and Seepage*, McGraw Hill, New York, 1965

Kacimov, A.R., and Yu.V. Obnosov, Steady water flow around parabolic cavities and through parabolic inclusions in unsaturated and saturated soils, *J. Hydrol.*, 238, 65-77, 2000.

Knight, J.H., J.R. Philip, and R.T. Waechter, The seepage exclusion problem for spherical cavities, *Water Resour. Res.*, 25, 29-37, 1989.

Philip, J.R., Numerical solution of equations of the diffusion type with diffusivity concentration-dependent, *Trans, Faraday Soc.*, 51, 885-892, 1955.

Philip, J.R., Numerical solution of equations of the diffusion type with diffusivity concentration-dependent. II, *Aust. J. Phys.*, 10, 29-42, 1957.

Philip, J.R., The seepage exclusion problem for sloping cylindrical cavities, *Water Resour. Res.*, 25, 1447-1448, 1989a.

Philip, J.R., Asymptotic solutions of the seepage exclusion problem for elliptic-cylindrical, spheroidal, and strip- and disc-shaped cavities, *Water Resour. Res.*, 25, 1531-1540, 1989b.

Philip, J.R., Some general results on the seepage exclusion problem, *Water Resour. Res.*, 26, 369-377, 1990.

Philip, J.R., J.H. Knight, and R.T. Waechter, Unsaturated seepage and subterranean holes: conspectus, and exclusion problem for circular cylindrical cavities, *Water Resour. Res.*, 25, 16-28, 1989a.

Philip, J.R., J.H. Knight, and R.T. Waechter, The seepage exclusion problem for parabolic and paraboloidal cavities, *Water Resour. Res.*, 25, 605-618, 1989b.

Polubarinova-Kochina, P.Ya., *Theory of Groundwater Movement* (translated from the Russian by J.M. Roger de Weist), Princeton University Press, Princeton, N.J., 1962.

Youngs, E.G., Moisture profiles during vertical infiltration, *Soil Sci.*, 84, 283-290, 1957.

Youngs, E.G., Soil physics and hydrology, *J. Hydrol.*, 100, 411-431, 1988.

Youngs, E.G., D.E. Elrick, and W.D. Reynolds, Comparison of steady flows from infiltration rings in "Green and Ampt" soils and "Gardner" soils. *Water Resour. Res.*, 29, 1647-1650, 1993.

E.G. Youngs, Institute of Water and Environment, Cranfield University, Silsoe, Bedfordshire MK45 4DT, England.

Column Flow in Stratified Soils and
Fingers in Hele-Shaw Cells: A Review

J.-Yves Parlange, Tammo S. Steenhuis, Ling Li,
D. A. Barry, and Frank Stagnitti

We examine the development of instability studies in the oil industry and in soil physics. The former being far more advanced has tended to mold the latter. This had some unfortunate consequences as oil studies tended to rely heavily on Hele-Shaw cells which provide a poor model for soils. In soil physics creation of column flow in soils has serious implications for infiltration and the transport of pollutants. It is to John Philip's credit that he recognized fairly early the practical importance of instability in soils.

INTRODUCTION

In his 1972 review [*Philip*, 1972] on "future problems of soil water research", John Philip listed "Stability of soil water flows" as the first in a list of five topics. He stated "till now almost all analyses of water movement in unsaturated soils have contained the implicit assumption that solutions of the flow equation are stable for flows in homogeneous soils. This appears to be very reasonable since the diffusion character of the equation suggests that disturbances will be damped and not magnified; but it remains an open and interesting question whether the possibility of capillary hysteresis affects flow stability in homogeneous soils. Hill and Parlange have performed elegant experiments on instability (fingering) during infiltration into a soil with a fine-textured layer over a coarser-textured one....contact angle problems in a surface layer may well be explained as the expression of instability....This work is certain to stimulate further studies of stability..."

This long quote covers much of what is discussed in the following. Although *Hill and Parlange* [1972] was in press at the time, Philip was aware of the study as the second author of that paper was on sabbatical at the Pye Laboratory.

Environmental Mechanics: Water, Mass and Energy Transfer in the Biosphere
Geophysical Monograph 129
Copyright 2002 by the American Geophysical Union
10.1029/129GM08

Even at that early stage, John Philip recognized the important implications of the work in modeling infiltration in the field as subsequently documented by *Starr et al.* [1978]. His statement on hysteresis was remarkably accurate as hysteresis is ubiquitous in fingering even though that recognition has been slow to develop. The interaction between contact angle problems and fingering remains a fundamental question and his suggestion that further studies would follow proved quite prophetic. Indeed shortly afterwards *Raats* [1973] and *Bridge and Collis-George* [1973] were the first to refer to Hill and Parlange in their respectively theoretical and experimental studies.

Philip [1972] also mentioned the earlier papers of *Saffman and Taylor* [1958] and *Wooding* [1969] which provide some theoretical background to fingering in stratified porous media as will be discussed later. In his important paper on fingering, *Raats* [1973] points out that "in almost all studies it has been tacitly assumed that small perturbations in flow patterns will tend to disappear. In other words, it is usually assumed that the flows are stable. Recent experiments by *Hill and Parlange* [1972] involving infiltration into layered soils suggest that this assumption is not always justified."

EARLY WORK

It is curious that the practical importance of instability in stratified soils was not recognized prior to 1972. As stated by *Baker and Hillel* [1990], "the phenomenon of fingering during infiltration into layered soils has attracted increasing interest since the work of *Hill and Parlange* [1972], who were among the first to demonstrate that fingering can occur in a fine-over coarse-textured profile". Most likely the earlier "demonstrations" had not obvious enough connection to infiltration in stratified soils, e.g. *Horton and Rogers* [1945]; *Lapwood* [1948]; *Heller* [1966]; *Elder* [1968]; and *Bachmat and Elrick* [1970] as they were concerned with instabilities caused by gradients of temperature or solute concentration in miscible liquids. Others were difficult to discover [*Engelberts and Klinkenberg*, 1951; *DeRoo*, 1952; and *Tabuchi*, 1961]. *Tabuchi* [1961] in particular drew a rough sketch of an embryonic finger starting to form in a layer of coarse glass beads underneath a layer of fine glass beads and obtained a necessary stability condition independently of *Saffman and Taylor* [1958]. *Smith* [1967] came close to our present understanding of fingering with "infiltration when the rate of rainfall is less than the infiltration capacity of the sand...For example, if a succession of drops from a burette falls upon a sand for several hours, it will be observed that these drops pass downwards in a filament which when established, does not spread. The balance of the sand is either dry or in an unsaturated condition...The problem arises largely in very light rainfall." The question of fingering in very light rainfall will be discussed briefly later on. Although not entirely clear, *Crosby et al.* [1968] referred to Smith's "ribbons" and possibly observed them in the case of a fine material overlying a coarse material..." and "this mechanism is the apparent explanation of the dry conditions found beneath the drain field in the Spokane Valley." An earlier paper by *Miller and Gardner* [1962] made similar points, but more emphasis was placed on the important point that "when most of the pores in a layer were larger than those in the surrounding soil, infiltration was temporarily inhibited...water must accumulate at a layer-soil interface until it is at a tension low enough to allow it to move into pores in the layer." Then "as the water passes through an initially dry sand layer, it characteristically wets the sand in only a few places and the remainder of the sand remains dry. Liquid movement through the sand is restricted to the water-filled channels." However the sand layers were only half a centimeter thick and eventually became uniformly wet.

Other papers put more emphasis on the presence of ribbons/tongues/fingers/columns, but as they were clearly caused by water repellency or air compressibility presumably these effects were seen as being of limited importance. For instance *Bond* [1964] states "rain penetrates into water repellent sands through narrow tongues, leaving the interven-

ing soil quite dry" and those "dry zones tend to persist". Bond also gave clear sketches of those tongues (looking more like columns). *Peck* [1965] also observed tongues in sand due to air compressibility and mentioned them briefly in two places: "After a short initial period of wetting it was observed that tongues developed in bounded columns of the sand and that in the long columns the tongues grew to be the dominant feature of each profile. This is an example of a density instability with analogies with that discussed by Saffman and Taylor". In the second mention the tongues are somewhat dismissed, i.e. "moisture profiles in bounded columns of the sand are not shown here because the profiles were dominated by the tongues, and the mean moisture contact...is not considered useful information" and no sketch of the tongues was presented.

In all those papers, with the exception of *Tabuchi* [1961] and *Bond* [1964] the formation of tongues was not central to the papers and thus were more easily missed. Indeed all those early papers are being "rediscovered" since the early 70's when the importance of fingering as a fundamental process for the analysis of infiltration of water in stratified soils was finally appreciated [*Philip*, 1972; and *Hill and Parlange*, 1972]. As mentioned by *Hill and Parlange* [1972] in the US alone "there are about 350 series in family groupings with fine or coarse silty, fine or coarse loamy layers over sandy or sandy-skeletal textures. In those soils, the wetting fronts following rains or irrigation should be unstable". The increasing concern about pollution of aquifers and the obvious mechanism provided by fingering, also mentioned by *Hill and Parlange* [1972], made it harder to ignore this important process. Undoubtedly as time passes more of those early papers will be rediscovered.

The early awareness of fingering in petroleum engineering noticed in several papers, e.g. *Hillel and Baker* [1988]; and *White et al.* [1976], had no parallel in soil physics, in part because of reference to oil and Hele-Shaw cells [*Saffman and Taylor*, 1958; *Chuoke et al.*, 1959; and *Wooding*, 1969]. Compounding the difficulty even further is that oil being viscous, viscosity became a dominant feature, or as *Saffman and Taylor* [1958] begin their paper: "when a viscous fluid filling the voids in a porous medium is driven forwards by the pressure of another driving fluid, the interface between them is liable to be unstable if the driving fluid is the less viscous of the two. This condition occurs in oil fields", then in their Hele-Shaw cell experiment they displaced glycerine by compressed air, not an obvious analogue to water displacing air in a stratified medium. In fact in the latter case viscosity is largely irrelevant as will become clear later on (although the term "viscous fingering" is sometimes still used incorrectly to describe the phenomenon). Of course this does not mean that some of the equations in *Saffman and Taylor* [1958] and the similar ones in *Chuoke et al.* [1959] cannot be properly reinterpreted. For instance the general necessary condition for instability, for water and air, neglecting both viscosity and density of air becomes

$$K > Q \qquad (1)$$

where Q is the water flux imposed by the upper layer of fine material, and K is some conductivity of water in the coarse layer underneath. Note that since conductivity is an increasing function of the water content, it is certainly necessary that

$$K_s > Q \qquad (2)$$

where K_s has its maximum value at saturation. Clearly with a Hele-Shaw cell saturation is always imposed and there is no ambiguity as to the value of K. *Hill and Parlange* [1972] also assumed that the fingers were saturated and thus took condition (2). Condition (2) is an obvious necessary condition since if Q was greater than K_s the whole area would have to be saturated to carry the water and no finger would be present. It is a fundamental contribution of *Hillel* [1987], *Hillel and Baker* [1988], and *Baker and Hillel* [1990] to have noticed that condition (2) is not constraining enough and condition (1) with (usually) a lower K must be satisfied. They associate this value with a "water entry" value for the water to penetrate the coarse layer. This condition provides a mechanism to explain the constriction of the water flow. Those points of entry were clearly observed by *Glass et al.* [1989]. *Hillel and Baker* [1988] further suggest that upon rewetting the water entry suction will be higher, resulting in drier fingers which was also observed by *Glass et al.* [1989] leading to interesting hysteresis phenomena [*Liu et al.*, 1995; and *Raats*, 1973]. For instance hysteresis is crucial by limiting lateral capillary diffusion of water which would otherwise remove the presence of fingers.

Glass et al. [1989] further observed that there were too many points of entry for all of them to become fingers. Then, merger takes place until only a few fingers with an optimal size, remain. These mergers took place just below the interface of the layers in the original experiments of *Hill and Parlange* [1972] and could not be observed. Because of a more homogeneous packing the merger in *Glass et al.* [1989] took place over a greater distance and could be clearly observed. The region where mergers take place is called the induction zone [*Hill and Parlange*, 1972].

Saffman and Taylor [1958] predicted the optimal width of the fingers in a Hele-Shaw cell by balancing the destabi-

lizing effects of gravity and the stabilizing effect of surface tension. The same formula was then "extended" by *Chuoke et al.* [1959] for a porous medium replacing the surface tension which loses its meaning for a diffuse wetting front by an "effective" surface tension. Well aware of this limitation, *Raats* [1973] and *Philip* [1975a] limited themselves to "Green and Ampt soils" with discontinuous wetting fronts and extended condition (1) for other situations, e.g. with nonwetting soils and compression of the air below the invading water.

It is interesting that *Raats* [1973] suggests that instability will take place for nonponding rainfall and a homogeneous, i.e. not stratified, soil, whereas *Philip* [1975a] does not. *Raats* [1973] states "infiltration of nonponding rainfall is very similar to infiltration of ponded water through a fine layer or crust as observed by *Hill and Parlange* [1972]". Indeed following this similarity one must look for some equivalent mechanism resulting in the concentration of flow provided by the points of entry in layered soils. One might speculate that when a large raindrop (large compared to the pore size) hits the soil surface with a positive pressure it will enter primarily through the largest pores and passages and if enough raindrops fall in the same neighborhood they might merge fast enough to form a finger? This could possibly explain why at low rainfall rates when raindrops are less likely to merge, fingers become rapidly wider and eventually the instability disappears [*Yao and Hendrickx*, 1996]. Another possible mechanism might be linked to some wettability effects. However, *Selker et al.* [1992a,b,c] and *Yao and Hendrickx* [1996] observed fingers under nonponding rainfall and wetting sand, although wetting problems may well have affected field and laboratory observations, e.g. *Hendrickx and Yao* [1996] and *Selker and Schroth* [1998]. As discussed by *Bond* [1964] we expect water repellent soils to exhibit fingering, see also *Bauters et al.* [1998, 2000]. Note that oil displaced by water will tend to leave some oily residue on the sand grains leading to contact angle problems [*Rimmer et al.*, 1996] which may well affect the oil flow [*DiCarlo et al.*, 1997, 2000; *Darnault et al.*, 1998; *Rimmer et al.*, 1998; and *Chao et al.*, 2000]. Thus, as suggested by *Philip* [1972] contact angle effects should always be considered at least as a contributing factor in the formation of fingers, especially for homogeneous soils.

Philip [1975a,b] was well aware that the limitations of a Green and Ampt soil "cast some doubt on the relevance of the model" but it "has the great advantage of being amenable to stability analysis. Unfortunately, formulations based on the Richards equation...are less so". Indeed he refers properly to the model as a "generalized Hele-Shaw cell" and suggested "stability studies of appropriate forms of the Richards equation", although "a general attack promises to be very difficult". *Philip* [1975b] then provided an estimate of the finger width, corrected later by *White et al.* [1976], who also presented some experimental results, again mostly with a Hele-Shaw cell, because "it satisfies the criteria of the delta-function model precisely, whereas a soil water system can at best approximate them". They also observed fingering with a coarse sand by increasing the air pressure ahead of the wetting front. They concluded the study with further doubt on the delta-function model and the need of more "work on the stability of actual diffuse fronts", as done by *Parlange and Hill* [1976].

ANALYSIS

Parlange and Hill [1976] derived an expression for the finger width, d, based on the analysis of Richards equation, i.e. a diffuse front, yielding

$$d = \pi \, \frac{S^2}{K(\theta - \theta_i)} \, \frac{1}{1 - Q/K} \qquad (3)$$

where S is the sorptivity given by

$$S^2 = \int_{\theta_i}^{\theta} D \left[\theta + \overline{\theta} - 2\theta_i\right] d\theta, \qquad (4)$$

where D is the soil-water diffusivity and θ_i is the initial water content, assumed small enough that the soil-water conductivity K at θ_i is negligible compared to its value at θ. The coefficient π is obtained for a two-dimensional finger as often observed in the laboratory, in the field where fingers are axisymmetric (more or less) the coefficient π should be replaced by 4.8 [*Glass et al.*, 1991].

Initially *Parlange and Hill* [1976] assumed that θ in Eq. (3) corresponds to saturation. However fingers are rarely saturated in soils (as they are in a Hele-Shaw cell). In fact their water content varies with depth. It was shown by *Selker et al.* [1992a,b] that θ varies with depth (measured from the interface between layers) according to the equation

$$z = \int_{\theta_o}^{\theta} \frac{D d\overline{\theta}}{K - v(\theta - \theta_i)}, \qquad (5)$$

where v is the constant downward speed of the fingers obtained after a short time, i.e. after all mergers have taken place and the fingers have reached a steady configuration.

θ_o is the value of θ at $z = 0$ and if we assume, following *Hillel and Baker* [1988], that the maximum value of θ corresponds to a water entry value θ_e, then, Eq. (5) gives

$$vt = \int_{\theta_o}^{\theta_e} \frac{D d\theta}{K - v(\theta - \theta_i)} \qquad (6)$$

which gives $\theta_o(t)$ when v and θ_e are known. In particular when $t \rightarrow \infty$ θ_o approaches an asymptotic value $\theta_{o\infty}$ with

$$K(\theta = \theta_{o\infty}) = v(\theta_{o\infty} - \theta_i). \qquad (7)$$

Note that between θ_e and θ_o all properties are measured on a drying curve of the matric potential. However, as the finger moves downwards within the sand there is a very narrow zone at the finger tip where the water content increases rapidly, thus operating on a wetting curve but reliable matric potential data are impossible to get in that region [*Liu et al.*, 1995; and *Selker et al.*, 1992a,b].

Going back to Eq. (3) it is not entirely clear which value of θ should be used to determine S and K. As noted by *Hillel and Baker* [1988], in agreement with Eqs. (1) and (3), we require $1 > Q/K$ or the total soil cross-section would be required to carry the water. This is true whenever $\theta_o \leq \theta \leq \theta_e$ is used. Indeed for steady state conditions the wetted fraction of soil is $F_\infty = Q/K_{o\infty}$. This might suggest using $\theta = \theta_{o\infty}$ in Eq. (3) however the fingers reach their thickness d when θ_o is somewhat above $\theta_{o\infty}$ but certainly less than θ_e. When $\theta_o \approx \theta_e$ as already mentioned the fingers are very narrow and mergers are required before d is obtained. Call θ_o^* this (unknown) value of θ_o to use in Eq. (3) and K_o^* the corresponding value of K_o.

First if $F^* = Q/K_o^*$ is much less than one, Q/K can be neglected in Eq. (3). *Glass et al.* [1991] looked at the impact of $[1 - Q/K]^{-1}$ on the value of d. By the time this is significant so many fingers are present that the impact of fingering, i.e. the fact that the flow bypasses most of the soil is lost. Thus for fingering to be important in practice, we require that Q/K be negligible in Eq. (3). Then we obtain a simpler equation for d,

$$d^* = \pi \, S_o^{*2} / K_o^* (\theta_o^* - \theta_i) \qquad (8)$$

Note that both S_o^{*2} and K_o^* are inversely proportional to the viscosity, thus d^* is independent of viscosity. The influence of viscosity can be felt only through $[1 - Q/K_o^*]$ and thus is irrelevant when fingering is important and F^* is small.

We are now going to show that d^* has only a very small dependence on θ_o^*. For coarse sands and as long as θ is not too close to zero and to θ_e, *Parlange and Hogarth* [1985] showed that the description of the soil-water conductivity of *Gardner* [1958] is realistic:

$$K = K_o^* \, exp \; \alpha\left(h - h_o^*\right) \qquad (9)$$

where α is more or less constant and h is the matric potential. Then Eq. (4) shows that for a coarse sand

$$S^{*2} \simeq 2\left(\theta^* - \theta_i\right) K_o^*/\alpha \qquad (10)$$

Hence we find the remarkably simple result,

$$d^* \simeq 2\pi/\alpha, \qquad (11)$$

i.e. the dependence on θ_o^* and θ_i has disappeared. This explains why as soon as θ_o is somewhat less than θ_e, d is essentially constant in time and space and is inversely proportional to Gardner's α. It must be remembered that Eq. (11) is a crude approximation. In particular it requires that $F^* << 1$. Under low flow rates, *Yao and Hendrickx* [1996] observed widening fingers and Eq. (3) should be used until $F^* = 1$ and the instability disappears.

DISCUSSION

We have limited ourselves so far to estimating d when only a liquid, e.g. water or oil, and a gas, e.g. air are present. It would be certainly important to extend the formula to the case of oil and water or even with three fluids, e.g. oil, water and air. One such extension, for oil and water, is based on a fundamentally wrong derivation. As the error was also somewhat easy to make, we shall briefly look at its cause; more details can be found in *Chandler et al.* [1998]. If, for instance water displaces oil, we consider the region where no oil is present and we call the lower boundary of this region, the "front". Ahead of the front there is a narrow zone where both oil and water are present and finally there is only oil. To solve the instability problem it is necessary to obtain the dependence of the front velocity on its curvature. If that dependence is neglected all wave lengths are found to be unstable, the smaller ones growing faster and only condition (1) is obtained [*Saffman and Taylor*, 1958]. The curvature effect stabilizes the small wave lengths and an estimate of d is obtained [*Parlange and Hill*, 1976]. To obtain the dependence of the front velocity on curvature, the flow in the narrow diffuse region, ahead of it, must be analyzed and to do so we must know the flow further ahead (here the pure oil). In the case of a gas [*Parlange and Hill*, 1976] this presented no problem as it was assumed that the air could move freely ahead and did not affect the flow of water. Usually this will not be the case when oil is displaced by

water, although it represents a limiting case [*Chandler et al.*, 1998]. *Glass et al.* [1991] instead impose that the diffuse front has zero thickness which is in fundamental contradiction with the existence of a diffuse front in a soil! Consequently the result must be seen as pure speculation for a soil (although it would hold for a Hele-Shaw cell). It is clear that any constraint on the oil flow will widen the fingers and hence assuming that the oil is free to move [*Chandler et al.*, 1998] provides the minimum estimate for finger size.

We have pointed out in several places that Hele-Shaw cells are not a very good analogue to a soil for finger analysis. In addition since fingers can only be saturated in a Hele-Shaw cell, there is no diffuse front and surface tension is affected by the curvature of the finger, rather than the curvature of meniscii between grains. There is another even more fundamental reason not to use the Hele-Shaw cells as a substitute for soils. All the experiments with Hele-Shaw cells show fingers connected with each other at their base [*Saffman and Taylor*, 1958; *Chuoke*, 1959; *Wooding*, 1969; *White et al.*, 1976, 1977; and *Tamai et al.*, 1987] and that base tends to move with the flow, i.e. the fingers are often very short. On the other hand with fingering in stratified soils, the fingers are separated below the induction region and hence their base does not move. Also the width of the fingers is constant whereas in Hele-Shaw cells, they tend to grow in width until they join at their base, indeed looking much like fingers in a hand, whereas in stratified soils they look more like "columns". It is unfortunate that following the influence of *Saffman and Taylor* [1958], *Hill and Parlange* [1972] kept the word "finger" and it is to be hoped that in the future the words "column flow" be adopted as more descriptive for stratified soils. We also believe that the differences in appearance are based on a fundamental physical difference in the flows. When the columns are formed in the stratified soil we expect some lateral diffusion, and drier sand between the columns can be sustained only because of hysteresis: to diffuse in the drier sand the columns are on a drying curve and on a wetting curve in between, so for the same potential differences in water content can be maintained [*DiCarlo*, 1999]. Certainly no hysteresis should normally be present in a Hele-Shaw cell when water and air are the two fluids so the fingers should become thicker until they finally merge. *Raats* [1973] also pointed out that whereas the tip is on a wetting curve the remainder of the column flow is on a drying curve.

CONCLUSION

In this short review we have tried to connect the histories of flow instability in porous media as they developed in petroleum studies and in soil physics. That connection was strongly influenced through the Hele-Shaw cell analogue, which turns out to have been at least misleading, both theoretically and experimentally, as fingers in the cells and column flow in stratified soils have very different properties and appearances. The presence of a fine textured layer over a coarse textured layer readily explains the appearance of points of entry at their interface resulting in column flow. In the case of a homogeneous soil the reasons for the concentration of flow are more speculative.

REFERENCES

Bachmat, Y., and D.E. Elrick, Hydrodynamic instability of miscible fluids in a vertical porous column, *Water Resour. Res.*, 6, 156-171, 1970.

Baker, R.S., and D. Hillel, Laboratory tests of fingering during infiltration into a layered soil, *Soil Sci. Soc. Am. J.*, 54, 20-30, 1990.

Bauters, T.W.J., D.A. DiCarlo, T.S. Steenhuis, and J.-Y. Parlange, Preferential flow in water-repellent sands, *Soil Sci. Soc. Am. J.*, 62,1185-1190, 1998.

Bauters, T.S. Steenhuis, D.A. DiCarlo, J.L. Nieber, L.W. Dekker, C.J. Ritsema, J.-Y. Parlange, and R. Haverkamp, Physics of water repellent soils, *J. Hydrol.*, 231-232:233-243, 2000.

Bond, R.D., The influence of the microflora on physical properties of soils. II. Field studies on water repellent sands, *Aust. J. Soil Res.*, 2, 123-131, 1964.

Bridge, B.J., and Collis-George, N., An experimental study of vertical infiltration into a structurally unstable swelling soil, with particular reference to the infiltration throttle, *Aust. J. Soil Res.*, 11, 121-132, 1973.

Chandler, D.G., Z. Cohen, E. Wong, D.A. DiCarlo, T.S. Steenhuis, and J.-Y. Parlange, Unstable fingered flow of water into a light oil, in *Hydrology Days*, edited by H. Morel-Seytoux, pp. 13-31, AGU Pub. 18, 1998.

Chao, W.-L., J.-Y. Parlange, and T.S. Steenhuis, An analysis of the movement of wetting and nonwetting fluids in homogeneous porous media, *Transport in Porous Media*, 41,121-135, 2000.

Chuoke, R.L., P. van Mears, and C. van der Poel, The instability of slow immiscible, viscous liquid-liquid displacements in porous media, *Trans. Am. Inst. Min. Eng.*, 216, 188-194, 1959.

Crosby, J.W., D.L. Johnstone, C.H. Drake, and R.L. Fenton, Migration of pollutants in a glacial outwash environment, *Water Resour. Res.*, 4, 1095-1113, 1968.

Darnault, C.J.G., J.A. Throop, D.A. DiCarlo, A. Rimmer, T.S. Steenhuis, and J.-Y. Parlange, Visualization by light transmission of oil and water contents in transient two-phase flow fields, *J. Contaminant Hydrol.*, 31, 337-348, 1998.

DeRoo, H.C., The soil and sub-soil geology of the Dreuthe table-land, *Boor en Spade*, 5, 102-118, 1952.

DiCarlo, D.A., T.W.J. Bauters, T.S. Steenhuis, J.-Y. Parlange, and B. Bierck, High-speed measurements of three-phase flow using synchrotron X rays, *Water Resour. Res.*, 33, 569-576, 1997.

DiCarlo, D.A., T.W.J. Bauters, C.J.G. Darnault, T.S. Steenhuis, and J.-Y. Parlange, Lateral expansion of preferential flow paths in sands, *Water Resour. Res.*, 35, 427-434, 1999.

DiCarlo, D.A., T.W.J. Bauters, C.J.G. Darnault, E. Wong, B.R. Bierck, T.S. Steenhuis, and J.-Y. Parlange, Surfactant-induced changes in gravity fingering of water through a light oil, *J. Contaminant Hydrol.*, 41, 317-334, 2000.

Elder, J.W., The unstable thermal interface, *J. Fluid Mech.*, 32, 69-96, 1968.

Engelberts, W.F., and L.J. Klinkenberg, Laboratory experiments on the displacements of oil by water from packs of granular materials, in *Proc. 3rd World Petr. Cong., Part II*, 544, The Hague, 1951.

Gardner, W.R., Some steady state solutions of unsaturated moisture flow equation with application to evaporation from a water table, *Soil Sci.*, 85, 228-232, 1958.

Glass, R.J., T.S. Steenhuis, and J.-Y. Parlange, Mechanism for finger persistence in homogeneous, unsaturated, porous media: Theory and verification, *Soil Sci.*, 148, 60-70, 1989.

Glass, R.J., J.-Y. Parlange, and T.S. Steenhuis, Immiscible displacement in porous media: Stability analysis of three-dimensional axisymmetric disturbances with applications to gravity driven wetting front instability, *Water Resour. Res.*, 27, 1947-1956, 1991.

Heller, J.P., Onset of instability patterns between miscible fluids in porous media, *J. Appl. Phys.*, 37, 1566-1579, 1966.

Hendrickx, J.M.H., and T-M. Yao, Prediction of wetting front stability in dry field soils using soil and precipitation data, *Geoderma*, 70, 265-280, 1996.

Hill, D.E., and J.-Y. Parlange, Wetting front instability in homogeneous soils, *Soil Sci. Soc. Am. Proc.*, 36, 697-702, 1972.

Hillel, D., Unstable flow in layered soils: A review, *Hydrol. Proc.*, 1, 143-147, 1987.

Hillel, D., and R.S. Baker, A descriptive theory of fingering during infiltration into layered soils, *Soil Sci.*, 146, 51-56, 1988.

Horton, C.W., and F.T. Rogers, Convection currents in a porous medium, *J. Appl. Phys.*, 16, 367-370, 1945.

Lapwood, E.R., Convection of a fluid in a porous medium, *Proc. Camb. Phil. Soc.*, 44, 508-521, 1948.

Liu, Y., J.-Y. Parlange, T.S. Steenhuis, and R. Haverkamp, A soil water hysteresis model for fingered flow data, *Water Resour. Res.*, 31, 2263-2266, 1995.

Miller, D.E., and W.H. Gardner, Water infiltration into stratified soil, *Soil Sci. Soc. Am. Proc.*, 16, 115-119, 1962.

Parlange, J.-Y., and D.E. Hill, Theoretical analysis of wetting front instability in soils, *Soil Sci.*, 122, 236-239, 1976.

Parlange, J.-Y., and W.L. Hogarth, Steady state infiltration: Consequences of α dependent on moisture content, *Water Resour. Res.*, 21, 1283-1284, 1985.

Peck, A.J., Moisture profile development and air compression during water uptake by bounded porous bodies. 3. Vertical columns. *Soil Sci.*, 100, 44-51, 1965.

Philip, J.R., Future problems of soil water research, *Soil Sci.*, 113, 294-300, 1972.

Philip, J.R., Stability analysis of infiltration, *Soil. Sci. Soc. Am. Proc.*, 39, 1042-1049, 1975a.

Philip, J.R., The growth of disturbances in unstable infiltration flows, *Soil Sci. Soc. Am. Proc.*, 39, 1049-1053, 1975b.

Raats, P.A.C., Unstable wetting fronts in uniform and nonuniform soils, *Soil Sci. Soc. Am. Proc.*, 37, 681-685, 1973.

Rimmer, A., J.-Y. Parlange, T.S. Steenhuis, C. Darnault, and W. Condit, Wetting and nonwetting fluid displacements in porous media., *Transport in Porous Media*, 25, 205-215, 1996.

Rimmer, A., D.A. DiCarlo, T.S. Steenhuis, B. Bierck, D. Durnford, and J.-Y. Parlange, Rapid fluid content measurement method for fingered flow in an oil-water-sand system using synchrotron X-rays, *J. Contaminant Hydrol.*, 31, 315-335, 1998.

Saffman, P.G., and G.I. Taylor, The penetration of a fluid into a porous medium or Hele-Shaw cell containing a more viscous liquid, *Proc. Soc. London, Ser. A*, 245, 312-331, 1958.

Selker, J., P. Leclerq, J.-Y. Parlange, and T.S. Steenhuis, Fingered flow in two dimensions. 1. Measurement of matric potential, *Water Resour. Res.*, 28, 2513-2521, 1992a.

Selker, J., J.-Y. Parlange, and T.S. Steenhuis, Fingered flow in two dimensions. 2. Predicting finger moisture profile, *Water Resour. Res.*, 28, 2523-2528, 1992b.

Selker, J., T.S. Steenhuis, and J.-Y. Parlange, Wetting front instability in homogeneous sandy soils under continuous infiltration, *Soil Sci. Soc. Am. J.*, 56, 1346-1350, 1992c.

Selker, J., and M.H. Schroth, Evaluation of hydrodynamic scaling in porous media using finger dimensions, *Water Resour. Res.*, 34, 1935-1940, 1998.

Smith, W.O., Infiltration in sand and its relation to ground water recharge. *Water Resour. Res.*, 3, 539-555, 1967.

Starr, J.L., H.C. DeRoo, C.R. Frink, and J.-Y. Parlange, Leaching characteristics of a layered field soil, *Soil Sci. Soc. Am. J.*, 42, 386-391, 1978.

Tabuchi, T., Infiltration and ensuing percolation in columns of layered glass particles packed in laboratory, *Trans. Agric. Eng. Soc. Japan*, 2, 27-36, 1961.

Tamai, N., T. Asaeda, and C.G. Jeevaraj, Fingering in two-dimensional, homogeneous, unsaturated porous media, *Soil Sci.*, 144, 107-112, 1987.

White, I., P.M. Colombera, and J.R. Philip, Experimental study of wetting front instability induced by sudden change of pressure gradient, *Soil Sci. Soc. Am. J.*, 40, 824-829, 1976.

White, I., P.M. Colombera, and J.R. Philip, Experimental studies of wetting front instability induced by gradual change of pressure gradient and by heterogeneous porous media, *Soil Sci. Soc. Am. J.*, 41, 483-489, 1977.

Wooding, R.A., Growth of fingers at an unstable diffusing interface in a porous medium or Hele-Shaw cell, *J. Fluid Mech.*, 39, 477-495, 1969.

Yao, T., and J.M.H. Hendrickx, Stability of wetting fronts in homogeneous soils under low infiltration rates, *Soil Sci. Soc. Am. J.*, 60, 20-28, 1996.

J.-Y. Parlange and Tammo S. Steenhuis, Department of Biological and Environmental Engineering, Riley-Robb Hall, Cornell University, Ithaca, NY 14853

Ling Li and D. A. Barry, School of Civil and Environmental Engineering, Crew Building, The University of Edinburgh, Edinburgh, Scotland EH9 3JN United Kingdom

Frank Stagnitti, School of Ecology and Environment, Deakin University, P.O. Box 423, Warrnambool 3280 Australia

Wetting Front Evolution in Randomly Heterogeneous Soils

Alexandre M. Tartakovsky

Department of Hydrology and Water Resources, University of Arizona
Tucson, Arizona

Shlomo P. Neuman

Department of Hydrology and Water Resources, University of Arizona
Tucson, Arizona

Daniel M. Tartakovsky

Group T-7, Los Alamos National Laboratory, Los Alamos, New Mexico

Philip [1975] was the first to investigate the stability of a wetting front in a stratified soil using rigorous hydrodynamic arguments. He based his analysis on the *Green and Ampt* [1911] model and treated permeability as a known function of depth. We adopt the same model to develop integro-differential equations for leading statistical moments of wetting front propagation in a three-dimensional, randomly heterogeneous soil. We solve these equations analytically for mean front position and mean pressure head gradient in one spatial dimension, to second order in the standard deviation of log conductivity. We do the same for second moments of front positions and pressure head gradient, which serve as measures of predictive uncertainty. To verify the accuracy of our solution, we compare it with the results of numerical Monte Carlo simulations.

INTRODUCTION

Unstable wetting fronts have been reported in many field tests and laboratory experiments, [e.g. *Chen et al.*, 1995]. The onset of instability can be understood in general terms within the context of linear theories developed for the immiscible displacement of one fluid by another in a Hele-Shaw cell [*Saffman* and *Taylor*, 1958] and in a macroscopically uniform porous medium [*Chuoke et al.*, 1959]; see *Neuman* and *Chen* [1996] for a recent extension.

In natural soils and rocks the phenomenon is strongly colored by systematic and random spatial variations in macroscopic medium properties. It has become common to treat medium properties as random fields [e.g. *Warrick et al.*, 1986; *Dagan and Neuman*, 1997]. A stochastic analysis of wetting front instability in randomly stratified soils has been published by *Chen and Neuman* [1996]. Their work is closely related to earlier deterministic analyses by *Raats* [1973] and especially *Philip* [1975]. Both Raats and Philip considered a simple one-dimensional model of a horizontal wetting front due to *Green and Ampt* [1911]. In this model, the wetting front forms a sharp interface between a uni-

Environmental Mechanics: Water, Mass and Energy Transfer in the Biosphere
Geophysical Monograph 129
Copyright 2002 by the American Geophysical Union
10.1029/129GM09

formly wet region above, and a relatively dry region of uniform volumetric water content below. Philip studied the effect of vertical permeability trends on the instability of a Green and Ampt front. He found that the front is unstable when pressure head immediately above it increases downward, but is stable otherwise. A continuous or sudden increase in permeability with depth across the wetting front tends to destabilize it.

Chen and Neuman [1996] derived a probabilistic criterion for the onset of wetting front instability during infiltration into a randomly stratified soil. They took the wetting front to form a sharp boundary and treated the natural log hydraulic conductivity, $Y = \ln K$, as a random multivariate Gaussian function of space. Whereas the mean, $\langle Y \rangle$, of this function may exhibit a spatial drift, its fluctuations $Y' = Y - \langle Y \rangle$ about the mean are statistically homogeneous with zero mean, $\langle Y' \rangle \equiv 0$, constant variance, σ_Y^2, and spatial correlation scale, l_Y. The authors obtained closed-form expressions for the probability of instability and for the mean critical wave number, both directly and via a first-order reliability method. They then used Monte Carlo simulations to verify their analytical solutions and to determine the mean maximum rate of incipient finger growth and corresponding mean wave number. Chen and Neuman found that random fluctuations in soil permeability may have either a stabilizing or a destabilizing effect on a wetting front, depending on the spatial trend, variance and spatial correlation scale of log hydraulic conductivity.

Shariati and Yortsos [2001] have addressed the effect of nonrandom stratification on the stability of miscible fronts without dispersion. Readers interested in additional references to unstable fronts in random and nonrandom heterogeneous media should consult their paper and that of *Chen and Neuman* [1996].

In this paper, we adopt the Green and Ampt model to develop integro-differential equations for leading statistical moments of wetting front propagation in a three-dimensional, randomly heterogeneous soil. We solve these equations analytically for the mean and variance of front depth and pressure head gradient in one spatial dimension, to second order in the standard deviation, σ_Y, of log conductivity. This renders our solution nominally valid for mildly heterogeneous soils with $\sigma_Y \ll 1$. Our analytical solution requires no assumptions about the statistical distribution of Y. To verify its accuracy, we compare it with the results of numerical Monte Carlo simulations. The comparison shows that the solution "works" for both mildly and moderately heterogeneous soils with σ_Y at least as large as 1.

PROBLEM DEFINITION

Consider a three-dimensional flow domain Ω_T enclosing a randomly heterogeneous soil (Fig. 1). Constant hydraulic head H is maintained at the soil surface, which acts as a Dirichlet boundary Γ_D at elevation $z = 0$. It causes a sharp wetting front γ to propagate downward, separating a wetted region Ω $(\Omega \subset \Omega_T)$ above the front from a non-wetted region below it. The wetted region is at constant volumetric water content θ_1 and the non-wetted region at constant residual water content θ_2. No flow takes place across the vertical Neumann boundaries Γ_N. Capillary pressure, $p_c(\mathbf{x}_\gamma)$, at any point $\mathbf{x}_\gamma$ on the front is equal to an effective entry pressure, p_e, of water into the unsaturated soil. Methods to determine p_e on the basis of moisture retention and relative conductivity characteristics of uniform and heterogeneous soils have been reviewed by *Chen and Neuman* [1996]. These authors found that, in a randomly stratified soil, variations in effective entry pressure have a negligible effect on the onset of wetting front instability. We therefore take p_e to be a deterministic constant.

As water content behind the front is constant, transient flow in Ω is governed by

$$\nabla \cdot \mathbf{q}(\mathbf{x},t) = 0 \qquad \mathbf{x} \in \Omega , \ t \geq 0 \qquad (1)$$

coupled with Darcy's law

$$\mathbf{q}(\mathbf{x},t) = -K(\mathbf{x})\nabla h(\mathbf{x},t) \qquad \mathbf{x} \in \Omega , \ t \geq 0 \qquad (2)$$

and subject to Dirichlet and Neumann boundary conditions

$$h(\mathbf{x},t) = H \qquad \mathbf{x} \in \Gamma_D \qquad (3)$$

$$\frac{\partial h(\mathbf{x},t)}{\partial x_\mathbf{x}} = 0 \qquad \mathbf{x} \in \Gamma_N \qquad (4)$$

where $\mathbf{q}$ is flux, $\mathbf{x} = (x_\mathbf{x}, y_\mathbf{x}, z_\mathbf{x})$ is a system of Cartesian coordinates with $z_\mathbf{x}$ defined to be positive downward, t is time, K is hydraulic conductivity, and $h = p_w / \rho_w g - z_\mathbf{x}$ is hydraulic head, p_w being pressure and ρ_w the density of water. The front γ, whose position is defined uniquely by a depth function $\xi(x_\mathbf{x}, y_\mathbf{x}, t) \equiv z_\mathbf{x} \,|\, \mathbf{x} \in \gamma$, is initially at the soil surface so that

$$\xi(x_\mathbf{x}, y_\mathbf{x}, 0) = \xi_0(x_\mathbf{x}, y_\mathbf{x}) = 0 \quad t = 0 \qquad (5)$$

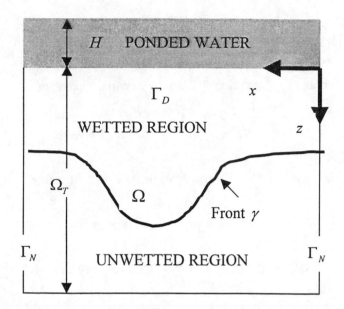

Figure1. Flow domain Ω (wetted region). Green's function defined on Ω_T that consists of wetted region and unwetted region, separated by front γ. Dirichlet boundary Γ_D with prescribed constant hydraulic head $h=H.$. No-flow Neumann boundary Γ_N.

At the front,

$$h(\mathbf{x},t) = a - \xi(x_{\mathbf{x}}, y_{\mathbf{x}}, t) \quad z_{\mathbf{x}} = \xi(x_{\mathbf{x}}, y_{\mathbf{x}}, t)), \, t \geq 0 \quad (6)$$

$$\mathbf{n}(\mathbf{x},t) \cdot \mathbf{q}(\mathbf{x},t) = \theta V_n(\mathbf{x},t) \quad z_{\mathbf{x}} = \xi(x_{\mathbf{x}}, y_{\mathbf{x}}, t), t \geq 0 \quad (7)$$

where $a = (p_o + p_e)/\rho_w g$, p_o is water pressure at the front due to excess air pressure (relative to atmospheric), g is acceleration due to gravity, $\theta = \theta_1 - \theta_0$, $\mathbf{n}$ is a unit normal pointing out of Ω, and V_n is the normal velocity of γ (positive when directed outward). The latter is defined as

$$V_n(\mathbf{x},t) = \frac{D\xi(x_{\mathbf{x}}, y_{\mathbf{x}}, t)}{Dt} n_z(\mathbf{x},t) \quad z_{\mathbf{x}} = \xi(x_{\mathbf{x}}, y_{\mathbf{x}}, t), t \geq 0 \quad (8)$$

where $D\xi / Dt$ is the Lagrangian velocity of a fixed point on the front and n_z is the vertical component of $\mathbf{n}$.

We treat $K(\mathbf{x})$ as a random field with mean $\langle K(\mathbf{x}) \rangle$, variance $\sigma_K^2(\mathbf{x})$ and spatial covariance function $C_K(\mathbf{x},\mathbf{y})$. We assume that the mean, variance and covariance of

$Y = \ln K$ can be estimated from, and conditioned on, field measurements of Y by geostatistical methods [e.g. Neuman, 1984]. Spatial drift and/or conditioning may render $Y(\mathbf{x})$, and therefore $K(\mathbf{x})$, statistically nonhomogeneous. As K is random, (1) – (8) are stochastic and their solution is uncertain. Our aim is to solve them in terms of leading moments of $h(\mathbf{x},t)$, its gradient and $\xi(x_{\mathbf{x}}, y_{\mathbf{x}}, t)$. For this, we represent all random quantities as the sum of their mean (designated by triangular brackets) and a zero-mean perturbation (designated by prime) about the mean.

MOMENT EQUATIONS

Integro-differential representation

Equations (1) - (8) are nonlinear due to the presence of a moving boundary, γ. To overcome this we introduce a deterministic Green function, defined as the solution of

$$\nabla_{\mathbf{y}} \cdot \left[\langle K(\mathbf{y}) \rangle \nabla_{\mathbf{y}} G(\mathbf{y};\mathbf{x}) \right] + \delta(\mathbf{y} - \mathbf{x}) = 0 \quad \mathbf{y}, \mathbf{x} \in \Omega_T \quad (9)$$

subject to homogeneous Dirichlet and Neumann boundary conditions

$$G(\mathbf{y};\mathbf{x}) = 0 \quad \mathbf{y} \in \Gamma_D \quad (10)$$

$$\nabla_{\mathbf{y}} G(\mathbf{y};\mathbf{x}) \cdot \mathbf{n}(\mathbf{y}) = 0 \quad \mathbf{y} \in \Gamma_N \quad (11)$$

As $G(\mathbf{y};\mathbf{x})$ is defined on the entire domain Ω_T including the wetted and nonwetted regions, it does not depend on front position. This idea was used by *Tartakovsky and Winter* [2001] to solve stochastic free surface problems without gravity. It allows writing an explicit, integro-differential expression for head,

$$\begin{aligned}
h(\mathbf{x},t) = &- \int_{\Omega(t)} K'(\mathbf{y}) \nabla_{\mathbf{y}} G(\mathbf{y};\mathbf{x}) \cdot \nabla_{\mathbf{y}} h(\mathbf{y},t) d\mathbf{y} \\
&- H \int_{\Gamma_D} \langle K(\mathbf{y}) \rangle \nabla_{\mathbf{y}} G(\mathbf{y};\mathbf{x}) \cdot \mathbf{n}(\mathbf{y}) d\mathbf{y} \\
&- \int_{\gamma(t)} \theta V_n(\mathbf{y},t) G(\mathbf{y};\mathbf{x}) d\mathbf{y} \\
&- \int_{\gamma(t)} \langle K(\mathbf{y}) \rangle \left[a - \xi(x_y, y_y, t) \right] \nabla_{\mathbf{y}} G(\mathbf{y};\mathbf{x}) \cdot \mathbf{n}(\mathbf{y},t) d\mathbf{y} \quad (12)
\end{aligned}$$

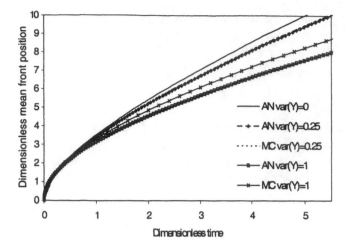

Figure 2. Dimensionless mean front positions, obtained analytically (AN) and with Monte Carlo simulations (MC) vs. Dimensionless time for different values of σ_Y^2 (var(Y)). (H-a)/l_Y=4.

Perturbation analysis

We expand K and all quantities that depend on it in powers of σ_Y, the standard deviation of $Y = \ln K$. Taking the ensemble mean of (12) yields, to leading orders of approximation,

$$h^{(0)}(\mathbf{x},t) = -HK_G \int_{\Gamma_D} \nabla_y G^{(0)}(\mathbf{y};\mathbf{x}) \cdot \mathbf{n}(\mathbf{y})\, d\mathbf{y}$$

$$- \int_{\gamma^{(0)}(t)} \theta V_n^{(0)}(\mathbf{y},t) G^{(0)}(\mathbf{y};\mathbf{x})\, d\mathbf{y}$$

$$-K_G \int_{\gamma^{(0)}(t)} \left[a - \xi^{(0)}(x_y, y_y, t)\right] \nabla_y G^{(0)}(\mathbf{y};\mathbf{x}) \cdot \mathbf{n}(\mathbf{y},t)\, d\mathbf{y} \quad (13)$$

$$\langle h(\mathbf{x},t)\rangle^{[2]} = K_G \int_{\Omega(t)^{(0)}} \mathbf{r}^{[2]}(\mathbf{y},t) \cdot \nabla_y G^{(0)}(\mathbf{y};\mathbf{x})\, d\mathbf{y}$$

$$- HK_G \int_{\Gamma_D} \nabla_y G^{(0)}(\mathbf{y};\mathbf{x}) \cdot \mathbf{n}(\mathbf{y})\, d\mathbf{y}$$

$$-K_G \int_{\Gamma_D} C_{Y\xi}^{[2]}(\mathbf{y}_\xi^{(0)}; x_y, y_y, t) \nabla_y h^{(0)}(\mathbf{y}_\xi^{(0)}, t) \cdot \nabla_y G^{(0)}(\mathbf{y}_\xi^{(0)};\mathbf{x})\, d\mathbf{y}$$

$$-\theta \int_{\langle \gamma(t)\rangle^{[2]}} \left[\langle V_n(\mathbf{y},t)\rangle^{[2]} - \frac{\sigma_Y^2}{2} V_n(\mathbf{y},t)^{(0)}\right] G^{(0)}(\mathbf{y};\mathbf{x})\, d\mathbf{y}$$

$$-K_G \int_{\langle \gamma(t)\rangle^{[2]}} \left[a - \langle \xi(x_y, y_y, t)\rangle^{[2]}\right] \nabla_y G^{(0)}(\mathbf{y};\mathbf{x}) \cdot \mathbf{n}(\mathbf{y},t)\, d\mathbf{y} \quad (14)$$

where superscripts (i) denote quantities that contain strictly i-th order terms, superscripts $[i]$ denote quantities that contain terms up to order i, $K_G = \exp\langle Y\rangle$ is the geometric mean of K, $\gamma^{(i)}$ and $\gamma^{[i]}$ are fronts corresponding to $\xi^{(i)}$ and $\xi^{[i]}$, $\mathbf{y}_\xi^{(0)} = \left[x_y, y_y, z_y = \xi^{(0)}(x_y, y_y, t)\right]^T$, $\mathbf{r}(\mathbf{x}) = -\langle Y'(\mathbf{x})\nabla h'(\mathbf{x})\rangle$ is a "residual" flux" given by

$$\mathbf{r}^{[2]}(\mathbf{x}) = K_G \int_{\Omega^{(0)}(t)} C_Y^{[2]}(\mathbf{x};\mathbf{y}) \nabla_y h^{(0)}(\mathbf{y},t) \cdot \nabla_y \nabla_x G^{(0)}(\mathbf{y};\mathbf{x})\, d\mathbf{y}$$

$$+ \int_{\gamma^{(0)}(t)} \theta C_{YV_n}^{[2]}(\mathbf{x};x_y, y_y, t) \nabla_x G^{(0)}(\mathbf{y};\mathbf{x})\, d\mathbf{y}$$

$$- K_G \int_{\gamma^{(0)}(t)} C_{Y\xi}^{[2]}(\mathbf{x};x_y, y_y, t) \nabla_y \nabla_x G^{(0)}(\mathbf{y};\mathbf{x}) \cdot \mathbf{n}(\mathbf{y})\, d\mathbf{y} \quad (15)$$

and $C_Y(\mathbf{x},\mathbf{y}) = \langle Y'(\mathbf{x})Y'(\mathbf{y})\rangle$,

$$C_{Y\xi}^{[2]}(\mathbf{x}; x_y y_y, t) = \left\langle Y'^{(1)}(\mathbf{x})\xi^{(1)}(x_y, y_y, t)\right\rangle$$

$$= K_G \int_{\Omega^{(0)}(t)} C_Y^{[2]}(\mathbf{x};\mathbf{z}) \nabla_z h^{(0)}(\mathbf{z})$$

$$\cdot \nabla_z G^{(0)}\left[\mathbf{z}; x_y, y_y, \xi^{(0)}(x_y, y_y, t)\right] d\mathbf{z}$$

$$+ \int_{\gamma^{(0)}(t)} \theta C_{YV_n}^{[2]}(\mathbf{x},\mathbf{z}) G^{(0)}\left[\mathbf{z}; x_y, y_y, \xi^{(0)}(x_y, y_y, t)\right] d\mathbf{z}$$

$$- K_G \int_{\gamma^{(0)}(t)} C_{Y\xi}^{[2]}(\mathbf{x}; x_z y_z, t)\mathbf{n}(\mathbf{z})$$

$$\cdot \nabla_z G^{(0)}\left[\mathbf{z}; x_y, y_y, \xi^{(0)}(x_y, y_y, t)\right] d\mathbf{z} \quad (16)$$

$$C_{YV_n}^{[2]}(\mathbf{x}; x_y, y_y, t) = \langle Y'(\mathbf{x})V_n^{(1)}(x_y, y_y, t)\rangle$$

$$= \frac{dC_{Y\xi}^{[2]}(\mathbf{x}; x_y y_y, t)}{dt} \cdot n_z(x_y, y_y, t) \quad (17)$$

From (6) it follows that

$$\langle h(\mathbf{x},t)\rangle^{[i]} = a - \langle \xi(x_x, y_x, t)\rangle^{[i]} \qquad z_x = \langle \xi(x_x, y_x, t)\rangle^{[i]} \quad (18)$$

where $\langle h(\mathbf{x},t)\rangle^{[i]}$ is given to zero and second order by (13) and (14), respectively. According to (8),

$$\langle V_n(\mathbf{x},t)\rangle^{[i]} = \frac{D\langle \xi(x_x, y_x, t)\rangle^{[i]}}{Dt} n_z(\mathbf{x},t) \quad (19)$$

The covariance

$$C_\xi^{[2]}(x_x,y_x,t;x_y,y_y,s) = \left\langle \xi'(x_x,y_x,t)\xi'(x_y,y_y,s)\right\rangle^{[2]}$$ of
front positions is obtained on the basis of (8) and (18) with
the aid of (11), (13) and (14),

$$C_\xi^{[2]}\left(x_x,y_x,t;x_y,y_y,s\right)=$$

$$\int_{\gamma^{(0)}(t)} \theta C_{\xi V_n}^{[2]}\left(x_y,y_y,s;x_z,y_z,t\right)G^{(0)}\left[\mathbf{z};x_x,y_x,\xi^{(0)}\left(x_x,y_x,t\right)\right]d\mathbf{z}$$

$$-K_G \int_{\gamma^{(0)}(t)} C_\xi^{[2]}\left(x_y,y_y,s;x_z,y_z,t\right)$$

$$\nabla_z G^{(0)}\left[\mathbf{z};x_x,y_x,\xi^{(0)}\left(x_x,y_x,t\right)\right]\cdot\mathbf{n}(\mathbf{z})d\mathbf{z}$$

$$+K_G \int_{\Omega^{(0)}(t)} C_{Y\xi}^{[2]}\left(\mathbf{z};x_y,y_y,s\right)\nabla_z h^{(0)}(\mathbf{z})$$

$$\cdot\nabla_z G^{(0)}\left[\mathbf{z};x_x,y_x,\xi^{(0)}\left(x_x,y_x,t\right)\right]d\mathbf{z} \qquad (20)$$

where

$$C_{\xi V_n}^{[2]}(x_y,y_y,s;x_z,y_z,t)=\frac{dC_\xi^{[2]}(x_y,y_y,s;x_z,y_z,t)}{dt}n_z(\mathbf{z}) \quad (21)$$

The variance $\sigma_\xi^{[2]}$ of front positions corresponds to the
limit of (20) as $(x_y,y_y,s)\to(x_x,y_x,t)$. Similar expressions
can be obtained for the covariance and variance of hydraulic head and its gradient.

Both the zero- and second-order problems are nonlinear
in mean head and front position. *Tartakovsky and Winter
[2001]* analyzed a gravity-free version of a similar problem
by using a somewhat different perturbation approximation.
Among others, their expression for second-order mean
head does not include a term containing $C_{Y\xi}^{(2)}$ as does the
third integral in our (14). The authors found that this
caused their second-order solution to overpredict the mean
position of the front.

Analytical solution in one dimension

We solve the above moment equations for front depth
$\xi(t)$ within a vertical depth interval $x\in(0,l)$ where
$l>\xi$. For this we take $Y(x)$ to have a constant mean,
$\langle Y\rangle$, and an exponential covariance,

$$C_Y\left(|x-y|\right)=\sigma_Y^2\exp\left(-\frac{|x-y|}{l_Y}\right) \qquad (22)$$

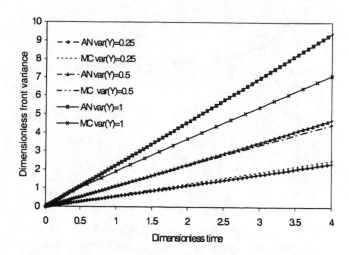

Figure 3. Dimensionless front variance obtained
analytically (AN) and with Monte Carlo simulations (MC)
vs. Dimensionless time for different values of σ_Y^2 (var(Y)).
(H-a)/l_Y=4.

where l_Y is the spatial autocorrelation scale of Y. The
auxiliary function $G_K(x,y)=\langle K\rangle G(x,y)$ satisfies

$$\frac{\partial^2 G_K(x,y)}{\partial x^2}+\delta(x-y)=0 \qquad 0\le x,y\le l \qquad (23)$$

subject to boundary conditions

$$G_K(x,y)=0 \qquad x=0 \qquad (24)$$
$$G_K(x,y)=0 \qquad x=l \qquad (25)$$

It is given by

$$G_K(x,y)=-(x-y)\mathrm{H}(x-y)+\frac{l-y}{l}x \qquad (26)$$

where H is the Heaviside function. Substituting (26) into
(12) and evaluating at $x=\xi$ yields

$$0=\theta\frac{d\xi(t)}{dt}\xi(t)+\langle K\rangle[a-\xi(t)]-\langle K\rangle H+\int_0^\xi K'(y)\frac{dh(y,t)}{dy}dy$$
$$(27)$$

Expanding the integral in Taylor series around $\langle\xi\rangle$ gives

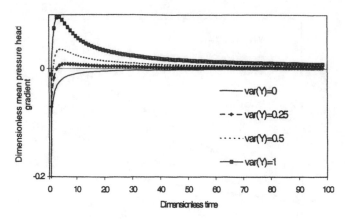

Figure 4. Dimensionless mean pressure head calculated analytically vs. Dimensionless time for different values of σ_Y^2 (var(Y)). (H-a)/l_Y=4.

$$0 = \theta \frac{d\xi(t)}{dt}\xi(t) + \langle K \rangle [a - \xi(t)] - \langle K \rangle H$$

$$+ \int_0^{\langle \xi \rangle} K'(y) \frac{dh(y,t)}{dy} dy + \xi'(t) K'(y) \frac{dh(y,t)}{dy}\bigg|_{y=\langle \xi \rangle} + \dots \quad (28)$$

The mean of (28) is, to leading orders of approximation,

$$\frac{\theta}{K_G} \frac{d\xi^{(0)}(t)}{dt} = 1 + \frac{H-a}{\xi^{(0)}(t)} \quad (29)$$

$$\theta \xi^{(0)}(t) \frac{d\langle \xi(t) \rangle^{(2)}}{dt} + \left(\theta \frac{d\xi^{(0)}(t)}{dt} - K_G \right) \langle \xi(t) \rangle^{(2)}$$

$$= -\frac{\theta}{2} \frac{d\left[\sigma_\xi^2(t)\right]^{[2]}}{dt} + \left[H - a + \xi^{(0)}(t)\right] K_G \frac{\sigma_Y^2}{2}$$

$$+ \int_0^{\xi^{(0)}} r^{[2]}(y,t) dy - C_{K\xi}^{[2]}(\xi^{(0)},t) \frac{dh^{(0)}(y,t)}{dy}\bigg|_{y=\xi^{(0)}} \quad (30)$$

Multiplying (28) by ξ' and taking the mean yields, to leading order,

$$\frac{\theta}{2} \xi^{(0)} \frac{d\left[\sigma_\xi^2(t)\right]^{[2]}}{dt} + \left(\theta \frac{d\xi^{(0)}}{dt} - K_G \right) \left[\sigma_\xi^2(t)\right]^{[2]}$$

$$= \frac{\xi^{(0)} + H - a}{\xi^{(0)}} \int_0^{\xi^{(0)}} C_{K\xi}^{[2]}(y,t) dy \quad (31)$$

Expressions (29) - (31) are ordinary differential equations subject to zero initial conditions that are easily

solved. As a prerequisite for solving (30) and (31), one must first solve an ordinary differential equation for $C_{K\xi}^{[2]}$,

$$\theta \xi^{(0)}(t) \frac{dC_{K\xi}^{[2]}(x,t)}{dt} + \left(\theta \frac{d\xi^{(0)}(t)}{dt} - K_G \right) C_{K\xi}^{[2]}(x,t)$$

$$= K_G^2 \left(1 + \frac{H-a}{\xi^{(0)}(t)} \right) \int_0^{\xi^{(0)}} C_Y^{[2]}(x,y) dy \quad (32)$$

subject to $C_{K\xi}^{[2]}(x,0) = 0$. Also required is the residual flux

$$r^{[2]}(x,t) = \frac{\theta}{K_G} \frac{\partial C_{K\xi}^{[2]}(x,t)}{\partial t} + K_G \sigma_Y^2 \frac{dh^{(0)}(x,t)}{dx} \quad (33)$$

Zero- and second-order mean head gradients are obtained from one-dimensional versions of (13) and (14), respectively, as

$$\frac{dh^{(0)}(x,t)}{dx} = -\frac{\xi^{(0)}(t) + H - a}{\xi^{(0)}(t)} \quad (34)$$

$$\frac{d\langle h(x,t) \rangle^{(2)}}{dx} = -\frac{\xi^{(0)}(t) + H - a}{\xi^{(0)}(t)} \frac{d\langle \xi(t) \rangle^{(2)}}{d\xi^{(0)}(t)}$$

$$+ \frac{\sigma_Y^2}{2} \frac{\xi^{(0)}(t) + H - a}{\xi^{(0)}(t)} + \frac{r^{[2]}(x,t)}{K_G} \quad (35)$$

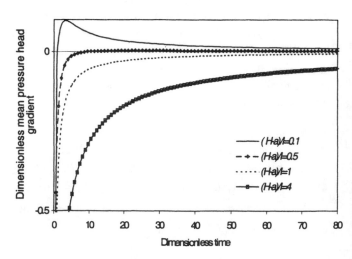

Figure 5. Dimensionless mean pressure head calculated analytically vs. dimensionless time for different values of (H-a)/l_Y. $\sigma_Y^2 = 1$.

and the second-order approximation of hydraulic head variance,

$$K_G \left\{ \sigma_{dh/dx}^2 \left[\xi^{(0)}(t) \right] \right\}^{[2]} = r^{[2]}(\xi^{(0)}, t) \frac{dh^{(0)}(x,t)}{dx} \Bigg|_{x=\xi^{(0)}}$$

$$+ \frac{\theta}{K_G} \frac{\partial}{\partial t} \left(C_{K\xi}^{[2]}(x,t) \frac{dh^{(0)}(x,t)}{dx} + \frac{\theta}{2} \frac{d\left[\sigma_\xi^2(t) \right]^{[2]}}{dt} \right) \Bigg|_{x=\xi^{(0)}} \quad (36)$$

Results and Comparison with Monte Carlo Simulations

To test our moment equations, we compare their solutions to sample moments obtained from 4000 Monte Carlo solution of (1) – (5). Figure 2 shows how dimensionless mean front depth $L = \xi / l_Y$ increases with dimensionless time $K_G t / \theta l_Y$ as a function of σ_Y^2 when $(H - a)/l_Y = 4$. The zero order moment solution is independent of variance and consistently overestimates the depth of the front. The second order moment solution is much closer to that obtained from Monte Carlo simulations and therefore more accurate. Accuracy is high for $\sigma_Y^2 = 0.25$ but deteriorates as σ_Y^2 increases. Mean front depth and its rate of advance diminish with increasing variance.

Figure 3 depicts dimensionless front variance σ_ξ^2 / l_Y^2 as a function of dimensionless time $K_G t / \theta l_Y$ for various values of σ_Y^2 when $(H - a)/l_Y = 4$. The moment solution compares favorably with Monte Carlo results when $\sigma_Y^2 \leq 0.5$ but de-

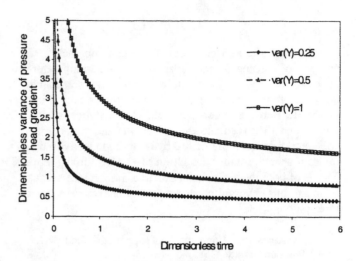

Figure 7. Dimensionless mean pressure head variance calculated analyticaly vs Dimensionless time for different values of $\sigma_Y^\div$ (var(Y)). $(H-a)/l_Y=4$

teriorates as σ_Y^2 increases. Dimensionless front variance grows in a near-linear fashion with dimensionless time, at a rate that increases with σ_Y^2.

According to *Philip* [1975], a front is stable when the gradient of pressure head immediately above it is negative, and unstable if this gradient is positive. For instabilities to develop, the front must undergo some slight initial perturbation. In the case of randomly heterogeneous media, such perturbations are introduced (among other causes) by random variations in soil properties. Figure 4 shows how mean pressure head gradient varies with dimensionless time for various σ_Y^2 when $(H - a)/l_Y = 4$. In a homogeneous soil with $\sigma_Y^2 = 0$, represented by the zero-order solution, the mean front in our example remains stable. Heterogeneity is seen to destabilize it at a dimensionless time that increases as σ_Y^2 goes down. Theory and the figure indicate that the gradient of mean pressure head tends asymptotically to zero with time. Figure 5 depicts the influence of $(H - a)/l_Y$ on the gradient of mean pressure head, and the onset of instability, when $\sigma_Y^2 = 1$. For the mean front to be stable, $(H - a)/l_Y$ must be sufficiently large. Otherwise the smaller is $(H - a)/l_Y$, the earlier does instability set in. An increase in the correlation scale l_Y has a destabilizing effect on the front when $\sigma_Y^2 = 1$ and $(H - a)/l_Y = 0.1$ (Figure 6).

Uncertainty in the prediction of mean pressure head gradient is largest at time zero, diminishes steeply at early time and more slowly at later time toward an asymptote that increases with σ_Y^2.

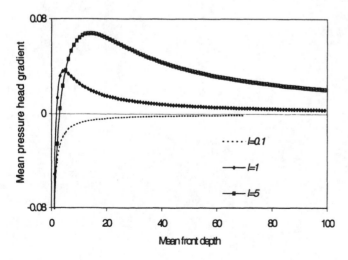

Figure 6. Mean pressure head calculated analytically vs. mean front depth for different l_Y. $(H-a)/l_Y = 0.1$, $\sigma_Y^\div = 0.5$.

SUMMARY

We proposed a new conditional moment approach to the analysis of wetting front propagation in randomly heterogeneous soils. We solved our moment equations analytically and found the result to be accurate for both mildly and moderately heterogeneous soils. Our analytical solution allows one to investigate the effect that the variance and spatial correlation scale of log hydraulic conductivity have on front behavior under the action of various forcing terms. This is of both theoretical and practical importance because in most soils, hydraulic conductivity behaves as a correlated random field.

Acknowledgments. This work was supported in part by EPA STAR Program Grant R826157-01-1.

REFERENCES

Chen G. and S. P. Neuman, Wetting front instability in randomly stratified soils, *Phys. Fluids A*, 8(2), 353-369, 1996.

Chen G., M. Taniguchi, and S. P. Neuman, An Overview of Instability and Fingering During Immiscible Groundwater Flow in Porous and Fractured Media, *NUREG/CR-6308*, prepared for U. S. Nuclear Regulatory Commission, Washington, D. C., May, 1995.

Chuoke R. L., P. van Meurs, and C. van der Poel, The instability of slow, immiscible, viscous liquid-liquid displacements in permeable media, *Trans. AIME*, 216, 188-194, 1959.

Dagan, G. and S.P. Neuman (editors), *Subsurface Flow and Transport: A Stochastic Approach*, Cambridge University Press, Cambridge, United Kingdom, 1997.

Green, W. H. and G. A. Ampt, Studies in soil physics, I. The flow of air and water through soil, *J. Agr. Sci.*, 4, 1-24, 1911.

Neuman, S.P., Role of Geostatistics in Subsurface Hydrology, 787-816 in *Geostatistics for Natural Resources Characterization*, Part 2, edited by G. Verly, M. David, A.G. Journal, and A. Marechal., D. Reidel Publ. Co., Dordrecht, Holland, 1984.

Neuman, S.P. and G. Chen, On instability during immiscible displacement in fractures and porous media, *Water Resour. Res.*, 32(6), 1891-1894, 1996.

Philip J. R., Stability analysis of infiltration, *Soil Sci. Soc. Amer. Proc.*, 39, 1042-1053, 1975.

Raats, P. A., Unstable wetting fronts in uniform and non-uniform soils, *Soil Sci. Soc. Am. Proc.*, 37, 681-685, 1973.

Shariati, M. and Y.C. Yortsos, Stability of miscible displacements across stratified porous media, *Phys. Fluids*, 13(8), 1-12, 2001.

Saffman P.G. and G.I. Taylor, The penetration of a fluid into a porous medium or Hele-Shaw cell containing a more viscous fluid, *Proc. Roy. Soc. London, Ser. A*, 245, 312-320, 1958.

Tartakovsky, D. M. and C. L. Winter, Dynamics of free surfaces in random porous media, *SIAM Journal of Applied Math.*, 61(6), 1857-1876, 2001.

Warrick A.W., Myers D.E., and Nielsen D. R., Geostatistical methods applied to soil science, 53-82 in *Methods of Soil Analysis, Part 1, Physical and Mineralogical Methods*, edited by A. Klute, Amer. Soc. Agronomy and Soil Science Society of America, Madison, Wisconsin, 1986.

S. P. Neuman, Department of Hydrology and Water Resources, University of Arizona, Tucson, Arizona 85721.

A. M. Tartakovsky, Department of Hydrology and Water Resources, University of Arizona, Tucson, Arizona 85721.

D. M. Tartakovsky, Group T-7, Los Alamos National Laboratory, Los Alamos, New Mexico 87545.

On Hydrostatics and Matristatics of Swelling Soils

C.D. Grant

Department of Soil and Water, Adelaide University, Adelaide, Australia

P.H. Groenevelt

Department of Land Resource Science, University of Guelph, Guelph, Canada

G.H. Bolt

Department of Environmental Science, Wageningen University, Wageningen, The Netherlands

We review and re-examine the development of the hydrostatics of swelling soils, starting with the pioneering work of John R. Philip. His early attempts to formulate the overburden potential of the soil water as an additional component to the pressure and gravitational potentials did cause some confusion, until it became clear that there is no additional component and that the actual pressure potential can be split into two parts: the (non-actual) pressure potential the soil water would have if the system were unloaded, and a remaining part, which may be called the overburden potential. Using fundamental thermodynamics it is then shown how the overburden potential can be calculated from the slopes of the bundle of load-pressure-dependent shrinkage curves. A set of experimental data published by *Talsma* is used to test the *Groenevelt-Bolt* equation of state for the bundle of shrinkage curves and for the numerical values of the overburden potentials for *Talsma*'s clay paste. Subsequently, the *Groenevelt-Bolt* equation is used to interpret the physical significance of the extremes of residual shrinkage curves, for which shrinkage data from the literature are used. The description of the static behavior of the matrix of swelling soils is termed "matristatics".

1. INTRODUCTION

When John R. Philip first turned his attention to swell/shrink media [*Philip*, 1968], primed by the kinetic experiments that were carried out and shown to him by *Smiles and Rosenthal* [1968], he entered this "muddy" scientific field of non-rigid porous media as a hydrodynamicist. He formulated the flux densities and the

Environmental Mechanics: Water, Mass and Energy Transfer in the Biosphere
Geophysical Monograph 129
Copyright 2002 by the American Geophysical Union
10.1029/129GM10

rates of volume change in clay-colloid pastes. Together with Smiles [*Philip and Smiles*, 1969], he continued his pioneering work, concentrating on the dynamics in non-rigid porous media. It was only after those two major contributions that Philip started to pay attention to the hydrostatics in swelling soils [*Philip*, 1969a,b,c, 1970a,b,c, 1971, 1972a,b,c], for which he received much inspiration from the pioneering work by Croney and Coleman (1961). Even in the early stages [*Philip*, 1969c], however, he preferred to deal with the hydrodynamics in swelling soils well before he arrived at the hydrostatics in swelling soils. He thus subdivided this field of science into "statics" and "dynamics". Indeed, the difficulties involved in the behavior of swelling (and shrinking) soils, in particular the behavior of water in such media, are so complex that a subdivision of the subject into these two sub-disciplines is even more urgent than it is for the subject of the behavior of water in rigid soils, and, of course, one should enter the field on the side of hydrostatics.

Similarly, when studying the volume change as a function of water content and load (overburden pressure, or envelope pressure), the starting point should be the description of the static (equilibrium) relationship between void ratio, e, and moisture ratio, ϑ, followed by the description of the rate of volume change.

The early emphasis on the dynamic behavior of water in non-rigid media caused some confusion [*Youngs and Towner*, 1970], which was manifest in *Philip*'s [1969a] early definition of the "overburden potential". He mistakenly suggested that the water in non-rigid soils had an additional component potential over and above the gravitational and pressure potentials, which he termed the "overburden potential". This is not so, and of course, in the early formulation of Darcy's Law in a material coordinate system [*Philip and Smiles*, 1969], the overburden potential does not appear.

In this paper we restrict ourselves to the statics of swelling media. The objective of this paper is to exploit the theory of *Groenevelt and Bolt* [1972] to analyze several sets of shrinkage data, provided in the literature, to obtain values of the overburden potential and to divide shrinkage curves into regions of 'structural' shrinkage, 'unsaturated proportional' shrinkage, and 'zero' shrinkage. We will first deal with the liquid phase, the hydrostatics, and subsequently with the solid phase, for which we now coin the term "matristatics".

2. HYDROSTATICS

2.1 The Water Retention Curves

The relation between the pressure (matric) potential of the soil water and the water content in swelling/shrinking soils is more complicated than in rigid soils. Leaving hysteresis (also more complicated in non-rigid soils) out of the picture, we will focus on the equilibrium relation between pressure potential and water content during a process of desorption, starting from a completely swollen state. In the case of a clay suspension such a 'completely swollen' state may not exist, because the system will tend to swell indefinitely in trying to bring the matric potential to zero. For practical cases in soil science, however, a completely swollen state does exist, and the pressure potential will find a zero value, such that, upon further addition of water, ponding will occur, and the pressure potential will turn positive. In such a completely swollen state, the system is located somewhere on the 1:1 line in a diagram of void ratio versus moisture ratio.

We characterize the system in the e-ϑ diagram, where e is the void ratio (volume of void or pore space per unit volume of solid space) and ϑ is the moisture ratio (volume of liquid-filled space per unit volume of solid space). We consider the liquid phase essentially free of salt, such that the extent of the electrical double layers is not influenced by the properties of the liquid phase. On the 1:1 line in the e-ϑ diagram the system is saturated.

We start with the system free of external constraint, free of an external pressure (load pressure or envelope pressure). The system then, when brought into contact with free water, will suck up water and swell on its way to our starting point, the completely swollen state. As long as the system sucks up water, it has a swelling pressure, Π, which is equal to the load pressure, P, that must be applied to the system in order for it to stop sucking up or releasing water. When the system finally stops sucking up water, and is thus in equilibrium with free water, the pressure potential of the soil water, p, is zero, and the system no longer swells. The swelling pressure is then also zero.

From this point, we start the desorption process by extracting water, *e.g.* by lowering the pressure potential, p, which then becomes negative. Water will leave the system and it will shrink. The swelling pressure, Π, will rise above zero. As long as the (shrinking) system stays saturated and thus on the 1:1 line, the pressure potential of the soil water and the swelling pressure will balance each other exactly. Thus:

$$^{0}p + \Pi = 0 , \qquad (1)$$

where ^{0}p is the pressure potential of the water at zero load pressure ($P = 0$).

The system will continue to shrink along the 1:1 line until air enters the system at $e = {}^{0}e^{*}$. This process of shrinking along the 1:1 line is often called 'normal' shrinkage, but we will refer to it as 'proportional' shrinkage. All along the saturated stretch of drying, the water content of the system decreases as the pressure potential decreases. This part of the desorption curve is an important ingredient in the study of the kinetics of desorption in clay pastes. The remaining part of the desorption curve of an unloaded shrinking soil (that part beyond air-entry where the e-ϑ relation is known as 'residual' shrinkage) most resembles the familiar water retention (or water release) curve in rigid soils. The only difference between the water retention curves is that the matrix-configuration of non-rigid soils continues to decline to varying degrees in response to the decreasing pressure potential in the liquid phase. The two parts of the desorption curve have to be determined by experiment. Good examples of such measurements of the unloaded desorption curve can be found in *Reeve and Hall* [1978] and in *Bronswijk and Evers-Vermeer* [1990]. These

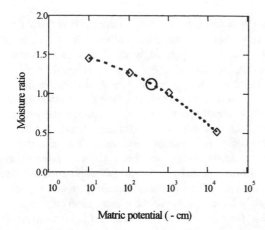

Figure 1. Water retention curve for *Bronswijk and Evers-Vermeer*'s [1990] Schermerhorn horizon C21g (52-77 cm), with the circle indicating the calculated point of air-entry.

authors did not plot the water retention curves – instead, they plotted the shrinkage curves (which we will focus on mainly). They did, however, indicate some values of the water potential along their shrinkage curves, and these provide the opportunity to construct their water retention curves.

When plotting (Fig. 1) the two parts of the desorption curve, $^0p(\vartheta)$ (*e.g.* for *Bronswijk and Evers-Vermeer*'s [1990] C21g horizon of their Schermerhorn soil), there are indications that at the air-entry point, the pressure potential, $^0p^*$, does not seem to show a discontinuity, while the slope of the shrinkage curve, $e(\vartheta)$, does show a discontinuity at air entry.

When the above-described desorption process is carried out while the system is under a constant load (envelope or overburden) pressure, P, the starting point, where $p = 0$, always occurs at a lower value of e (and thus at a value of Π greater than zero) compared to the unloaded state, and the value of Π will be equal to P. The desorption process will again lead the system down the 1:1 line, until air enters the system at $e = e^*(P)$, smaller than $^0e^*$. All along this saturated stretch of (loaded) proportional shrinkage before air-entry we have [*Groenevelt and Kay*, 1981]:

$$p + \Pi = P \ , \qquad (2)$$

which is essentially the equation of *Terzaghi* [1925].

Again, along the stretch of proportional shrinkage, the water content decreases with the decreasing water potential. Beyond the air-entry point the water content keeps dropping with the water potential. This part of the water release curve is completely unknown in classical soil physics because, for rigid soils, the water retention curve is independent of the load pressure.

The desorption curve for a shrinking medium under constant load pressure, P, has to be determined by experiment, and differs from the desorption curve at $P = 0$. Thus, each swelling soil has to be characterized by a bundle of water retention curves, each curve belonging to a particular value of P and which cannot cross over the curves for other pressures. No such bundle of measured water retention curves for real soils is available in the literature, although a few individual points of such a bundle, measured in the field, are provided by *Talsma* [1977a].

2.1.1 Examples of water retention curves for swelling soils. In contrast to rigid soils, water retention curves of swelling soils consist of two parts: a saturated part and an unsaturated part. For studies concerning the structural stability of swelling soils, it is of interest to know whether the slope of the retention curve at the transition point between the two parts (*i.e.* at the air-entry point) changes abruptly or not. To provide a preliminary answer to this question we examine two of the thirty-one shrinkage curves provided in the literature for different soil clods from different soil horizons by *Bronswijk and Evers-Vermeer* [1990]. They identified on their e-ϑ (shrinkage) curves, the values of ϑ at which the pressure potentials reached 0, 10, 100, 1000, and 16000 cm of water tension. They did not, however, measure the pressure potential of the water at precisely the point of air-entry, $^0p^*$. To make such an observation during the shrinkage process, of course, requires constant, long-term 'baby-sitting'. All thirty-one shrinkage curves presented by *Bronswijk and Evers-Vermeer* concern 'real' soils and each of them has a value of ϑ for which the pressure potential reaches zero, representing the completely swollen state.

The soil clods tested by *Bronswijk and Evers-Vermeer* where taken at specified depths in the soil profile. *In situ*, they existed under an overburden (load) pressure, P, but this was removed when they were brought into the laboratory. We assume the samples were exposed to free water in the laboratory for long enough to swell to equilibrium and thus represent the unloaded ($P = 0$) state. To answer the question about continuity of the differential water capacity at air-entry, we choose two of the thirty-one shrinkage curves, such that two of the five points of the water retention curve occupy the saturated part and two fall in the unsaturated part. The first water retention curve is shown in Fig. 1 for the Schermerhorn horizon C21g (52-77 cm). The best estimate of the moisture ratio at air entry, $^0\vartheta^*$, from the shrinkage curve is 1.12, indicated by the circle in Fig. 1. This then leads to $^0p^* = -350$ cm, so that at air-entry the water is at 350 cm suction. The water retention curve in Fig. 1 suggests that, for the present soil, the slopes of the saturated part (left of the circle) and the unsaturated part (right of the circle) do not change abruptly.

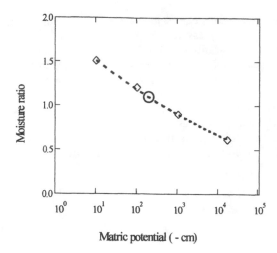

Figure 2. Water retention curve for *Bronswijk and Evers-Vermeer*'s [1990] Oosterend horizon A11(0-22cm), with the circle indicating the calculated point of air-entry.

The second shrinkage curve we inspect here is the Oosterend soil horizon A11(0-22cm) shown in (Fig. 2). The best estimate of the moisture ratio at air-entry, $^0\vartheta^*$, from the shrinkage curve is 1.09, indicated by the circle in Fig. 2. This then leads to the pressure potential at air-entry being $^0p^* = -189$ cm. Again, Fig. 2 suggests that, for this soil, the slopes of the saturated part (left of circle) and the unsaturated part (right of circle) do not change abruptly. The implications of this in terms of soil structural stability have not yet been investigated, and should be.

2.2 The Overburden Potential

As indicated earlier, the overburden potential is not an additional component potential to be added to the pressure (or matric) and gravitational component potentials (and the osmotic, electrical and thermal component potentials). There is no 'additional' form of energy 'carried' by the water that, in some way, emanates from the overburden. We make these introductory statements in view of the enormous impact J.R. Philip's work has had, and will continue to have, on the discipline of Soil Physics. Works, in which a statement such as: "This approach was used recently [*Philip* 1969a] to establish that Φ, the total potential governing vertical equilibrium and movement of water in a swelling soil, is made up of three components, the moisture (or matric) potential, Ψ, the gravitational potential, $-z$, and the overburden potential, Ω, so that $\Phi = \Psi - z + \Omega$" [*Philip*, 1970a, p.294] have caused much confusion. As argued by *Youngs and* Towner [1970], this statement is incorrect.

In the previous section we dealt with the water retention curve of a non-loaded ($P = 0$), non-rigid soil. When the soil is under a constant load (overburden) pressure, it has a

different water retention curve. At the starting point ($p = 0$), in the fully swollen state, the value of ϑ is smaller than when $P = 0$, increasingly so, when P increases. At the air-entry point, the value of ϑ (*i.e.* ϑ^*), is smaller than $^0\vartheta^*$, increasingly so with greater P. From the above, it can be expected that this will result in a bundle of water retention curves that cannot cross over and are more or less parallel.

As indicated above, the shrinkage curves for different load pressures cannot cross over, but at the same time they cannot be constructed from the unloaded water retention curve. Direct measurements must be taken with the system under a load. These measurements, however, do not necessarily have to involve the observation of the pressure potential (tension) of the water. This is courtesy of a thermodynamic relationship involving the differentials of two intensive variables (p and P) and two extensive variables (e and ϑ). The relationship follows strictly from basic thermodynamics (Gibbs equation) and a fundamental law of calculus (Maxwell relation). A precise derivation of this thermodynamic relation was presented by *Groenevelt and Bolt* [1972, Eqns (6) and (12)], and reads:

$$(\delta p/\delta P)_\vartheta = (\delta e/\delta\vartheta)_{P,T,p_a} \tag{3}$$

On the left-hand side of this equation one finds the differential behavior of the pressure potential (at constant moisture ratio) with changing load pressure. On the right-hand side of Eqn (3) one finds the slope of the shrinkage curve at constant load pressure (which should be measured at constant temperature and air pressure, p_a). This opens the possibility to calculate, from a bundle of measured shrinkage curves, the value of the pressure potential, $p(P,\vartheta^\#)$, when the system is under a load, P, at a certain moisture ratio, $\vartheta^\#$, if the value of the pressure potential is known at that moisture ratio in the unloaded state, $^0p(\vartheta^\#)$. The original condition on the left-hand side was that the ratio of the densities of the water and solid phases (ρ_w/ρ_s) remains constant. Together with the assumption that water and mineral matter are incompressible, this condition is equivalent to constant ϑ but is not equivalent to the condition of constant volumetric water content, θ. Thus, when the value of $^0p(\vartheta)$ is known, the value of $p(\vartheta,P)$ can be calculated from:

$$\int_0^P (\partial p/\partial P)\,dP = \int_0^P (\partial e/\partial\vartheta)\,dP \tag{4}$$

Each of the above partial differentials has the same conditions as in Eqn (3). The numerical value of these definite integrals is by definition the overburden potential, Ω. Thus,

$$p(P,\vartheta) = {}^0p(\vartheta) + \Omega \ , \tag{5}$$

Table 1. Moisture ratios and void ratios from *Talsma*'s [1977b] three shrinkage curves.

P = 0 kPa		P = 6.3 kPa		P = 11.2 kPa	
ϑ	e	ϑ	e	ϑ	e
0.25	1.30	0.20	1.16	0.20	1.04
0.41	1.32	0.25	1.17	0.25	1.05
0.57	1.40	0.33	1.17	0.33	1.05
0.65	1.45	0.43	1.21	0.49	1.11
0.73	1.45	0.61	1.28	0.67	1.18
0.79	1.49	0.68	1.31	0.87	1.28
0.87	1.54	0.86	1.40	1.05	1.35
1.14	1.61	1.03	1.44	1.17	1.39
1.22	1.63	1.17	1.49	1.49	1.49
1.30	1.65	1.26	1.51		
1.37	1.68	1.54	1.61		
1.45	1.75	1.64	1.64		
1.63	1.80				
1.75	1.85				
1.91	1.92				

where $p(P,\vartheta)$ is the *in situ* pressure potential of the water while the system sits under the overburden, and ${}^{0}p(\vartheta)$ is the pressure potential that same system would attain (after equilibration) if the overburden (load) were removed while maintaining the mass ratio of water and solid material.

Only one useful bundle of shrinkage curves could be found in the literature, and it consists of only three curves. Furthermore, it is not for a natural soil – but a remoulded clay, and no information whatsoever is provided about pressure potentials. It was published in 1977b by *T. Talsma*, a close associate of J.R. Philip. Because of its rarity and its high quality, we will analyze *Talsma*'s data set in detail. In particular, we will scrutinize the calculated overburden potential curves.

2.2.1 Talsma's Data

We present in Table 1 the complete set of data reported in a single graph by *Talsma* in his 'Note' published in 1977b. Strictly speaking the unloaded shrinkage line is represented by two data sets under small loads (0.02 kPa and 0.14 kPa). These loads were necessary for technical reasons but there is quite a bit of 'wobble' in the data, indicating that the experimental difficulties were considerable. For this reason we ignore here the data for 0.14 kPa and adopt the data for the smaller load, 0.02 kPa, as the 'unloaded' shrinkage curve. We extracted the numerical values of the data points from *Talsma*'s graph with the greatest possible precision. There were three curves for the void ratio versus moisture ratio, each applicable to a different load pressure: $P = 0$ kPa, $P = 6.3$ kPa, and $P = 11.2$ kPa. At a wet bulk density in the field of 1.5 Mg/m^3, these overburden pressures would occur respectively, at the soil surface, at 42 cm below the soil surface, and at 75 cm below the soil surface. These locations could be said to represent the top, the middle and the bottom of a 'typical' horticultural root zone. There was

a total of 36 data points: 15 for the unloaded curve ($P = 0$), 12 for the load pressure of 6.3 kPa, and 9 for the load pressure of 11.2 kPa (Table 1). Each curve had one point on the 1:1 line. All 36 points were fitted to *Groenevelt and Bolt's* [1972] single constitutive equation, which consists of several sub-models and reads as follows:

$$\vartheta(e) = [k_2(e^n P - k_1)[k_2 P + \ln[(e-\varepsilon)k_3^{-1} + \exp(-k_1 k_2 \varepsilon^{-n})]]^{-1}]^{1/n} \quad (6)$$

where ϑ, e and P are defined above, ε is the void ratio at air-entry, and k_1, k_2, k_3, n are curve-fitting parameters.

The curve-fitting of Eqn (6) to all 36 points in Table 1 (using a 'solve block' in Mathcad [*Mathsoft* 1998] to produce the least sum of squared errors) provides an excellent fit for the three shrinkage curves. For convenience, however, we use here the rounded values for the four fitting parameters, *viz.* $k_1 = 1$ bar, $k_2 = 2$ bar^{-1}, $k_3 = 2$ and n = 0.7. This produces a result (Fig. 3), which is sufficiently accurate for the calculation of the $\Omega(\vartheta)$ curve, relating the overburden potential to the moisture ratio. Of course, all variables and parameters that have the dimension of pressure, P, p, Π, k_1 and $(k_2)^{-1}$ are expressed in the same units (here we use bar).

Equation (6) shows the moisture ratio as the explicit variable and the void ratio as the implicit variable - it is not a trivial matter to make *the void ratio* the explicit variable. We therefore proceed to differentiate $\vartheta(e)$ at constant P and subsequently invert the result to obtain the desired slope of the shrinkage curves, $\alpha_P(e)$, at constant P. These slopes are plotted in Fig.4a and can be seen to be of very similar shape. In fact, by shifting the $\alpha_0(e)$ curve to the left by 0.16 units of e, and by shifting the $\alpha_{112}(e)$ to the right by 0.11 units of e, the three curves nearly coincide (Fig.4b).

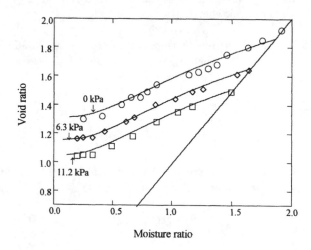

Figure 3. *Talsma*'s [1977b] shrinkage curves for the three load pressures: P=0 (circles), P=6.3 kPa (diamonds) and P=11.2 kPa (squares) fitted to Eqn (6).

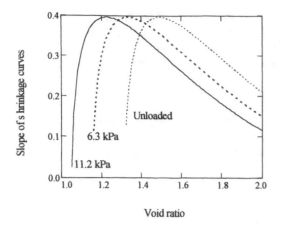

Figure 4a. Slopes of *Talsma*'s three shrinkage curves as a function of the void ratio (dotted line = Unloaded; dashed line = 6.3 kPa; solid line = 11.2 kPa).

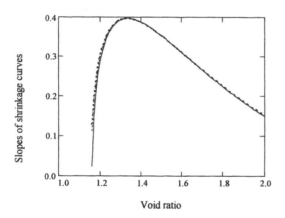

Figure 4b. *Talsma*'s 3 slope curves with $\alpha_0(e)$ and $\alpha_{11.2}(e)$ shifted to coincide with $\alpha_{6.3}(e)$.

This similarity shows that the curves are near to parallel, which means that $(\delta e/\delta \vartheta)_P$ at a fixed value of ϑ is nearly independent of P. Thus at constant ϑ,

$$\int_0^P \frac{\delta e}{\delta \vartheta} \, dP = \alpha_0 P \qquad (7)$$

is approximately correct. Thus it can be said (again approximately) that:

$$p(P, \vartheta) = {}^0p(\vartheta) + \alpha_0 P \qquad (8)$$

The relation in Eqn (8) was first proposed by *Croney and Coleman* [1961]. For *Talsma*'s clay paste the largest value of α_0 is 0.4. Thus, at the highest value of P, 11.2 kPa, the highest value of the overburden potential that occurred

during *Talsma*'s experiments was $\Omega = 0.4 \times 11.2 = 4.48$ kPa, or approximately 45 cm water pressure.

2.3 The *Groenevelt-Bolt* Equation of State

In the 1972 special issue of *Soil Science* dedicated to L.A. Richards, *Groenevelt and Bolt* published an equation of state (Eqn 6, above) for a hypothetical bundle of shrinkage curves relating the void ratio to the moisture ratio at different load pressures. At that time, however, there were no reliable experimental data available to test the model. In 1974 *Groenevelt and Parlange* presented a thermodynamic theory to describe the transition from the saturated to the unsaturated part of the shrinkage curve at the point of air entry. In their analysis of the unloaded shrinkage curve for an Australian clay soil they wrote [*Groenevelt and Parlange* 1974]:

> "We note that at the air entry point the experimental data suggest that the e-ϑ curve is not tangent to the line $e = \vartheta$. It is important to know whether a discontinuity in $\delta e/\delta \vartheta$ at the air entry point is physically acceptable, or whether this apparent discontinuity must be regarded as an experimental error. We show here that the discontinuity is perfectly acceptable from a thermodynamic point of view and can be associated with a second order phase transition in the sense of Ehrenfest."

In 1976 *Sposito and Giraldez* confirmed the above:

> "Following a heuristic suggestion by Groenevelt and Parlange, a thermodynamic description is given of a postulated order-disorder transformation in a swelling soil. This transformation is shown to correspond to a phase transition of the second order at the air entry point, wherein the onset of a microscopic ordering process among the clay crystals in the soil occurs and there is a consequent discontinuity in the slope of the swelling curve".

In 1977, *Talsma* published his one but last paper as member of the staff of CSIRO Division of Environmental Mechanics, Canberra, in the form of a short Note. His data show undeniable and very pronounced discontinuities in the slopes of the 3 shrinkage curves at air entry; their magnitudes being 0.267, 0.292 and 0.310 respectively - all well below unity. Of course, these low values are partly a consequence of the boundary conditions (acrylic containing rings) imposed by *Talsma* [1977b].

In 1981 *Groenevelt and Kay* published an extensive analysis of *Talsma*'s data examining the different sub-models of the *Groenevelt-Bolt* equation, and attempting to calculate the effective stress and the χ factor of *Bishop and Blight* [1963]. In 1983, *Giraldez and Sposito* attempted to apply the same equation to *Talsma*'s data "by minimizing the sum of squares of differences between calculated and measured values with the Rosenbrock algorithm". They commented that "the fitting of the model required

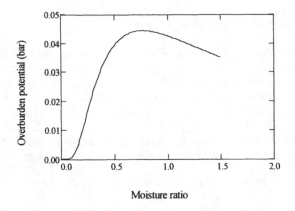

Figure 5. Overburden potential as a function of moisture ratio for *Talsma*'s clay paste at load pressure $P=0.112$ bar from Eqn (12).

considerable computational effort", which is of course no longer a problem now that high-speed optimization subroutines are readily available for personal computers. Their more serious complaint was that "on physical grounds, the model parameters, k_1, k_2, k_3 and n, should remain constant as the pressure changes in a given soil". Of course these parameters should remain constant, and they do, when the optimization is carried out for the entire bundle of shrinkage curves at once (not individually), or for the submodels first, followed by the application of the derived values for the parameters to the entire bundle [*c.f. Groenevelt and Kay* 1981].

It is indeed opportune to analyze the different submodels. For example, from the first sub-model of *Groenevelt and Bolt* [1972, Eqn.(16)] combined with their Eqn. (3), one can calculate the pressure potential of the water, p^*, at the air entry point:

$$p^* = P - k_1\varepsilon^{-n} \qquad (9)$$

Thus, even though *Talsma* did not provide any information whatsoever about pressure potentials in the water, it can be calculated (from Eqn 9, above) that at *Talsma*'s three air entry points, the pressure potentials of the water were -648, -644, and -644 cm water pressure, respectively. These values are considerably lower than those we found for the two soils of *Bronswijk and Evers-Vermeer* [1990] identified in Figs 1 and 2. This, of course, is due to the different boundary conditions. *Talsma* used soil cores, whereas *Broenswijk and Evers-Vermeer* used soil clods. The boundary conditions, together with the initial conditions (such as the size of the sample), have a large influence on the location and the shape of the shrinkage curves.

It should be noted from the above calculations that the pressure potential in the water at the air entry point appears to be independent of the load pressure, P. The sub-models

proposed by *Groenevelt and Bolt* in 1972 were first attempts to formulate the basic physical phenomena – these of course are open to refinement and adaptation.

Finally, returning one more time to the overburden potential, by accepting the value of $\delta e/\delta \vartheta$ at fixed ϑ to be independent of the load pressure P, we can write:

$$\Omega = P\,(\delta e/\delta\vartheta)_\vartheta \qquad (10)$$

We now concentrate on the unloaded shrinkage curve. For $P=0$ the *Groenevelt-Bolt* equation of state can be written with the void ratio as the explicit variable:

$$e(\vartheta) = \varepsilon + k_3[\exp(-k_0/\vartheta^n) - \exp(-k_0/\varepsilon^n)] \qquad (11)$$

where $k_0 = k_1 k_2$ and $\varepsilon = {}^0\varepsilon$. The overburden potential can thus be expressed in terms of the moisture ratio (Fig. 5):

$$\Omega = P[k_3\,(k_0/\vartheta^n)\,(n/\vartheta)\,\exp(-k_0/\vartheta^n)] \quad (bar), \qquad (12)$$

At values of $\vartheta > 1.49$, which is the air-entry point of the curve for $P = 0.112$ (bar), the situation is slightly different. For a value of ϑ on the 1:1 line (say $\vartheta^\#$) upon unloading, the value of p will first change in proportion to the change in P, according to Eqn (2). This continues as long as the system stays saturated (or as long as the system stays on the line where the slope of the e-ϑ curve is unity). Upon further unloading, when p reaches $p^\#$ (*i.e.* value of pressure potential at air entry for the system under load, $P^\#$), all further change in p follows Eqn (10), as shown here:

$$\Omega(\vartheta^\#) = (P-P^\#) + P^\#(\delta e/\delta\vartheta)^\# \qquad (13)$$

Equations (6) to (13) have a bearing on the following analysis of some British swelling soils (with structural shrinkage) and some Norwegian swelling soils (without structural shrinkage).

3. MATRISTATICS

We turn our attention now to the 'matristatics' of swelling media, which we define here as the interpretation of solid-phase behavior from the standard shrinkage curve. By differentiating Eqn (11) one obtains the slope of the shrinkage curve as a function of the moisture ratio, $\sigma(\vartheta)$:

$$\sigma(\vartheta) = nk_0k_3[\vartheta^{-(n+1)}\exp(-k_0/\vartheta^n)] \qquad (14)$$

This slope function, $\sigma(\vartheta)$, usually has two inflection points, the locations of which can be identified by differentiating Eqn (14) to produce the curvature function:

$$\kappa(\vartheta)= nk_0k_3\exp(-k_0\vartheta^{-n})\,[(-\vartheta^{-n}\,n - \vartheta^{-n} + nk_0\vartheta^{-2n})\,/\,\vartheta^2]\,, \qquad (15)$$

Table 2. Moisture ratios and void ratios for two soils of contrasting soil structure [after *Reeve and Hall* 1978].

Faulkbourne Bw (Excellent soil structure)		Ragdale (Very poor soil structure)	
ϑ	e	ϑ	e
0.837	0.837	0.825	0.825
0.690	0.784	0.760	0.776
0.659	0.774	0.641	0.666
0.549	0.727	0.580	0.612
0.430	0.610	0.546	0.575
0.348	0.549	0.485	0.517
0.267	0.491	0.450	0.490
0.183	0.476	0.419	0.456
0.152	0.478	0.357	0.405
		0.319	0.387
		0.287	0.382
		0.267	0.368
		0.247	0.368
		0.229	0.368

and finding its extremes. These extremes can be used to distinguish regions of the unsaturated shrinkage curve, and give them physical significance (examples shown below).

While there is a paucity of published data suitable for this sort of analysis, the results of *Reeve and Hall* [1978] and *Olsen and Haugen* [1998] provide a good starting point. We present briefly here an evaluation of the slopes and curvatures at both the wet and dry ends of their shrinkage curves. For the wet end we use data from undisturbed soil cores and clods published by *Reeve and Hall* [1978], and for the dry end we use data from natural soil clods published by *Olsen and Haugen* [1998].

3.1 Wet end of Shrinkage Curves

We present in Table 2 the shrinkage data for two soils of contrasting structure as scored in the field by *Reeve and Hall* [1978]. The Faulkbourne soil (high structure-score) exhibits 'structural' shrinkage, while the Ragdale soil (low structure-score) does not.

An examination of the curvatures for these two soils using Eqn (15) enables the separation of 'structural' versus 'normal' shrinkage [*Groenevelt and Grant* 2001a]. The point of maximum curvature marks the end of structural shrinkage and the beginning of normal shrinkage (Fig. 6a,b). The magnitude of structural shrinkage can thus be calculated unambiguously as the difference between the void ratio at air-entry and the void ratio at the point of maximum curvature. *Groenevelt and Grant* [2001a], in their analysis of the data presented by *Reeve and Hall* [1978], showed that at this point the calculated volumetric air content is a good indicator of the relative quality of soil structure in the field (as scored by *Reeve and Hall*).

3.2 Dry end of Shrinkage Curves

For this analysis, we examine two shrinkage curves published by *Olsen and Haugen* [1998] for a topsoil and a subsoil, the data for which are shown in Table 3. The curves were fitted to Eqn (11), differentiated to obtain the slopes using Eqn (14), and subsequently differentiated to obtain the curvatures using Eqn (15).

The point of maximum curvature (inflection point) at the dry end of the shrinkage line can be used to unambiguously

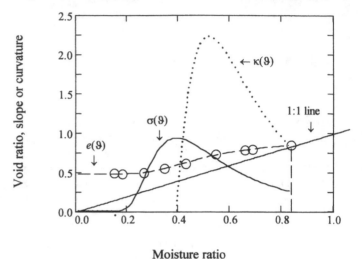

Figure 6a. Void ratio (dashed curve), slope (solid curve) and curvature (dotted curve) for shrinkage data (circles) of *Reeve and Hall* [1978] for their Faulkbourne Bw soil. The vertical broken line indicates the moisture ratio at air entry during shrinkage.

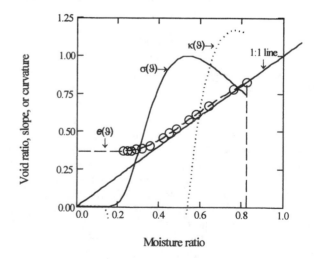

Figure 6b. Void ratio (dashed curve), slope (solid curve) and curvature (dotted curve) for shrinkage data (circles) of *Reeve and Hall* [1978] for their Ragdale soil. The vertical broken line indicates the moisture ratio at air entry during shrinkage.

Table 3. Moisture ratios and void ratios for *Olsen and Haugen*'s [1998] topsoil and subsoil.

Topsoil		Subsoil	
ϑ	e	ϑ	e
0.78	0.98	0.89	0.96
0.73	0.94	0.83	0.90
0.66	0.90	0.75	0.83
0.49	0.85	0.51	0.73
0.47	0.84	0.30	0.71
0.43	0.84	0.20	0.70
0.40	0.84	0.13	0.69
0.32	0.83	0.00	0.68
0.27	0.83		
0.23	0.82		
0.19	0.82		
0.15	0.81		

identify the shrinkage limit, which historically has been a rather coarse, manual procedure. Shrinkage limits calculated by this procedure are compared in Table 4 with values measured by *Olsen and Haugen* [1998]. Further analysis of the data presented by *Olsen and Haugen* shows that the maximum curvature (inflection point) at the wet end represents the lower plastic limit quite well. Lower plastic limits calculated by this procedure are compared in Table 4 with *Olsen and Haugen*'s measured values.

3.3 Matristatics of Talsma's Clay Paste

Finally, we wish to analyze *Talsma*'s shrinkage curves from the point of view of the behavior of the matrix. Applying Eqn (14) one finds the slope values of *Talsma*'s unloaded shrinkage curve as given in Fig. 7a, using $k_0 = 2$, $k_3 = 2$, and $n = 0.7$. From this figure, or from Eqn (15), one then finds the maximum slope ($\sigma_{max} \approx 0.40$) occurs at $\vartheta \approx 0.76$ (where the curvature, $\kappa(\vartheta)=0$) as presented in Fig. 7b. Thus, according to Eqn (7), $\vartheta \approx 0.76$ is the moisture ratio where, for a given load pressure, P, the overburden potential is maximal. At this point, the percentage of the load pressure, P, carried by the matrix is minimal.

By differentiating Eqn (15) and setting the differential equal to zero one finds the two extremes at:

$$\vartheta_1 = (2nk_0)^{1/n} [3(n+1)+[9(n+1)^2 - 4(n+1)(n+2)]^{0.5}]^{-1/n} \quad (16)$$

$$\vartheta_2 = (2nk_0)^{1/n} [3(n+1)-[9(n+1)^2 - 4(n+1)(n+2)]^{0.5}]^{-1/n} \quad (17).$$

The first extreme, $\vartheta_1 \approx 0.23$, is the inflection point at the dry end and is an unbiased estimate of the 'shrinkage limit' [*Groenevelt and Grant* 2001b]. The second extreme, $\vartheta_2 \approx 1.30$, is the inflection point at the wet end, and separates the region of 'structural' shrinkage from that of 'unsaturated, proportional' shrinkage (*ibid*). In the middle between these two extremes, where the curvature is zero at $\vartheta \approx 0.76$, the

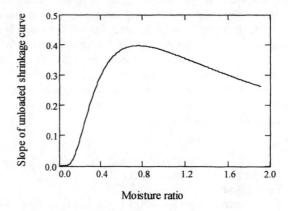

Figure 7a. Slope values for *Talsma*'s unloaded shrinkage curve as a function of the moisture ratio.

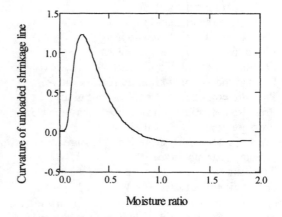

Figure 7b. Curvature values for *Talsma*'s unloaded shrinkage curve as a function of the moisture ratio.

Table 4. Comparison of measured and predicted consistency limits for *Olsen and Haugen*'s [1998] data.

Soil	Gravimetric Water Contents (kg/kg)			
	Shrinkage limits calculated from maximum curvatures at dry end	Shrinkage limits measured by *Olsen & Haugen*	Lower plastic limits calculated from maximum curvatures at wet end	Lower plastic limits measured by *Olsen & Haugen*
Topsoil	0.21	0.20	0.28	0.29
Subsoil	0.21	0.20	0.29	0.27

matrix carries the smallest percentage of the total load or overburden.

4. CONCLUSIONS

Hydrostatics in swelling soils, as pioneered by JR Philip, is extended here to calculate the water potential in soils under loaded conditions. This has significant implications for understanding subsoil constraints to root growth in heavy-textured and texture-contrast soils where horizons change volume with wetting and drying [*Groenevelt et al.* 2001]. Our analysis and interpretation of shrinkage lines using their maximum curvatures (inflection points), provide a powerful tool to separate the different stages of the shrinkage process. This tool works equally well for data obtained from soil cores as for data obtained from soil clods. It is expected that this tool will be useful in soil physics and soil mechanics.

List of Symbols

C Differential water capacity (bar^{-1})
e Void ratio (dimensionless)
$^0e^*$ Void ratio with zero load pressure at air entry (dimensionless)
$e^*(P)$ Void ratio with load pressure, P, at air entry (dimensionless)
k_1 Fitting parameter (bar)
k_2 Fitting parameter (bar-1)
k_3 Fitting parameter (dimensionless)
n Fitting parameter (dimensionless)
p Pressure potential of soil water (bar, negative)
p^* Pressure potential at air entry (bar, negative)
^0p Pressure potential at $P = 0$ (bar, negative)
$^0p^*$ Pressure potential with zero load pressure at air entry
p_a Pressure potential of soil air (bar)
P Load pressure, overburden pressure, or envelope pressure (bar, positive)
T Temperature
z Gravitational potential (bar)
$\alpha_P(e)$ Slope of shrinkage curve with moisture ratio as explicit variable and void ratio as implicit variable, $\vartheta(e)$, at constant load pressure, P.
ε Void ratio on shrinkage line at air entry (dimensionless)
$^0\varepsilon$ Void ratio on unloaded shrinkage line at air entry (dimensionless)
θ Volumetric water content (dimensionless)
Φ Total water potential (bar)
Ψ Matric potential (bar, or cm in Figs 1 and 2, negative)
ϑ Moisture ratio (dimensionless)
$\vartheta^\#$ Moisture ratio linking loaded and unloaded states (dimensionless)
$^0\vartheta^*$Moisture ratio, zero load at air entry (dimensionless)
ρ_w, ρ_s Density of water, density of particles (g cm^{-3})
Π Swelling pressure (bar)

σ Differential overburden potential (dimensionless)
Ω Overburden potential (bar)

Acknowledgment. Adelaide University supported the senior author to undertake this work while on study leave at the University of Guelph, Canada during 2000.

REFERENCES

Bishop, A.W., and G.E. Blight, Some aspects of effective stress in saturated and partly saturated soils, *Geotechnique, 13*, 177-197, 1963.

Bronswijk, J.J.B., and J.J. Evers-Vermeer, Shrinkage of Dutch clay soil aggregates, *Netherlands J. Agric. Sci., 38*, 175-194, 1990.

Croney, D., and J.D. Coleman, Pore pressure and suction in soil. In *Pore Pressure and Suction in Soils*, pp. 31-37, Butterworths, London, 1961.

Giraldez, J.V., and G. Sposito, A general soil volume change equation: 2. Effect of load pressure, *Soil Sci. Soc. Am. J., 47*, 422-425, 1983.

Groenevelt, P.H., and G.H. Bolt, Water retention in soil, *Soil Sci., 113*, 238-245, 1972.

Groenevelt, P.H., and C.D. Grant, Re-evaluation of the structural properties of some British swelling soils, *European J. Soil Sci., 52 (4)*, (in press), 2001a.

Groenevelt, P.H., and C.D. Grant, Curvature of shrinkage lines in relation to the consistency and structure of a Norwegian clay soil, *Geoderma* (in press), 2001b.

Groenevelt, P.H., C.D. Grant, and S. Semetsa, A new procedure to determine soil water availability, *Australian J. Soil Res., 39*, 577-598, 2001.

Groenevelt, P.H., and B.D. Kay, On pressure distribution & effective stress in unsaturated soils, *Canadian J. Soil Sci., 61*, 431-443, 1981.

Groenevelt, P.H., and J.-Y. Parlange, Thermodynamic stability of swelling soils, *Soil Sci. 118*, 1-5, 1974.

Mathsoft 1998. 'Mathcad 8 Professional Academic'. Mathsoft Inc. Cambridge MA. USA. (<http://www.mathsoft.com>).

Olsen, P.A., and L.E. Haugen, A new model of the shrinkage characteristic applied to some Norwegian soils, *Geoderma, 83*, 67-81, 1998.

Philip, J.R., Kinetics of sorption and volume change in clay-colloid pastes, *Australian J. Soil Res., 6*, 249-267, 1968.

Philip, J.R., Moisture equilibrium in the vertical in swelling soils. I. Basic theory, *Australian J. Soil Res., 7*, 99-120, 1969a.

Philip, J.R.. Moisture equilibrium in the vertical in swelling soils. II. Applications, *Australian J. Soil Res., 7*, 121-141, 1969b.

Philip, J.R., Hydrostatics and hydrodynamics in swelling soils, *Water Resources Res., 5*, 1070-1077, 1969c.

Philip, J.R., Hydrostatics in swelling soils and soil suspensions: unification of concepts, *Soil Sci., 109*, 294-298, 1970a.

Philip, J.R., Flow in porous media, *Ann. Rev. Fluid Mech. 2*, 177-204, 1970b.

Philip, J.R., Reply to note by E.C.Youngs and G.D.Towner on "Hydrostatics and hydrodynamics in swelling soils", *Water Resources Res., 6*, 1248-1251, 1970c.

Philip, J.R., Hydrology of swelling soils. In: *Salinity and Water Use*, edited by T. Talsma and J.R. Philip, Proc. of a Symposium held at the Australian Academy of Science, Canberra, pp. 95-107, Macmillan, London, 1971.

Philip, J.R., Hydrostatics and hydrodynamics in swelling media, in *Fundamentals of Transport Phenomena in Porous Media*, Proc. IAHR Symposium, Haifa, 1969, pp. 341-355, Elsevier, Amsterdam, 1972a.

Philip, J.R., Recent progress in the theory of irrigation and drainage of swelling soils, *Proc. 8th Congr. Int. Comm. Irrig. Drain*, pp. c13-c28, Varna, Bulgaria, 1972b.

Philip, J.R., Future problems of soil water research, *Soil Sci., 113*, 294-300, 1972c.

Philip, J.R., and D.E. Smiles, Kinetics of sorption and volume change in three-component systems, *Australian J. Soil Res., 7*, 1-19, 1969.

Reeve, M.J., and D.G.M. Hall, Shrinkage in clayey subsoils of contrasting structure. *J. Soil Sci., 29*, 315-323, 1978.

Smiles, D.E., and M.J. Rosenthal, The movement of water in swelling materials, *Australian J. Soil Res., 6*, 237-248, 1968.

Sposito, G., and J.V. Giraldez, Thermodynamic stability and the law of corresponding states in swelling soils. *Soil Sci. Soc. Am. J. 40*, 352–358, 1976.

Talsma, T., Measurement of the overburden component of total potential in swelling field soils. *Australian J. Soil Res., 15*, 95-102, 1977a.

Talsma, T., A note on the shrinkage behaviour of a clay paste under various loads, *Australian J. Soil Res., 15*, 275-277, 1977b.

Terzaghi, K., Erdbaumechanik auf bodenphysikalisher Grundlage, Deuticke, Vienna, 1925.

Youngs, E.G., and G.D. Towner, Comments on "Hydrostatics and hydrodynamics in swelling soils" by J.R. Philip, *Water Resources Res., 6*, 1246-1247, 1970.

C.D. Grant, Department of Soil and Water, Adelaide University, Waite Campus, PMB No.1 Glen Osmond, SA. 5064. Australia (e-mail: cameron.grant@adelaide.edu.au)

P.H. Groenevelt, Department of Land Resource Science, University of Guelph, Guelph, Ontario, N1G 2W1, Canada (e-mail: pgroenev@lrs.uoguelph.ca)

G.H. Bolt (retired), Department of Environmental Science/Soil Quality, Wageningen University and Research Centre, 6700 HB Wageningen Netherlands (e-mail:jerry.bolt@bodsch.benp.wau.nl)

Water and Solute Transfer in Porous Media

CSIRO Land and Water, Canberra, Australia

John Philip's classical approach to environmental problems was based on a flux equation whose elements permit measurement, and material continuity. This approach identifies a small set of material properties that are necessary and sufficient for analysis and can be measured at the scale of application. Philip warned, on a number of occasions, against quantitative prediction of material properties at one scale using insights developed at another. At the same time he encouraged simplifications where observation permitted them and he strongly objected to unnecessary complication when these insights are applied to practical problems. He was disappointed when simplifications and insights, revealed in some of his quasi-analytical solutions to flow equations, were not incorporated in computer models. This paper illustrates some of these issues in relation to 1-dimensional flow of liquid in swelling systems and, specifically, to approximations that simplify analysis of some chemical engineering unit processes. It also illustrates approaches to hydrodynamic dispersion that derive from early insights of Philip. In particular, it draws attention to benefits of solid- and water-based space-like coordinates in analysing hydrodynamic dispersion and reaction during unsteady flow in unsaturated soils and some benefits in field management.

1. INTRODUCTION

John Philip's approach to environmental processes was based on classical continuum mechanics and, in the case of water, on the equation of *Richards* [1931] who combined Darcy's law and insights of *Buckingham* [1907], among others, with a continuity equation for the water. This approach identifies material properties necessary and sufficient to solve problems to which the equation applies. These, together with appropriate initial and boundary conditions reveal behaviour, through the logic of the mathematics, that permits test of theory or even behaviour that

Environmental Mechanics: Water, Mass and Energy Transfer in the Biosphere
Geophysical Monograph 129
Copyright 2002 by the American Geophysical Union
10.1029/129GM11

may be inaccessible to experiment [*Philip*, 1957a]. He emphasized the importance of matching the scale of analysis to that of application and insisted on identifying the key material properties at that scale. Examples in his *Advances in Hydroscience* classic [*Philip*, 1969], which follows the general structure of texts such as *Carslaw and Jaeger* [1959] and *Crank* [1975], illustrate the power of the approach.

The approach provided a guide to test theory and to measure material properties; it provided a basis for practical analysis; it provided a useful set of simplifications and approximations to theory; and it underpins the test of more complicated "trans-scientific" excursions [*Philip*, 1980]. It also provided a recipe to formulate a new but analogous theory of 1-dimensional flow in swelling materials.

John Philip supported William of Ockham's thesis that the simplest explanation of a phenomenon must be the most attractive, despite the assertion [*Bronowski*, 1979] that it represents no more than an aesthetic pleasure akin to sacrificing your queen to permit a knight to mate. He worried, too, about models that overlooked the efficiencies and physical insights that simplified and illuminated his analyses and also about modeling that was not frequently and searchingly tested experimentally [*Philip*, 1991].

This paper first illustrates some of these issues in relation to water flow in swelling materials and then explores an approach to hydrodynamic dispersion and chemical reaction in soils, which was encouraged by Philip's insights.

2. FLOW IN SWELLING MATERIAL

Origins of theory of water flow in non-swelling soils are described well by *Philip* [1969, 1974]. Extension of the approach to swelling systems was precipitated by analyses of *Raats and Klute* [1968a, b] and experiments of *Smiles and Rosenthal* [1968]. John Philip's interest in this area illustrates his style and is worth recalling. Margaret Rosenthal and I performed a series of experiments where we measured the evolving water content profiles following imposition of a step change in water content at the outflow end of an initially uniform column of clay in a pressure membrane cell. Our analysis was formulated in rudimentary solid-based space-like coordinates and applied mathematicians at Sydney University recommended we approach John for advice. Within a week of our discussing the work with him, he had produced an analysis but in physical space and time [*Philip*, 1968]. Margaret and I were astonished at the precision of his approach but more irritated at the facility with which he analyzed issues that we had battled over for months. Characteristically, he expressed his "considerable debt to Dr Smiles and Miss Rosenthal for interesting me in this topic"; he also assured us that there was no future in using material coordinates for this class of problem. He then produced the flurry of papers on vertical equilibrium and flow in swelling materials cited in *Philip* [1970]. Within three months, however, he recognized the significance of papers arising from Peter Raats' doctoral thesis [*Raats and Klute*, 1968a, b] and, in retrospect, that of *McNabb* [1960], and revised his opinion of the use of material coordinates. He also realized that a solid-based space-like coordinate resulted in an equation of exactly the same form as Richards' equation. This made much of the analysis in his 1969 review paper on theory of infiltration accessible to flow in swelling material.

Principal features of the analysis are set out in *Philip* [1970], and papers he cited there. Briefly, 1-dimensional water flow in swelling soils requires that we recognise that water flows with, and relative to, the solid particles. Darcy's law describes flux of water relative to the particles

[*Zaslavsky*, 1965] in response to a space gradient of hydraulic head (manometric pressure head plus gravitational head). The solid flux times the water content (per unit amount of solid) defines the component of the total flux advected with the moving particles. Because both the solid and the liquid move during non-steady flow, material balance equations are necessary for both. Substitution of the equation for the solid in that for the water, however, leads to a balance equation for the water expressed in a coordinate, m, based on the distribution of the solid. A flow equation (1) analogous to that of *Richards* [1931] then emerges when Darcy's law is combined with the material balance equation expressed in the solid based coordinates.

$$\frac{\partial \vartheta}{\partial t} = \frac{\partial}{\partial m}\left(k_m(\vartheta)\frac{\partial \psi}{\partial m} \right) + \frac{\partial}{\partial m}\left(k_m^*(\vartheta) \right). \quad (1)$$

In (1), ϑ is the moisture ratio (volume of water per unit volume of solid), t is time, k_m is hydraulic conductivity in the m-coordinate, ψ is the water potential, and $k_m^*(\vartheta)$ is a conductivity like term that encapsulates $k_m(\vartheta)$, overburden and gravity [*Kim et al.*, 1992, *Philip*, 1969; *Smiles*, 2000a]. The hydraulic conductivities $k_m(\vartheta)$ and $k(\theta_w)$ in 'material' and 'physical' space are related by

$$k_m(\vartheta) = k(\theta_w)\theta_s, \quad (2)$$

with θ_w the volume fraction of the water and θ_s the volume fraction of the solid. In the saturated system discussed below,

$$k_m^*(\vartheta) = (1-\rho)k_m(\vartheta), \quad (3)$$

with ρ the specific gravity of the solid.

Gravity enters the analysis both because it contributes explicitly to the hydraulic head and implicitly to the manometric pressure, p_w, defined by

$$p_w = \psi(\vartheta) + \alpha \int_z^T \gamma \, dz, \quad (4)$$

where the first term on the right is the *unloaded* moisture potential and the second term is the overburden potential, calculated as the integral of γ, the wet specific gravity of the soil, from the point of elevation, z, to the soil surface at T, [*Groenevelt and Bolt*, 1972; *Philip*, 1970]. The overburden term incorporates the consequences of vertical displacement of the profile as water content changes [*Philip*, 1969]. The α term moderates the effect of the total load (including any surface load) according to an average of the slope of the $e(\vartheta, p)$ relationship which describes the volume/ water content/ pressure properties of the material. [*Croney et al.,* 1958; *Croney and Coleman*, 1961; *Groenevelt and Bolt*, 1972] in which the void ratio, e, parameterizes the volume, ϑ parameterizes the water content, and p is the load.

Solutions of Richards' equation, and hence equation (1), are well-known and advection of water associated with solid movement is implicitly dealt with in the solid-based coordinate; *Philip* [1968] exemplifies mathematical complications if the advection term remains an explicit component of the flow equation.

In common with non-swelling soil theory, use of this equation requires that the $k_m(\vartheta)$ and $\psi(\vartheta)$ characteristics of the material be well-defined. But, in addition, swelling soil theory requires the solid specific gravity, ρ, and the $e(\vartheta, p)$ relationship   These characteristics are readily measured although they do not generally permit prediction from properties at other scales. *Smiles* [2000a] identifies most of the workers who have used the approach in soil science since 1970

Analogous problems are also encountered in civil engineering where studies date back to those of *Terzaghi* [1923] and in chemical engineering where analogous theory was formulated, particularly in Japan and the United States [see for example, *Atsumi et al.*, 1973; *Gibson et al.*, 1967; *Leu*, 1986; *Shirato et al.*, 1986].

Engineering provides useful insights into theory and experimental techniques that would be very useful in soil science but are not often applied and *vice versa*.

Two topical examples illustrate Philip's contentions that the simplest approach to a phenomenon that is fully consonant with observation is best. One of these examples deals with basic issues of flow; the other illustrates economies that arise from insights that *Philip* [1969] offers into solutions of Richards' equation.

2.1. Flow Laws in Swelling Materials

In the analysis of water flow giving rise to equation (1), transfer of the solid is linked by continuity with the flux of the water and Darcy's law describes the flux of water relative to the solid. But an approach based on Darcy's law and continuity is not unique; a school of thought has focussed on the flux law for the solid in swelling systems [eg *Landman and Russel*, 1993; *Nakano et al.*, 1986]. This approach has been aired recently in the soils literature (*Angulo-Jaramillo et al.*, 1997]. It derives from Stokes' flow and the notion that the settling of independent particles might be integrated to describe the settling of an interacting particle swarm.

There is also an analogy with diffusion consequent on Brownian motion of dispersed colloid particles. Regrettably, neither a Stokes' flow model nor Brownian motion applies easily across the range of materials from a diffuse colloidal suspension to an unsaturated *non-swelling* soil. Darcy's law, however, explicitly describes the "inverse" process of water flow relative to the solid and we can measure both $\psi(\vartheta)$ and $k_m(\vartheta)$ across almost this entire range. Furthermore, the complementary material balance equations for the solid and the liquid make Darcy's law sufficient to describe fully the transfer of both components in the system so a flux law for the solid is unnecessary [*Kirby and Smiles*, 1999].

Tests of this approach initially used the analysis of *Bruce and Klute* [1956] and focussed on non-steady flow resulting from constant potential boundary conditions. A rapidly converging iterative method, based on an approach of *Parlange* [1972], then permitted approximate but accurate ways to solve and test the water flow equation for a wide range of other important initial and boundary conditions. *Smiles* [1986] summarised a set of these tests. Further experimental series such as *Smiles et al.* [1985] and *Kirby and Smiles* [1989] tested flows on slurries over ranges of solution salt concentration and temperature under transient and steady state flow conditions. They found no reason to reject the basic premises at least for smectites. For example, steady state measurements of $k_m(\vartheta)$ corresponded well with values calculated from the diffusivity, $D_m(\vartheta)$ measured using the method of *Bruce and Klute* [1956] and $\psi(\vartheta)$ [*Kirby and Smiles*, 1989] using the definition of diffusivity, which is often used for non-hysteretic flow.

$$D_m(\vartheta) = k_m(\vartheta) \frac{d\psi}{d\vartheta}. \qquad (5)$$

The measurements also corresponded well with estimations calculated by differentiating measured sorptivity/pressure functions [*Philip*, 1957b; *Smiles and Harvey*, 1974].

Other experiments explored effects of temperature and found flow to be consistent with the temperature dependence of the kinematic viscosity of water, within limits of experimental measurement. These experiments put to rest concerns about curious effects related to water structure close to clay mineral surfaces in clay slurries in the ranges $4 < \vartheta < 40$ and $277 < T/K < 302$. Furthermore, there was no statistically significant temperature dependence in $\psi(\vartheta)$ of these clays. It is important to note, however, that in the range of suctions imposed, the clay remained saturated and surface tension effects would not have been evident. Concerns about the temperature dependence of flow in unsaturated soils and clays [see for example, *Stoffregen et al.*, 1997] remain unanswered.

Steady- and non-steady state, and static equilibrium experiments also explored effects of solution salt concentrations. For example, measured $\psi(\vartheta)$ functions were qualitatively consistent with double layer theory but that theory failed to predict quantitative changes in $\psi(\vartheta)$ accompanying changes in the equilibrium NaCl solution concentration in the concentration range 0-0.1M. This prompted *Philip and Smiles* [1982] to reiterate the difficulties of trying to transcend scales of discourse even in materials as structurally simple as clay slurries. At the same time, measurements of $\psi(\vartheta)$ and $k_m(\vartheta)$ are unambiguous and are sufficient to fully describe flow under many circumstances at that scale. Furthermore, we can remain quite agnostic about the origins or details of the interactions that give rise to them. It is sufficient that these characteristics exist and are measurable.

These studies also demonstrated a hitherto unobserved unique relation between k and ψ over 3 orders-of-magnitude of solution salt concentration [*Smiles*, 1986]. The numerical benefits of this relation are evident but its basis in theory has yet to be elucidated.

Confusion may arise in test of theory and experimental artifacts may prejudice interpretation of results. For example, volume change requires that experimental constraints be considered carefully and experiments that fail to recognize their importance will be misleading. The macroscopic theory originally proposed by *Philip and Smiles* [1969] envisaged essentially 1-dimensional (vertical) volume change. It recognized 3-dimensional cracking but considered that only the vertical component would affect energetics of flow. Experiments therefore sought to ensure lateral constraint but free vertical movement. *Bridge et al.* [1970] and *Collis-George and Bridge* [1973] describe experiments and strategies to deal with these requirements. *Smiles and Colombera* [1975] used another simple technique with success in a limited range of overburden constraint. Spurious results emerge if these conditions are not met and full development of theory will almost certainly have to appeal to methods of soil mechanics where the triaxial cell has been designed to deal with such problems [*Kirby and Smiles*, 1999].

Analysis of steady state experiments must also recognise that flow must be related in a non-linear way to the imposed water pressure difference. The phenomenon was used by *Kirby and Smiles* [1989] to measure $k_m(\psi)$ 'directly'.

2.2. *Physically Based Simplifications of Theory*

In John Philip's series solutions to the Richards' equation in one-dimension and multi-dimensional absorption, the leading term in each case is exactly the solution for one-dimensional absorption. This reveals the importance of the 1-dimensional diffusion solution during early stages of flow. It is also consistent with his truncated 2-parameter infiltration equation

$$i(t) = S t^{1/2} + A t , \qquad (6)$$

in which $i(t)$ is the cumulative infiltration and he defined the sorptivity, S [*Philip*, 1957b]. S is the fundamental integral property of the first term of the series solution. Philip recognised the inconsistencies between A in equation (6) and the second and subsequent terms in his series solution; he returned to this issue in *Philip* [1987]. Despite these theoretical difficulties, the importance of the first term in the series solution and in equation (6) led him to identify t_{grav} as the time before which one does not need to consider the effect of gravity in infiltration. Similarly, in 2- and 3-dimensional axisymmetric flow, he identified t_{geom} as the time after which geometry needs to be considered. Both t_{grav} and t_{geom} are estimated by comparing S with the second coefficient in the series (or A in the case of equation (6)), which is related to the hydraulic conductivity. The better than order-of-magnitude estimate proves very useful in non-swelling water flow theory; it is also helpful in analysis of flow in swelling materials.

The effect of gravity

The effect of gravity in swelling systems much exercised John Philip. Among other things, he recognised that where the solid specific gravity is greater than one, infiltration into saturated swelling systems results in an *increase* in the gravitational potential energy of the system. This contrasts with a *decrease* in potential energy that accompanies infiltration in non-swelling soil. Thus, vertical infiltration in these materials is analogous to capillary rise in non-swelling ones. The mathematical expression of this effect is seen for the saturated system in equation (3), where the $(1-\rho)$ term changes the sign of $k_m^*(\vartheta)$. An average slope of the $e(\vartheta, p)$ relationship moderates this effect in unsaturated swelling soils but *Philip* [1995] continued to argue for a practical approximate approach that sets α as the slope of the unloaded

$e(\vartheta, p)$ relationship despite early criticism of this approximation by Youngs and others. Experiments by *Smiles* [1974] exemplify the issue during infiltration of water into a saturated clay and reasonable limits to t_{grav} for that system were calculated. A limited set by of experiments of *Smiles and Colombera* [1975] in the laboratory and *Perroux and Zegelin* [1984] in the field, offer support.

The hydraulic properties of a potentially acid sulfate soil provide a recent example. The material was recovered from a depth of about 1.5m to provide preliminary hydraulic conductivity and potential measurements needed to predict consequences of loading and drainage. Pressure membrane cell equilibrium and outflow were used to measure $\psi(\vartheta)$ (Figure 1) and $\psi(S)$ (Figure 2) on this material.

The linear relationships observed in Figures 1 and 2 have previously been observed for much wetter materials. There is no reason, in principle, that they should be so but the observation facilitates calculations of $D_m(\vartheta)$ (Figure 3) and then $k_m(\vartheta)$ (Figure 4) using equations (7) and (5) [*Smiles and Harvey* [1974]:

$$D_m(\vartheta) = \frac{1}{2(\vartheta_0 - \vartheta_i)}\left(\frac{\partial S^2}{\partial \vartheta} - \frac{r S^2}{(\vartheta_0 - \vartheta_i)}\right). \quad (7)$$

In (7) the subscripts refer to initial (i) and boundary (o) values of ϑ and r is the power of the flux concentration relation [*Philip*, 1972] expressed in the form,

$$F(\vartheta) = \left(\frac{\vartheta - \vartheta_i}{\vartheta_o - \vartheta_i}\right)^r, \quad (8)$$

with $0 < r < 1$ depending on the flow regime.

$D_m(\vartheta)$ shown in Figure 3 is a decreasing function of ϑ. This is quite different from the non-swelling soil experience but consistent with other swelling material data as application of the analysis of *Bruce and Klute* [1956] to the data of Figure 9 reveals. The phenomenon makes desorption from swelling systems similar to absorption by a non-swelling soil and, because cumulative outflow tends to be proportional to $t^{1/2}$ for relatively long times, desorptivity is measured accurately.

Figure 4 shows that, compared to $k(\theta_w)$ in rigid soils, $k_m(\vartheta)$ is a comparatively weakly increasing function. This is a common observation. In addition, $k_m(\vartheta)$ is not matched well using material analogues of equations, such as that of

Kozeny/Carman/Hatch discussed by *Dallavalle* [1948] (his equations 13.32 *et seq.*), which seek to relate k to θ_w.

These data permit us to estimate the time during which gravity might be neglected during vertical flow that might follow the imposition of a surface load. An example relates to the time during which gravity free flow controls consolidation of a road or other pavement laid on such a material. Using the approach based on equation (6) and setting $A \approx (1-\rho)(k_m(\vartheta_0) - k_m(\vartheta_i))/2$ [*Smiles* 1974], it emerges that $t^{1/2}$ behavior will be 10 times greater than that due to gravity for more than 17 years. This is in accord with observations of a road base on a similar soil where settlement

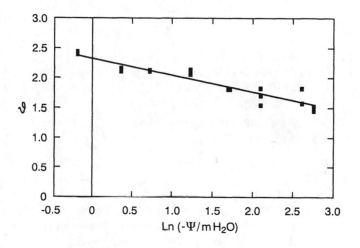

Figure 1. Water content, ϑ, water potential, ψ, characteristic of an acid sulfate soil

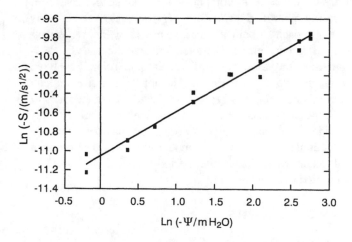

Figure 2. Sorptivity-water potential, $S(\psi)$ function for an acid sulfate soil.

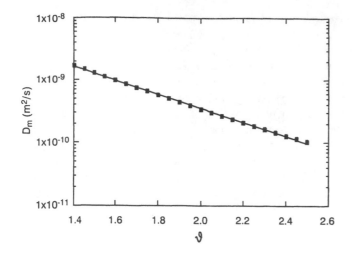

Figure 3. Moisture diffusivity-water content function, $D_{\mathrm{m}}(\vartheta)$, derived from data shown in Figures 1 and 2 using equation (7) and setting r = 0.5 [*Parlange*, 1975].

was still proceeding linearly with respect to $t^{1/2}$ after more than 2 years [*Smiles* 1973].

The effect of geometry

The notion of t_{geom} has been applied to centrifugal filtration. This process is energetically advantageous because both the water and the solid potential energies diminish in the centripetal force field and equilibrium water contents in the filter cake are much less than they would be, for example, in a ponded sediment in the same force field [*Smiles*, 1976]. But analyses have been complicated [*Bear et al.*, 1984] or seriously simplified [*Tiller and Hysung*, 1993]. An early time approach to both the axi-symmetric cylindrical aspects and the gravity effect, however, reduces the analysis to 1-dimension and, for the case of wet clays, to a simply analysed diffusion form [*Smiles*, 1999]. The approach derives directly from that of *Philip* [1969] and provides reliable estimates of the time for which the approximation is valid. It also provides a simple bridge to constant pressure filtration methods that can be used to model the centrifugal problem and again to estimate the period of time for which a simplified approach is valid.

2.3. General Comments

The use of the hydraulic head

John Philip's approach to the components of the water potential in a vertical swelling system was mechanistic and classical. It identified (equation 4) the manometric pressure, p_{w}, as the sum of the overburden potential and the unloaded water potential, ψ (or effective stress). Only the

total stress and p_{w} permit direct measurement in the field, however, and the unloaded water potential remains an inferred property. In this sense Philip was not as pragmatic as *Miller* [1975] who argued for the use of p_{w} rather than the inferred unloaded water potential, ψ, as a basis from which to define material properties and flow. Miller's point derived from the notion and use of the hydraulic head as the driving force for all water flow, with manometric pressure and gravity the two measurable components. *Smiles* [2000a] identifies examples where this issue is not recognised. Among other things, it results in a *SSSA* [1997] Glossary definition of ψ that is incorrect for swelling soils and is generally inapplicable.

Effects of cracking

These were considered early by *Raats* [1969, 1984] and *Philip and Smiles* [1969] who recognized that definition and measurement of a mass based space-like coordinate is not prejudiced by cracks. All that is required is that the cumulative mass of solid be defined in terms of a unit area of cross section that includes cracks. Relations between aggregate and profile volume change, however, have confused and complicated analysis. It is easy to measure the volume change of aggregates but the consequence of this swelling on the profile is less obvious. The thesis of *Bronswijk* [1990] provides an exhaustive set of data and a clear distinction between the effect of aggregate volume change on vertical profile displacement. The results of cracking on flow remain a problem yet to be dealt with.

A classical approach based on material balance and Darcy's law remains the most useful first approach to water flow in swelling soils. The approach has been well-tested

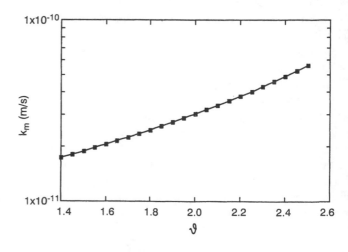

Figure 4. Material conductivity-water content function, $k_{\mathrm{m}}(\vartheta)$, derived from Figures 3 and 1 using equation (3).

in saturated systems such as clay suspension, mine tailings and sediments and many of its elements are systematically used in civil and chemical engineering. It is not widely used in soil science however, and papers still offer needlessly complicated explanations and inept experiments. Furthermore, definitions in standard texts are incorrect. The Glossary of Soil Science Terms [SSSA 1997], for example, has no definition of overburden that draws attention to its effect on the water potential. In addition, the definitions on page 103 of that Glossary totally ignore the consequences of swelling and the notion of effective stress that have been of operational concern in civil engineering for more than 60 years [Terzaghi 1943].

3. HYDRODYNAMIC DISPERSION

In the mid-1970s John Philip, David Elrick, John Knight and I realised that absorption experiments like those of Bruce and Klute [1956] provided a systematic way to study hydrodynamic dispersion in soil during unsteady unsaturated. Explicitly, it emerged [Smiles et al., 1978] that both the water and the soil solution concentrations preserved similarity in terms of distance divided by the square root of time during absorption of a solution by a relatively dry soil. This approach for the first time, permitted systematic exploration of dispersion during unsaturated soil water flow. It was developed in the context described in subsection 4.3.1 of Raats [2002] and it provided a flexible alternative to the Danckwerts/Saffman/Taylor focus on capillary flow that underpins the 'breakthrough' experiments on which previous examination of the phenomenon in soils was based. This is because material coordinates based on the distribution of the water improve description of the advection of the solutes in a setting of water contents that may vary in space and time.

For example, it put in context the significance of the water content and the pore water velocity dependence of the dispersion coefficient and showed that then current approaches to dispersion might be simplified. In particular it put realistic constraints on the conditions of flow that are of concern in soil science and thence to the relevance of the velocity dependence of the dispersion coefficient during 'natural' flow of water in soil. The analysis was subsequently extended to flux boundary conditions and Smiles et al. [1981], contemporaneously with Wilson and Gelhar [1981], showed how the approach is simplified by using space-like water based coordinates. For absorption, this approach identifies the origin of the coordinate system as the notional piston front that would exist if the absorbed water completely displaced the water originally present. Bond's and co-workers' studies in the 1980s extended the

approach to chemical reaction. A set of papers in 1990 summarizes insights arising from the approach [Bond and Phillips, 1990a, b, c].

The approach is not extensively applied although recent studies illustrate its versatility and power in critical analyses. We focus on two of these.

3.1. Dispersion and Reaction During Soil Water Absorption

Selection of a low-level radioactive waste repository in central Australia requires that we identify regolith materials that restrict the movement of radionuclides that might be released inadvertently. The task is complicated because material is sampled using reverse cycle impact drilling. This procedure destroys field structure although the clay mineral content appears unchanged. On the presumption that clay content provides a useful first estimate of the chemical activity, we examined effects of clay content and its variation between samples on the retardation properties of the material. In principle, retardation of the nuclide relative to the water can be predicted from the slope of the adsorption isotherm measured in "batch experiments" [Bolt, 1976; Bond and Phillips, 1990c, Freeze and Cherry, 1980; Hashimoto et al., 1964], Lai and Jutinak, 1972] but absorption experiments actually measure that retardation. Smiles [2001] describes such experiments, where a uniform, relatively dry, horizontal column of soil absorbed aqueous solution containing radionuclides from a source at constant water potential. Experiments were terminated after different specified times and the columns sectioned. Water soluble salts, tritium and Co-60 concentrations were measured in each section and graphed as a function of distance from the inflow surface divided by the square root of time at which the profiles are measured.

Figure 5 shows the mass fraction of water, θ_g, graphed as a function of $M = mt^{-1/2}$. Here, m is the cumulative mass of soil solid (per unit area of column cross section) measured from the inlet end of the column and t is the time at which the column was sectioned. The self-similar water content profiles revealed in this figure indicate that Darcy's Law is observed, that the column is uniform with initially uniform water content, and that the step-function water potential boundary condition is realized. The hatched areas are equal and the vertical line near 9×10^{-2} g/cm^{2}s$^{1/2}$ is a "front" that would exist if absorbed water displaced water originally present in its entirety. It is important to note that our use of the "piston front" here and later does not imply physical piston displacement but simply that during Darcy flow in soil the hydrodynamics and physical chemistry of the process result in something that resembles a "front".

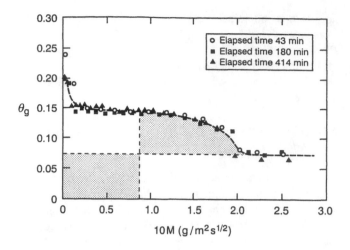

Figure 5. Water content (θ_g) vs M ($= mt^{-1/2}$) profiles measured during absorption by a relatively dry soil. m is a mass-based space-like coordinate. Experiments were terminated at the times shown in the figure.

Figure 6 shows concentrations of Co-60, water soluble salts and H-3 as functions of M. These data also preserve similarity consistent with step changes at $t = 0$ in each of these sets. Similarity also implies that chemical equilibrium effectively exists in the columns. Within experimental eror, the H-3 and soluble salt 'fronts' coincide at $M \approx 9\times10^{-2}$ g/cm^2.s$^{1/2}$. The Co-60 'front' is at $M < 1\times10^{-2}$ g/cm^2.s$^{1/2}$. It is retarded relative to the other tracers because the Co-60 is adsorbed (principally by the clay) and in an environment dominated by the absorbed solution rather than that originally present, and which is characterized by concentrations shown in $M>1\times10^{-1}$ g/cm^2.s$^{1/2}$.

It is interesting to consider these data in a space that is defined by the distribution of the water and satisfies its material balance. This coordinate is defined, for this case, by

$$G(m,t) = \int_0^M \theta_g \, dM - \int_{\theta_g(i)}^{\theta_g(0)} M \, d\theta_g \qquad (9)$$

with $\theta_g(i)$ and $\theta_g(0)$ the initial and boundary water contents of Figure 1. The second integral on the right hand side is Philip's sorptivity, S. The origin of this coordinate system ($G = 0$) is the notional piston front between the solution originally present and the invading one [*Smiles et al.*, 1981].

Figure 7 relates M to G using the data of Figure 5 and equation (9). $G = 0$ when $M \approx 9\times10^{-2}$ g/cm^2.s$^{1/2}$.

Figure 8 shows Co-60, soluble salts and concentrations of H-3 presented in this water-based coordinate system. The advancing "front" for H-3 and the retreating "front" for water soluble salts coincide with $G = 0$ and with the piston

front identified in Figure 5. They are distributed about it because of diffusion relative to the water. This indicates that the anions, in this case, do not react with the soil, and the tritiated water behaves like water. The delay of the Co-60 profile, relative to $G = 0$, is related to the retardation factor of *Hashimoto et al.* [1964]. It reflects the average slope, β, of the adsorption isotherm for Co-60 in the chemical environment of the soil during this process. In this system, the distribution coefficient, β, equals the ratio of the location of the Co "front" in water space (shown in Figure 4) to its location in mass based space (Figure 2) [*Smiles*, 2001]. Note that β is the same as the K_d of *Freeze and Cherry* [1980] but its units reflect the molal convention we use. Here, $\beta \approx 1.6$ g$_{water}$/g$_{solid}$.

The water-based coordinate is revealing and useful for anions and for H-3. When cation solution concentrations are small relative to the amount on the exchange capacity, as is the case with the Co-60, then the soil-solid based coordinate may be more illuminating. A subsequent experimental set where acid washed sand was added to the soil demonstrated this. These experiments are described by *Smiles* [2001] also. The addition of sand increased the sorptivity of the soil but decreased local adsorption of Co-60 associated with the clay mineral. The increase in S together with reduced local adsorption resulted in a greater penetration of solutes into the columns; increased S also increased the total amount of each solute in the column. These effects were virtually eliminated by using a space-like coordinate defined by the cumulative mass of clay (rather than soil solid as in Figure 6) and expressing the Co-60 (and other concentrations per unit mass of clay; and by 'normalizing' that space with regard to S. The latter device

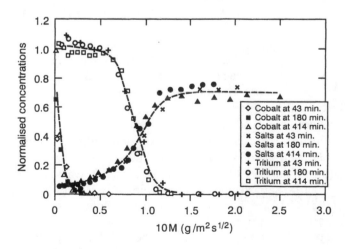

Figure 6. Normalized concentrations of soluble salts, tritium and Co-60, corresponding to the water content profiles of Figure 5, and graphed vs M

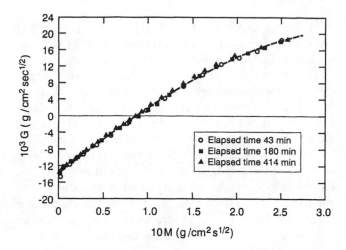

Figure 7. Relation between M, the solid-based coordinate of Figure 5, and G, a water based coordinate calculated by applying equation (9) to the data of Figure 5.

involves dividing equation (9) throughout by the sorptivity integral on the right hand side.

The use of a clay-based coordinate was foreshadowed by *Raats* [1997] and by *Smiles* [1997]. The qualitative efficacy of such an approach is not surprising, and we routinely estimate, for example, gypsum requirement on the basis of the cumulative exchange capacity of some arbitrary depth of topsoil. This analysis quantifies that approach.

Different clay minerals will behave differently but for the experiments of *Smiles* [2001], extension to a coordinate based on cumulative cation exchange capacity or surface charge density offers no further insights. The use of these approaches to predict behaviour systematically and quantitatively across clay mineral types and concentrations would seem a profitable area for study.

It is not necessary that both the water and solute invasion be associated with step-function, or constant-flux, boundary conditions. A water-based coordinate can be similarly used in the field where redistribution of water or water extraction by plant roots occurs as *Raats* [1975] and *Wilson and Gelhar* [1981] describe.

In conclusion, this procedure determines the fate of a nuclide during non-steady invasion of solution containing that nuclide into a particular soil. It provides a check on estimates based on "batch " (*in vitro*) exchange isotherms but it is more realistic because the measurement is made *in-vivo* and in association with the cation and anion suite in the neighbourhood of the chemical reaction during the 'chromatographic' process. It makes no presumption about the exchange environment required for an *in vitro* measurement. A corollary of the procedure is that it permits determination of actual isotherms involving all cations and ani-

ons in the displacement process, should that be required. This latter determination requires measurement of the principal soluble and exchangeable cations throughout the column. Each point on the Co-60 curve in Figures 6 and 8 then represents one point on an exchange isotherm [*Bond and Phillips*, 1990c] in the presence of the other soil cations at that point.

Solute profiles, such as those shown in Figure 6, are generally presented in physical space but their presentation in a solid based M-space is a trivial extension where the bulk densities of the columns remain constant. They are potentially more general than those presented in physical space, however, because they are strain-independent [*Raats and Klute*, 1968a; *Smiles and Rosenthal*, 1968]. The next example illustrates this versatility.

3.2. Dispersion and Reaction During Desorption From Clay

In the process of slip casting, a porous mould absorbs water from wet clay to produce a cast that is subsequently air dried and kiln-fired to make a ceramic object. Preliminary experiments using a plaster-of-Paris mould were consistent with theory of flow between different porous media but the structure of the sodic clay used as the slip was changed by penetration of Ca into the clay from the mould. We then conducted constant pressure filtration experiments that sectioned the clay "cast" and showed that Ca moved from a gypsum 'membrane' at the outflow surface into the clay against the flow of the water. This is different from conventional breakthrough and absorption experiments where both the water and the solute tend to go in the same direction. Experimental detail for the sequence is provided in Smiles [2000b] and the papers cited there.

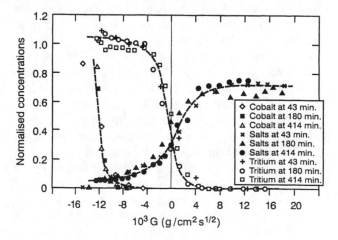

Figure 8. Profiles of soluble salts, tritium and Co-60 shown in Figure 6, regraphed in terms of G shown in Figure 7.

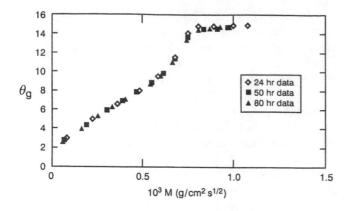

Figure 9. Water content, (θ_g) vs M (= $mt^{-1/2}$) profiles measured during desorption of wet clay under constant pressure. Experiments were terminated at the times shown in the figure.

Figures 9-12 are analogues of Figures 5-8 for this system. Figure 9 shows $\theta_g(M)$ for experiments terminated after approximately 24, 50 and 80 hours. These profiles are self-similar in $M = mt^{-1/2}$ but note that the system is water saturated (ie $\theta_g = \rho e$) so the (one-dimensional) volume change is exactly equal to the sorptivity, S. Changes in structure in the clay consequent on the change in solution type and concentration also preserve similarity in terms of M. *Smiles et al.*,[1996] argued that this observation is consistent with the requirement that the potential and conductivity characteristics of the material follow unique paths across a surface that takes into account the changing physical-chemistry of the clay. It also requires that chemical equilibration and consequent structural response are rapid compared with the rate of movement of the water.

At the same time the change in structure was not as great as we would have expected had the clay been laid down, *de novo*, in equilibrium with the cation suite shown in Figure 10. Explicitly, ϑ at $M = 0$ was more than twice as great as would then have been the case.

Figure 10 shows profiles of total (water soluble plus exchangeable) Ca and Na concentrations in M-space with similarity again observed in the Ca and (displaced) Na profiles. Cation exchange capacity (CEC) is also shown.

Because the cation exchange capacity in this system is great, additional cation in solution exceeds it by only about 10%. At the same time, because the total amount of cation is presented, part of the decrease in Na is the result of the water content decrease in the system in $M < 8 \times 10^{-4}$ g/cm^2 s$^{1/2}$. This also affects the total amount of Ca in M-space.

Figure 11 shows the relation between M and the water based G-coordinate defined by equation (9). Because S is now negative, the origin of the coordinate lies in the 'free' solution expressed from the clay column and at a material distance equal to S from it.

Figure 12 shows the data of Figure 10 graphed in G-space. The concentrations are expressed here per unit mass of water to facilitate material balance calculations.

The cation exchange capacity reflects the distribution of clay in the space-like water based G-coordinate. Consider, for example, the total concentration of the Ca at $G = -S$ (recall that S is negative here) where $M = 0$. The vertical dashed line in Figure 12 indicates this plane. The Ca concentration here greatly exceeds the concentration of CaSO$_4$ at $G < S$ (27 mmols/L) shown to the left of this line. The Ca concentration in the solution just inside the column must approximate that just outside, however, so that the 'excess' (≈ 0.25-0.027 mmol/g$_{water}$) is adsorbed on the clay. The approximately constant value of Na in G-space close to $G = S$ arises because the accumulation of clay 'compensates' for the decreasing exchangeable Na associated with it at $M = 0$. *Smiles et al.* [1996] and *Smiles* [2000b] present more detail. Further analysis of this type of data awaits a good way to discriminate between the water soluble and adsorbed phases in "stiff" clay. At the same time the methods offer an approach to dispersion, chemical reaction and structural change in clay liners which is of great importance in retention ponds used to contain noxious industrial and mining wastes. The approach also provides a way to explore the significance of surface diffusion relative to diffusion in the liquid phase in colloidal systems and clay soils and of cation exchange on the physical properties of clay systems.

4. CONCLUDING REMARKS

The paper is both eclectic and parochial and focuses on experiments and analysis based in insight and influence of John Philip. For this reason, perhaps insufficient attention is paid to thermodynamic approaches to the hydrostatics of swelling soils exemplified by *Groenevelt and Bolt* [1972]

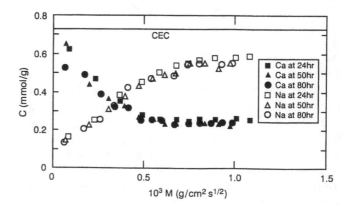

Figure 10. Profiles, in M-space, of total concentrations (mmols/g$_{clay}$) of calcium and sodium corresponding to the water content profiles of Figure 9

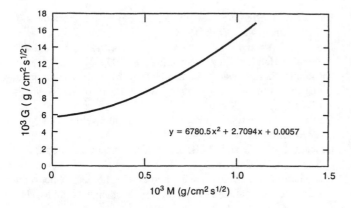

Figure 11. Relation between the solid based M-coordinate of Figure 9 and G, a water based space-like coordinate calculated using equation (7) with Figure 9 data.

and *Sposito* [1975], and its extension to flow, which is reviewed to some extent by *Giraldez and Sposito*, [1985]. It also passes over consequences of cracking in relation to hydrodynamic dispersion and flow.

The paper exemplifies John Philip's strong view that the ultimate aim of environmental science is field understanding and management. In view of its scale and complexity, however, the natural environment is generally too complicated to permit application of deterministic theory and modeling may be the only way to provide insights for sound planning. At the same time, sensible and physically based simplifications of theory provide practical and economical bases from which to design and to check the more empirically based approaches that are often our only recourse for prediction at the scale and complexity of the environment [*Philip*, 1980]. The theory should not be needlessly complicated. Soundly based physical simplifications often provide powerful and economical methods to explore likely outcomes of more expensive field engineering. The approach does not seek to exclude design based or more formal modeling of systems in the large, but it complements that modeling.

Water flow in swelling systems provides examples that identify unnecessarily complicated analyses and the care required to ensure that artifacts arising from a failure to recognize basic premises of theory do not corrupt experiments. Swelling systems also illustrate practical and economical benefits of physically based simplifications that derive from Philip's approach to infiltration. Thus, the approach exemplified in Figures 1-4 has been used in this laboratory for industrial and mining materials as different as piggery effluents, phosphate slimes and peanut butter.

Analysis of hydrodynamic dispersion during unsteady, unsaturated flow, precipitated by Philip, illustrates the way a novel but rational approach provides information to assist

in field design. Thus the nuclide experiments provide reliable material characterization to help rank field sites for repository selection at modest cost when only structurally disturbed samples are available. Study of dispersion and reaction during desorption from clays provides entirely new insights into processes that occur during slip casting as well as in clay barriers used to line retention ponds. These are critical guides to field behavior and to subsequent and more costly fieldwork. The study also offers new methods to explore basic issues of transfer in soils and similar materials.

NOTATION

α	weighting factor for the overburden potential
γ	wet specific gravity of soil system
θ_g	water mass fraction
θ_s	solid volume fraction
θ_w	water volume fraction
ϑ	volume of water per unit volume of solid
ρ	solid specific gravity
ψ	matric potential
A	coefficient in the Philip infiltration equation
$D_m(\vartheta)$	soil water diffusivity defined in material space
g	water based space-like coordinate
G	water based space-like coordinate divided by $t^{1/2}$
i	cumulative infiltration
$k(\theta_w)$	hydraulic conductivity in space
$k_m(\vartheta)$	hydraulic conductivity in material space
m	solid based space-like coordinate
M	$m/t^{1/2}$
p_w	manometric pressure
S	sorptivity
z	vertical coordinate positive upwards
t	time

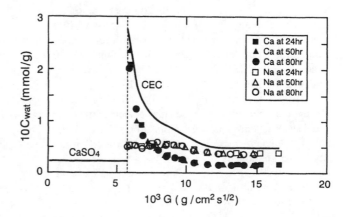

Figure 12. Profiles of total concentrations (mmols/g_{water}) of calcium and sodium regraphed in terms of the G-space illustrated in Figure 11.

REFERENCES

Angulo-Jaramillo, R., M. Vauclin, R. Haverkamp, and P. Gérard-Marchant, Dual gamma-ray scanner and instantaneous profile method for swelling unsaturated materials, in *"Characterization and measurement of the hydraulic properties of unsaturated porous media"* edited by M. Th. van Genuchten, F. J. Leij, and L. Wu, pp. 459-466, University of California, Riverside, 1997 .

Atsumi, K., T. Akiyama, and S. Miyagawa, Expression of thick slurry of titanium dioxide in water, *J. Chem. Engng Japan* 6 236-240, 1973.

Bear, J., M. Y. Corapcioglu, and J. Balakrishna, Modeling of centrifugal filtration in unsaturated deformable porous media, *Adv. Water Resources* 7, 150-167, 1984.

Bolt, G. H. Transport and accumulation of soluble soil components, in 'Soil Chemistry. A. Basic Elements'. Edited by G. H. Bolt and M. G. M. Bruggenwert, pp. 126-140. Elsevier, Amsterdam, 1976.

Bond, W. J., and I. R. Phillips, Ion transport during unsteady water flow in an unsaturated clay soil, *Soil Sci. Soc. Amer. J.,* 54, 636-645, 1990a.

Bond, W. J., and I. R. Phillips, Cation exchange isotherms obtained with batch and miscible-displacement techniques, *Soil Sci. Soc. Amer. J.,* 54, 722-728, 1990b.

Bond, W. J., and I. R. Phillips, Approximate solutions for cation transport during unsteady, unsaturated soil water flow, *Water Resour. Res.* 26, 2195-2205, 1990c.

Bridge, B. J., N. Collis-George, and R. Lal, The effect of wall lubricants and column confinement on the infiltration behaviour of a swelling soil in the laboratory, *Aust. J. Soil Res.,* 8, 259-272, 1970.

Bronowski, J., *The common sense of science.* Harvard UP. Cambridge MA, 1979.

Bronswijk, H., *Magnitude, modeling and significance of swelling and shrinkage processes in clay soils.* Doctoral Thesis, Wageningen Agricultural University Wageningen, The Netherlands 1991.

Bruce R. R., and A. Klute, The measurement of soil moisture diffusivity, *Soil Sci. Soc. Amer. Proc.,* 20, 458-462, 1956.

Buckingham, E., *Studies on the movement of soil moisture.* USDA Bureau of Soils Bulletin No 38, Washington, 1907.

Carslaw, H. S. and J. C. Jaeger, *Conduction of heat in solids*, 2nd Edn, Oxford University Press, London, 1959.

Collis-George, N., and B. J. Bridge, The effect of height of sample and confinement on the moisture characteristic of an aggregated swelling clay soil, *Aust. J. Soil Res.,* 11, 107-120, 1973.

Crank, J. *The mathematics of diffusion*, 2nd Edn, Oxford University Press, London, 1975.

Croney, D., and J. D. Coleman, Pore pressure and suction in soil, in: *"Pore pressure and suction in soil"*, pp. 31-37, Butterworths, London, 1961.

Croney, D., Coleman, J. D., and Black, W. P. M. Studies of the movement and distribution of water in soil in relation to highway design and performance. DSIR Road Research Laboratory Report RN/3209/DC.JDC.WPMB April, 1958

Dallavalle, J. *Micromeritics: the technology of fine particles*, 2nd Edn, Pitman, New York, 1948.

Freeze, R. A., and J. A. Cherry 'Groundwater', pp. 402-405, Prentice-Hall Inc: Englewood Cliffs NJ, 1979.

Gibson, R. E., England, G. L., and Hussey, M. J. L., The theory of one-dimensional consolidation of saturated clays, *Géotechnique* 17, 261-273, 1967.

Giraldez, J. V., and G. Sposito, Infiltration in swelling soil. *Water Resour. Res.,* 21, 33-44, 1985.

Groenevelt, P. H., and G. H. Bolt, Water retention in soil, *Soil Sci.,* 113, 238-245, 1972.

Hashimoto, I., K. B. Deshpande, and H. C. Thomas, Péclet numbers and retardation factors for ion exchange columns, *Ind. Engng Chem. Fund.,* 3, 213-218, 1964.

Kim, D. J., H. Vereecken, and J. Feyen, Comparison of multidisciplinary approaches and unification of concepts on the movement of water and soil in deformable porous media, *Soil Sci.,* 156, 141-148, 1992.

Kirby, J. M., and D. E. Smiles, Hydraulic conductivity of aqueous bentonite pastes, *Aust. J. Soil Res.,* 26, 561-574, 1989.

Kirby, J. M., and D. E. Smiles, Comment on "Dual-energy synchrotron X ray measurements of rapid soil density and water content changes in swelling soils during infiltration" by Patricia Garnier et al., *Water Resour. Res.,* 35, 3585-3587, 1999.

Landman K. A. and W. B. Russel, Filtration at large pressures for strongly flocculated suspensions, *Phys. Fluids*, A5, 550-560.

Lai, S-H., and J. L. Jurinak, Cation adsorption in one-dimensional flow through soils: A numerical solution, *Water Resour. Res.,* 8, 99-107, 1972.

Leu, W., Principles of compressible cake filtration, in 'Encyclopedia of Fluid Mechanics', edited by N. P. Cheremisinoff, pp. 865-904, Gulf Publishing Company: Houston, Texas, 1986

McNabb, A., A mathematical treatment of one dimensional soil consolidation, *Q. Appl. Math.* XVIII, 337-347, 1960.

Miller, E. E., Physics of swelling and cracking soils, *J. Colloid and Interface Sci.,* 52, 434-443, 1975.

Nakano, M., Y. Amemiya, and K. Fuji, Saturated and unsaturated hydraulic conductivity of swelling clays, *Soil Sci.,* 141, 1-6, 1986.

Parlange, J-Y., Theory of water movement in soils: I. One dimensional absorption, *Soil Sci.,* 111, 134-137, 1972.

Parlange, J-Y., Determination of soil water diffusivity by sorptivity measurements, *Soil Sci. Soc. Amer. J.,* 39, 1011-1012, 1975.

Perroux, K. M. and S. J. Zegelin, Constant flux infiltration in swelling soil, in *"The properties and utilization of cracking clay soils"*, edited by J. W. McGarity, E. H. Hoult, and H. B. So, *Reviews in Rural Science 5*, pp. 150-154, University of New England: Armidale 1984.

Philip, J. R., The role of mathematics in soil physics, *J. Aust. Inst. Agric. Sci.,* 23, 293-301, 1957a.

Philip, J. R., Theory of infiltration: 4. Sorptivity and algebraic infiltration equations, *Soil Sci.,* 84, 257-264, 1957b.

Philip, J. R., Kinetics of sorption and volume change in clay-colloid pastes, *Aus. J. Soil Res.,* 6, 249-267, 1968.

Philip, J. R., Theory of infiltration, *Adv. Hydrosci.,* 5, 215-296, 1969.

Philip, J. R., Flow in porous media, *Ann. Review Fluid Mech.*, 2, 177-204, 1970.

Philip, J. R., On solving the unsaturated flow equation: 1. The flux-concentration relation, *Soil Sci.*, 116, 328-9335, 1972.

Philip, J. R., Fifty years progress in soil physics, *Geoderma* 12, 265-280, 1974.

Philip, J. R., Field heterogeneity: some basic issues, *Water Resour. Res.*, 16, 443-448, 1980.

Philip, J. R., The infiltration joining problem, *Water Resour. Res.*, 23, 2239-2245, 1987.

Philip, J. R., Soils, natural science and models, *Soil Sci.*, 151, 91-98, 1991.

Philip, J. R., Phenomenological approach to flow and volume change in soils and other media, *Appl. Mech. Reviews* 48, 650-658, 1995.

Philip, J. R., and D. E. Smiles, Kinetics of sorption and volume change in three-component systems, *Aust. J. Soil Res.*, 7, 1-19, 1969.

Philip, J. R., and D. E. Smiles, Macroscopic analysis of the behaviour of colloidal suspensions, *Adv. Coll. Interface Sci.*, 17, 83-103, 1982.

Raats, P. A. C., Axial fluid flow in swelling and shrinking porous rods. Abstracts 40th Annual Meeting of the Society of Rheology, St Paul, Minnesota, USA, 1969.

Raats, P. A. C., Distribution of salts in the root zone, *J. Hydrol.*, 27, 237-248, 1975.

Raats, P. A. C., Mechanics of cracking soils, in *Symposium on water and solute movement in heavy clay soils*, edited by J. Bouma and P. A. C. Raats, ILRI, Wageningen, The Netherlands pp23-38, 1984.

Raats, P. A. C., Kinematics of subsidence of soils with a non-conservative solid phase. Proceedings on CD-ROM of a symposium on *"New concepts and theories in soil physics"* at the 16th World Congress of Soil Science 20-26 August 1998, Montpellier, France, 1997.

Raats, P. A. C., Multidimensional flow of water in saturated and unsaturated soils, in *"Heat and mass transfer in the environment"*, edited by P. A. C. Raats, D. E. Smiles and A. W. Warrick, pp 000-000, AGU Geophysical Monograph Series, American Geophysical Union, Washington DC 2002.

Raats, P. A. C., and A. Klute, Transport in soils: the balance of mass, *Soil Sci. Soc. Amer. Proc.*, 32, 161-6, 1969a.

Raats, P. A. C., and A. Klute, Transport in soils: the balance of momentum, *Soil Sci. Soc. Amer. Proc.*, 32, 452-6, 1969b.

Richards, L. A., Capillary conduction of liquids through porous mediums, *Physics 1*, 318-333, 1931.

Shirato, M., T. Murase, M. Iwata, and T. Kurita, Principles of expression and design of membrane compression-type filter press operation, in *'Encyclopedia of Fluid Mechanics'*, edited by N. P. Cheremisinoff, pp. 905-964, Gulf Publishing Company: Houston, Texas, 1986

Smiles, D. E. Examination of settlement data for an embankment on a wet light clay, *Aust. Road Res.*, 5, 55-59, 1973

Smiles, D. E. Infiltration into a swelling material, *Soil Sci.*, 117, 140-147, 1974.

Smiles, D. E. Sedimentation and filtration equilibria, *Separ Sci.*, 11, 1-16, 1976.

Smiles, D. E. Principles of constant-pressure filtration, in *'Encyclopedia of Fluid Mechanics'*, edited by N. P. Cheremisinoff, pp. 791-824, Gulf Publishing Company: Houston, Texas, 1986

Smiles, D. E., Water balance in swelling materials: some comments, *Aust. J. Soil Res.*, 35, 1143-1152, 1997.

Smiles, D. E. Centrifugal filtration in particulate systems, *Chem. Engng Sci.*, 54, 215-224, 1999.

Smiles, D. E., Hydrology of swelling soils: a review, *Aust. J. Soil Res.*, 38, 501-521, 2000a.

Smiles, D. E., Material coordinates and solute movement in consolidating clay, *Chem. Engng Sci.*, 55, 773-781, 2000b.

Smiles, D. E., Hydrology of swelling soils: a review, *Aust. J. Soil Res.*, 38, 502-521, 2000a [corrigendum: 40, 1467, 2001].

Smiles, D. E., C. J. Barnes, and W. R. Gardner, Water relations of saturated bentonite: some effects of temperature and salt concentration, *Soil Sci. Soc. Amer. J.*, 49, 66-69, 1985.

Smiles, D. E., and P. M. Colombera, The early stages of infiltration into a swelling soil, in *'Heat and Mass Transfer in the Biosphere: 1. Transfer Processes in the Plant Environment'*, edited by D. A. de Vries and N. H. Afgan, pp. 77-85, Scripta Book Company: Washington DC, 1975.

Smiles, D. E., and A. G. Harvey, Measurement of moisture diffusivity of wet swelling systems, *Soil Sci.*, 116, 391-399, 1974.

Smiles, D. E., J. M. Kirby, and I. P. Little, Hydrodynamic dispersion and chemical reaction in sodium bentonite during filtration in the presence of calcium sulphate, *Chem. Engng Sci.*, 51, 3647-3655, 1996.

Smiles, D. E., K. M. Perroux, S. J. Zegelin,, and P. A. C. Raats, Hydrodynamic dispersion during constant rate absorption of water by soil, *Soil Sci. Soc. Amer. J.*, 45, 453-458, 1981.

Smiles, D. E., J. R. Philip, J. H. Knight, and D. E. Elrick, Hydrodynamic dispersion during absorption of water by soil, *Soil Sci. Soc. Amer. J.*, 42, 229-234, 1978.

Smiles, D. E., and M. J. Rosenthal, The movement of water in swelling materials, *Aust. J. Soil Res.*, 6, 237-248, 1968.

Sposito, G., On the differential equation for the equilibrium moisture profile in swelling soil, *Soil Sci. Soc. Amer. J.*, 39, 1053-1056, 1975.

SSSA *Glossary of Soil Science Terms*, Soil Science Society of America, Inc: Madison, WI, 1997.

Stoffregen, H., C. Wessolek, M. Renger, and R. Plagge, Effects of temperature on hydraulic conductivity, in *"Characterization and measurement of the hydraulic properties of unsaturated porous media"* edited by M. Th. van Genuchten, F. J. Leij, and L. Wu, pp. 497-506, University of California, Riverside, 1997.

Terzaghi, K. Die Berechnung der Durchlassigkeitsziffer des Tones aus dem Verlauf der Hydrodynamischen Spannungserscheinungen, *Akademie der Wissenschaften in Wein, Sitzungberichte, Mathematisch-Naturewissenschaftliche Klasse*, Part IIa, *132*, 3-4, 125-138, 1923.

Terzaghi, K. *Theoretical soil mechanics*, John Wiley, New York, 1943.

Tiller, F. M., and N. B. Hsyung, Unifying the theory of thickening, filtration and centrifugation, *Wat. Sci. Technol.*, 28, 1-9, 1993.

Wilson, J. L., and L. W. Gelhar, Analysis of longitudinal dispersion in unsaturated flow, 1, The analytical method, *Water Resour. Res.*, 17, 122-130, 1981.

Zaslavsky, D. Saturated and unsaturated flow equation in an unstable medium, *Soil Sci.,* 98, 317-321, 1964.

Dr D. E. Smiles, CSIRO Land and Water, PO Box 1666, Canberra ACT 2601, Australia.

Equilibrium Moisture Profiles in Consolidating, Sulfidic, Coastal Clay Soils

Ian White

Centre for Resource and Environmental Studies, Australian National University, Canberra, Australia

Large areas of soft sulfidic coastal clay soils are being developed around the world. These soils can be up to 40m deep, have shallow watertables and volumetric water contents approaching 80%. They pose significant problems and frequently require costly consolidation. Swelling soil theory, developed for industrial slurries is directly applicable to these soils. Here, simple analytic expressions for equilibrium moisture contents in consolidating soils are derived using an analytic moisture characteristic that is consistent with both diffuse double layer theory and measurements on coastal clay soils. Consolidation from applied surface loads, falling watertables and increases in the soil solution electrolyte concentration are considered. The total buoyant specific solid volume per unit surface area is identified as an important scaling parameter. Effective consolidation requires that applied surface loads exceed this parameter. Predictions of chemical consolidation are based on the untested assumption that relative results for clay slurries are directly applicable to these soils. The predictions show that modest increases in electrolyte concentrations can produce significant consolidation, in some cases comparable with that produced by an applied surface load of 40 tonnes/m^2. Increasing the charge of the electrolytes has a dramatic impact on consolidation. These predictions remain to be tested.

1. INTRODUCTION

John Philip was unhappy with the macroscopic, phenomenological theory of water in swelling soils and had planned to revisit the topic in his "retirement". His early foray into swelling soils was marked by uncharacteristic flaws in his otherwise impeccable physical intuition. Like so many of his works, this foray was catalysed by intriguing experimental results, in this case the perceptive observations and analysis of *Smiles and Rosenthal* [1968] on clay slurries. *Philip's* [1968] first paper on swelling materials,

Environmental Mechanics: Water, Mass and Energy Transfer in the Biosphere
Geophysical Monograph 129
Copyright 2002 by the American Geophysical Union
10.1029/129GM12

however, eschewed their use of Lagrangean, material coordinate systems. Instead, he used the needlessly confusing and mathematically more complex Eulerian physical space coordinates. It took *Raats and Klute* [1968] to finally convince him to abandon this cumbersome coordinate system [*Smiles*, 1995].

In a subsequent flurry of papers on the hydrostatics and hydrodynamics of swelling soils [*Philip*, 1969a; 1969b; 1969c], Philip assumed a unique relation between void ratio and moisture ratio of swelling soils, independent of overburden pressure. This led him to conclude, at static equilibrium, that air at atmospheric pressure could exist well below the water table. *Youngs and Towner* [1970] pointed out that this is not physically possible and *Philip* [1970a] acknowledged his error. Despite these flaws, John Philip's early papers on swelling materials illustrate many of their unique properties.

Shortcomings in the macroscopic theory of the hydrology of swelling soils have been clearly identified by *Smiles* [2000]. These principally occur when swelling systems contain air at atmospheric pressure. It is generally agreed that the Darcian theory works well for water-saturated, slurries encountered in industrial and mining processes. There is, however, a perception that real soils do not behave like saturated clay materials [*Youngs*, 1995]. This perception appears mistaken. There are about 10^8 ha of soils around the world, partly used for agricultural production and development, for which saturated swelling soil theory appears appropriate [*Kim et al.*, 1992; *Smiles*, 1997; *White et al.*, 2001a]. These soils are deposits of marine- or estuarine-origin, sulfidic, clay soils [commonly called acid sulfate soils] laid down mostly during the last sea level rise, from about 11000 years ago [*Pons*, 1978]. They have watertables at or near the surface, contain up to 80% by volume of water, and exist in deposits up to 40 m thick [*White et al.*, 1997]. Techniques and theory developed for slurries work equally well for these soils [*White et al.*, 2001a].

Major developments on these very soft coastal soils usually require consolidation, which is a critical, time-consuming and expensive procedure. Failures, particularly in roadway embankments, can cause costly problems. In some coastal areas, roadways have settled by as much as 4m over 40 years. Their rate of consolidation is consistent with swelling soil theory [*Smiles*, 1973]. Frequently the ultimate possible consolidation is unknown. In this work, we will explore the prediction of equilibrium moisture profiles in soft, sulfidic, coastal clay soils consolidated one-dimensionally by applied surface loads or by falling water tables or by changes in soil solution electrolytes.

Terzaghi [1923] introduced material coordinates and recognised that consolidation requires water flow through the soil. *Croney* [1952], concerned with road construction in soft soils, developed an iterative scheme for estimating moisture profiles in soft soils under surface loads using the soils' measured moisture characteristics. *Croney and Coleman* [1961] derived the conditions for thermodynamic equilibrium in soils under applied surface loads and self-weight [see also *Groenevelt and Bolt*, 1972; *Sposito*, 1975a; 1975b].

Philip [1970b] treated the hydrostatics of dilute slurries as an equilibrium between sedimentation and Brownian motion. Smiles and coworkers have explored extensively both the statics and dynamics of consolidating clay suspensions and industrial slurries under a range of conditions including centrifugation [see *Smiles*, 2000]. This body of work has demonstrated that the macroscopic, Darcian theory was valid for these 'difficult materials [*Smiles*, 1976a]. Smiles predicted equilibrium profiles during both sedimentation, when water accumulated above the slurry, and filtration, with drained from beneath the slurry, for red mud slurries generateg in the processing of bauxite. *Talsma*

[1974; 1977a; 1977b] and *Talsma and van der Lelij* [1976] measured moisture profiles and the impact of overburden loads on swelling clays, in both the laboratory and the field. Talsma concluded that moisture and solid profiles of unsaturated clay soils in the field may be consistent with equilibrium theory, but the contribution of overburden to the total soil water potential was not very large. *Stroosnijder* [1976] also measured the impact of overburden on shrinkage and *Kim et al.* [1992] considered the contribution of overburden to water movement in marine-origin clay soil.

'Quick clays' exhibit dramatic changes in their volume and rheological properties on the addition of electrolytes to the clays [*Rosenqvist*, 1952; 1966; *Mitchell*, 1986]. *Smiles et al.* [1985] and *Smiles* [1995] examined the impacts of additions of NaCl and saturated $CaSO_4$ solutions on the moisture characteristic of bentonite slurries. The large systematic changes they measured were not quantitatively predictable by double layer theory. Their work suggests that increasing the soil solution electrolyte concentration may be an effective way of aiding consolidation.

This work follows that of *Smiles* [1976b] but considers the impact of applied surface loads. An analytic form of the moisture characteristic will be used to derive simple expressions for consolidating soil profiles under a range of loads. The contribution of soil solution electrolyte concentration to equilibrium moisture content profiles and consolidation of sulfidic clay soils is also investigated.

2. EQUILIBRIUM MOISTURE PROFILES IN SWELING SOILS

2.1 Matric Potential Profiles

Water potentials here are expressed in work per unit weight of water so that the unit of potential will be metres of water. The total potential, Φ, (hydraulic head) for water in soil is the sum of the gravitational potential, z, (the elevation relative to an arbitrary datum, here defined positive upwards) and the 'manometric pressure' of water in the soil, p_w (the pressure a tensiometer measures).

$$\Phi = z + p_w \qquad (1)$$

This relationship holds for both swelling and non-swelling soils and the watertable is defined as the surface where $p_w = 0$ [*Croney*, 1952; *Philip*, 1969a; b]. In non-swelling soils $p_w = \psi$, with ψ the matric potential. In swelling soils the weight of the overlying soil, together with any applied surface load, P_T, is carried partly by the solid structure and partly by the liquid phase of the soil. Both loads together make up the mechanical load, $P(z)$ [or the total normal stress, σ] at position z [here the arbitrary datum will be taken as the base of the swelling soil deposit]. The component of potential in the water phase due to $P(z)$, the overburden potential, Ω, must be included in the total potential [*Croney*, 1952; *Croney and Coleman*, 1961; *Philip*, 1969a; b; *Youngs and Towner*, 1970; *Groenevelt and Bolt*, 1972; *Sposito*, 1975a].

$$\Phi = z + \psi + \Omega = z + p_w \qquad (2)$$

In swelling soils, ψ in (2) is the unloaded matric potential [*Smiles* 2000]. In engineering terms, $\psi = -\sigma'$, with σ' the effective interparticle stress. The manometric pressure of water relative to atmospheric pressure in swelling systems is $p_w = \psi + \Omega$ and the position of the watertable is the surface where $p_w = 0$ or $\psi = -\Omega$.

We will be concerned here with saturated swelling systems and therefore avoid complications with the partitioning of the overburden between the water and solid phases that occur in three-phase systems [*Smiles*, 2000]. For saturated, swelling soils in the normal shrinkage region at or beneath the watertable, $\Omega = P(z)$, and, at equilibrium,

$$\begin{aligned}\Phi &= z + \psi(\vartheta) + P(z) \\ &= z + \psi(\vartheta) + \left[P_T + \int_z^{Z_T} \gamma(\vartheta, P(z')) dz' \right] = Z_W \end{aligned} \qquad (3)$$

Here ϑ is the moisture ratio [ratio of volume of soil water to volume of solid], Z_T, represents the position of the soil surface and Z_W, a constant, is the position of the watertable. The wet specific gravity of the soil, γ is:

$$\gamma = (\vartheta + \gamma_s)/(\vartheta + 1) = \theta_w + \theta_s \gamma_s \qquad (4)$$

with γ_s the specific gravity of the solid and θ_w, θ_s the volume fractions of the solid and water phases.

If the watertable is shallow, such that the capillary fringe extends to the soil surface and the soil lies within the normal shrinkage range, as often happens in coastal, sulfidic clay soils, then (3) is valid for the whole depth of the profile. *Croney* [1952] suggested that equilibrium moisture profiles $\vartheta(z)$ could be found from (3) by an iterative procedure in which a constant ϑ, consistent with field observation, is chosen for all depths as an initial estimate. The profile of $P(z)$ is then calculated for this initial moisture distribution, and a first estimate of the unloaded matric potential profile is found from (3). The measured moisture characteristic $\psi(\vartheta)$ is then used to determine the next estimate of $\vartheta(z)$ and the process is repeated. Usually one or two iterations are sufficient. If, however, the moisture characteristic

is known, then direct solution is possible [*Philip*, 1969c; *Smiles*, 1976b].

Consolidation of soft soil profiles usually involves the application of surface loads and the removal of excess water at the surface, often via vertical wick drains. As the soil consolidates it is more appropriate and convenient to work in a material coordinate system $m(z, \vartheta)$ [*Smiles*, 1974; Smiles, 1976a; Smiles, 2000] which conserves the solid volume per unit area in the soil profile:

$$m(z, \vartheta) = \int_0^z [1/(1 + \vartheta)] dz' = \int_0^z \theta_s dz' \qquad (5)$$

The physical space coordinate is recovered from $\vartheta(m)$ profiles using:

$$z = \int_0^m (1 + \vartheta) dm' = \int_0^m (1/\theta_s) dm' \qquad (6)$$

In this material coordinate system, the equilibrium equation (3) becomes [*Smiles*, 1974; 1976b]:

$$\psi(m) = -[W + P_T + (\gamma_s - 1)(m_T - m)] \qquad (7)$$

Here m_T is the total volume of solid material per unit area in the soil profile and $W = Z_T - Z_W$ is the depth to the watertable from the consolidating soil surface. Equation (7) can be rewritten as:

$$\begin{aligned}\frac{\psi}{(\gamma_s - 1)m_T} &= -\left[\frac{P_T + W}{(\gamma_s - 1)m_T} + \left(1 - \frac{m}{m_T} \right) \right] \\ \psi^* &= -\left[P_T^* + W^* + 1 - \frac{m}{m_T} \right]\end{aligned} \qquad (8)$$

with $\quad \psi^* = \psi /[(\gamma_s - 1)m_T]$

and $\quad P_T^* + W^* = (P_T + W)/[(\gamma_s - 1)m_T]$.

The total 'buoyant' specific volume of solid per unit area in the deposit, $(\gamma_s - 1)m_T$ in (8), provides a scaling length for both unloaded matric potential and the applied load and watertable depth. It also follows from (7) and (8) that watertable depth and surface load act in concert and are interchangeable. Since during consolidation, watertables are frequently at the soil surface, we shall set $W = 0$ and only consider consolidation brought about by imposed surface loads. It will be recognised, however, that the impact of applied surface loads P_T is identical to that of lowering the

Table 1. Parameters in the moisture characteristic (10) parameters for two saturated coastal clay soils together with the specific gravity of their solid phase [*White et al.*, 2001a].

Soil	A	B	γ_s	Soil Solution
Netherlands; marine-origin	3.24	0.25	2.57	seawater
Eastern Australian estuarine-origin	2.34	0.30	2.55	4% seawater [22 M/m³ Cl⁻]

position of the watertable, W (the specific influence of W is treated in *Smiles* [1976b] and *White et al.*,[2001b]).

2.2 Moisture Content Profiles in Material Space

In order to use the equilibrium matric potential profiles to predict equilibrium moisture content profiles, knowledge of the soil's moisture characteristic, $\vartheta(\psi)$, is required. For dilute clay suspensions in which double layer interactions are negligible, the balance between sedimentation and Brownian motion requires [*Philip*, 1970b]:

$$\vartheta + 1 = h/\psi \qquad (9)$$

where h is a constant at a fixed temperature. It seems unlikely in coastal clay soils, with $\vartheta < 4$, that double-layer interactions can be ignored. Simple double layer theory suggests that the moisture characteristic for swelling clays should be of the form [*Sposito*, 1984; *Smiles et al.*, 1985]:

$$\vartheta = A - B \ln|\psi| \qquad (10)$$

A and B are constants for a given clay matrix, soil solution electrolyte concentration and temperature. The moisture characteristics of clay and industrial slurries [*Smiles and Harvey*, 1973; *Smiles*, 1975; 1976a; 1995] and soft clay soils of estuarine or marine origin [*White et al.*, 2001a] in the normal shrinkage range are much better described by (10) than (9). The actual magnitudes of the dependencies of A and B on soil solution concentration, however, are not given quantitatively by double layer theory [*Smiles et al.*, 1985], and (10) must be regarded as semi-empirical.

The moisture profile in material space follows from (7) and (10).

$$\vartheta = A - B \ln[P_T + (\gamma_s - 1)(m_T - m)] \qquad (11)$$

The form of the moisture characteristic (10) imposes a problem at the surface of unloaded soils with watertable at the surface where (11) has a physically unrealistic singularity.

In order to use (11), the parameters in the moisture characteristic must be known. Table 1 lists the parameters for two soils, one a marine-origin clay soil from the Netherlands [*Kim et al.*, 1992], the other an estuarine soil from eastern Australia [*White et al.*, 2001a], whose moisture characteristics have been fitted to (10) as shown in Figure 1. We shall use these soils for illustrative calculations.

Equation (11) can be rewritten as:

$$(\vartheta - \vartheta_T)/B = \ln\left[P_T^*/\left(P_T^* + 1 - m/m_T\right)\right] \qquad (12)$$

with $\vartheta_T = A - B \ln(P_T)$ the moisture ratio at the soil surface. Moisture profiles, parameterised in this way, depend only on the relative position in the soil profile in material space and on the normalised surface loading P_T^*.

The normalised moisture ratio difference $(\vartheta - \vartheta_T)/B$ is plotted in Figure 2 as a function of the normalised material coordinate, m/m_T, for selected, dimensionless applied surface loads, P_T^* normalised using $(\gamma_s - 1)m_T$ as in (8).

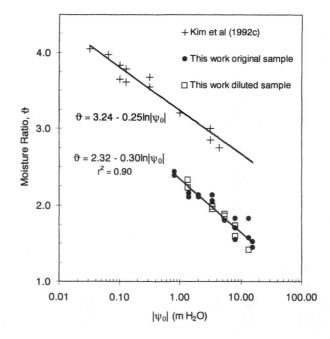

Figure 1. Moisture characteristics for then eastern Australian estuarine sulfidic clay soil and results estimated from *Kim et al.* [1992] for a Netherlands marine-origin clay soil both fitted to equation (10) [*White et al.*, 2001a].

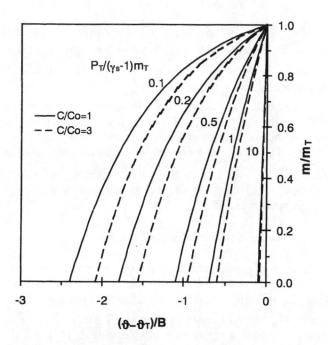

Figure 2. Reduced moisture ratio profiles in reduced material space for a range of normalised surface loads. The dashed curves show the impact of increasing the soil 1:1 electrolyte concentration three-fold.

It is obvious from (12) and Figure 2 that when $P_T^* \approx 10$ the moisture ratio at all depths approaches that at the soil surface and an approximately constant moisture content is reached at all depths. It is also apparent from (12) that, in order to produce a uniform consolidation in an embankment traversing a landscape with variable soft soil depths, the applied surface loading must be proportional to the total "buoyant" specific volume of solid per unit area, $(\gamma_s - 1)m_T$, at any location.

The moisture ratio difference on the left hand side of (12) can also be normalised with regard to the moisture ratio difference between the soil surface and its base, $(\vartheta_0 - \vartheta_T)/B = \ln\left[(P_T^*+1)/P_T^*\right]$. The resultant normalised moisture ratio $(\vartheta - \vartheta_T)/(\vartheta_0 - \vartheta_T)$ depends only on P_T^* and m/m_T :

$$\frac{\vartheta - \vartheta_T}{\vartheta_0 - \vartheta_T} = \frac{\ln\left[\left(P_T^* + 1 - m/m_T\right)/P_T^*\right]}{\ln\left[\left(P_T^* + 1\right)/P_T^*\right]} \quad (13)$$

While (13) is independent of the parameters of the moisture characteristic (10), its functional form is determined by that of (10). Figure 3 shows the plot of profiles of the normalised moisture ratio (13) in relative material space as a function of normalised applied surface loading. It can be

seen at applied surface loads, $P_T^* > 1$, the moisture ratio profile approaches $(\vartheta - \vartheta_T)/(\vartheta_0 - \vartheta_T) = 1 - m/m_T$.

2.3 Consolidation and the Relation between Physical and Material Coordinates

The relation between the material and space coordinates can be found by substituting for ϑ from (11) in (6):

$$z = m\left\{1 + A - B\ln\left[(\gamma_s - 1)m_T\right]\right\}$$
$$+ Bm_T\left[\left(P_T^* + 1 - \frac{m}{m_T}\right)\ln\left(P_T^* + 1 - \frac{m}{m_T}\right)\right] \quad (14)$$
$$- Bm_T\left[\left(P_T^* + 1\right)\ln\left(P_T^* + 1\right) + \frac{m}{m_T}\right]$$

The relation between the position of the surface of the swelling soil (relative to its base), which is the thickness of the swelling soil deposit under any surface load, $Z_T(P_T^*)$, and the total 'buoyant' volume of solid per unit area in the entire soil profile follows from (14) when $m = m_T$:

$$Z_T(P_T^*) = Z_T(0)$$
$$- m_T B\left[\left(P_T^* + 1\right)\ln\left(P_T^* + 1\right) - P_T^* \ln P_T^*\right] \quad (15a)$$

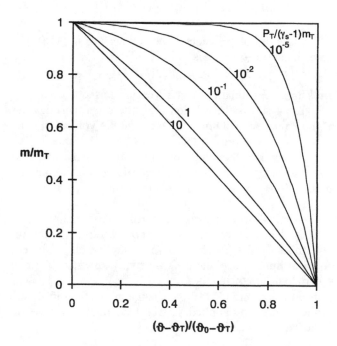

Figure 3. Normalised moisture ratio profiles in reduced material space for a range of normalised surface loads.

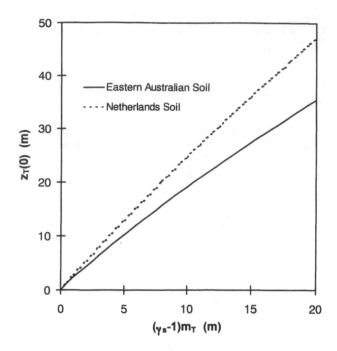

Figure 4. Dependence of the thickness of the two soils in Table 1 on the total 'buoyant' specific volume of solid per unit area.

where $Z_T(0)$ is the position of the soil surface with no imposed surface load:

$$Z_T(0) = m_T\left[1 + A - B\{\ln[(\gamma_s - 1)m_T] - 1\}\right] \qquad (15b)$$

The rate of compression of the soil surface in (15a) with normalised surface load is

$\partial Z_T(P_T^*)/\partial P_T^* = -m_T B \ln(1 + 1/P_T^*)$, and shows a diminishing consolidation with increasing surface loads.

Figure 4 shows the relation (15b) between the thickness of the unloaded swelling soil layer and the volume of solid contained in the profile for the two soils in Table 1. The marine-origin Netherlands soil clearly has thicker deposits than the estuarine soil for the same volume of solid per unit area. This is consistent with values of the A parameter in Table 1

Figure 5 shows an example of the consolidation of an initially 10 m thick deposit by applied surface loads (or watertable depths) for the eastern Australian clay soil. The practical limit of applied loads is usually less than 10m and watertables less than 3 m.

It follows from (15) that the change in position of the soil surface under any applied surface load relative to the unloaded soil surface is:

$$\frac{Z_T(0) - Z_T(P_T^*)}{m_T B} = (P_T^* + 1)\ln(P_T^* + 1) - P_T^* \ln(P_T^*) \qquad (16)$$

The relative consolidation, $Z_T(0) - Z_T(P_T^*)$ normalised by $m_T B$, depends only on the normalised applied surface load. For large surface loads the right hand side of (16) approaches $\ln(P_T^* + 1)$.

Figure 6 shows the normalised, relative consolidation as a function of the normalised surface load. The logarithmic dependence of relative consolidation on applied load is apparent. The most rapid changes in consolidation occur for $P_T^* < 10$. Note that the applicability of the moisture characteristic (10) may be limited to the normal shrinkage range, probably to $\vartheta \geq 1$.

2.4 Moisture Content Profiles in Physical Space

In order to predict equilibrium moisture profiles in physical space, the total volume of solid per unit surface area, m_T, for in a given depth of an unloaded swelling soil deposit is calculated iteratively from (15b). The moisture profile in material space is then calculated from (11) and the corresponding position in physical space is determined from (14) for any applied surface load. Figure 7 shows the predicted moisture profiles in physical space for an initially 10m thick deposit of the eastern Australian soil, for an unloaded profile and for P_T^* of 1 and 10. Here, to remove the

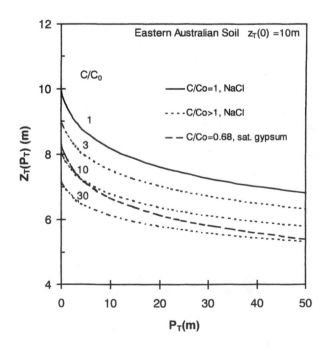

Figure 5. Consolidation of an eastern Australian soil deposit initially 10 m thick as a function of applied surface load. Also shown is the impact of increasing the NaCl concentration and using saturated gypsum.

singularity at the surface for unloaded soil we have taken the watertable depth as 0.1 m.

The significant consolidation of the soil surface and the dewatering of the profile brought about by applied surface loads is apparent in Figure 7. A normalised surface load of 1 in Figure 7. corresponds to a surface load of 4.9 tonnes/m^2.

3. ELECTROLYTE CONSOLIDATION OF SWELLING SOILS

It has been recognised for more than 50 years that the volume rheological properties of "quick" clay soils can be dramatically altered by the addition of electrolytes to the soil water [*Rosenqvist*, 1952; 1966; *Mitchell*, 1986]. Here we will examine, in an approximate sense, the dependence of equilibrium water content profiles in soft swelling soils on soil solution electrolyte concentration, C. This approach is based on the recognition that the parameters A and B in the moisture characteristic (11) are functions of soil solution electrolyte concentration [*Smiles et al.*, 1985; *Smiles*, 1995] so that (10) becomes:

$$\vartheta(C,\psi) = A(C) - B(C)\ln|\psi| \quad (17)$$

Double layer theory for parallel clay platelets [*Sposito*, 1984; *Smiles et al.*, 1985] suggests:

$$A(C) = B(C)\ln\left(aZ_i^2 \sum_i C_i\right) \quad (18)$$

$$B(C) = \frac{bS}{\left(Z_i^2 \sum_i C_i\right)^{1/2}}$$

In (18) a and b are constants for a given soil at a fixed temperature, Z_i and C_i are the valency and concentration of the ith electrolyte species and S is the specific surface area of the soil. The constant a incorporates the surface electric potential on the charged surfaces. For one-to-one electrolytes, *Smiles et al.* [1985] showed that (18) can be rewritten as:

$$\frac{A(C)}{A_0} = \frac{B(C)}{B_0}\frac{\ln(aC)}{\ln(aC_0)} \quad (19)$$

$$\frac{B(C)}{B_0} = \left(\frac{C_0}{C}\right)^{1/2}$$

In (19), A_0 and B_0 are values of A and B at an arbitrary reference 1:1 electrolyte concentration, C_0. From (18), it would appear that using a reference concentration to normalise parameters removes any specific soil dependence from the B parameter term and most of that from the A term.

Smiles et al. [1985] measured the concentration dependence of A and B for dilute bentonite slurries when the electrolyte was predominantly NaCl. They increased the electrolyte concentration 36-fold and found only qualitative agreement with simple diffuse double layer theory. Instead, it can be shown that their data fit remarkably well:

$$\frac{A(C)}{A_0} = \frac{B(C)}{B_0} = 1 - 0.12\ln\left(\frac{C}{C_0}\right) \quad (20)$$

The fact that both the A and B parameters have the same relative dependence suggests that any specific soil dependence has been removed. We will assume here, for illustrative calculations, that the relationship (20) also holds for estuarine-origin soil in Table 1. With this assumption, the relative moisture profile in material space given by (12) becomes:

$$(\vartheta_T - \vartheta)/B_0 = \left[1 - 0.12\left(\frac{C}{C_0}\right)\right]\ln\left[\left(P_T^* + 1 - m/m_T\right)/P_T^*\right] \quad (21)$$

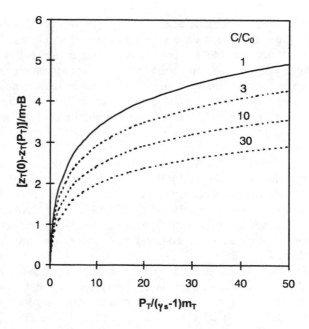

Figure 6. Dependence of reduced relative consolidation on the normalised surface load for four soil solution concentrations of 1:1 electrolytes.

Figure 2 also shows the effect on the relative moisture profile of a three-fold increase the 1:1 electrolyte concentration in the soil solution for a range of applied normalised surface loads. It can be seen that the added electrolyte produces a larger relative change in the moisture ratio of the profile at lower applied pressures. The normalised moisture ratio $(\vartheta - \vartheta_T)/(\vartheta_0 - \vartheta_T)$ in (13) is independent of moisture characteristic parameters and therefore independent of electrolyte concentration.

The effect of electrolytes on the position of the soil surface follows directly by substitution for A and B from (20) into (15). Figure 5 also shows the impact of increasing the one-to-one electrolyte concentration on the position of the soil surface for the eastern Australian soil for up to a 30-fold increase in soil solution concentration. It can be seen that a 30-fold increase in soil solution concentration (to about seawater) produces approximately the same consolidation as a very large applied surface load of 40 m (40 tonnes/m^2).

The change in depth of the unloaded soil profile at electrolyte concentration C, $z_T(C,0)$, relative to that at concentration C_0, $z_T(C_0,0)$ follows from (15b) and (20):

$$\frac{z_T(C_0,0) - z_T(C,0)}{m_T B_0}$$

$$= 0.12 \ln\left(\frac{C}{C_0}\right)\left(1 + \frac{A_0}{B_0} - \ln[(\gamma_s - 1)m_T]\right) \quad (22)$$

The estimated impact of increasing soil electrolyte concentration on the relative change in position of the unloaded soil surface for the eastern Australian estuarine soil is shown in Figure 8. Here the reference concentration is taken as the ambient solution concentration in Table 1 and the reference values of A and B are therefore those in Table 1. It can be seen in Figure 8 that substantial consolidation can be achieved by even relatively modest increases in soil electrolyte concentration.

The change in soil water content profiles with both imposed surface loading and a three-fold increase in soil one-to-one electrolyte concentration are also shown in Figure 7 for a 10m thick deposit of the eastern Australian soil. Again the effectiveness of a relatively modest increase in soil solution concentration on consolidation and dewatering is evident.

Thus far we have considered the impact of added 1:1 electrolytes. Diffuse double layer theory suggests that ions with higher charge should have a much greater impact on consolidation than 1:1 electrolytes [see (18)]. *Smiles* [1995] compared the moisture characteristic for bentonite slurries in saturated gypsum solutions with those in NaCl solutions.

The comparison shows the 2:2 electrolyte has a dramatic impact on the moisture characteristic. The results can be summarised as:

$$\frac{A(C,Z)}{A_0} = \frac{B(C,Z)}{B_0} = 1 - 0.12Z \ln\left(\frac{Z^2 C}{C_0}\right) \quad (23)$$

Here, Z is the valency of the 2:2 electrolyte, C_0 is the reference concentration of the 1:1 electrolyte and A_0 and B_0 are the parameter values at that concentration. The dependence shown in (23) is even greater than that expected from (20).

If it is assumed that this dependence is also applicable to other soils then we can use (23) to predict the impact of substituting the soil water solution with a 2:2 electrolyte, such as CaSO$_4$. Figure 5 shows the consolidation that is predicted to occur with a saturated gypsum solution. With 4% seawater initially present in the pore water under field conditions, the value of C/C_0 on substitution with saturated gypsum is only 0.68. None-the-less, Figure 5 suggest that the gypsum solution can bring about consolidations equivalent to 1:1 electrolytes with $C/C_0 \approx 10$ at low applied surface loads and $C/C_0 \approx 30$ at high surface loads.

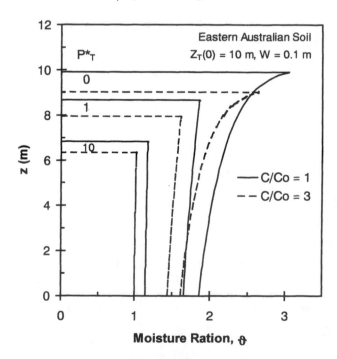

Figure 7. Consolidation and dewatering of a 10 m deep soil profile of the eastern Australian soil with watertable at a depth of 0.1 m. The impact of three applied surface loads as well as three-fold increase in the soil solution 1:1 electrolyte concentration from its ambient concentration, C_0.

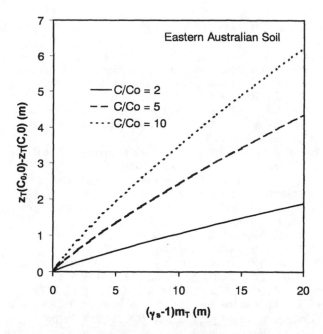

Figure 8. Change in the height of the eastern Australian soil relative to the height for the unloaded soil with the ambient solution concentration in Table 1 at three increased soil solution concentrations for a 1:1 electrolyte.

4. DISCUSSION AND CONCLUSIONS

There are large regions of soft, sulfidic coastal soil with very shallow watertables to which the macroscopic theory of water equilibrium and movement, developed for two-phase, liquid-solid swelling systems, is applicable. Infrastructure development on these soils normally requires consolidation. The application of equilibrium swelling soil theory to them helps identify critical processes in their consolidation and can be used to predict their ultimate consolidation. Here, simple analytical expressions have been derived for equilibrium moisture profiles in profiles subject to surface loads, based on a form for the moisture characteristic for swelling clay systems, $\vartheta = A - B \ln|\psi|$, which is consistent with simple, diffuse double layer theory [*Smiles et al.,* 1985]. It is expected that the validity of this form may be restricted to the normal shrinkage range and experimental examination of its applicability to other soils and other moisture ratios is warranted.

Here the total 'buoyant' specific volume of solid per unit area in the deposit, $(\gamma_s - 1) m_T$, is identified as an appropriate scaling length for both unloaded matric potential, the applied surface load and watertable depth. Simple forms for normalised equilibrium moisture profiles in material space were produced that are independent of the particular soil parameters of the moisture characteristic. These again dem-

onstrate the great conceptual simplicity of the use of material coordinates. Both depth to watertable and surface loads are additive and interchangeable in the analysis. For normalised applied surface loads $P_T^* > 1$, these profiles approach a simple linear form in normalised material space. When the normalised applied surface loads, $P_T^* > 10$, moisture contents are almost constant with depth in the soil profile.

In order to achieve uniform consolidation under structures, such as road embankments, traversing swelling soil deposits whose depths vary across the landscape, as occur in old backswamp areas in estuarine areas in eastern Australia, the applied surface loads should be proportional to $(\gamma_s - 1) m_T$. The analysis shows that there is diminishing consolidation with increased surface loads. For the soils considered here, the appropriate range of normalised surface load is in the range 1 to 10. This appears to impose practical limits on the depth of saturated swelling soil deposits over which structures can be built using only surface applied loads. With the soils considered here that depth appears to be about 15 m.

The impact of soil solution electrolyte concentration on equilibrium moisture profiles and consolidation was also investigated. There is little information on which to base an analysis. Speculative predictions here have been based on the dependence of the measured moisture characteristics of bentonite slurries on solution concentration and composition [*Smiles et al.,* 1985; *Smiles,* 1995]. We have suggested here that, from double layer theory and because $A(C)/A_0 = B(C)/B_0$, we may have removed the soil dependence completely by using a reference concentration. Unfortunately, the concentration dependence of $A(C)/A_0$ and $B(C)/B_0$ are not that expected from simple, diffuse double layer theory. This requires additional research.

Predictions of equilibrium profiles, based on the assumption that $A(C)/A_0$ and $B(C)/B_0$ for the coastal soils have the same dependencies as bentonite slurries, show significant dewatering and consolidation produced by relatively modest increases in the soil solution concentration of 1:1 electrolytes. The consolidation, for example, produced by a thirty-fold increase in solution concentration of 1:1 electrolyte in the eastern Australian soil (to approximately seawater) is equivalent to a massive surface load of about 40 m. For 2:2 electrolytes (for which there is less data) dramatic consolidations are predicted at small concentrations.

These predictions suggests that methods to alter the soil solution concentration could greatly assist consolidation in coastal, sulfidic soils, provided methods of injecting electrolytes are developed. When the sulfides in these coastal sulfidic soils oxidise they produce high concentrations of dissolved aluminium and iron in acidic pore water solutions [*Kittrick et al.,* 1986]. The 'ripening' of these soils by the presence of these ions and their exchange with the clay complex results in considerable consolidation. One method of increasing consolidation could be to be promote oxidation at depth in the profile through vertical wick drains. However, the impact of acid exported from oxidised profiles can be devastating for estuarine ecosystems [*Sammut et al.,* 1996] and extreme care would be needed if this procedure was adopted.

This work has dealt with the ultimate consolidation and moisture profiles at equilibrium in coastal clay soils. The rate at which that equilibrium is approached is also extremely important. Practical experience suggests that the settling of embankments on coastal soils occurs over decades [*Smiles and Poulos,* 1973; White *et al.,* 2001a]. The substantial increases observed in the rate of dewatering of clays caused by adding electrolytes [*Smiles et al.,* 1985; Smiles, 1995] suggests that consolidation can be speeded up significantly. This remains to be tested both in the laboratory and the field.

Predictions in this work on the impact of soil solution electrolytes on consolidation at equilibrium are admittedly highly speculative. They depend on the transferability of results for bentonite slurries to coastal clay soils. Experimental testing of both the assumptions and the predictions is required.

John Philip hoped that a macroscopic theory of swelling soils could be constructed from a rigorous, three-dimensional theory of colloids [*Philip,* 1970d]. The work presented here would not have pleased him. It is speculative and empirically-based. Simple representations of colloid theory do not appear to describe the behaviour of even bentonite slurries. There is still much to be done in this area.

Acknowledgments. Dr D.E. Smiles, CSIRO Land and Water Canberra, is thanked for many stimulating discussions and comments. Assoc. Prof. M.D. Melville, School of Geography, University of NSW, Sydney is thanked for advice and comments. Support from the Water Research Foundation of Australia, the Australian Research Council, under ARC Large Grant A39917105 and the NSW ASSPRO are gratefully acknowledged.

REFERENCES

Croney, D., The movement and distribution of water in soils, *Géotechnique, 3,* 1-16, 1952.

Croney, D., and J. D. Coleman, Pore pressure and suction in soil, in *Pore Pressure and Suction in Soil,* pp. 31-37, Butterworths, London, 1961.

Dent, D. L., *Acid Sulphate Soils: a Baseline for Research and Development,* No. 39. Internat. Instit. for Land Reclamation and Improvement, Wageningen, the Netherlands, 1986.

Groenvelt, P. H., and G. H. Bolt, Water retention in soil, *Soil Sci., 113,* 238-245, 1972.

Kim, D. J., H. Vereecken, J. Feyen, D. Boels, and J. J. B. Bronswijk, On the characterisation of properties of an unripe marine clay soil II. A method on the determination of hydraulic properties, *Soil Sci., 154,* 59-71, 1992.

Kittrick, J.A., D.S.Fanning, and L.R. Hosner. *Acid sulfate weathering,* SSSA Spec. Publ. 10. Soil Science Society of America, Madison, Wisconsin, 1982.

Mitchell, J. K., Practical problems from surprising soil behaviour, *J. Geotech. Engng., 112,* 259-289, 1986.

Philip, J. R., Kinetics of sorption and volume change in clay-colloid pastes, *Aust. J. Soil Res., 6,* 249-267, 1968.

Philip, J. R., Hydrostatics and hydrodynamics in swelling soils, *Water Resour. Res., 5,* 1070-1077, 1969a.

Philip, J.R., Moisture equilibrium in the vertical in swelling soils, 1, Basic theory, *Aust. J. Soil Res., 7,* 99-120, 1969b.

Philip, J.R., Moisture equilibrium in the vertical in swelling soils, 1, Applications, *Aust. J. Soil Res., 7,* 121-141, 1969c.

Philip, J. R., Reply, *Water Resour. Res., 6,* 1248-1251, 1970a.

Philip, J.R., Hydrostatics in swelling soils and soil suspensions: Unification of concepts, *Soil Sci., 109,* 294-298, 1970b.

Philip, J.R., Diffuse double layer interactions in one-, two-and three dimensional particle swarms. *J. Chem. Phys., 52,* 1387-1398, 1970c.

Pons, L. J. Outline of the genesis, characteristics, classification and improvement of acid sulphate soils, in *Proceedings of the International Symposium on Acid Sulphate Soils 13-29 Aug. 1972 Wageningen,* edited by H. Dost,. Pub. No. 18, Vol. 1, pp.3-27, Internat. Instit. Land Reclamation and Improvement, Wageningen, the Netherlands, 1973.

Raats, P.A.C. and A. Klute. Transport in soils: the balance of mass, *Soil Sci. Soc. Amer. Proc., 32,* 161-169, 1968.

Rosenqvist, T., Consideration of the sensitivity of Norwegian quick-clays, *Geotechnique, 3,* 195-200, 1953.

Rosenqvist, T., Considerations on the sensitivity of Norwegian quick clays –A Review, *Engineering Geology, 1,* 445-450, 1966.

Sammut, J., I. White, and M.D. Melville,. Acidification of an estuarine tributary in eastern Australia due to drainage of acid sulfate soils. *Marine and Freshwater Research, 47.* 669-684, 1996.

Smiles, D.E., An examination of settlement data for an embankment on a light wet clay, *Aust Road Res., 5,* 55-59, 1973.

Smiles, D.E., On the validity of the theory of flow in saturated swelling material, *Aust. J. Soil Res., 14,* 389-395, 1976a.

Smiles, D.E., Sedimentation and filtration equilibrium. *Separation Sci., 11,* 1-16, 1976b.

Smiles, D.E., Principles of constant pressure filtration, in *Encyclopedia of Fluid Mechanics,* edited by N. P. Cheremisinoff, pp 791-824, Gulf Publishing Company, Houston, Texas, USA, 1986.

Smiles, D.E., Liquid flow in swelling soils, *Soil Sci. Soc. Amer. J., 59,* 313-318 1995.

Smiles, D.E., Water balance in swelling materials: some comments, *Aust. J. Soil Res., 35,* 1143-52, 1997.

Smiles, D.E., Hydrology of swelling soils: a review, *Aust. J. Soil Res., 38,* 501-521, 2000.

Smiles, D.E., and A.G. Harvey, Measurement of the moisture diffusivity of swelling systems, *Soil Sci., 116,* 391-399, 1973.

Smiles, D.E. and H. G. Poulos, The one-dimensional consolidation of columns of soil of finite length, *Aust. J. Soil Res.*, 7, 285-291, 1969.

Smiles, D.E., and M.J. Rosenthal, The movement of water in swelling materials, *Aust J. Soil Res.*, 6, 237-248, 1968.

Smiles, D.E., C. J. Barnes and W. R. Gardner, Water relations of saturated bentonite: Some effects of temperature and solution salt concentration, *Soil Sci Soc. Amer. J.*, 49, 66-69, 1985.

Sposito, G., A thermodynamic integral equation for the equilibrium profile in swelling soil, *Water Resour. Res.*, 11, 499-500, 1975a.

Sposito, G., On the differential equation for the equilibrium moisture profile in swelling soil. *Soil Sci. Soc. Amer. Proc.*, 39, 1053-1056, 1975b.

Sposito, G., *The Surface Chemistry of Soils*. Oxford University Press, New York, 1984.

Stroosnijder, L. Infiltratie en herverdeling van water in grond. *Versl. Landouwk. Onderz.* No. 847, 1976.

Talsma, T., Moisture profiles in swelling soils, *Aust J. Soil Res.*, 12, 71-75, 1974.

Talsma, T., and A. van der Lelij, Infiltration and water movement in an in situ swellig soil during prolonged ponding, Aust *J. Soil Res.*, 14, 337-349, 1976.

Talsma, T., Measurement of the overburden component of total potential in swelling field soils. *Aust J. Soil Res.*, 15, 95-102, 1977a.

Talsma, T., A note on the shrinkage behaviour of a clay paste under various loads. *Aust J. Soil Res.*, 15, 275-277, 1977b.

Terzaghi, K., Die Berechnung der Durchlassidkeitsziffer des Tones aus den Verlauf der Hydrodynamischen Spannungserscheinungen, *Akad. Der Wissenschaften in Wein, Sitzungberichte, Mathematisch-Naturewissenschaftliche Klasse*, Part IIa, 132, 3-4, 125-138, 1923.

White, I., M.D. Melville, B.P. Wilson, and J. Sammut, Reducing acid discharge from estuarine wetlands in eastern Australia, *Wetlands Ecol. and Management*, 5, 55-72, 1997.

White, I., D.E. Smiles, S. Santomartino, P. van Oploo, B.C.T. MacDonald, and T.D. Waite. Dewatering and the hydraulic properties of soft, sulfidic coastal clay soils. *Water Resour. Res.* Submitted, 2001a.

White, I., B.C.T. MacDonald, and M.D. Melville. Modelling shallow groundwater in soft, sulfidic, coastal sediments, in *Proceedings, Modsim 2001, ANU, Canberra, 10-13 December 2001, Canberra, edited by F. Ghassem* (in press).

Youngs, E.G., Developments in the theory of infiltration, *Soil Sci. Soc. Amer. J.*, 59, 307-313, 1995.

Youngs, E. G. and G. D. Towner, Comments on 'Hydrodynamics in swelling soils' by J. R. Philip, *Water Resour. Res.*, 6, 1246-1247, 1970.

Ian White, Centre for Resource and Environmental Studies, Australian National University, Canberra, ACT, 0200, AUSTRALIA

Solute Transport in Infiltration-Redistribution Cycles in Heterogeneous Soils

S. C. Lessoff

Dept. of Fluid Mechanics and Heat Transfer, Tel Aviv University, Ramat Aviv, Israel

P. Indelman

Dept. of Civil Engineering, Technion, Haifa, Israel

G. Dagan

Dept. of Fluid Mechanics and Heat Transfer, Tel Aviv University, Ramat Aviv, Israel

An analytic model of transient unsaturated infiltration is presented. The model is based on the column conceptualization of flow and transport in unsaturated soil which is expanded here to incorporate repeated infiltration and redistribution stages. The transport of reactive solute is modeled by assuming three mechanisms: advection by gravitational water flow, equilibrium sorption and linear decay. Solutions of the flow and transport equations are derived for multiple infiltration-redistribution cycles and for Dirac and finite pulse solute applications. Expressions are derived for average moisture content and for average concentration regarding the soil saturated conductivity a random value.

1. INTRODUCTION

Flow and transport in the unsaturated zone has long been a field of active research, yet practical engineering and agricultural applications still rely on painstaking calibration of empirical models. This is because the nonlinear inter-relationships between unsaturated flow properties crucially complicate the solution of practical problems. Lacking analytic solutions, the characterization of important processes becomes difficult or unfeasible. Significant efforts have been undertaken to solve analytically the flow equations for homogeneous soils. Most of the results that have been obtained are for one-dimensional vertical flow under different simplified assumptions. These solutions are available in many publications [e.g., *Philip*, 1969; *Gardner*, 1958; *VanGenuchten*, 1980]. Additionally, the governing equations were solved numerically for one- two- and three-dimensional flows. Although many interesting and important solutions have been achieved, these studies clearly show how much efforts should be undertaken to arrive at results relevant to practical applications.

Natural soils reveal spatial heterogeneity in their properties (primarily conductivity) and this heterogeneity may significantly change the nature of flow and transport. Therefore, treating soil properties as randomly

Environmental Mechanics: Water, Mass and Energy Transfer in the Biosphere
Geophysical Monograph 129
Copyright 2002 by the American Geophysical Union
10.1029/129GM13

distributed has gained wide acceptance. Modeling flow and transport with random soil properties requires considering the flow equations as stochastic ones. Obviously, taking into account spatial heterogeneity renders flow problems in unsaturated soils even more complicated with infinitesimal hopes for analytical general solutions. Two main avenues have been suggested in the literature for characterizing flow and transport in stochastic unsaturated domains: numerical simulation and stochastic-analytical solution of the flow and transport equations under simplifying assumptions. The present work addresses the second analytic approach.

The advantage of the numerical approach is that general problems can be simulated. Many computer codes have been developed and successfully applied to both theoretical studies and field test data. Heterogeneity is often simulated using a Monte-Carlo approach. Recently numerical solutions have been sought to stochastic transformations of the equations of flow and transport [e.g., *Zhang and Winter*, 1998; *Osnes*, 1998]. The review of the numerical approach is out of scope of this paper and can be found in many publications [e.g., *Russo*, 1993; *Burr et al.*, 1994].

Many analytical studies derive expressions for the statistical moments of pressure head in unsaturated media using a perturbation approach similar to that originally developed for saturated uniform flows [e.g., *Dagan*, 1989]. For example solutions have been derived by *Andersson and Shapiro* [1983] characterizing one-dimensional steady flow, by *Yeh et al.* [1985a, b] characterizing three-dimensional steady-state flow and by *Mantoglou and Gelhar* [1983] characterizing three-dimensional transient flow. Perturbation solutions require the assumption that changes in the mean head occur on a spatial scale that is large in comparison to the characteristic scale of random head variability. However, this assumption is not generally valid for unsaturated flows [*Indelman et al.*, 1993].

An approach free of the assumption of stationary mean head was suggested by *Dagan and Bresler* [1979] and *Bresler and Dagan* [1979]. Their column model represents shallow three-dimensional infiltration by one-dimensional flow in a random collection of homogeneous columns. For various problems of solute transport with constant uniform moisture content, analytical solutions were obtained by solving a one-dimensional convective-diffusion equation in each column and averaging over the columns [*Warrick and Nielsen*, 1980; *Bresler and Dagan*, 1981; *Dagan and Bresler*, 1988; *Destouni and Cvetkovic*, 1991; *Dagan*, 1993]. Recently, *Indelman et al.* [1998] applied the column model to two

cycle unsteady, unsaturated flow and this paper extends their developments to multiple cycles, arbitrary initial solute location, and a finite initial solute pulse.

In this paper we develop simple analytical expressions characterizing flow and transport under conditions typical to agricultural fields and field experiments. The typical field conditions to which we refer are as follows. Conservative or reactive substances are transported downward into the soil by water applied at the soil surfaces. Flow and transport occur by repeated two stage infiltration redistribution cycles.

The first stage of the cycle, infiltration, occurs when water is applied to the dry soil. We neglect the temporal variability of water flux during wetting events. This assumption is appropriate when the variability of the water application rate occurs on time scales much smaller than the time scale of soil drainage [*Rodriguez-Iturbe et al.*, 1999]. The duration of the infiltration stage is determined by the time of water application and the time required for all of the applied water to infiltrate into the soil.

The second stage of the cycle, redistribution, occurs during prolonged dry periods. Redistribution is characterized by a no flux conditions at the soil surface.

The plan of the paper is as follows. The first introductory section outlines the problem, method of solution and structure of the paper. The second section presents the solution to flow in a homogeneous column for the two aforementioned stages of infiltration and redistribution. In the third section, conservative and reactive solute transport is solved in a single column for instantaneous and finite pulse sources, for an arbitrary initial solute depth and for multiple infiltration-redistribution cycles. In the fourth section we solve the average moisture content and the average concentration in a random collection of columns representing a heterogeneous field. Finally in the fifth section we summarize the limitations and capabilities of the model presented.

2. FLOW MODEL

Application of the column model characterize transport in unsaturated soils was first suggested by *Dagan and Bresler* [1979] and developed in series of publications [e.g., *Bresler and Dagan*, 1983a, b; *Dagan and Bresler*, 1983; *Destouni and Cvetkovic*, 1991; *Cvetkovic and Destouni*,1989]. The applicability of the model was investigated by *Protopapas and Bras*, [1991] and *Or and Rubin*, [1993], who compared results based on the column model with two-dimensional solutions. According to the column model, the three-dimensional flow domain is regarded as a random series of vertical columns. In each column the one-dimensional flow equation for moisture content θ is given by

$$\frac{\partial \theta}{\partial t} + \frac{\partial q}{\partial z} = 0 \qquad (1)$$

with the infiltration flux

$$q = -K_s K_r(\theta)\left[\frac{\partial \Psi}{\partial z} - 1\right] \qquad (2)$$

In (2) K_s and K_r are soil saturated and relative conductivities, respectively, Ψ is the pressure head and the axis z is directed vertically downward with the origin at the soil surface. Combining (2) with functional dependence of θ on Ψ produces a nonlinear system which, in general, can not be solved analytically.

As is seen from (2) the flow of water into the soil is driven by the two terms in the right-hand-side, namely, the pressure gradient and gravity. For long periods of irrigation the pressure gradient can be neglected as compared to gravity, [Or and Rubin, 1993; Green and Ampt, 1911] i.e. flow is gravitational. We may then simplify the flow equations by combining (1) with (2) to get

$$\Delta\frac{\partial S}{\partial t} + K_s\frac{\partial K_r(S)}{\partial z} = 0 \qquad (3)$$

where $S = (\theta - \theta_r)/\Delta$ is the soil saturation, θ_s and θ_r are the saturated and residual moisture contents of the soil, respectively and $\Delta = \theta_s - \theta_r$. Relative conductivity is characterized by a power dependence on saturation

$$K_r(\theta) = S^{1/\beta} \qquad (4)$$

where $\beta = const.$

Following the approach of Indelman et al. [1998], we consider the water flow as consisting of two stages: an infiltration stage during which a volume of water per unit surface area W is imbibed through the soil surface and a redistribution stage during which no water is imbibed through the soil surface. To simplify calculations we assume that the initial moisture content of the soil is at residual $\theta(z,0) = \theta_r$. Note that the approach developed below is applicable to soils with initial water content between residual θ_r and field capacity θ_c, i.e. $\theta(z,0) = \theta_0$ with $\theta_r \leq \theta_0 \leq \theta_c$, under the assumption that $K_r(\theta_0) \approx 0$. The soil is assumed heterogeneous in horizontal plane only by regarding K_s as independent of z and a random stationary function of (x,y). Vertical random variability of K_s can be taken into account using the approach developed by Indelman et al. [1993].

However, the derivations become too complicated for the present objective.

The exact solution of (1) and (2) would yield a water content distribution in which θ grows steeply down from the soil surface to a 'plateau' and then steeply decreases to the initial value at depth [Bresler, 1973]. We solve (3) to characterize the 'plateau' structure following Dagan and Bresler [1979], i.e. the solution is sought in the form

$$S(z,t) = S_f(t)H\left[z_f(t) - z\right] \qquad (5)$$

where H is the Heaviside step function, $z_f(t)$ is the location of the wetting front and S_f is the saturation at the wetting front.

2.1. Infiltration Stage

For simplicity we assume that, during the infiltration stage, water is applied uniformly over the soil surface of each column at the constant rate r (L/T) such that the time of irrigation is equal to $t_{ir} = W/r$. It is convenient to introduce the dimensionless conductivity $\hat{K}_s = K_s/r$ and the parameter γ

$$\gamma = \begin{cases} 1/\hat{K}_s & \text{if } r < K_s \\ 1 & \text{if } r \geq K_s \end{cases} \qquad (6)$$

The surface boundary condition accounting for ponding is as follows

$$\hat{q}(\hat{z},\tau)|_{z=0} = 1 \quad \text{if } r < K_s \qquad (7)$$

and

$$\theta|_{\hat{z}=0} = \theta_s \quad \text{otherwise}$$

where $\hat{q} = q/r$, $\tau = tr/W$, and $\hat{z} = z/W$ are nondimensional flux, time and depth respectively. The solution of (3) for (7) is given by the step function (5) where $z_f = q_f\tau/(S_f\Delta)$ with q_f and S_f being the infiltration flux and saturation at the wetting front location respectively. The moisture content during infiltration is shown schematically in Figure 1a. From (3)-(7) we get that during infiltration:

$$\hat{q}_f = \gamma\hat{K}_s \quad ; \quad S_f = \gamma^\beta \qquad (8)$$

The infiltration flux $\hat{q}(\hat{z})$ is now determined from (1), (4) and (8) as follows

$$\hat{q}(\hat{z},\tau) = \hat{q}_f H\left[\hat{z}_f(\tau) - \hat{z}\right] \qquad (9)$$

where $\hat{z}_f$ is the location of the wetting front which moves linearly in time

$$\hat{z}_f(\tau) = \hat{q}_f\tau/(S_f\Delta) = \gamma^{1-\beta}\hat{K}_s\tau/\Delta \qquad (10)$$

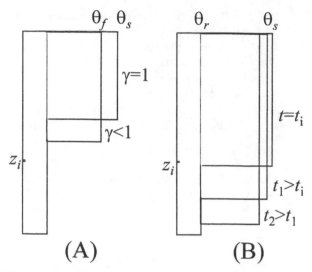

Figure 1. A schematic illustration of the soil moisture content during infiltration in two columns with different values of K_s (a) and redistribution in a column at 3 times (b).

The duration of the infiltration stage τ_{inf} depends on the rate of infiltration and the column conductivity and is defined by $\tau_{\mathrm{inf}} = 1/(\gamma \hat{K}_s)$. The depth of water infiltration at the end of the stage is $\hat{z}_{\mathrm{inf}} = \gamma^{-\beta}/\Delta$.

2.2. Redistribution Stage

During redistribution, the initial and boundary conditions are

$$S(\hat{z}, \tau_{\mathrm{inf}}) = \gamma^\beta H(\hat{z}_{\mathrm{inf}} - \hat{z}), \quad \hat{q}\,(0,\tau)|_{\tau > \tau_{\mathrm{inf}}} = 0 \quad (11)$$

The water flux and velocity of the wetting front is determined by (2) for gravitational flow ($q = K_s K_r$), (4) and by water mass conservation at the front location $[(\theta_f - \theta_r)\partial z_f/\partial t = q_f]$

$$\hat{q}_f(\tau) = \hat{K}_s S_f^{1/\beta}(\tau); \quad \frac{d\hat{z}_f}{d\tau} = \frac{\hat{K}_s S_f^{1/\beta - 1}}{\Delta} \quad (12)$$

Substituting the global water mass balance during redistribution $\hat{z}_f = (S_f \Delta)^{-1}$ into (12) yields a differential equation for the moisture content at the wetting front location

$$\frac{d}{d\tau} [S_f(\tau)]^{-1} = \hat{K}_s S_f^{1/\beta - 1}(\tau) \quad (13)$$

The moisture content during redistribution is shown schematically in Figure 1b. The solution of (13) which satisfies the initial condition $S_f(\tau_{\mathrm{inf}}) = \gamma^\beta$ is given by

$$S_f(\tau) = \gamma^\beta \Theta^{-\beta}(\tau) \quad (14)$$

where

$$\Theta(\tau) = 1 + \frac{1}{\beta}\left(\gamma \hat{K}_s \tau - 1\right) \quad (15)$$

The wetting front location is

$$\hat{z}_f(\tau) = \frac{1}{S_f(\tau)\Delta} = \frac{\gamma^{-\beta}}{\Delta} \Theta^\beta(\tau) \quad (16)$$

The location of the water front for one set of parameters is illustrated in Figure 2. The water flux $\hat{q}(\hat{z}, \tau)$ results from (1), (5), (12) and (14) for $\tau > \tau_{\mathrm{inf}}$

$$\hat{q}\,(\hat{z}, \tau) = \hat{K}_s \hat{z} \Delta \left(\frac{\gamma}{\Theta(\tau)}\right)^{\beta+1} H\,[\hat{z}_f(\tau) - \hat{z}] \quad (17)$$

Summarizing, the moisture content distribution, as depicted in Figure 1, is determined by

$$S(\hat{z}, \tau) = S_f(\tau) H\,[\hat{z}_f(\tau) - \hat{z}] \quad (18)$$

with

$$S_f(\tau) = \gamma^\beta \begin{cases} 1 & \tau \le 1/(\gamma \hat{K}_s) \\ \Theta^{-\beta}(\tau) & \tau > 1/(\gamma \hat{K}_s) \end{cases} \quad (19)$$

and

$$\hat{z}_f(\tau) = \frac{1}{\Delta} \begin{cases} \gamma^{1-\beta} \hat{K}_s \tau & \tau \le 1/(\gamma \hat{K}_s) \\ [\Theta(\tau)/\gamma]^\beta & \tau > 1/(\gamma \hat{K}_s) \end{cases} \quad (20)$$

The reader is reminded that γ (6) contains a switch discriminating between ponded and unsaturated conditions, $\tau = tr/W$ is non-dimensional time, $\tau_{\mathrm{inf}} = 1/(\gamma \hat{K}_s) = r/(\gamma K_s)$ is the time at the end of the infiltration stage, Θ is given by (15).

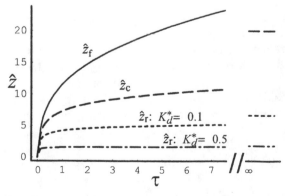

Figure 2. The location of the front over time for the water and solute (conservative and reactive). In all cases $\theta_r = 0.05$, $\Delta = 0.35$, $\hat{K}_s = 3$, and $\beta = 1/3$

3. TRANSPORT IN A SINGLE COLUMN

Neglecting pore-scale dispersion, advective transport of a conservative solute in a column is described by the equation

$$\frac{\partial C}{\partial t} + \frac{\partial (uC)}{\partial z} = 0 \qquad (21)$$

where C is the solute concentration (mass solute per mass soil solids) and $u = q/\theta$. Equation (21) assumes that mass transfer into immobile water is much faster than large scale advection.

The characteristics of equation (21) are defined by the equation:

$$\frac{dz_c(t)}{dt} = u(z,t) \qquad (22)$$

where $z_c(t)$ is the location of a solute concentration characteristic. Note that the concentration along the characteristics is not constant but satisfies the equation

$$\frac{dC}{dt} = -C\frac{\partial u}{\partial z} \qquad (t > t_i) \qquad (23)$$

For $u = q/\theta$, combining (23) with (1) shows that $C/\theta =$ const. on the characteristics. The general solution for concentration at any time is

$$C\left[z_c(t),t\right] = C_0\left[z_0\right]\frac{\theta\left[z_c(t),t\right]}{\theta_0} \qquad (24)$$

where C_0 and θ_0 are the concentration and water content at depth z_0 at time t_0 and $z_c(t)$ is the solution of (22) for the initial condition $z_c(t_0) = z_0$. The proportionality of moisture content (mass water per mass soil solids) to concentration (24) results from the definition of concentration as dissolved solute per mass soil solids. As stated above, due to the non-linearity of (1) and (2), it is not possible to solve (22) in the general case.

3.1. Solution to Gravitational Transport

For gravitational flow $q = K_s K_r$ and dry initial conditions $\theta(z,0) = \theta_r$, we determine $\hat{u} = \hat{q}/\theta$ in (22) from (9) and (17)-(20), as follows:

$$\hat{u}(\hat{z},\tau) = \frac{\gamma \hat{K}_s}{\theta_r + \gamma^\beta \Delta} H\left[\hat{z}_f(t) - \hat{z}\right] \qquad (25)$$

during the infiltration stage ($\tau \leq \tau_{\text{inf}}$), and

$$\hat{u}(\hat{z},\tau) = \frac{\gamma^{\beta+1}\Delta \hat{K}_s \hat{z}\,\Theta^{-1}(\tau)}{\theta_r \Theta^\beta(\tau) + \gamma^\beta \Delta} H\left[\hat{z}_f(\tau) - \hat{z}\right] \qquad (26)$$

during redistribution stage ($\tau > \tau_{\text{inf}}$).

With $\hat{u}(\hat{z},\tau)$ given by (25) and (26), equation (22) is solved (Appendix A) for the initial solute location $\hat{z}_c(0) = \hat{z}_0$ giving

$$\hat{z}_c(\tau) = \hat{z}_0 H\left(\hat{z}_0 - \hat{z}_f\right) + \left[\frac{\gamma \hat{K}_s \tau + \hat{z}_0 \theta_r}{\theta_r + \gamma^\beta \Delta} H(\tau_{\text{inf}} - \tau)\right.$$

$$\left. + \frac{1 + \hat{z}_0 \theta_r}{\theta_r + \gamma^\beta \Delta \Theta^{-\beta}(\tau)} H(\tau - \tau_{\text{inf}})\right] H\left(\hat{z}_f - \hat{z}_0\right) \qquad (27)$$

It can be seen that until the wetting front reaches the solute location (i.e. $\hat{z}_f(\tau) < \hat{z}_0$), the second term on the right hand side of (27) is zero and $z_c(\tau) = z_0$. After the wetting front reaches the solute location, the first term on the right hand side in (27) becomes zero. Then, the first term inside the square brackets in (27) controls during infiltration (if the wetting front reaches the solute location during the infiltration period) and the second term describes advection during redistribution.

Solutions (20) and (27) show the principal difference between water flow and conservative solute transport in the unsaturated zone. It is seen (Figure 2) that the solute moves slower than the wetting front. Indeed, in its propagation, the water fills the pore space determined by the front water content $S_f(\tau)\Delta$ (19), whereas the solute occupies additionally the pore space of the residual water θ_r. If the duration of the infiltration stage was not limited, the solute would propagate behind the water front unboundedly downward. For limited volume of applied water, during the redistribution stage both the water flux and the water content decrease for fixed $\hat{z}$. This leads to the finite depth of solute propagation. Indeed, it follows from (20) and (27) that the water front propagates unboundedly downward $\lim_{\tau \to \infty} \hat{z}_f(\tau) = \infty$, whereas the solute front reaches the finite depth $\lim_{\tau \to \infty} \hat{z}(\tau) = 1/\theta_r + \hat{z}_0$ provided $\theta_r \neq 0$. Thus, the maximum depth of penetration of a solute pulse starting at the surface is shown on the far right in Figure 2, whereas there is no maximum depth of water penetration.

3.2. Finite Pulse

To illustrate the transport of a finite solute pulse let us assume that an initially dry column $[\theta(\hat{z},0) = \theta_r]$ has mass per unit surface area M of solute initially distributed over a thin layer of depth l_0 ($l_0 \ll W/\theta_s$) i.e.

$$C(z,0) = \begin{cases} M/(\rho_b l_0) & z < l_0 \\ 0 & z \geq l_0 \end{cases} \qquad (28)$$

where ρ_b is the bulk density of the soil. For $\hat{l}_0 = l_0/W$, and after the wetting front has passed the initial location of the plume ($\hat{z}_f > \hat{l}_0$) we can derive expressions for the depth to the top $\hat{z}_-(\tau)$ and bottom limits $\hat{z}_+(\tau)$ of the solute pulse as follows

$$\hat{z}_-(\tau) = \frac{\gamma \hat{K}_s \tau}{\theta_r + \gamma^\beta \Delta}, \quad \hat{z}_+(\tau) = \frac{\gamma \hat{K}_s \tau + \hat{l}_0 \theta_r}{\theta_r + \gamma^\beta \Delta} \qquad (29)$$

during infiltration $(\tau \leq \tau_{\text{inf}})$, and

$$\hat{z}_-(\tau) = \frac{1}{\theta_r + \gamma^\beta \Theta^{-\beta}(t)\Delta}, \quad \hat{z}_+(\tau) = \frac{1 + \hat{l}_0 \theta_r}{\theta_r + \gamma^\beta \Delta \Theta^{-\beta}(t)} \qquad (30)$$

during redistribution $(\tau > \tau_{\text{inf}})$.

We define a spreading coefficient $\xi(\tau)$ as the current length of the plume divided by its initial length. For the initial conditions (28), after the wetting front has passed the plume $(\hat{z}_{\text{f}} > \hat{l}_0)$, ξ is given by

$$\xi(\tau) = \frac{\hat{z}_+(\tau) - \hat{z}_-(\tau)}{\hat{l}_0} = \frac{\theta_r}{\theta(\tau)} \qquad (31)$$

The meaning of ξ can be better understood by considering the compression spreading cycle of a plume during a single wetting-drying cycle. When the wetting front reaches the top of the plume, $z_{\text{f}} = z_-$, combining (9) or (17) for infiltration and redistribution respectively with (22) shows that the top of the plume begins to move downward while the bottom of the plume remains stationary. This compresses the plume. Simultaneously, the average moisture content between z_- and z_+ increases. Because the concentration per unit mass soil is proportional to the moisture content (24), the average concentration increases while the plume contracts preserving the solute mass. Equations (20) and (22) require that the water front moves faster than the top of the plume $dz_{\text{f}}/dt > dz_-/dt$. Therefore, the top of the plume lags behind the water front. When the water front reaches the bottom of the plume $z_- < z_{\text{f}} = z_{\cdot+}$, and if the flow is still in the infiltration stage, then (25) requires that u is independent of depth. Therefore z_+ and z_- move at the same rate and the plume extent remains fixed. Later, during the redistribution stage, u increases with depth (26). Therefore the bottom of the plume descends into the soils faster than the top of the plume $dz_-/dt < dz_+/dt$ and the plume expands. While the plume is expanding, the soil moisture content is reduced thus reducing the concentration per soil mass (24) and preserving the solute mass balance. Asymptotically as the soil dries $(\theta(\tau) \to \theta_r$ as $\tau \to \infty)$ the plume extent returns to $\hat{l}_0$ and the concentration returns to C_0. Note that the spreading described here is a function of the mass balance only and does not include the effects of local dispersion. In a heterogeneous collection of columns there will also be stochastic spreading of the mean plume as is discussed in the sequel.

3.3. Reactive Solute

Transport with equilibrium sorption and first order decay obeys the mass balance equation

$$\frac{\partial C}{\partial t} + \frac{\partial(u_r C)}{\partial z} = -\lambda C \qquad (32)$$

where u_r is the retarded advection velocity of the reactive solute and λ is the first order decay rate. For linear-equilibrium sorption, $u_r = q/\bar{\theta}$ where $\bar{\theta} = \theta + K_d^*$ is the solute capacity of the soil, $K_d^* = K_d \rho_b$ with K_d the partition coefficient.

Assuming K_d^* independent of z and given the initial moisture profile $\theta(z, 0) = \theta_r$ and the initial condition $C(z, 0) = C_0(z)$, the solution of (32) by the method of characteristics gives

$$C[z_r(t), t] = C_0[z_0] e^{-\lambda t} \frac{\bar{\theta}[z_r(t), t]}{\bar{\theta}_0} \qquad (33)$$

where $z_r(t)$ is the solution of the equation of the characteristic for the initial condition $z_r(t_0) = z_0$ and $\bar{\theta}_0 = \theta_0 + K_d^*$.

For the two cycle gravitational infiltration redistribution flow described by (18)-(20) we compute $\hat{u}_r = \hat{q}/\bar{\theta}$ from (9) and (17). Defining $\hat{z}_r = z_r/W$, we get the expression for $\hat{z}_r(\tau)$ by replacing θ_r in (27) with $\bar{\theta}_r$. The behavior of $\hat{z}_r(\tau)$ is illustrated for a few values of K_d^* in Figure 2. The spreading coefficient for a reactive plume is given by $\xi(\tau) = \bar{\theta}_r/\bar{\theta}(\tau)$ which is larger than ξ for conservative plumes.

Thus, we have derived approximate solutions of the problems of water flow and of transport of conservative and reactive solutes in an unsaturated soil column for both infiltration and redistribution. For soils of finite residual water content, the maximum depth of contaminant penetration is bounded by $W/\bar{\theta}_r$. This is in contrast to the water infiltration front which propagates unboundedly with time (Figure 2).

3.4. Multiple Cycles

Typically the above infiltration redistribution cycle is repeated either at regular intervals (e.g., sprinkler irrigation) or at irregular intervals (e.g., due to rainfall). When the time interval between consecutive wetting events is much shorter than the characteristic time of redistribution, the wetting events can be treated as one event with the application rate averaged over the entire time [Rodriguez-Iturbe et al., 1999].

If the time interval between consecutive wetting events is much longer than the characteristic time of redistribution, we can assume that transport due to previous events has essentially reached its asymptotic maximum depth and therefore, transport at any given time is essentially due only to the latest wetting event. In this case, fluid flow is given by (18)-(20) with the surface fluid flux condition (W, r) of the current wetting event, while transport is given by (33) with the initial condition corresponding to the asymptotic long time state of the previous wetting event.

For example consider an infinitely thin initial layer of solute $C_0(z) = M\delta(z)/\rho_b$ where δ is the Dirac function. The solute undergoes homogeneous linear-equilibrium sorption and first-order decay. Starting from the surface, the plume is transported by N wetting events where the applied fluid volume of each event is W. After the wetting front from the Nth wetting event passes the plume, the concentration profile will be

$$C(\hat{z}) = \frac{M}{\rho_b W} \delta \left(\frac{[\hat{z} - \hat{z}_r(\tau_N)]\overline{\theta}}{\overline{\theta}_r} \right) \exp\left(-\hat{\lambda}\tau\right) \frac{\overline{\theta}}{\overline{\theta}_r} \quad (34)$$

where τ_N is the time since the beginning of the Nth wetting event and $\hat{z}_r(\tau_N)$ is given by (27) with θ_r and z_0 replaced by $\overline{\theta}_r$ and $\hat{z}_0^{(N)}$ respectively with

$$\hat{z}_0^{(N)} = (N-1)/\overline{\theta}_r \quad (35)$$

4.   AVERAGING OF FLOW AND TRANSPORT

In non-dimensional form, the model developed above contains the following input parameters: (i) three soil parameters affecting flow (the saturated conductivity $\hat{K}_s$, the mobile moisture capacity of the soil Δ and the power constant β) (ii) an additional parameter to describe conservative transport (the residual moisture content of the soil θ_r) and (iii) two additional parameters characterizing reactions in the soil (rate of losses $\hat{\lambda}$ and the effective distribution coefficient K_d^*). Although each of these parameters may be considered as uncertain, K_s is generally regarded as varying over a few orders of magnitude. Due to this great magnitude of variability we treat K_s as the single heterogeneous parameter and model it as a random function of horizontal coordinate x and y whose probability density function (p.d.f.) and cumulative probability function (c.d.f.) are $p(K_s)$ and $P(K_s)$ respectively. We limit the study primarily to deriving ensemble means of flow and transport variables only. The mean value of any chosen variable $A(z, t; K_s)$ is defined by

$$\langle A(z, t)\rangle = \int_0^\infty dK_s p(K_s) A(z, t; K_s) \quad (36)$$

4.1.   Mean Water Content

The mean water content is given by

$$\langle \theta(\hat{z}, \tau)\rangle = \theta_r + \langle S(\hat{z}, \tau)\rangle \Delta \quad (37)$$

where $\langle S\rangle$ is determined by substituting (18)-(20) into (36). We define the functions resulting from (15) for $\gamma = 1$ and $\gamma < 1$ respectively

$$\begin{aligned} \Theta(\tau; \hat{K}_s) &= 1 + \left(\hat{K}_s\tau - 1\right)/\beta \\ \Theta_1(\tau) &= 1 + (\tau - 1)/\beta \end{aligned} \quad (38)$$

The calculations show (Appendix B) that $\langle S\rangle$ is expressed in terms of the following three functions

$$\begin{aligned} P(x) &= \int_0^x d\hat{K}_s p(\hat{K}_s) \quad &(39) \\ \overline{P}(x) &= \int_0^x d\hat{K}_s p(\hat{K}_s)\Theta^{-\beta}(\tau; \hat{K}_s) \\ \widetilde{P}(x) &= \int_x^\infty d\hat{K}_s p(\hat{K}_s)\left(\frac{1}{\hat{K}_s}\right)^\beta \end{aligned}$$

With these definitions the expression of $\langle S\rangle$ is given by

$$\langle S(\hat{z}, \tau)\rangle = \widetilde{P}\left[\left(\frac{\hat{z}\Delta}{\tau}\right)^{1/\beta}\right] H(\hat{z}\Delta - \tau) + \quad (40)$$
$$\left[P(1) - P\left(\frac{\hat{z}\Delta}{\tau}\right) + \widetilde{P}(1)\right] H(\tau - \hat{z}\Delta)$$

for $\tau \leq 1$, and

$$\begin{aligned} \langle S(\hat{z}, \tau)\rangle &= H(1 - \hat{z}\Delta)[\Theta_1^{-\beta}(\tau)\widetilde{P}(1) \\ &\quad + P(\tau^{-1}) - P(\hat{z}\Delta/\tau) + \overline{P}(1) - \overline{P}(\tau^{-1})] \\ &\quad + H(\hat{z}\Delta - 1) H[\tilde{z}(\tau) - \hat{z}] \\ &\quad \left\{\overline{P}(1) - \overline{P}[\tilde{K}_s(\tau)] + \Theta_1^{-\beta}(\tau)\widetilde{P}(1)\right\} \\ &\quad + H[\hat{z} - \tilde{z}(\tau)]\Theta^{-\beta}(\tau, 1)\widetilde{P}\left[(\hat{z}\Delta)^{1/\beta}/\Theta_1(\tau)\right] \end{aligned} \quad (41)$$

for $\tau > 1$ where

$$\tilde{K}_s(\tau) = \left\{1 + \beta\left[(\hat{z}\Delta)^{1/\beta} - 1\right]\right\}/\tau; \quad \tilde{z}(\tau) = \Theta_1^\beta(\tau)/\Delta \quad (42)$$

We see in (39) that the calculation of the mean water content requires a one-dimensional integration.

Figure 3 shows the mean saturation profile for an ergodic set of flow tubes (a) at the end of water application $(\tau = 1)$ and (b) after the end of water application $(\tau = 1.6)$. Both times are near the time that infiltration would have stopped for a profile with homogenous K_s

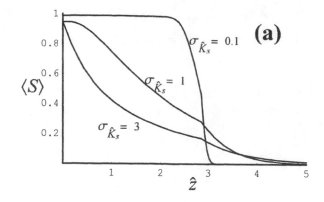

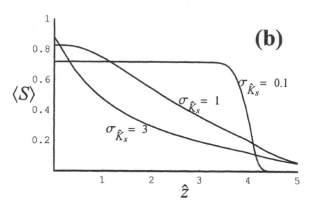

Figure 3. The average moisture content profile (a) at the end of water application ($\tau = 1$); and (b) after the end of water application ($\tau = 1.6$). In both cases the field parameters are $\langle \hat{K}_s \rangle = 1$, $\beta = 1/3$, and $\Delta = 0.35$.

equal to the distribution mean. Therefore for the field with low K_s variance, almost all of the applied water has infiltrated. On the other hand, in the field with the high K_s variance, not all of the applied water has infiltrated because a large portion of flow tubes have K_s significantly lower than the population mean. Therefore, the total infiltrated water volume is less for the high variability field. As expected, in the field with greater conductivity variability, the wetting front is more dispersed.

4.2. Transport of an Instantaneous Pulse

For simultaneous instantaneous solute injection with initial condition $C_0(\hat{z}) = M/(W\rho_b)\delta(\hat{z})$, the mean solute concentration $\langle C(\hat{z}, \tau) \rangle$ results from (34) and (36) as follows

$$\langle C(\hat{z},\tau) \rangle = \frac{M\,e^{-\hat{\lambda}\tau}}{\rho_b W} \int_0^\infty d\hat{K}_s p(\hat{K}_s) \qquad (43)$$
$$\delta\left\{[\hat{z} - \hat{z}_r(\tau_N)]\,\bar{\theta}(\tau, K_s)/\bar{\theta}_r\right\}\bar{\theta}(\tau, \hat{K}_s)/\bar{\theta}_r$$

Substituting (33) into (43) leads to the following expressions of the mean solute concentrations (see Appendix C)

$$\langle C(\hat{z},\tau) \rangle = \frac{M\,e^{-\hat{\lambda}\tau}}{\rho_b W} \left[\frac{\bar{\theta}_s}{\tau} p\left(\frac{\bar{\theta}_s \hat{z}}{\tau} \right) H\left(\frac{\tau}{\bar{\theta}_s} - \hat{z} \right) \right. \qquad (44)$$
$$\left. + \frac{\partial \hat{K}_{s2}}{\partial \hat{z}} p(\hat{K}_{s2}) H(\bar{\theta}_s \hat{z} - \tau) H(\tau - \bar{\theta}_r \hat{z}) \right]$$

for $\tau \leq 1$ and

$$\langle C(\hat{z},\tau) \rangle = \frac{M\,e^{-\hat{\lambda}\tau}}{\rho_b W} \left[\frac{\bar{\theta}_s}{\tau} p\left(\frac{\bar{\theta}_s \hat{z}}{\tau} \right) H(1 - \bar{\theta}_s \hat{z}) \right.$$
$$+ \frac{\partial \hat{K}_{s1}}{\partial \hat{z}} p(\hat{K}_{s1}) H(\bar{\theta}_s \hat{z} - 1) H\left[\hat{z}_p(\tau) - \hat{z} \right] \qquad (45)$$
$$\left. + \frac{\partial \hat{K}_{s3}}{\partial z} p(\hat{K}_{s3}) H\left[\hat{z} - \hat{z}_p(\tau) \right] H(1 - \bar{\theta}_r \hat{z}) \right]$$

for $\tau > 1$. Here $\hat{z}_p(\tau) = \left[\bar{\theta}_r + \Delta\Theta_1^{-\beta}(\tau) \right]^{-1}$ and

$$\hat{K}_{s1} = \frac{1}{\tau}\left[1 + \beta\left(\frac{\hat{z}\Delta}{1 - \bar{\theta}_r \hat{z}} \right)^{1/\beta} - \beta \right] \qquad (46)$$

$$\hat{K}_{s2} = \left(\frac{\hat{z}\Delta}{\tau - \bar{\theta}_r \hat{z}} \right)^{1/\beta} ;\ \hat{K}_{s3} = \frac{1}{\Theta_1(\tau)}\left(\frac{\hat{z}\Delta}{1 - \bar{\theta}_r \hat{z}} \right)^{1/\beta}$$

The behavior of the mean concentration is illustrated in Figure 4. At the end of infiltration ($\tau = 1$) the mean concentration profile is bimodal. The first peak in the $\tau = 1$ curve near the top of the profile is due to the large number of columns with low conductivity. In all of these columns, the velocity is exceedingly slow and solute remains near the top of the profile. The second peak represents transport for $\hat{K}_s$ near the mean $\hat{K}_s$ value of 1. During redistribution, the solute in the low conductivity columns is strongly detained at low S. This causes the large spreading of the plume in Figure 4 at $\tau = 10$. At large times ($\tau = 1000$) the plume again becomes concentrated because in all of the columns solute approaches the maximum infiltration depth ($W/\bar{\theta}_r$) which is not affected by the heterogeneity of $\hat{K}_s$. Figure 5 shows the effect the variance of $\hat{K}_s$ on transport at $\tau = 20$. As the variance of $\hat{K}_s$ increases, the plume becomes more dispersed. When the variability of conductivity is very high ($\sigma = 6$), the concentration profile becomes bimodal. As above (Figure 4), the two peaks represent a large number of tubes with disparate low values of $\hat{K}_s$ whose solute all remain near the top of the

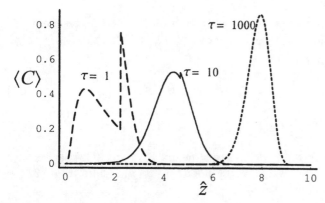

Figure 4. The average concentration profile for a conservative solute at the end of water application ($\tau = 1$) and at two times after water application stops ($\tau > 1$). In all cases the field parameters are $\langle \hat{K}_s \rangle = 1$, $\sigma_{\hat{K}_s} = 1.0$, $\beta = 1/3$, $\Delta = 0.35$ and $\theta_r = 0.1$.

profile and a large number of columns whose conductivity is near the mean value of $K_s = 1$.

5. SUMMARY

We present an analytic model of transient flow and transport in unsaturated soil. The basic simplification of the model is the assumption that flow is essentially vertical which is most appropriate for shallow domains whose vertical extent is much less than the horizontal scale of heterogeneity or whose vertical extent is much less than their horizontal extent. We assume gravitational flow. The analytic solution to flow derived here applies to homogeneous columns. Care must be taken in applying the model to real soils with heterogeneity of hydraulic and transport parameters along the flow path. The model neglects the effects of local dispersion which implies advection dominated transport. We assume that the mass transfer rate between mobile and immobile phases is fast in relation to the rate of transport so that there is local equilibrium.

Using simple analytical expressions, the infiltration redistribution model succeeds in explaining the often observed phenomenon that, when water and solute are simultaneously applied to an unsaturated column, the water drains through the column, but the solute is retained. Water draining through the column only samples the mobile porosity of the soil. On the other hand, solute mixes through all of the soil water. Therefore, at any time, the mobile water content above the water front equals the applied water $S_f \Delta z_f = W$. Similarly, at any time, the total soil solute capacity above the solute front equals the solute capacity of the applied water

$(S_f \Delta + \bar{\theta}_r) z_r = W$. As the soil dries, S goes to 0. For a finite volume of applied water (W is finite), the wetting front descends unboundedly while the solute front assymptotically approaches the depth $W / \bar{\theta}_r$.

The model presented here may also be valuable for explaining the time behavior of moisture content and solute in the root zone. This is of significant value because the root zone is the most active zone of biological activity in the subsurface. Due to the simplicity of the analytical solution derived, the model presented here may be used to predict the effect of transient flow coupled to spatially variable processes. For example, we use model to derive analytical expressions for ergodic transport in a heterogeneous field with random heterogeneity of hydraulic conductivity transverse to the direction of flow.

APPENDIX A: DERIVATION OF $\hat{z}(\tau)$ FOR CONSERVATIVE SOLUTE

We define the solute capacity above the front as:

$$Q(\tau) = \int_0^{\hat{z}_c(\tau)} \theta(\tau, \hat{z}) \, d\hat{z} \qquad (A1)$$

To derive (27) we find a differential form for Q as follows:

$$\begin{aligned}
\frac{dQ}{d\tau} &= \theta[\tau, \hat{z}_c(\tau)] \frac{d\hat{z}_c}{d\tau} + \int_0^{\hat{z}_c(\tau)} \frac{\partial \theta(\tau, \hat{z})}{\partial \tau} d\hat{z} \\
&= \theta[\tau, \hat{z}_c(\tau)] u(\tau, \hat{z}_c(\tau)) - \int_0^{\hat{z}_c(\tau)} \frac{\partial q(\tau, \hat{z})}{\partial \hat{z}} d\hat{z} \\
&= q[\tau, \hat{z}_c(\tau)] - q[\tau, \hat{z}_c(\tau)] + q[\tau, 0] = q[\tau, 0]
\end{aligned} \qquad (A2)$$

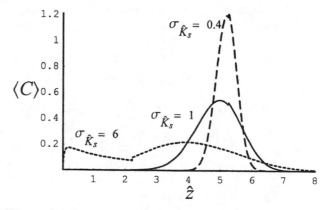

Figure 5. The average concentration profile of a conservative solute for three values of the variance of $\hat{K}_s$. In all cases the field parameters are $\langle \hat{K}_s \rangle = 1$, $\theta_r = 0.1$, $\beta = 1/3$, $\Delta = 0.35$ and $\tau = 20$.

due to our assumption of a square wetting front (5),

$$\hat{z}_c(\tau) = Q(\tau)/\theta_f(\tau) \qquad (A3)$$

Combining the initial condition $Q(\tau = 0) = \hat{z}_0\theta_r$ with (A2) and (A3) gives (27).

APPENDIX B: DERIVATION OF THE MEAN WATER CONTENT

The mean saturation $\langle S \rangle$ (18)-(20) can be represented as a sum $\langle S \rangle = \langle S_1 \rangle + \langle S_2 \rangle$ where

$$\langle S_1 \rangle = \int_0^1 d\hat{K}_s\, p(\hat{K}_s) S_f(\tau) H(\hat{z}_f(\tau) - \hat{z}) \quad (B1)$$

$$\langle S_2 \rangle = \int_1^\infty d\hat{K}_s\, p(\hat{K}_s) S_f(\tau) H(\hat{z}_f(\tau) - \hat{z}) \quad (B2)$$

We calculate the two means (B1) and (B2) separately.

For (B1) $\gamma = 1$, hence substituting for $\hat{z}_f$ yields

$$\langle S_1 \rangle = \int_0^1 d\hat{K}_s\, p(\hat{K}_s) \left\{ H(1 - \hat{K}_s\tau) H\left(\frac{\hat{K}_s\tau}{\hat{z}\Delta} - 1\right) \right.$$
$$\left. + \Theta^{-\beta}(\tau;\hat{K}_s) H(\hat{K}_s\tau - 1) H\left[\frac{\Theta^\beta(\tau;\hat{K}_s)}{\hat{z}\Delta} - 1\right] \right\} (B3)$$

For times $\tau \leq 1$ the system is definitely in the infiltration stage and (B3) becomes

$$\langle S_1 \rangle = [P(1) - P(\hat{z}\Delta/\tau)] H(\tau - \hat{z}\Delta) \qquad (B4)$$

where $P(x)$ is the c.d.f. of $\hat{K}_s$. For $\tau > 1$ (B3) becomes

$$\langle S_1 \rangle = \int_0^{1/\tau} d\hat{K}_s\, p(\hat{K}_s) H\left(\frac{\hat{K}_s\tau}{\hat{z}\Delta} - 1\right) \qquad (B5)$$
$$+ \int_{1/\tau}^1 d\hat{K}_s p(\hat{K}_s)\Theta^{-\beta}(\tau;\hat{K}_s) H\left[\frac{\Theta^\beta(\tau;\hat{K}_s)}{\hat{z}\Delta} - 1\right]$$
$$= \left[P\left(\frac{1}{\tau}\right) - P\left(\frac{\hat{z}\Delta}{\tau}\right) + \overline{P}(1) - \overline{P}\left(\frac{1}{\tau}\right)\right] H(1 - \hat{z}\Delta)$$
$$+ \left\{\overline{P}(1) - \overline{P}\left[\tilde{K}_s(\tau)\right]\right\} H(\hat{z}\Delta - 1) H[\tilde{z}(\tau) - \hat{z}]$$

with $\overline{P}(x)$, $\tilde{K}_s$ and $\tilde{z}$ given by (39) and (42) of the text.

For (B2) $\gamma = 1/\hat{K}_s$ and we have

$$\langle S_2 \rangle = \int_1^\infty d\hat{K}_s \frac{p(\hat{K}_s)}{\hat{K}_s^\beta} \left\{ H\left[\frac{\hat{K}_s^\beta\tau}{\hat{z}\Delta} - 1\right] H(1 - \tau) \right.$$
$$\left. + \Theta_1(\tau)^{-\beta} H\left[\frac{(\Theta_1(\tau)\hat{K}_s)^\beta}{\hat{z}\Delta} - 1\right] H(\tau - 1) \right\}$$
$$= \left\{ \tilde{P}(1) H\left(\frac{\tau}{\hat{z}\Delta} - 1\right) + \tilde{P}\left[\left(\frac{\hat{z}\Delta}{\tau}\right)^{1/\beta}\right] H\left(\frac{\hat{z}\Delta}{\tau} - 1\right) \right\}$$
$$H(1 - \tau) + \Theta_1^{-\beta}(\tau) \left\{ \tilde{P}(1) H\left[\frac{\Theta_1^\beta(\tau)}{\hat{z}\Delta} - 1\right] \right.$$
$$\left. + \tilde{P}\left[\frac{(\hat{z}\Delta)^{1/\beta}}{\Theta_1(\tau)}\right] H\left[\frac{\hat{z}\Delta}{\Theta_1^\beta(\tau)} - 1\right] \right\} H(\tau - 1) \qquad (B6)$$

with $\tilde{P}(x)$ given by (39) of the text.

APPENDIX C: DERIVATION OF THE MEAN SOLUTE CONCENTRATION FOR INSTANTANEOUS INPUT

The mean concentration (43) can be represented as a sum $\langle \hat{C}(\hat{z},\tau) \rangle = M e^{-\hat{\lambda}\tau} \left[\hat{C}_1(\hat{z},\tau) + \hat{C}_2(\hat{z},\tau) \right] / (\rho_b W)$ where

$$\hat{C}_1(\hat{z},\tau) = \int_0^1 d\hat{K}_s p(\hat{K}_s)\delta\left[\hat{z} - \hat{z}_c(\tau)\right] \quad (C1)$$

$$\hat{C}_2(\hat{z},\tau) = \int_1^\infty d\hat{K}_s p(\hat{K}_s)\delta\left[\hat{z} - \hat{z}_c(\tau)\right] \quad (C2)$$

We calculate the two means (C1) and (C2) separately.

For (C1) $\gamma = 1$. Therefore $\hat{z}_c(\tau)$ results from (27) as follows

$$\hat{z}_c(\tau) = \frac{\hat{K}_s\tau}{\theta_s}H(1 - \hat{K}_s\tau) + \qquad (C3)$$
$$\frac{1}{\theta_r + \Delta\Theta^{-\beta}(\tau;\hat{K}_s)}H(\hat{K}_s\tau - 1)$$

When $\tau \leq 1$ introducing $\hat{u} = \hat{K}_s\tau/\theta_s$ produces

$$\hat{C}_1(\hat{z},\tau) = \frac{\theta_s}{\tau}\int_0^{\tau/\theta_s} d\hat{u}\, p\left(\frac{\hat{u}\theta_s}{\tau}\right)\delta(\hat{z} - \hat{u})$$
$$= \frac{\theta_s}{\tau}p\left(\frac{\hat{z}\theta_s}{\tau}\right) H(\tau - \theta_s\hat{z}) \qquad (C4)$$

For $\tau > 1$ introducing a new variable by $\hat{v} = 1/[\theta_r + \Delta\Theta^{-\beta}(\tau;\hat{K}_s)]$ we have

$$\hat{C}_1(\hat{z},\tau) = \frac{\theta_s}{\tau}\int_0^{1/\theta_s} d\hat{u}\, p\left(\frac{\hat{u}\theta_s}{\tau}\right)\delta(\hat{z} - \hat{u})$$
$$+ \int_{1/\theta_s}^{\hat{z}_p(\tau)} d\hat{v}\, p(\hat{K}_{s1})\frac{\partial\hat{K}_{s1}}{\partial\hat{z}}\delta(\hat{z} - \hat{v})$$
$$= \frac{\theta_s}{\tau}p\left(\frac{\hat{z}\theta_s}{\tau}\right) H(1 - \theta_s\hat{z}) \qquad (C5)$$
$$+ \frac{\partial\hat{K}_{s1}}{\partial\hat{z}}p(\hat{K}_{s1}) H(\theta_s\hat{z} - 1) H(\hat{z}_p(\tau) - \hat{z})$$

where $\hat{K}_{s1}$ and $\hat{z}_p$ are given by (46).

If $\hat{K}_s > 1$ then $\gamma = 1/\hat{K}_s$ and $\hat{z}_c(\tau)$ is given by

$$\hat{z}_c(\tau) = \frac{\tau}{\theta_r + \Delta(1/\hat{K}_s)^\beta}H(1 - \tau) \qquad (C6)$$
$$+ \frac{1}{\theta_r + \Delta(\hat{K}_s)^{-\beta}\Theta_1^{-\beta}(\tau)}H(\tau - 1)$$

Denoting

$$A = \left\{ \begin{array}{l} \tau \\ 1 \end{array} \right. ; \quad D = \Delta\left\{ \begin{array}{ll} 1 & \text{for } \tau \leq 1 \\ \Theta_1^{-\beta}(\tau) & \text{for } \tau > 1 \end{array} \right. \qquad (C7)$$

and substituting (C6) into (C2) yields

$$\hat{C}_2(\hat{z},\tau) = \int_1^\infty d\hat{K}_s p(\hat{K}_s)\delta\left(\hat{z} - \frac{A}{\theta_r + D/\hat{K}_s^\beta}\right) \quad (C8)$$

Introducing a new variable $\chi = A/[\theta_r + D(1/\hat{K}_s)^\beta]$ transforms (C8) to

$$\langle \hat{C}(\hat{z}, \tau) \rangle = \frac{M}{\rho_b W} \int_{\frac{A}{\theta_r + D}}^{\frac{A}{\theta_r}} d\chi \, p\left[K_s^*(\chi)\right] \frac{\partial K_s^*(\chi)}{\partial \chi} \delta(\hat{z} - \chi)$$

$$= \frac{M}{\rho_b W} p\left[K_s^*(\hat{z})\right] \frac{\partial K_s^*(\hat{z})}{\partial \hat{z}} H\left(\hat{z} - \frac{A}{\theta_r + D}\right) H\left(\frac{A}{\theta_r} - \hat{z}\right) \tag{C9}$$

with

$$K_s^*(\hat{z}) = \left(\frac{D\hat{z}}{A - \theta_r \hat{z}}\right)^{1/\beta} \tag{C10}$$

Expressions (C4), (C5) and (C9) lead to (44) of the text.

Acknowledgments. The support of the Chinese-Israeli Fund for Scientific and Strategic Research and Development provided by the Israel Ministry of Science, Culture and Sports and by the Chinese Ministry of Science and Technology is gratefully acknowledged.

REFERENCES

Andersson, J., and A. M. Shapiro, Stochastic analysis of one-dimensional steady state unsaturated flow: A comparison of Monte Carlo and perturbation methods, *Water Resour. Res.*, *25*, 121–133, 1983.

Bresler, E., Simultaneous transport of solutes and water under transient unsaturated flow conditions, *Water Resour. Res.*, *9*, 975–986, 1973.

Bresler, E., and G. Dagan, Solute dispersion in unsaturated heterogeneous soil at field scale II: Applications, *Soil Science Soc. of Am. J.*, *43*, 476–472, 1979.

Bresler, E., and G. Dagan, Convective and pore scale dispersive solute transport in unsaturated heterogeneous fields, *Water Resour. Res.*, *17*, 1683–1693, 1981.

Bresler, E., and G. Dagan, Unsaturated flow in spatially variable fields 2. Application of water models to various fields, *Water Resour. Res.*, *19*, 421–428,, 1983a.

Bresler, E., and G. Dagan, Unsaturated flow in spatially variable fields 3. Solute transport models and their application to two fields, *Water Resourc. Res.*, *19*, 429–435, 1983b.

Burr, T. D., E. A. Sudicky, and R. L. Naff, Nonreactive and reactive solute transport in three-dimensional heterogeneous porous media: Mean displacement, plume spreading and uncertainty, *Water Resour. Res.*, *30*, 791–815, 1994.

Cvetkovic, V. D., and G. Destouni, Comparison between resident and flux-averaged concentration models for field-scale solute transport in the unsaturated zone, in *Contaminant Transport in Groundwater*, edited by H. E. Kobus and W. Kinzelbach, pp. 245–250, A. A.Balkemia, Rotterdam Netherlands, 1989.

Dagan, G., *Flow and Transport in Porous Formations*, Springer-Verlag, New-York, 1989.

Dagan, G., The Bresler-Dagan model of flow and transport: Recent theoretical developments, in *Water Flow and Solute Transport in Soils, Developments and Applications*, edited by G. Dagan and D. Russo, vol. 20 of *Advances Series in Agricultural Sciences*, pp. 13–32, Springer-Verlag Advanced Series in Agricultural Sciences, Berlin Heidelberg New York, 1993.

Dagan, G., and E. Bresler, Solute dispersion in unsaturated heterogeneous soil at field scale I: Theory, *Soil Science Soc. of Am. J.*, *43*, 461–467, 1979.

Dagan, G., and E. Bresler, Unsaturated flow in spatially variable fields 1. Derivation of models of infiltration and redistribution, *Water Resourc. Res.*, *19*, 413–420, 1983.

Dagan, G., and E. Bresler, Variability of yield of an irrigated crop and its causes 1. Statement of the problem and methodology, *Water Resour. Res.*, *24*, 381–387, 1988.

Destouni, G., and V. Cvetkovic, Field scale mass arrival of sorptive solute into the groundwater, *Water Resour. Res.*, *27*, 1315–1325, 1991.

Gardner, W. R., Some steady state solutions of unsaturated moisture flow equations with application to evaporation from a water table, *Soil Science*, *85*, 228–232, 1958.

Green, W. H., and G. A. Ampt, Studies on soil physics, part 1, the flow of air and water through soils, *J. Agric. Sci.*, *4*, 1–24, 1911.

Indelman, P., D. Or, and Y. Rubin, Stochastic analysis of unsaturated steady flow through bounded heterogeneous formations, *Water Resour. Res.*, *29*, 1141–1148, 1993.

Indelman, P., I. Touber-Yasur, B. Yaron, and G. Dagan, Stochastic analysis of water flow and pesticides transport in a field experiment, *Journal of Contaminant Hydrology*, *32*, 77–97, 1998.

Mantoglou, A., and L. W. Gelhar, Stochastic modeling of large-scale transient unsaturated flow systems, *Water Resour. Res.*, *23*, 37–46, 1983.

Osnes, H., Stochastic analysis of velocity spatial variability in bounded rectangular heterogeneous aquifers, *Adv. in Water Resources*, *21*, 203–215, 1998.

Or, D., and Y. Rubin, Stochastic modelling of unsaturated flow in heterogeneous media with water uptake by plant roots: Tests of the parallel columns model under two-dimensional flow conditions, *Water Resour. Res.*, *29*, 4109–4119, 1993.

Philip, J. R., Theory of infiltration, *Advances In Hydrosciences*, *5*, 215–196, 1969.

Protopapas, A. L., and R. L. Bras, The one-dimensional approximation for infiltration in heterogeneous soils, *Water Resour. Res.*, *27*, 1019–1027, 1991.

Rodriguez-Iturbe, I., A. Porporato, L. Ridolfi, V. Isham, and D. R. Cox, Probabilistic modeling of water balance at a point: The role of climate, soil and vegetation, *Proc. R. Soc. Lond. A*, *455*, 3789–3805, 1999.

Russo, D., Stochastic modeling of macrodispersion for solute transport in a heterogeneous unsaturated porous formation, *Water Resour. Res.*, *29*, 383–397, 1993.

VanGenuchten, M. T., A closed-form equation for predicting the hydraulic conductivity of unsaturated soils, *Soil Sci. Am. J.*, *44*, 892–898, 1980.

Warrick, A. W., and D. R. Nielsen, Spatial variability of soil physical properties, in *Applications of Soil Physics*, edited by D. Hillel, pp. 319–344, Academic Press, New York, NY, 1980.

Yeh, T. C., L. W. Gelhar, and A. L. Gutjahr, Stochastic analysis of unsaturated flow in heterogeneous soils: 1 statistically isotropic media, *Water Resour. Res.*, *21*, 447–456, 1985a.

Yeh, T. C., L. W. Gelhar, and A. L. Gutjahr, Stochastic analysis of unsaturated flow in heterogeneous soils: 2 statistically anisotropic media with variable alpha, *Water Resour. Res., 21*, 457–464, 1985b.

Zhang, D., and C. L. Winter, Nonstationary stochastic analysis of steady state flow through variably saturated, heterogeneous media, *Water Resour. Res., 34*(5), 1091–1100, 1998.

G. Dagan and S. C. Lessoff, Dept. of Fluid Mechanics and Heat Transfer, Tel Aviv University, Ramat Aviv 69978, Israel (e-mail: dagan@eng.tau.ac.il; slessoff@eng.tau.ac.il)

P. Indelman, TECHNION - Israel Institute of Technology, Faculty of Civil Engineering, Technion City, Haifa 32000, Israel (e-mail: indelman@techunix.technion.ac.il)

Analytical Solutions for Two-Dimensional Solute Transport with Velocity-Dependent Dispersion

Philip Broadbridge, R. Joel Moitsheki and Maureen P. Edwards

Institute for Mathematical Modelling and Computational Systems, University of Wollongong, New South Wales, Australia.

A form of the solute transport equation is transformed from Cartesian to streamline coordinates. Symmetry analysis of this equation with a point water source reveals a 5-parameter symmetry group. Exploitation of the rich symmetry properties of the equation leads to a number of associated reduced partial differential equations - that is, partial differential equations where the number of independent variables has been reduced by one. Using further symmetry reductions and other transformation techniques, the construction of new solutions for non-radial solute transport on a background of radial water flow is possible.

1. INTRODUCTION

Throughout his working life, John Philip devoted much of his energy to analysing macroscopic models for real-world problems in environmental mechanics. During the 1990s, one of those areas that captured his attention was solute transport [*Philip,* 1994, 1996]. This field is of immense practical interest since regional soil contamination and salinisation has become one of our most serious environmental problems. The time scale of these regional processes is of the order of several decades. Therefore, for the purposes of environmental management, it takes too long to experimentally determine the outcomes of agricultural and industrial practices. Predictions must be made by mathematical modelling or by designing small physical models whose results may be sensibly scaled up to predict outcomes at the field scale. A full theory of solute transport will require understanding of microscopic transport processes

Environmental Mechanics: Water, Mass and Energy Transfer in the Biosphere
Geophysical Monograph 129
Copyright 2002 by the American Geophysical Union
10.1029/129GM14

in fluctuating fluid flow fields in networks of tortuous channels (for a review, see e.g. [*Jury,* 1988]). Nevertheless, macroscopic transport models, described in terms of partial differential equations, will remain important for efficiently predicting solute transport at the field scale. In practical problems, it is normal to solve the relevant partial differential equations by approximate numerical methods. However, we face the serious problem that available numerical packages have significant disagreements in their prediction of solute dispersion [*Woods et al.,* 1998]. Therefore, exact solutions are very important not only because they provide insight but also because they are needed as validation tests for numerical schemes.

Solute dispersion is complicated even at the macroscopic level because the dispersion coefficient increases with fluid velocity, which in general is varying in space and time. The fluid velocity vector field cannot be an arbitrary smooth function of space and time; it must conform to the established laws of fluid flow in porous media. Although passive scalar transport in solvent-conducting porous media has been intensively studied by many people for many years, realistic exactly solvable models with spatially varying dispersion coefficient are very rare. Perhaps the most notable effort in this direction has been that of *Moench* [1989], who obtained the Laplace transform for solute concentration during transport from an injection well to a pumped withdrawal well after approximating the flow as being radial towards the withdrawal well. In this case, the Laplace transform must be inverted numerically but this has been achieved with demonstrated accuracy [*Moench*, 1991]. Most other good approximate and exact analytic results have similarly focussed on radial transport in two or three dimensions [*Hoopes and Harleman*, 1967; *Eldor and Dagan*, 1972; *Tang and Babu*, 1979; *Hsieh*, 1986; *Novakowski*, 1972; *Fry et al.*, 1993]. Simple one-dimensional, axisymmetric or spherically symmetric geometries are most insightful in our analysis of solute dispersion. However, in these simplest cases the transport process is automatically represented as a partial differential equation in two dependent variables, including the time variable t and only one space variable r. Such equations do not provide a genuine test for numerical simulations of two-dimensional flows. *Zoppou and Knight* [1996] made some progress in this direction by producing the point source solution for dispersion in a background of hyperbolic water streamlines bounded by a wedge. The main drawback of this solution is that it required a special form of anisotropy in the velocity dependence of the dispersion tensor.

Over his many years of research in environmental mechanics, John Philip's powers of physical intuition and deductive reasoning were such that he rarely needed more than undergraduate calculus to solve seemingly difficult applied problems. However, we find it necessary to use some slightly more sophisticated algebraic tools to progress towards our aim; that is to construct exact two dimensional solutions to velocity-dependent isotropic dispersion that are not axially symmetric. To this end, we select a form of the solute transport equation that has a particularly large symmetry group of invariance transformations, depending continuously on a number of real parameters. In Section 2, we formulate the class of partial differential equations that we will study. We will briefly outline the requisite Lie symmetry theory in Section 3. In Section 4, we summarise the results of our symmetry analyses. This leads to a number of new exotic solutions for non-radial solute transport on a background of radial water flow. These are discussed further in Section 5.

2. SOLUTE TRANSPORT EQUATIONS

Here, we concentrate on macroscopic deterministic models based on local conservation laws [e.g. see

Wierenga 1995]. In the most complete formulations, the dispersion tensor may have anisotropic dependence on various components of pore velocity [*Bear*, 1979]. However, for the purposes of exact analysis, we assume here that dispersion is isotropic, with the single dispersion coefficient being a function of pore water speed $v = |\mathbf{V}|/\theta$. Here θ is the volumetric water concentration in the soil. The solute flux density $\mathbf{J}$ is the sum of three components,

$$\mathbf{J} = -\theta D_0 \nabla c - \theta D_e(v) \nabla c + c\mathbf{V}, \qquad (1)$$

due to molecular diffusion, dispersion and convection respectively. $\mathbf{V}$ is the volumetric Darcian water flux. The dispersion coefficient D_e is found to be an increasing function of pore water speed. It has often been convenient to model this function as a power law, $D_e = D_1 v^m$, with $1 \le m \le 2$. This has some experimental support [*Salles et al.*, 1993]. When we combine (1) with the equation of continuity for mass conservation,

$$\frac{\partial(c\theta)}{\partial t} + \nabla \cdot \mathbf{J} = \mathbf{0}, \qquad (2)$$

we obtain the convection-dispersion equation,

$$\frac{\partial(c\theta)}{\partial t} = \nabla \cdot [\theta D(v) \nabla c] - \nabla \cdot (c\mathbf{V}), \qquad (3)$$

where $D(v) = D_0 + D_e(v)$. For the remainder of this article, we shall consider two dimensional steady flow of water in saturated soils. These satisfy $\theta = \theta_s$, along with Darcy's law $\mathbf{V} = -K_s \nabla \Phi$, where θ_s is the water content at saturation, Φ is the total hydraulic pressure head and K_s is the hydraulic conductivity at saturation. For flow saturated soils, the equation of continuity $\nabla \cdot \mathbf{V} = 0$ combined with Darcy's law implies Laplace's equation

$$\nabla^2 \Phi = 0, \qquad (4)$$

and (3) takes the form

$$\frac{\partial c}{\partial t} = \nabla \cdot [D(v) \nabla c] + k \nabla \Phi \cdot \nabla c, \qquad (5)$$

where $k = K_s/\theta_s$ and $v = |k\nabla\Phi|$. Since in two dimensional Darcian saturated flow, the velocity $\mathbf{v}$ is irrotational and the fluid is incompressible, there exists a stream function $\psi(x,y)$ which is a harmonic function conjugate to the pore velocity potential which is $\phi(x,y) = K_s\Phi$. Thus, $\mathbf{V} = (-\frac{\partial\phi}{\partial x}, -\frac{\partial\phi}{\partial y}) = (-\frac{\partial\psi}{\partial y}, \frac{\partial\psi}{\partial x})$. After the transformation from Cartesian coordinates (x,y) to streamline coordinates (ϕ, ψ), convection takes place in the direction of constant ψ, and the convection-dispersion equation transforms to [*Hoopes and Harleman*, 1967; *Ségol*, 1994]:

$$\frac{1}{v^2}\frac{\partial c}{\partial t} = \frac{\partial}{\partial\phi}\left[D(v)\frac{\partial c}{\partial\phi}\right] + \frac{\partial}{\partial\psi}\left[D(v)\frac{\partial c}{\partial\psi}\right] + \frac{\partial c}{\partial\phi}. \quad (6)$$

This form of the solute transport equation happens to be more amenable to symmetry analysis.

In most dispersion problems of interest, molecular diffusion is negligible compared to dispersion, and we approximate $D(v)$ by $D_e(v) = v^p$ with $1 \le p \le 2$. For radial water flows from a line source of strength q, in terms of the radial coordinate r, the Darcian flux is $\mathbf{V} = q/r$, and the pore velocity is $\mathbf{v} = \mathbf{V}/\theta_s$, for which the velocity potential is $\phi = -(q/\theta_s)\log\ r$ and the stream function is $\psi = -(q/\theta_s)\arctan(y/x)$. In this case, Equation (5) takes the form

$$\frac{\partial c}{\partial t} = \nabla\cdot\left[D_1\frac{(q/\theta_s)^p}{r^p}\nabla c\right] + \frac{(q/\theta_s)}{r}\frac{\partial c}{\partial r}. \quad (7)$$

Note that the gradient operator here is not simply radial as we are allowing solute concentration to depend on the polar angle. Equation (7) may be non-dimensionalised and rescaled so that all coefficients of proportionality are unity. Consider dimensionless quantities $C = c/c_s$, $T = t/t_s$ and $(X,Y,R) = (x,y,r)/l_s$, where s-subscripted parameters represent suitable concentration, time and length scales. The unique choice of time scale t_s and length scale l_s that will normalize Equation (7) is

$$t_s = D_1^{2/p}(\frac{q}{\theta_s})^{1-2/p},$$

$$l_s = D_1^{1/p}(\frac{q}{\theta_s})^{1-1/p}.$$

Then Equation (7) rescales to

$$\frac{\partial C}{\partial T} = \nabla\cdot\left[\frac{1}{R^p}\nabla C\right] + \frac{1}{R}\frac{\partial C}{\partial R}. \quad (8)$$

Now we will summarize the techniques of symmetry reduction before we apply them to find forms of the convection-dispersion equation that allow exact solution.

3. ALGEBRAIC TECHNIQUES FOR SYMMETRY REDUCTION

The theory and applications of continuous symmetry groups were founded 120 years ago by Lie [1880]. There are many modern readable accounts of this theory but we will mention only a few here [Ovsiannikov, 1982; Bluman and Kumei, 1989; Hill, 1992; Ibragimov, 1995]. Given a continuous one-parameter symmetry group, in

most practical cases we may reduce the number of independent variables by one. The most familiar symmetry is the rotational symmetry that enables us to reduce (x,y) to the single radial variable r. For example, consider the nonlinear diffusion equation, familiar to all who study porous media,

$$\frac{\partial\theta}{\partial t} = \nabla\cdot[D(\theta)\nabla\theta].$$

This equation is invariant under the group of plane rotations

$$x' = x\ \cos(\epsilon) - y\ \sin(\epsilon),$$
$$y' = x\ \sin(\epsilon) + y\ \cos(\epsilon),$$
$$\theta' = \theta.$$

This Lie group of transformations depends continuously on the group parameter ϵ which is the rotation angle. The invariants of this transformation group are θ and the radial coordinate $r = (x^2 + y^2)^{1/2}$. Rotationally invariant solutions satisfy a reduced P.D.E. for $\theta(r,t)$.

The next most familiar example is the scaling symmetry. For example, the nonlinear diffusion equation is invariant under the Boltzmann scaling symmetry

$$x' = xe^\epsilon,$$
$$t' = te^{2\epsilon},$$
$$\theta' = \theta.$$

The group invariants are θ and $\phi = x/t^{1/2}$. Invariant solutions satisfy an O.D.E. for $\theta = f(\phi)$ or equivalently $x = t^{1/2}g(\theta)$. Philip [1957,1969] used this form as a starting point for the infiltration series to be used when gravity is not ignored in flow of water in unsaturated soils.

The identity transformation, which must belong to any group, is conventionally labeled by $\epsilon = 0$ (e.g. for rotation by angle $\epsilon = 0$ or scaling by factor $e^\epsilon = e^0 = 1$). As well as the familiar geometric one-parameter groups, there may be additional more complicated symmetry groups that apply only to special subclasses within the class of governing equations. For example, if $D(\theta) = \theta^{-4/3}$, then the one dimensional nonlinear diffusion equation is invariant under the non-obvious symmetry group [e.g. Galaktionov et al., 1988]

$$x' = \frac{x}{1+\epsilon x},$$
$$\theta' = \theta(1+\epsilon x)^{-3},$$
$$t' = t.$$

Fortunately, we need only use infinitesimal symmetry techniques, since Lie's fundamental result is that the whole of the one parameter group can be determined from the transformation laws up to first degree in ϵ.

If we know the coefficients $\tau(t, x, \theta)$, $\xi(t, x, \theta)$ and $\eta(t, x, \theta)$ of the infinitesimal transformations

$$t' = t + \epsilon\tau + O(\epsilon^2),$$

$$x' = x + \epsilon\xi + O(\epsilon^2),$$

$$\theta' = \theta + \epsilon\eta + O(\epsilon^2),$$

then the full group may be written formally as

$$f' = e^{\Gamma} f(t, x, \theta),$$

where Γ is the infinitesimal symmetry operator $\Gamma = \tau\frac{\partial}{\partial t} + \xi\frac{\partial}{\partial x} + \eta\frac{\partial}{\partial \theta}$. The exponential of Γ may be defined as the usual power series.

Given a governing partial differential equation, we may derive determining relations for the coefficients $\tau(t, x, \theta)$, $\xi(t, x, \theta)$ and $\eta(t, x, \theta)$ in order for Γ to be a symmetry operator. These determining relations are linear partial differential equations and they can be generated and solved automatically by many computer algebra programs.

Conveniently, the coefficients of Γ lead directly to the form of invariant solutions, obtained by solving the invariant surface condition

$$\tau\frac{\partial\theta}{\partial t} + \xi\frac{\partial\theta}{\partial x} = \eta,$$

for example by the method of characteristics.

If we begin with a P.D.E. in three independent variables, such as Equation (6), then a single symmetry will allow us to reduce to a P.D.E. in two independent variables. If the resultant reduced P.D.E has an additional symmetry, then it may be reduced further to an O.D.E. The reduced P.D.E. is guaranteed to have an inherited symmetry if the original P.D.E. has two independent symmetries satisfying $[\Gamma_1, \Gamma_2] = \gamma\Gamma_1$, with γ constant. Successive reductions can then take place by Γ_1 followed by a reduced form of Γ_2. In more complicated situations, successive reductions can take place by a chain of symmetries for the original P.D.E., provided that set of symmetries has the algebraic structure of a solvable Lie algebra [e.g. *Olver*, 1986]. Even if the original P.D.E with a symmetry does not have a solvable Lie symmetry algebra, an additional hidden symmetry may still show up for the reduced P.D.E. In practice, these may be found by successively reapplying the symmetry-finding procedure to each reduced equation.

4. SYMMETRY REDUCTIONS OF SOLUTE TRANSPORT EQUATIONS

Ultimately, we wish to obtain exact solutions to the system of Equations (4) and (5) . However, if we look for Lie symmetries of the entire system (4) and (5), then we will find nothing more than rescaling of c, translation in t, and translations and rotations in (x, y). These are the only conformal maps that leave not only (4) but (5) invariant. Nevertheless, we may hope to find symmetries that leave the single Equation (5) invariant when $\Phi(x, y)$ is a special solution of Laplace's equation. This may lead to useful reductions and solutions of (5) even if $\Phi(x, y)$ itself is not an invariant solution of Laplace's equation. For this purpose, we could carry out a symmetry classification of the single equation (5), treating $\Phi(x, y)$ as a free coefficient function. Given the class of functions $\Phi(x, y)$ that lead to extra symmetries, we could later select from these, solutions of Laplace's equation.

In order to generate and solve the symmetry determining relations, we have used the freely available program *DIMSYM* [*Sherring*, 1993], that is written as a subprogram for the computer algebra package *REDUCE* [*Hearn*, 1985]. The only point symmetries for the general equation (5) are combinations of translations in T, rescaling of C and linear superposition. The symmetry operators are linear combinations of $\Gamma_1 = \frac{\partial}{\partial T}$, $\Gamma_2 = C\frac{\partial}{\partial C}$, and $\Gamma_{\infty} = h(X, Y, T)\frac{\partial}{\partial C}$, where $h(X, Y, T)$ is any particular solution of (5). The output of *DIMSYM* indicates special algebraic and differential equations among the free functions $D(v)$ and $\Phi(x, y)$, which, if satisfied, may lead to additional special symmetries. In fact, we have found that *DIMSYM* more easily finds special symmetric cases when the general convection-dispersion equation is expressed in terms of streamline coordinates, as in (6). Not surprisingly, even when the pore velocity is non-uniform, many special symmetries arise when D is constant. Even this simpler case is directly applicable for modelling convection and molecular diffusion, or as a first approximation to dispersion. This case was studied more extensively in an earlier paper [*Broadbridge et al.*, 2000]. Not all symmetric cases have yet been determined, but useful additional symmetries certainly occur when the water velocity is radial or when it represents strained flow along hyperbolic streamlines bounded by a wedge. From an arbitrary initial condition, we showed how to construct exact solute concentration profiles in terms of Laguerre polynomials, modified Bessel functions and confluent hypergeometric functions when the water flow had hyperbolic streamlines bounded by a wedge. For the case of hyperbolic strained flow, additional symmetries do not occur for any velocity-dependent dispersion coefficient $D(v)$. For radial water flow, with power-law dispersion coefficient $D(v) = v^p$, additional symmetries occur only for the cases $p = 0$, $p = -2$ and $p = 2$. This choice $p = 2$ is in accord with *Taylor's* [1953] theory of dispersion by fluctuations of a fluid velocity field and it seems to be a reasonable model for dispersion in porous media [*Philip*, 1994; *de Gennes*, 1986]. For this case, the solute transport equation may be rescaled to Equation (8) with $p = 2$. For the relevant normalised point water source, $\phi = -\log R$, ψ is simply the clockwise polar angle coordinate $-\arctan(Y/X)$ and $v = e^\phi$. In this case, besides the generic symmetries Γ_1, Γ_2 and Γ_∞, Equation (6) has three additional independent symmetries

$$\Gamma_3 = -\left(\frac{T^2}{4} + \frac{T}{2} - \frac{T}{4}e^{-2\phi} + \frac{1}{16}e^{-4\phi}\right) C\frac{\partial}{\partial C}$$

$$+ T^2\frac{\partial}{\partial T} - \frac{T}{2}\frac{\partial}{\partial \phi},$$

$$\Gamma_4 = \left(-\frac{T}{4} + \frac{1}{8}e^{-2\phi}\right) C\frac{\partial}{\partial C} + T\frac{\partial}{\partial T} - \frac{1}{4}\frac{\partial}{\partial \phi} \quad \text{and}$$

$$\Gamma_5 = \frac{\partial}{\partial \psi}.$$

The optimal system is $\{\Gamma_3\,;\,\Gamma_3+\Gamma_1\,;\,\Gamma_3+\alpha\Gamma_5\,;\,\Gamma_3+\Gamma_4+\alpha\Gamma_5\,;\,\Gamma_4\,;\,\Gamma_4+\alpha\Gamma_1\,;\,\Gamma_4+\alpha\Gamma_1+\beta\Gamma_5\,;\,\Gamma_5\,;\,\Gamma_2\,;\,\Gamma_5+\alpha\Gamma_2\}$. Wherever they appear in this list, α and β represent arbitrary constants. In Table 1, we list the canonical invariants and the reduced P.D.E.s associated with each of these symmetries.

We are able to construct a variety of exact non-radial solutions in terms of elementary functions, Bessel functions, and Kummer's function [*Abramowitz and Stegun*, 1972] using subsequent symmetry reductions, and other transformation techniques. In Section 5, we construct some well behaved solutions that are invariant under the complicated symmetry Γ_3. Solutions that are invariant under other symmetries will be derived elsewhere.

5. INVARIANT NON-RADIAL SOLUTIONS

Consider the Γ_3-invariant solutions of the form

$$C = \exp\left(-\frac{T}{4} - \frac{\log T}{2} + \frac{1}{4}e^{-2\phi} - \frac{1}{16T}e^{-4\phi}\right) \times F(\rho, \gamma),$$

with F satisfying the P.D.E. listed in Table 1. Since

$$[\Gamma_3, \Gamma_4] = -\Gamma_4,$$

the reduced equation inherits the symmetry Γ_4, which now takes the form

$$\Gamma_4 = \rho\frac{\partial}{\partial \rho} + F\frac{\partial}{\partial F},$$

leading to the reduction

$$F = \rho g(\gamma), \quad \text{with} \quad g''(\gamma) + 8g(\gamma) = 0.$$

In terms of the original variables, this leads to the solution

$$C = k_3 + k_4 \cos(2\sqrt{2}\psi + k_5)\sqrt{T}R^{-2}$$

$$\times \exp\left(-\frac{T}{4} + \frac{R^2}{4} - \frac{R^4}{16T}\right). \tag{9}$$

For convenience of interpretation, we have neglected the analogous solutions wherein sine functions replace cosine functions, and we have added the constant solution k_3. The solution (9) has the concentration boundary condition $C = k_3$ at $\psi = 2^{-3/2}(k_5 - \frac{\pi}{2})$ and the zero flux boundary condition $\mathbf{J} \cdot \mathbf{n} = 0$, where $\mathbf{n}$ is the outward (circumferential) normal vector at $\psi = 2^{-3/2}(k_5 - \pi)$. The solution is depicted schematically in Figure 1. Liquid at the lower radial boundary is maintained at concentration k_3; for example this may be the equilibrium saturated concentration where the liquid contacts a salt block. The liquid and the solute are contained by a barrier at the upper radial boundary. The concentration is initially at the uniform equilibrium value. For some time, water with a lower concentration of solute flows in from the origin, flushing the interior and reducing its solute concentration. After some time, the inflowing water again becomes saturated with solute and the interior again approaches its initial concentration. At each point, the concentration reaches its minimum value at time

$$T = 1 + (1 + \frac{1}{4}R^4)^{1/2}.$$

Unfortunately, it is common for symmetry solutions not to have easily interpretable boundary conditions because they are very special solutions with few parameters that can be adjusted to satisfy boundary conditions. Sometimes, the number of free parameters may be greatly increased because the first reduced equation happens to be equivalent to a standard constant-coefficient linear equation with many solutions obtainable by linear transforms or series methods. For example, the P.D.E. obtained by reduction under Γ_3 transforms to the negative Helmholtz equation

$$\frac{\partial^2 V}{\partial \chi^2} + \frac{\partial^2 V}{\partial v^2} - V = 0, \tag{10}$$

Table 1. Reduced P.D.E.s

Γ_i	Reduced P.D.E.
Γ_3	$4\rho^2 \frac{\partial^2 F}{\partial \rho^2} + 8\rho \frac{\partial F}{\partial \rho} + \frac{\partial^2 F}{\partial \gamma^2} = 0$ with $C = \exp\left(-\frac{T}{4} - \frac{\log T}{2} + \frac{1}{4}e^{-2\phi} - \frac{1}{16T}e^{-4\phi}\right) \times F(\rho,\gamma)$, $\rho = Te^{2\phi}$, $\gamma = \psi$
Γ_4	$16\rho^2 \frac{\partial^2 F}{\partial \rho^2} + (24\rho - 1) \frac{\partial F}{\partial \rho} + \frac{\partial^2 F}{\partial \gamma^2} = 0$ with $C = \exp\left(-\frac{T}{4} + \frac{1}{4}e^{-2\phi}\right) \times F(\rho,\gamma)$, $\rho = Te^{2\phi}$, $\gamma = \psi$
$\Gamma_3 + \Gamma_1$	$4\rho^2 \frac{\partial^2 F}{\partial \rho^2} + 8\rho \frac{\partial F}{\partial \rho} + \frac{\partial^2 F}{\partial \gamma^2} + \left(\frac{1}{4\rho^2} + \frac{1}{16\rho^4}\right) F = 0$ with $C = \exp\left(-\frac{T}{4} - \frac{1}{4}\tan^{-1} T - \frac{\log(T^2+1)}{4} + \frac{1}{4}e^{-2\phi} - \frac{T}{16(T^2+1)}e^{-4\phi}\right) \times F(\rho,\gamma)$, $\rho = \sqrt{T^2+1}\, e^{2\phi}$, $\gamma = \psi$
$\Gamma_3 + \Gamma_5$	$4\rho^4 \frac{\partial^2 F}{\partial \rho^2} + 8\rho^3 \frac{\partial F}{\partial \rho} + \frac{\rho^2}{\alpha^2} \frac{\partial^2 F}{\partial \gamma^2} + \frac{\partial F}{\partial \gamma} = 0$ with $C = \exp\left(-\frac{T}{4} - \frac{\log T}{2} + \frac{1}{4}e^{-2\phi} - \frac{1}{16T}e^{-4\phi}\right) \times F(\rho,\gamma)$, $\rho = Te^{2\phi}$, $\gamma = \frac{\psi}{\alpha} + \frac{1}{T}$
$\Gamma_3 + \Gamma_4 + \alpha\Gamma_5$	$4\rho^2 \frac{\partial^2 F}{\partial \rho^2} + 8\rho \frac{\partial F}{\partial \rho} + \frac{\partial^2 F}{\partial \gamma^2} - \alpha^2 \frac{\partial F}{\partial \gamma} + \left(\frac{1}{4\rho^2} + \frac{1}{16\rho^4}\right) F = 0$ with $C = \exp\left(-\frac{T}{4} - \frac{1}{4}\tan^{-1} T - \frac{\log(T^2+1)}{4} + \frac{1}{4}e^{-2\phi} - \frac{T}{16(T^2+1)}e^{-4\phi}\right) \times F(\rho,\gamma)$, $\rho = \sqrt{T^2+1}\, e^{2\phi}$, $\gamma = \tan^{-1} T - \frac{\psi}{\alpha}$
$\Gamma_4 + \alpha\Gamma_1$	$16\rho^2 \frac{\partial^2 F}{\partial \rho^2} + (24\rho - 1) \frac{\partial F}{\partial \rho} + \frac{\partial^2 F}{\partial \gamma^2} - \frac{\alpha}{\rho} F = 0$ with $C = \exp\left(-\frac{T}{4} + \frac{\alpha}{4}\log(T+\alpha) + \frac{1}{4}e^{-2\phi}\right) \times F(\rho,\gamma)$, $\rho = (T+\alpha)\, e^{4\phi}, \gamma = \psi$
$\Gamma_4 + \alpha\Gamma_5$	$16\rho^2 \frac{\partial^2 F}{\partial \rho^2} + (24\rho - 1) \frac{\partial F}{\partial \rho} + \frac{\gamma^2}{\alpha^2} \frac{\partial^2 F}{\partial \gamma^2} + \left(\frac{1}{\alpha^2} - \frac{1}{\rho^2}\right)\gamma \frac{\partial F}{\partial \gamma} = 0$ with $C = \exp\left(-\frac{T}{4} + \frac{1}{4}e^{-2\phi}\right) \times F(\rho,\gamma)$, $\rho = (T+\alpha)\, e^{4\phi}$, $\gamma = Te^{-\frac{\psi}{\alpha}}$
$\Gamma_4 + \alpha\Gamma_1 + \beta\Gamma_5$	$16\rho^2 \frac{\partial^2 F}{\partial \rho^2} + (24\rho - 1) \frac{\partial F}{\partial \rho} + \frac{\gamma^2}{\beta^2} \frac{\partial^2 F}{\partial \gamma^2} + \left(\frac{\gamma}{\beta^2} - \frac{\gamma}{\rho}\right)\gamma \frac{\partial F}{\partial \gamma} + \frac{\alpha}{4\rho} F = 0$ with $C = \exp\left(-\frac{T}{4} + \frac{\alpha}{4}\log(T+\alpha) + \frac{1}{4}e^{-2\phi}\right) \times F(\rho,\gamma)$ $\rho = (T+\alpha)\, e^{4\phi}$, $\gamma = (T+\alpha)e^{-\frac{\psi}{\beta}}$
$\Gamma_5 + \alpha\Gamma_2$	$\frac{\partial^2 F}{\partial \rho^2} - e^{-4\rho} \frac{\partial F}{\partial \gamma} + \left(2 + e^{-2\phi}\right) \frac{\partial F}{\partial \rho} + \alpha^2 F$ with $C = \exp(\alpha\,\psi) \times F(\rho,\gamma)$, $\rho = \phi$, $\gamma = T$

under the transformation

$$F = U(\chi, \upsilon)e^{-\chi} \;\; ; \;\; \rho = e^{2\chi} \;\; ; \;\; \psi = \upsilon \;\; ; \;\; V = \frac{\partial U}{\partial \upsilon}.$$

Equation (10) has many available solutions since it has been intensively studied in several physical applications including quasilinear steady unsaturated flow [e.g. *Philip*, 1985b; *Waechter and Philip*, 1985]. If we impose zero-concentration initial conditions and zero normal flux boundary conditions on a wedge,

$$C = 0 \;\; \text{at} \;\; T = 0,$$

$$\mathbf{J} \cdot \mathbf{n} = 0 \;\; \text{at} \;\; \psi = \psi_0, \psi_1,$$

these transform to

$$V \to 0, \;\; \chi \to -\infty, \tag{11}$$

$$V = 0, \;\; \upsilon = \psi_0, \psi_1. \tag{12}$$

Without loss of generality, we take $\psi_0 = 0$. By separation of variables within the self adjoint equation (10), we obtain a general Fourier series solution. Expressed in the original variables, this solution is, for $R > T^{1/2}$,

$$C = T^{-1/2} \exp\left(-\frac{1}{4}T + \frac{1}{4}R^2 - \frac{1}{16T}R^4\right)$$

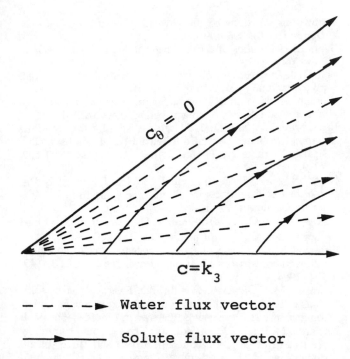

$c_\theta = 0$

$c = k_3$

- - - - ▶ Water flux vector

————————▶ Solute flux vector

Figure 1. Schematic representation of solute streamlines for solution given by Equation (9).

$$\times \sum_{n=0}^{\infty} A_n (TR^{-2})^{\alpha_n} \cos(n\pi\psi/\psi_1), \qquad (13)$$

and for $R < T^{1/2}$,

$$C = T^{-1/2} \exp\left(-\frac{1}{4}T + \frac{1}{4}R^2 - \frac{1}{16T}R^4\right)$$

$$\times \sum_{n=0}^{\infty} A_n (TR^{-2})^{-1-\alpha_n} \cos(n\pi\psi/\psi_1), \qquad (14)$$

where $\alpha_n = \frac{1}{2}[\sqrt{(n\pi/\psi - 1)^2 + 1} - 1]$.

Figure 2 is a polar plot of this solution for solute contained in a right angled wedge ($\psi_1 = \pi/2$), and with a step profile at T=1,

$$C(\psi, 1) = 1.0 \text{ for } \psi > \pi/4,$$

$$C(\psi, 1) = 0.8 \text{ for } \psi < \pi/4.$$

The dimensionless total solute content is

$$\int_0^\infty \int_0^{\psi_1} C(R, \psi, T)\, R\, dR d\psi = A_0 \pi \left[1 + \operatorname{erf}\left(\frac{\sqrt{T}}{2}\right)\right].$$

This shows that an amount $A_0\pi$ is deposited instantaneously at the origin and that an equal amount is injected continuously over time, with a decreasing supply rate. Note that total solute content does not depend on A_n with $n > 0$. These coefficients have no effect on total solute content or on mean solute transport rate across a circular arc $R = $ constant and $0 < \psi < \psi_1$. However, they dictate the variability of concentration and flux on polar angle. Although the source is isotropic with respect to water flux, it is not isotropic with respect to solute flux. Notionally, the continuous source represents a discharge pipe that is covered with a filter of variable strength.

6. CONCLUSIONS

In a letter dated 10th. February 1998, John Philip responded to congratulations on his admission to the Order of Australia:

> "...It's encouraging that the system can, in the end, tolerate the occasional oddball (though longevity may help). Of course I'm pleased; but, in a deep sense, the same reward has been the sheer fun of any (non-) labours over these 52 years."

We too have experienced a little of the fun in tackling an exacting mathematical problem in environmental mechanics. The symmetry analysis has produced a rich array of variable reductions and exact solutions for non-radial solute transport on a background of radial water flow. As far as we are aware, these are the only known solutions for non-radial two-dimensional isotropic velocity-dependent dispersion. However, we know that the outcomes fall a little short of the late John Philip's ideals of exact solutions for key nonlinear boundary value problems with direct testable practical implications. In symmetry analysis of complicated partial differential equations, it is unusual that the boundary conditions are directly interpretable. In the solutions that we have displayed, we maintain constant-concentration boundary conditions or zero-flux boundary conditions that are indeed interpretable. However, the solute injection rates at the source must be special functions of time. These may provide some insight on the effect of varying water velocity in dispersion but they are more likely to be important as bench tests for two-dimensional numerical schemes.

While the solute transport equations are linear, they have highly variable coefficients. We have found that this has been more troublesome than the nonlinear-

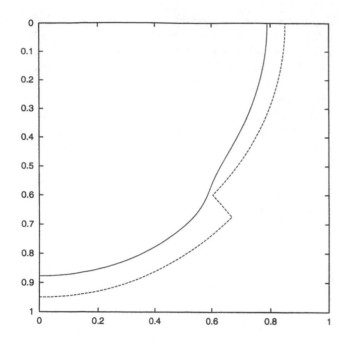

Figure 2. Polar plot for analytic solution of solute concentration given by Equation (13) showing concentration as a function of polar angle for given values of R and T. The step function represents the imposed condition at $(R, T) = (1, 1)$. The other curve shows smoothing at a later time; $(R, T) = (1.2, 1.2)$.

ity in symmetry classifications of the general nonlinear Richards equation for unsaturated water transport [*Oron and Rosenau*, 1986; *Sposito*, 1990; *Edwards*, 1994; *Edwards and Broadbridge*, 1994; *Yung et al.*, 1994; *Edwards and Broadbridge*, 1995; *Baikov et al.*, 1997]. However, the scope for symmetry analysis of linear P.D.E.s such as the solute transport analysis has greatly broadened following the initially surprising result of *Broadbridge and Arrigo* [1999] that every solution of any second or higher order linear P.D.E. is invariant under some classical Lie symmetry. This gives us hope that we may be able to incorporate boundary conditions from the outset of symmetry analyses.

Acknowledgments. R. Joel Moitsheki is grateful to the University Postgraduate Award, University of Wollongong and to the National Research Foundation of South Africa for financial assistance.

REFERENCES

Abramowitz, M. and I. A. Stegun (Eds.), *Handbook of Mathematical Functions*, 1046 pp., Dover Publications, New York, 1972.

Baikov, V. A., Gazizov, R. K., Ibragimov, N. H., and V. F. Kovalev, Water redistribution in irrigated soil profiles:

Invariant solutions of the governing equation, *Nonlinear Dynamics*, 13, 395-409, 1997.

Bear, J., *Hydraulics of Groundwater*, 569 pp., McGraw-Hill, New York, 1979.

Bell, R. J. *Introductory Fourier Transform Spectroscopy*, 329 pp., Academic, San Diego, Calif., 1972.

Bluman, G. W., and S. Kumei, *Symmetries and Differential Equations*, 412 pp., Springer-Verlag, New York, 1989.

Broadbridge, P., and D. J. Arrigo, All solutions of standard symmetric linear partial differential equations have classical Lie symmetry, *J. Math. Anal. Applic.* 234, 109-122, 1999.

Broadbridge, P., J. M. Hill, and J. M. Goard, Symmetry Reductions of Equations for Solute transport in Soil, *Nonlinear Dynamics*, 22, 15-27, 2000.

de Gennes, P. G., Hydrodynamic dispersal in unsaturated porous media, *J. Fluid Mech.*, 136, 189-200, 1983.

Edwards, M. P., Classical symmetry reductions of nonlinear diffusion-convection equations, *Phys. Lett.*, A 190, 149-154, 1994.

Edwards, M. P., and P. Broadbridge, Exact transient solutions to nonlinear-diffusion convection equations in higher dimensions, *J. Phys. A: Math. Gen.*, 27, 5455-5465, 1994.

Edwards, M. P., and P. Broadbridge, Exceptional symmetry reductions of Burgers' equation in two and three spatial dimensions, *J. Appl. Math. & Physics (ZAMP)*, 46, 595-622, 1995.

Eldor, M., and G. Dagan, Solutions of hydrodynamic dispersion in porous media, *Water Resour. Res.*, 8, 1316-1331, 1972.

Fry, V. A., Istok, J. D., and R. B. Guenther, An analytical solution to the solute transport equation with rate-limited desorption and decay, *Water Resour. Res.*, 29, 3201-3208, 1993.

Galaktionov, V. A., Dorodnitsyn, V. A., Elenin, G. G., Kurdyumov, S. P., and A. A. Samarskii, A quasilinear heat equation with a source: Peaking, localization, symmetry, exact solutions, asymptotics, structures, *J. Sov. Math.*, 41, 1222-1292, 1988.

Hearn, A. C., Reduce User's Manual Version 3.2, *Rand Publication CP78*, 130 pp., The Rand Corporation, Santa Monica, California, 1985.

Hill, J. M., *Differential Equations and Group Methods for Scientists and Engineers*, 161 pp., CRC Press, Boca Raton, Florida, 1992.

Hoopes, J. A. and D. R. F. Harleman, Dispersion in radial flow from a recharge well, *J. Geophys. Res.*, 72, 3595-3607, 1967.

Hsieh, P. A., A new formula for the analytical solution of the radial dispersion problem, *Water Resour. Res.*, 22, 1597-1605, 1986.

Ibragimov, N. H. (Ed.), *CRC Handbook of Lie Group Analysis in Differential Equations*, Vol. 2, 546 pp., CRC Press, Boca Raton, Florida, 1995.

Jury, W. A., Solute transport and dispersion, in *Flow and Transport in the Natural Environment: Advances and Applications*, edited by W. L. Steffen and O. T. Denmead, pp. 1-16, Springer-Verlag, Berlin, 1988.

Lie, S., Theorie der Transformationsgruppen, *Math. Ann.*, 16, 441-528, 1880.

Moench, A. F., Convergent radial dispersion: A Laplace transform solution for aquifer tracing testing, *Water Resour. Res.*, 25, 439-447, 1989.

Moench, A. F., Convergent radial dispersion: A note on evaluation of the Laplace transform solution , *Water Resour. Res.*, 27, 3261-3264, 1991.

Novakowski, K. S., The analysis of tracer experiments conducted in divergent radial flow fields, *Water Resour. Res.*, 28, 3215-3225, 1992.

Olver, P. J., *Applications of Lie Groups to Differential Equations*, 513 pp., Springer-Verlag, New York, 1986.

Oron, A., and P. Rosenau, Some symmetries of the nonlinear heat and wave equations, *Phys. Lett.*, A 118, 172-176, 1986.

Ovsiannikov, L. V., *Group Analysis of Differential Equations,* 416 pp., Academic Press, New York, 1982.

Philip, J. R., The theory of infiltration: 1. The infiltration equation and its solution, *Soil Sci.*, 83, 345-357, 1957.

Philip, J. R., Theory of infiltration, *Adv. Hydrosci.*, 5, 215-296, 1969.

Philip, J. R., Scattering functions and infiltration, *Water Resour. Res.*, 21, 1889-1894, 1985b.

Philip, J. R., Some exact solutions of convection-diffusion and diffusion equations, *Water Resour. Res.*, 30, 3545-3551, 1994.

Philip, J. R., Reply to Wang and Yeh, *Water Resour. Res.*, 32, 489-490, 1996.

Salles, J., J.-F. Thovert, R. Delannay, L. Presons, J.-L. Auriault, and P. M. Adler, Taylor dispersion in porous media: Determination of the dispersion tensor, *Phys. Fluids,* A 5, 2348-2376, 1993.

Ségol, G., *Classic Groundwater Simulations: Proving and Improving Numerical Models*, 531 pp., Prentice-Hall, Englewood Cliffs, New Jersey, 1994.

Sherring, J., DIMSYM: Symmetry determination and linear differential equations package, *Research Report www.latrobe.edu.au/www/mathstats/Maths/Dimsym/*, 55 pp., Latrobe University Mathematics Dept., Melbourne, 1993.

Sposito, G., Lie group invariance of the Richards equation, in *Dynamics of Fluids in Hierarchical Porous Media*, edited by J. H. Cushman, pp. 327-347, Academic Press, London, 1990.

Tang, D. H., and D. K. Babu, Analytical solution of a velocity dependent dispersion problem, *Water Resour. Res.*, 15, 1471-1478, 1979.

Taylor, G. I., Dispersion of soluble matter in solvent flowing through a tube, *Pro. R. Soc. London A.*, 219, 186-203, 1953.

Waechter, R. T., and Philip, J. R., Steady two- and three-dimensional flows in saturated soil:The scattering analog,*Water Resour res*, 21, 1875-1887, 1985.

Wierenga, P. J., Water and solute transport and storage, in *Handbook of Vadose Zone Characterization and Monitoring,* edited by L. G. Wilson, G. E. Lorne, and S. J. Cullen, pp. 41-60, Lewis Publishers, Boca Raton, Florida, 1995.

Woods, J., C. T. Simmons, and K. A. Narayan, Verification of black box groundwater models, in *EMAC98: Proceedings of the Third Biennial Engineering Mathematics and Applications Conference*, edited by E. O. Tuck and J. A. K. Stott, pp. 523-526, Institution of Engineers Australia, Adelaide, 1998.

Yung, C. M., Verburg, K., and P. Baveye, Group classification and symmetry reductions of the non-linear diffusion-convection $u_t = (D(u)u_x)_x - K'(u)u_x$, *Internat. J. Nonlin. Mechs.*, 29, 273-278, 1994.

Zoppou, C., and J. H. Knight, Analytical solution of a spatially variable coefficient advection-diffusion equation in one-, two- and three- dimensions, *Applied Mathematical Modelling*, 23, 667-685, 1998.

P. Broadbridge, Maureen P. Edwards and R. Joel Moitsheki, Institute for Mathematical Modelling and Computational Systems, University of Wollongong, NSW 2522, Australia.   (e-mail: phil_broadbridge@uow.edu.au; maureen_edwards@uow.edu.au; rjm02@uow.edu.au)

Stability Criteria for the Vertical Boundary Layer Formed by Throughflow Near the Surface of a Porous Medium

C. J. van Duijn, G. J. M. Pieters

Department of Mathematics and Computer Science, Eindhoven University of Technology, Eindhoven, The Netherlands

R. A. Wooding

CSIRO Land and Water, Canberra, Australia

A. van der Ploeg

MARIN, Wageningen, The Netherlands

We consider gravitational instability of a saline boundary layer formed by evaporation induced upward throughflow at a horizontal surface of a porous medium. Two paths are followed to analyse stability: the energy method and the method of linearised stability. The energy method requires constraints on saturation and velocity perturbations. The usual constraint is based on the integrated Darcy equation. We give a fairly complete analytical treatment of this case and show that the corresponding stability bound equals the square of the first root of the Bessel function J_0. This explains previous numerical investigations by *Homsy & Sherwood* [1975, 1976]. We also present an alternative energy method using the pointwise Darcy equation as constraint, and we consider the time dependent case of a growing boundary layer. This alternative energy method yields a substantially higher stability bound which is in excellent agreement with the experimental work of *Wooding et al.* [1997a, b]. The method of linearised stability is discussed for completeness because it exhibits a different stability bound. The theoretical bounds are verified by two-dimensional numerical computations. We also discuss some cases of growing instabilities. The presented results have applications to the theory of stability of salt lakes and the salinization of groundwater.

1. INTRODUCTION

Consider a semi-infinite porous medium with a horizontal upper boundary. If a uniform upward flow exists within the medium and through the boundary, and if appropriate boundary conditions apply, a spatially one-dimensional boundary layer may be created and sustained by the outflow. For in-

Environmental Mechanics: Water, Mass and Energy Transfer in the Biosphere
Geophysical Monograph 129
Copyright 2002 by the American Geophysical Union
10.1029/129GM15

stance, if the surface is maintained at a temperature different from that of the medium and the saturating fluid, a thermal boundary layer is formed with an equilibrium thickness proportional to the ratio of thermal diffusivity to upflow rate. Similarly, a boundary layer is formed by dispersing solute if the solute concentration at the boundary differs from the concentration of the solution issuing from the medium.

Such flows occur naturally in areas of groundwater discharge. These may be characterised by very low flow rates, leading to boundary layers of significant thickness. An upflow of warm or hot groundwater has been postulated for some shallow geothermal areas (*Wooding* [1960]). As the surface is relatively cold, a thermal boundary layer of cool water is formed below the surface. A reversal of this situation relative to gravity may arise for *in situ* coal gasification (*Homsy & Sherwood* [1975]), where a hot reaction surface forms a boundary layer at the lower horizontal boundary of a cooler permeable layer. Boundary layers are also formed in semi-arid regions containing extensive areas of groundwater discharge (*Gilman & Bear* [1996]; *Wooding et al.* [1997a]). The groundwater contains salt. After throughflow induced by evaporation, the salt remains behind at the surface to form saline deposits (salt lakes). These salt lakes may be 'dry' at the surface under the influence of evaporation, or may contain standing water (ponding), perhaps varying seasonally between the two states.

In each of these examples the fluid in the horizontal groundwater boundary layer differs in density from the fluid in the adjacent permeable medium, and the question of the gravitational stability of the boundary layer arises. *Wooding* [1960] treated the case of a constant-pressure (ponded) boundary by linearised stability theory. *Jones & Persichetti* [1986] applied linear analysis to a permeable layer with all combinations of boundary condition and throughflow direction. *Nield* [1987] obtained approximate stability criteria by variational means. *Gilman & Bear* [1996] treated the linearised stability of a horizontal unsaturated layer (vadoze zone) overlying a shallow water table. *Wooding et al.* [1997a, b] discussed saturated groundwater movement with dry or ponded conditions at the surface, and used both experimental and numerical methods to simulate the unstable behaviour of a boundary layer growing from an initial salinity discontinuity at the surface, and including the margin, of a dry salt lake.

In an important step, *Homsy & Sherwood* [1975, 1976] pointed out that the presence of throughflow contributes non-symmetric (odd-order) terms to the stability equations. The linear, time-independent part of the stability equations is not self-adjoint, and linear stability analysis is applicable only when the system is definitely unstable. Subcritical instabilities of finite amplitude are possible at Rayleigh numbers below the critical value derived using linear theory (*Davis* [1971], *Straughan* [1992]).

In the present work we are concerned with this aspect and also with the stability of a growing boundary layer. For simplicity we consider only the dry lake case in a vertical upflow, in which we assume that a rapidly established saturated surface layer exists yielding a steady boundary condition for the salt concentration. We will employ both the energy (variational) method and the method of linearised theory.

1.1. Stability of the Equilibrium Saline Boundary Layer

In applying the energy method we follow two approaches. The first one is the 'standard approach' as outlined, for example, by *Homsy & Sherwood* [1975, 1976] or by *Straughan* [1992]. In this approach one incorporates an integral constraint in the class of admissible perturbations, which is based on continuity and the integrated Darcy equation. The Euler–Lagrange equations with boundary conditions can be combined into a second order eigenvalue problem with time as a parameter. One of the goals of this paper is to demonstrate that at equilibrium, when the boundary layer has reached its large time profile, this eigenvalue problem can be solved in terms of Bessel functions yielding

$$R_{E_1} = 5.7832 \qquad (1.1)$$

as a value of the Rayleigh number below which the system is definitely stable; note that $\sqrt{R_{E_1}}$ is the first root of the Bessel function J_0 (*Abramowitz & Stegun* [1972, p. 409]).

In a second approach we deviate from Homsy & Sherwood and consider a different maximum problem. Using the same functional, we replace the integral constraint with an exact differential relation which is now based on continuity and the 'pointwise' Darcy equation. This yields a sixth order eigenvalue problem which we solve numerically by the Jacobi–Davidson method. With the given boundary conditions we find approximately

$$R_{E_2} = 8.590 \qquad (1.2)$$

as the largest Rayleigh number below which the system is definitely stable. The close agreement of this result with the numerical results of *Pieters* [2001] and the experimental results of *Wooding et al.* [1997a, b] is discussed in Section 5.

For completeness we also consider the linearised stability analysis of the equilibrium boundary layer. This yields a fourth order eigenvalue problem. Using again the Jacobi–Davidson method we find approximately

$$R_L = 14.35 \qquad (1.3)$$

as a critical Rayleigh number above which the system is definitely unstable.

Given the physical parameters of the system a value for the Rayleigh number R_s results. This value may fall within one of three ranges: definitely stable for $R_s \leqslant R_{E_i}$ ($i = 1, 2$), definitely unstable for $R_s > R_L$, and possibly unstable to disturbances of finite amplitude (leading to subcritical instabilities) when $R_{E_i} < R_s \leqslant R_L$.

Homsy & Sherwood [1976] considered throughflow in a finite slab. Their numerical results for large thickness of the slab approximately give the critical Rayleigh numbers (1.1) and (1.3).

1.2. *Time Dependent Growth of the Saline Boundary Layer*

Problems of fluid instability with impulsively-generated (time-dependent) density profiles have been discussed, in particular, by *Homsy* [1973], who used the energy method to treat global stability of fluid layers, and *Caltagirone* [1980], who compared the stability behaviour using linear and energy methods and also used finite-difference computations for a horizontal porous layer with a sudden rise in surface temperature. These studies, however, did not involve a superimposed throughflow.

Our case involves a dispersive boundary layer in an upflow, and we shall identify approximate parameter values where instability is likely to occur. Section 3 explains the stability analysis for a growing boundary layer. Here time t appears as a parameter. In the early stages of development, the layer is sufficiently thin to be stabilised by the given boundary conditions. However, the monotonic increase in layer thickness with time will be accompanied by decreasing stability of the system as the influence of the boundary diminishes. This is shown in Figures 2 and 3.

Figure 2 shows a family of curves in the a, R plane, a denoting the horizontal wavenumber, for increasing values of t. The curves are obtained with the energy method based on the differential constraint. For a given time $t > 0$, corresponding to an instantaneous state of the growing boundary layer, let $R_E(t)$ denote the minimum of the corresponding curve. Similarly, Figure 3 shows a family of curves obtained with the linearised stability method. Now, let $R_L(t)$ denote the minimum of the curve corresponding to time t.

We now have the following refinement with respect to the equilibrium case. If $R_s \leqslant R_E(\infty) = R_{E_2}$, the layer will attain a stable equilibrium profile. If, however, $R_s > R_{E_2}$ we can determine a time t_E^s, corresponding to $R_s = R_E(t_E^s)$, and conclude the stability of the growing boundary layer for $t < t_E^s$. On the other hand, if $R_s > R_L$ we can nominate an elapsed time t_L^s corresponding to $R_s = R_L(t_L^s)$ and conclude the instability of the layer for $t > t_L^s$. These observations follow from the nature of the curves in Figures 2 and 3. The curves in Figure 2 are upper bounds for regions of stable (a, R) combinations, whereas the curves in Figure 3 are lower bounds for regions of unstable (a, R) combinations.

The shape, i.e. number of 'salt-fingers' or critical wavenumber, of growing instabilities depends substantially upon the perturbations present during the initial stable period. This is investigated numerically in Section 4, where we use a finite element approach based on the stream function formulation. If initial perturbations are periodic and sufficiently small we observe growing instabilities in the theoretically predicted range. This is shown in Figure 5. Other perturbations are considered as well. Some qualitative features of the

computational results are explained in terms of the stability bounds. In particular the stochastic case meets the theory quite satisfactory (see Sub-subsection 4.3-*a*).

In Section 5 we present conclusions and discuss experimental Hele–Shaw results (*Wooding et al.* [1997a, b]) in terms of our theoretical findings. Theory and experiment are reproduced in Figure 10, showing excellent agreement.

The results presented in this contribution are taken from three extensive technical reports, *Van Duijn et al.* [2001a, b] and *Pieters* [2001]. These reports are available upon request from the authors at Eindhoven University of Technology.

2. PROBLEM FORMULATION

Following *Wooding et al.* [1997a], we consider a uniform isotropic porous medium occupying the three dimensional halfspace $\Omega = \{(x, y, z) : -\infty < x, y < \infty, \ z > 0\}$, where z points vertically downwards. The medium is saturated with a fluid of variable density ρ: i.e. water with dissolved salt. Along the upper boundary $\{z = 0\}$ we prescribe density and fluid flow corresponding to a 'dry lake bed', with a sufficient rate of evaporation to remove all free surface water and a rapid buildup of salt. Yet it is assumed that the salt solution everywhere and always fills up the pore space. If ρ_r denotes the fluid density in 'natural circumstances' (i.e. far away from the outflow boundary) and ρ_m the maximum density at the outflow boundary, we have $\rho_r \leqslant \rho \leqslant \rho_m$ throughout the flow domain Ω. Here ρ_m may represent the fluid density in an overlying pond or the density of the salt-saturated solution.

The flow equations in terms of the Boussinesq approximation (*Bear* [1972], *Nield & Bejan* [1992], *Wooding et al.* [1997a]) are given by:

Fluid incompressibility

$$\text{div } \boldsymbol{q} = 0 \; ; \qquad (2.1)$$

Darcy's law

$$\frac{\mu}{\kappa} \boldsymbol{q} + \text{grad } p - \rho g \boldsymbol{e}_z = \boldsymbol{0} \; ; \qquad (2.2)$$

Salt transport

$$\phi \frac{\partial \rho}{\partial t} + \text{div } (\rho \boldsymbol{q}) = \mathbb{D} \Delta \rho \; . \qquad (2.3)$$

Here $\boldsymbol{q}$ denotes fluid discharge, μ fluid viscosity, κ medium permeability, p fluid pressure, g gravity constant, ϕ porosity and $\mathbb{D}$ an appropriately defined dispersivity or diffusivity. Further, $\boldsymbol{e}_z$ denotes the unit vector in z-direction, pointing downwards.

These equations are considered in Ω subject to the boundary conditions

$$\boldsymbol{q} = -E \boldsymbol{e}_z \quad \text{and} \quad \rho = \rho_m \quad \text{at } z = 0 \qquad (2.4)$$

and initial condition

$$\rho\big|_{t=0} = \rho_r \quad \text{in } \Omega \,. \qquad (2.5)$$

Here E denotes the evaporation rate.

We recast the problem in dimensionless form by setting

$$S = \frac{\rho - \rho_r}{\rho_m - \rho_r} \quad \text{and} \quad U = \frac{q}{u_c} \,, \quad u_c = \frac{(\rho_m - \rho_r)g\kappa}{\mu} \,, \qquad (2.6)$$

and by introducing the thickness of the equilibrium boundary layer $\mathbb{D}/E$ and $\phi\mathbb{D}/E^2$, respectively, as scales for length and time. This yields

$$\text{div } U = 0 \,, \qquad (2.7)$$

$$U + \text{grad } P - S e_z = 0 \,, \qquad (2.8)$$

$$\frac{\partial S}{\partial t} + R_s U \cdot \text{grad } S = \Delta S \,. \qquad (2.9)$$

in Ω and for all $t > 0$, subject to

$$U = U_0 := -\frac{1}{R_s} e_z \quad \text{and} \quad S = 1 \quad \text{at } z = 0 \quad (2.10)$$

and

$$S\big|_{t=0} = 0 \quad \text{in } \Omega \,. \qquad (2.11)$$

Here $P = (p - \rho_r g \frac{\mathbb{D}}{E} z)/(\rho_m - \rho_r)g\frac{\mathbb{D}}{E}$ represents departures of the dimensionless pressure from hydrostatic conditions and R_s the system Rayleigh number

$$R_s = \frac{(\rho_m - \rho_r)g\kappa}{\mu E} = \frac{u_c}{E} \,. \qquad (2.12)$$

The main purpose of this paper is to investigate the stability properties of the flow problem defined by (2.7)–(2.12). More specifically, we will consider the stability of the ground state implied by the uniform initial condition (2.11). Because of the constant boundary data (2.10), this ground state can be determined explicitly. It is characterized by the uniform upflow

$$U = U_0 \quad \text{in } \Omega \qquad (2.13)$$

and the growing boundary layer, for $z > 0$,

$$S = S_0(z,t) = \frac{1}{2}e^{-z}\text{erfc}\left[\frac{z-t}{2\sqrt{t}}\right] + \frac{1}{2}\text{erfc}\left[\frac{z+t}{2\sqrt{t}}\right] \,, \qquad (2.14)$$

satisfying

$$S_0(z,t) \to e^{-z} \quad \text{as } t \to \infty \,. \qquad (2.15)$$

The corresponding pressure $P = P_0$ is found by integrating Darcy's law (2.8). The stability analysis is based on the expansion

$$S = S_0 + s \,, \quad U = U_0 + u \quad \text{and} \quad P = P_0 + p \,, \qquad (2.16)$$

with $u = (u, v, w)$, and where S, U and P satisfy equations (2.7)–(2.9) and boundary conditions (2.10). In the next section we study the corresponding perturbation equations. In the analysis we drop the subscript s on R_s and denote the Rayleigh number by R. This is to distinguish between R as an eigenvalue in the equations and its value R_s for the actual physical system.

3. ANALYSIS OF PERTURBATION EQUATIONS

Based on experimental observations of early instabilities we assume that the perturbations are periodic in the horizontal x, y plane. Further we require that the perturbations vanish at and far below the outflow boundary:

$$s = u = 0 \quad \text{at } z = 0, \infty \,, \qquad (3.1)$$

expressing that $\{S, U\}$ and $\{S_0, U_0\}$ both satisfy (2.10) and behave similarly at large depth.

3.1. Perturbation Equations

Substituting (2.16) into equations (2.7)–(2.9) and writing R instead of R_s, yields the system (in Ω and for all $t > 0$)

$$\text{div } u = 0 \,, \qquad (3.2)$$

$$u + \text{grad } p - s e_z = 0 \,, \qquad (3.3)$$

$$\frac{\partial s}{\partial t} - \frac{\partial s}{\partial z} + Rw\frac{\partial S_0}{\partial z} + Ru \cdot \text{grad } s = \Delta s \,. \qquad (3.4)$$

As in *Lapwood* [1948] we note that equations (3.2) and (3.3) can be combined to give for s and w the linear relation

$$\Delta w = \Delta_\perp s \quad \text{in } \Omega \,, \qquad (3.5)$$

where $\Delta_\perp$ denotes the horizontal Laplacian

$$\frac{\partial^2}{\partial x^2} + \frac{\partial^2}{\partial y^2} \,.$$

This relation plays a crucial role in various parts of the stability analysis.

Because of the assumed x, y-periodicity, we may restrict the analysis of equations (3.2)–(3.4) to the periodicity cell

$$\mathcal{V} = \left\{(x,y,z) : |x| < \pi/a_x, \ |y| < \pi/a_y, \ 0 < z < \infty\right\} \,, \qquad (3.6)$$

where a_x and a_y are the, as yet unspecified, horizontal wavenumbers. We call

$$a := \sqrt{a_x^2 + a_y^2} \qquad (3.7)$$

the horizontal wavenumber of the periodicity cell $\mathcal{V}$.

There are two well-known paths to carry out the stability analysis: the variational energy method and the method of linearised stability. Some important references in this respect are *Wooding* [1960], *Nield* [1987], *Straughan* [1992] and *Homsy & Sherwood* [1976]. Because of the existing throughflow ($U = U_0$), the energy method and the linearized stability method yield different stability bounds on the Rayleigh number. Therefore we will discuss both approaches.

3.2. Variational Energy Method

In the energy method one estimates the time derivative of the L^2-norm of the saturation perturbation. In particular, the aim is to find the largest R-interval for which

$$\frac{d}{dt}\int_{\mathcal{V}} s^2 < 0 . \tag{3.8}$$

Here and in integrals below we disregard the infinitesimal volume elements in the notation. The related maximum R-value clearly will depend on the wavenumber a and, because $S_0 = S_0(z,t)$, on time t. Once (3.8) is established, it follows that the L^2-norm of the velocity perturbation is bounded as well, since (*Van Duijn et al.* [2001b])

$$\int_{\mathcal{V}} |\boldsymbol{u}|^2 \leqslant \int_{\mathcal{V}} s^2 . \tag{3.9}$$

This is a direct consequence of (3.2) and (3.3).

To investigate (3.8), we multiply (3.4) by s and integrate over $\mathcal{V}$. Using (3.2) we find the identity

$$\frac{d}{dt}\frac{1}{2}\int_{\mathcal{V}} s^2 = -\int_{\mathcal{V}} |\text{grad } s|^2 - R\int_{\mathcal{V}} sw\frac{\partial S_0}{\partial z} . \tag{3.10}$$

Thus if R is chosen such that the right-hand side of (3.10) is negative for all perturbations satisfying a given constraint, then stability is guaranteed.

It is our aim to investigate the consequences of two different constraints. In the first we consider perturbations satisfying (3.2) and the integrated Darcy equation:

$$\int_{\mathcal{V}} |\boldsymbol{u}|^2 - \int_{\mathcal{V}} sw = 0 . \tag{3.11}$$

This approach is a modification of that used by *Homsy & Sherwood* [1976]. While they considered a stationary ground state only and solved the corresponding eigenvalue problem numerically, we are in the position to deal with time evolution of the primary profile as well. However, we shall not pursue the time dependence for this constraint. Instead we give a complete analytical treatment of the case where the ground state is given by the equilibrium case (2.15) for all $t > 0$. This analysis explains quite elegantly some of the previously obtained numerical results.

In the second constraint, we consider perturbations satisfying the differential expression (3.5). We shall treat the

time dependent ground state and show that this differential constraint significantly improves integral constraint (3.11).

3.2.1. Integral constraint.   Identity (3.10) and constraints (3.2), (3.11) lead to the maximum problem

$$\frac{1}{R} = \sup_{(s,\boldsymbol{u})\in\mathbf{H}} \frac{-\int_{\mathcal{V}} \frac{\partial S_0}{\partial z}sw}{\int_{\mathcal{V}} |\text{grad } s|^2} \tag{3.12}$$

with

$$\mathbf{H} = \big\{(s,\boldsymbol{u}) : x,y\text{-periodic with respect to } \mathcal{V},$$
$$s = \boldsymbol{u} = 0 \text{ at } z = 0,\infty,$$
$$\text{div } \boldsymbol{u} = 0 \text{ and } \int_{\mathcal{V}} |\boldsymbol{u}|^2 = \int_{\mathcal{V}} sw\big\} .$$

The corresponding Euler–Lagrange equations are

$$\begin{cases} -2\Delta s + R\frac{\partial S_0}{\partial z}w - \mu w = 0 , \\ 2\mu\boldsymbol{u} - \text{grad } \pi + R\frac{\partial S_0}{\partial z}s\boldsymbol{e}_z - \mu s\boldsymbol{e}_z = \mathbf{0} , \\ \text{div } \boldsymbol{u} = 0 \quad\text{and}\quad \int_{\mathcal{V}} |\boldsymbol{u}|^2 = \int_{\mathcal{V}} sw , \end{cases}$$

where μ (constant in space) and π are Lagrange multipliers. Applying the scaling

$$\boldsymbol{u} := \frac{\lambda}{\sqrt{R}}\boldsymbol{u} , \quad \mu = \frac{R}{\lambda^2} \quad\text{and}\quad p = -\frac{1}{2}\frac{\lambda}{\sqrt{R}}\pi ,$$

one finds

$$\begin{cases} \frac{\sqrt{R}}{2}\left(\frac{1}{\lambda} - \lambda\frac{\partial S_0}{\partial z}\right)w + \Delta s = 0 , & (3.13) \\ \frac{\sqrt{R}}{2}\left(\frac{1}{\lambda} - \lambda\frac{\partial S_0}{\partial z}\right)s\boldsymbol{e}_z - \boldsymbol{u} - \text{grad } p = \mathbf{0} , & (3.14) \\ \text{div } \boldsymbol{u} = 0 , & (3.15) \\ \int_{\mathcal{V}} |\boldsymbol{u}|^2 = \frac{\sqrt{R}}{\lambda}\int_{\mathcal{V}} sw . & (3.16) \end{cases}$$

These equations were also found by Homsy & Sherwood with a slightly different interpretation of the parameter λ.

Note that (3.14) has a structure similar to Darcy's law.

As before, (3.14) and (3.15) can be combined to give

$$\Delta w = \frac{\sqrt{R}}{2}\left(\frac{1}{\lambda} - \lambda\frac{\partial S_0}{\partial z}\right)\Delta_\perp s . \tag{3.17}$$

Further, multiplying (3.14) by $\boldsymbol{u}$, integrating the result over $\mathcal{V}$, and using (3.16) yields the useful identity

$$\lambda^2 = \frac{\int_{\mathcal{V}} sw}{-\int_{\mathcal{V}} \frac{\partial S_0}{\partial z}sw} . \tag{3.18}$$

Finally, multiplying (3.13) by s, integrating the result over $\mathcal{V}$, and using (3.18) gives

$$\int_{\mathcal{V}} |\operatorname{grad} s|^2 = \frac{\sqrt{R}}{\lambda} \int_{\mathcal{V}} sw \,. \qquad (3.19)$$

Next we introduce the periodicity. Setting $s := as$, with a given by (3.7), we find from (3.13) and (3.17) the equations (with D signifying d/dz)

$$\left(D^2 - a^2\right) s + \frac{a\sqrt{R}}{2}\left(\frac{1}{\lambda} - \lambda\frac{\partial S_0}{\partial z}\right) w = 0\,, \qquad (3.20)$$

$$\left(D^2 - a^2\right) w + \frac{a\sqrt{R}}{2}\left(\frac{1}{\lambda} - \lambda\frac{\partial S_0}{\partial z}\right) s = 0\,, \qquad (3.21)$$

for $0 < z < \infty$. Note that in these equations t appears as a parameter through the ground state. We seek non-trivial solutions subject to the homogeneous conditions (3.1) and the constraint (3.18).

As a first observation we note that (3.20), (3.21) and the boundary conditions imply $s = w$. Hence we are left with the second order boundary value problem (for $0 < z < \infty$)

$$\begin{cases} \left(D^2 - a^2\right) s + \dfrac{a\sqrt{R}}{2}\left(\dfrac{1}{\lambda} - \lambda\dfrac{\partial S_0}{\partial z}\right) s = 0\,, & (3.22) \\ s(0) = s(\infty) = 0\,, & (3.23) \end{cases}$$

subject to the constraint (replacing w by s in (3.18))

$$\lambda^2 = \frac{\displaystyle\int_0^\infty s^2}{-\displaystyle\int_0^\infty \frac{\partial S_0}{\partial z} s^2}\,. \qquad (3.24)$$

Identity (3.19) rewrites into

$$\int_0^\infty (Ds)^2 = \left(\frac{a\sqrt{R}}{\lambda} - a^2\right)\int_0^\infty s^2\,. \qquad (3.25)$$

This expression and equation (3.22), using $\partial S_0/\partial z \to 0$ as $z \to \infty$, imply that nontrivial solutions only exist in the parameter range

$$1 < \frac{\sqrt{R}}{a\lambda} < 2\,. \qquad (3.26)$$

So far we have not used the explicit form of S_0. In the analysis below we confine ourselves to the equilibrium case (2.15), where S_0 is a simple decaying exponential. Introducing the new parameters

$$\delta = \frac{\sqrt{R}}{a\lambda} \quad (\text{with } 1 < \delta < 2)\,, \quad \alpha = \sqrt{\frac{2R}{\delta}}\,,$$

$$\beta = \beta(a,\delta) = 2a\sqrt{1 - \frac{\delta}{2}}\,, \qquad (3.27)$$

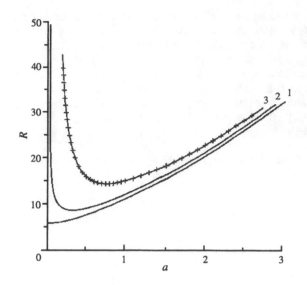

Figure 1. Comparison of estimates involving lowest eigenvalue R_1 versus wavenumber a for the equilibrium boundary layer. Curve 1: Energy method using integral constraint. Curve 2: Energy method using differential constraint. Curve 3: Linearised stability method using Jacobi–Davidson (solid curve) and Frobenius expansion (crossed points).

and the transformation

$$\xi = \alpha e^{-z/2}\,, \quad f(\xi) = s(z)\,, \qquad (3.28)$$

we find for f a boundary value problem involving the Bessel equation

$$\xi^2 f'' + \xi f' + (\xi^2 - \beta^2)f = 0 \quad \text{on } 0 < \xi < \alpha\,, \qquad (3.29)$$

with

$$f(0) = f(\alpha) = 0\,. \qquad (3.30)$$

Here primes denote differentiation with respect to ξ. A solution of (3.29) satisfying the first condition in (3.30) is

$$f(\xi) = J_\beta(\xi)\,, \qquad (3.31)$$

with J_β denoting the Bessel function of the first kind, order β. Next we fix $a > 0$ and consider

$$J_{\beta(a,\delta)}(\xi_1) = 0 \quad \text{for } 1 < \delta < 2\,, \qquad (3.32)$$

where $\xi_1 = \xi_1(a,\delta)$ is the first positive zero of J_β. Then setting $\alpha = \xi_1$ in the second equation of (3.27), we obtain the first eigenvalue R_1 for the given values of a and δ:

$$R_1 = R_1(a,\delta) = \frac{1}{2}\delta\big(\xi_1(a,\delta)\big)^2 \quad \text{for } 1 < \delta < 2\,. \qquad (3.33)$$

Keeping a fixed, we now turn to the integral constraint (3.24). In the transformed variables it reads

$$\frac{1}{\delta} = 2a^2 \frac{\int_0^{\xi_1} \frac{1}{\xi} J_\beta^2(\xi) d\xi}{\int_0^{\xi_1} \xi J_\beta^2(\xi) d\xi} \ . \qquad (3.34)$$

The question now arises whether there exists a unique number $\delta_a \in (1,2)$ such that $\delta = \delta_a$ satisfies (3.34). This would result in the first eigenvalue

$$R_1(a) := R_1(a, \delta_a) \quad \text{for } a > 0 \ . \qquad (3.35)$$

The proof involves some technical details which are given in *Van Duijn et al.* [2001b]. The energy stability curve in the a, R plane is plotted as curve 1 in Figure 1. If perturbations are x, y-periodic with wavenumber a and if $R_s < R_1(a)$, then the ground state (at equilibrium) is stable in the L^2-sense. The construction implies

$$R_1(0) = \lim_{a \downarrow 0} R_1(a) = 5.78318 \cdots \quad (\text{first zero of } J_0)^2 \ . \qquad (3.36)$$

Homsy & Sherwood used a numerical shooting method to solve the eigenvalue problem. They found (3.36) approximately as a stability bound.

3.2.2. Differential constraint.

In a second approach we want to achieve (3.8) for perturbations satisfying the diffential constraint (3.5). This leads to a maximum problem in which (3.12) is considered for the space of perturbations

$$\widetilde{\mathbf{H}} = \big\{ (s, w) : x, y\text{-periodic with respect to } \mathcal{V},$$
$$s = w = 0 \text{ at } z = 0, \infty, \text{ and } \Delta w = \Delta_\perp s \text{ in } \mathcal{V} \big\} \ .$$

This maximum problem results in an eigenvalue problem which has a much higher complexity than the eigenvalue problem related to (3.11). In fact it leads to a sixth order differential equation in terms of w, for which no explicit solution is known. However, one expects to have a more accurate description, yielding larger Rayleigh numbers, in particular since (3.5) is based on the pointwise Darcy equation. This statement is made precise in Appendix A.

Now the Euler–Lagrange equations read (*Van Duijn et al.* [2001b]):

$$\left(D^2 - a^2\right) s = \frac{R}{2} \frac{\partial S_0}{\partial z} w + \frac{a^2}{2} \pi \ , \qquad (3.37)$$

$$\left(D^2 - a^2\right) \pi = -R \frac{\partial S_0}{\partial z} s \ , \qquad (3.38)$$

with $\pi(0) = 0$ as natural boundary condition, and

$$\left(D^2 - a^2\right) w = -a^2 s \ . \qquad (3.39)$$

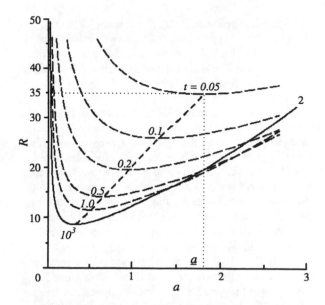

Figure 2. Stability curves for equilibrium boundary layer according to the energy method with differential constraint. Dashed curves show lowest eigenvalue R_1 versus wavenumber a prior to equilibrium, treating time as parameter. Numerical values are calculated by the Jacobi–Davidson method. Short-dashed curve traces minima of stability curves with increasing $t > 0$. Solid curve 2 is taken from Figure 1 (equilibrium case).

These equations need to be solved for $0 < z < \infty$ and they contain time t (through $S_0 = S_0(z, t)$) as parameter. Eliminating π from equations (3.37) and (3.38) yields a fourth order equation in s and w, and the further elimination of s using (3.39) leads to the sixth order w equation

$$\left(D^2 - a^2\right)^3 w + \frac{a^2 R}{2} \left\{ \left(D^2 - a^2\right) \left(\frac{\partial S_0}{\partial z} w\right) \right.$$
$$\left. + \frac{\partial S_0}{\partial z} \left(D^2 - a^2\right) w \right\} = 0 \ . \qquad (3.40)$$

The corresponding boundary conditions for this equation are

$$w(\infty) = 0 \ , \qquad (3.41)$$

implying that all higher order derivatives vanish as well at $z = \infty$, and

$$w(0) = D^2 w(0) = D^4 w(0) = 0 \ . \qquad (3.42)$$

The first two conditions are obvious. The third one is a consequence of $\pi(0) = 0$; this condition implies $D^2 s(0) = 0$ from (3.37), which is then used in (3.39). In terms of the variables w, s and π, we have the homogeneous conditions

$$w = s = \pi = 0 \quad \text{at} \quad z = 0, \infty. \qquad (3.43)$$

The eigenvalue problem (3.40), (3.41) and (3.42), or equivalently (3.37)–(3.39) subject to (3.43), was solved numerically by the Jacobi–Davidson method. This method is briefly described in *Van Duijn et al.* [2001b]. Detailed information is given in *Fokkema et al.* [1999] and *Sleijpen & Van Der Vorst* [1996].

For a given wavenumber $a > 0$ and time $t > 0$, let $R_E(a, t)$ denote the smallest positive eigenvalue. The dashed curves in Figure 2 show the numerical approximations of the curves $\{(a, R) : a > 0, \ R = R_E(a, t)\}$ for increasing values of t. Note that these curves essentially move downwards, except for large a and t. At large time they converge to the equilibrium curve, corresponding to (2.15). This limit case is also shown in Figure 1 (curve 2). The results obtained with the differential constraint are superior to the results obtained with the integral constraint. In particular, the minimum of curve 2 is $R = 8.590$ approximately, which is significantly higher than the minimum of about $R = 5.78$ of curve 1.

To interpret the results of the time dependent case, we set

$$R_E(t) := \min_{a>0} R_E(a, t) \quad \text{for } 0 < t < \infty \qquad (3.44)$$

and we recall the Rayleigh number of the physical system R_s, given by (2.12).

If $R_s < R_E(\infty) =: R_E$, which we denoted by R_{E_2} in the introduction, the boundary layer is definitely stable for all $t > 0$. However, if $R_s > R_E$, we can only conclude that the boundary layer is stable for $0 < t < t_E^s$, where t_E^s is determined by $R_s = R_E(t_E^s)$. When $t > t_E^s$ no direct conclusions can be drawn. The appearance and form of the growing instabilities critically depends on the choice of initial perturbations. This is further investigated in Section 4.

3.3. Linearised Stability

In the method of linearised stability one disregards the higher order terms in (3.4) and considers the approximate linear saturation equation

$$\frac{\partial s}{\partial t} - \frac{\partial s}{\partial z} + Rw\frac{\partial S_0}{\partial z} = \Delta s \quad \text{in } \Omega \qquad (3.45)$$

for $t > 0$. We shall seek nontrivial solutions of this equation together with (3.5), subject to the homogeneous boundary conditions (3.1). In case of a stationary ground state one looks for solutions having an exponential growth rate in time. Since here, the ground state (2.14) depends on time as well, such a construction is only possible under the assumption that the rate of change of the ground state is small compared with the growth rate of infinitesimal perturbations (the frozen profile approach). Hence, for given $t > 0$, we consider instead of (3.45) the approximate equation

$$\frac{\partial s}{\partial \tau} - \frac{\partial s}{\partial z} + Rw\frac{\partial S_0}{\partial z}(z, t) = \Delta s \quad \text{in } \Omega \qquad (3.46)$$

for $\tau > 0$ and sufficiently small. In fact we have two time scales: a large time scale for the evolving ground state and a small time scale for the perturbation. Now again t appears as a parameter in the equation, as in the case of the energy methods. From here on the procedure is quite standard (*Wooding* [1960], *Nield & Bejan* [1992]). Applying again the x, y- periodicity, taking σ as the exponential growth rate and setting

$$s, w = s, w(z) \exp\big(\sigma\tau + i(a_x x + a_y y)\big), \qquad (3.47)$$

we find from (3.5) and (3.46) the coupled set of second order equations

$$\begin{cases} (D^2 - a^2)\, w = -a^2 s\,, & (3.48) \\ \left(D^2 + D - a^2 - \sigma\right) s = R\dfrac{\partial S_0}{\partial z}(z, t)w\,. & (3.49) \end{cases}$$

The corresponding eigenvalues now depend on a, t and on the growth rate σ. An analysis as in *Van Duijn et al.* [2001a] shows for the smallest positive eigenvalue $R_1(a, t, \sigma)$:

$$R_1(a, t, \sigma) \lesseqgtr R_1(a, t, 0) \quad \text{if and only if } \sigma \lesseqgtr 0\,. \quad (3.50)$$

These inequalities imply the following. Let R_s be sufficiently close to $R_1(a, t, 0)$. If $R_s > R_1(a, t, 0)$, then there exists a $\sigma > 0$ such that $R_s = R_1(a, t, \sigma)$. In other words, if $R_s > R_1(a, t, 0)$, there exists a growing infinitesimal perturbation which implies that the boundary layer is unstable. If $R_s < R_1(a, t, 0)$ no definite statement about stability can be made. Only certain infinitesimal perturbations now decay. Others, and in particular large perturbations, may still grow in time.

As a consequence it suffices to analyse equations (3.48), (3.49) for the case of neutral stability $\sigma = 0$. Eliminating s and setting $\sigma = 0$, gives for w the fourth order eigenvalue problem

$$\left(D^2 + D - a^2\right)\left(D^2 - a^2\right) w = -a^2 R\frac{\partial S_0}{\partial z}(z, t)w \quad (3.51)$$

for $0 < z < \infty$, with

$$w(0) = D^2 w(0) = 0 \quad \text{and} \quad w(\infty) = 0\,. \qquad (3.52)$$

The equilibrium case ($t = \infty$ and $S_0 = e^{-z}$) can be treated by a semi-analytical technique based on a Frobenius expansion in terms of descending exponentials (*Van Duijn et al.* [2001b], *Wooding* [1960]). As a result one finds an accurate approximation to the lowest eigenvalue $R_1(a)$ for any wavenumber $a > 0$. In Figure 1, point values of $R_1(a)$ have been plotted as crosses, showing excellent agreement with solid curve 3 – the numerical solution of the eigenvalue problem (3.51), (3.52) using the Jacobi–Davidson method. We find

$$R_L := \min_{a>0} R_1(a) = R_1(a_c) = 14.35 \qquad (3.53)$$

with

$$a_c = 0.759 \qquad (3.54)$$

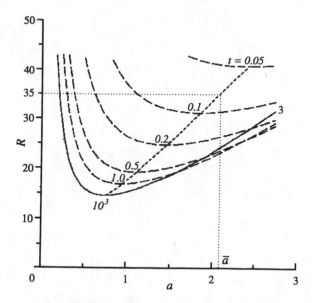

Figure 3. Stability curves for equilibrium boundary layer according to the linearised stability method. Dashed curves show lowest eigenvalue R_1 versus wavenumber a prior to equilibrium, treating time as parameter. Numerical values are calculated by the Jacobi–Davidson method. Solid curve 3 is taken from Figure 1 (equilibrium case).

approximately. These numbers, in good agreement with the numerical results of *Homsy & Sherwood* [1976], are characteristic of the linearised stability method.

To study the instability of the growing boundary layer we need to consider the eigenvalue problem for each finite $t > 0$. Let $R_L(a,t)$ denote the smallest positive eigenvalue. Again we used the Jacobi–Davidson method to find accurate numerical approximations. These results are shown in Figure 3 where the dashed curves indicate $R_L(a,t)$ for increasing values of t. Note again that these curves essentially move downwards, except for large a and t. As $t \to \infty$ convergence towards the equilibrium curve $R_1(a)$ is attained.

As before, we set

$$R_L(t) := \min_{a>0} R_L(a,t) \,. \qquad (3.55)$$

If $R_s > R_L(\infty) =: R_L$, an estimate for the onset time of instability is found by the crossover time t_L^s determined by $R_s = R_L(t_L^s)$. In other words, the boundary layer becomes unstable for $t > t_L^s$. If $R_s \approx R_L$, the boundary layer becomes unstable when it is close to its equilibrium profile.

For $R_s < R_L$ no definite statement about stability is possible. The linearised stability analysis only implies that infinitesimal small perturbations vanish for $R_s < R_L$. Subcritical instabilities originating from large perturbations may still grow in time. This is a consequence of the uniform upflow, implying that the eigenvalue problem is not self-adjoint (*Homsy & Sherwood* [1976]).

4. GROWING INSTABILITIES

The theoretical stability bounds tell us how the system will respond to periodic perturbations of the initial state $S = 0$. For $R_s < R_E$ we expect decaying perturbations and for $R_s > R_L$ growing instabilities or salt-fingers. In this section we verify this behaviour by means of numerical experiments. We also investigate the response of the system to non-periodic (Figures 7,8) and a combination of periodic (Figure 9) initial perturbations.

In the numerical experiments we consider the two-dimensional truncated flow domain

$$\Omega_H^L := \{(x,z) : -L < x < L, \ 0 < z < H\} \,. \qquad (4.1)$$

In this definition the quantities H and L are scaled with respect to the length scale $\mathbb{D}/E$. The truncated flow domain needs additional boundary conditions for the velocity U and the saturation S: we set $S = 0$ and $U = U_0$ at $z = H$ and we impose no-flow and no salt transport along the lateral boundaries.

We solve equations (2.7)–(2.9) in terms of the saturation S and the stream function Ψ, where

$$U = \left(-\frac{\partial \Psi}{\partial z}, \frac{\partial \Psi}{\partial x} \right) \,. \qquad (4.2)$$

Following *de Josselin de Jong* [1960] we obtain the system

$$\begin{cases} \dfrac{\partial S}{\partial t} + R_s \left(\dfrac{\partial \Psi}{\partial x} \dfrac{\partial S}{\partial z} - \dfrac{\partial \Psi}{\partial z} \dfrac{\partial S}{\partial x} \right) = \Delta S \,, & (4.3) \\[2mm] \Delta \Psi = \dfrac{\partial S}{\partial x} \,, & (4.4) \end{cases}$$

in Ω_H^L and for all $t > 0$. The corresponding boundary conditions result directly from the imposed saturation and flow behaviour.

4.1. The Numerical Method

Let $t^n = n\Delta t$, $n = 1, 2, \cdots, N$, N sufficiently large, and let S^n denote the saturation at $t = t^n$. The corresponding stream function is found from

$$\Delta \Psi = \frac{\partial S^n}{\partial x} \quad \text{in } \Omega_H^L \,. \qquad (4.5)$$

This problem is discretised by the finite (linear) element method. The corresponding matrix equation is iteratively solved using the conjugate gradient method. The numerical approximation of (4.5) is denoted by Ψ_h^n.

Next we consider

$$\frac{\partial S}{\partial t} + R_s \left(\frac{\partial \Psi_h^n}{\partial x} \frac{\partial S}{\partial z} - \frac{\partial \Psi_h^n}{\partial z} \frac{\partial S}{\partial x} \right) = \Delta S \,, \qquad (4.6)$$

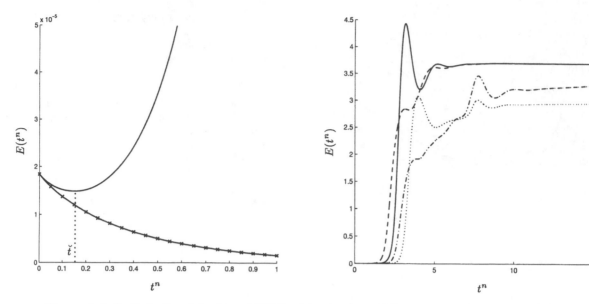

Figure 4. Left: Early-time behaviour of $E(t^n)$ originating from (4.8): solid line ($R_s = 35$), crossed solid line ($R_s = 5$). Right: Long-time behaviour of energy $E(t^n)$ for $R_s = 35$. Solid line corresponds to perturbation (4.8), dotted line to (4.9), dash-dotted line to (4.10) and finally dashed line to (4.11).

in Ω_H^L and for $t > t^n$. Again we use a finite element discretisation, together with an upwind discretisation for the convective part. Now the corresponding linear system is iteratively solved with the bi-conjugate gradient stabilized method. For the time integration we use the implicit Euler scheme. The numerical solution of (4.6) is denoted by S_h^{n+1}. Replacing S^{n+1} in equation (4.5) by the approximation S_h^{n+1}, the above cycle is repeated for subsequent time steps.

The numerical method does not involve automatic time-step adaptation nor does it include algorithms for local mesh refinement. This implies that the time-step and mesh are fixed during the computations. The mesh consists of square elements of length h. Motivated by the convergence behaviour of the scheme (see *Pieters* [2001] for technical details), we use $\Delta t = 0.005$ and $h = 0.15625$ for respectively the time-step and element length, and we fix $H = 5$ and $L = 25$.

4.2. Stability Criterion

To decide whether the system is stable or unstable, a stability criterion is required. Inspired by the energy method we consider the functional

$$E(t^n) := \int_{\Omega_H^L} |\boldsymbol{u}_h^n(x, z, t)|^2 \, \mathrm{d}x\mathrm{d}z \,, \qquad (4.7)$$

where $\boldsymbol{u}_h^n$ is the numerical approximation of the velocity perturbation $\boldsymbol{u}$. It is found by substracting the ground state U_0

from the numerical solution $U_h^n = (-\partial \Psi_h^n/\partial z \,, \ \partial \Psi_h^n/\partial x)$.

Numerical observations (*Pieters* [2001]) show that $E(t^n)$ either decreases from a positive value $E(0)$ towards zero for large n, or $E(t^n)$ first decreases, reaches a minimum and then strongly increases away from $E(0)$. Based on these observations, we call the ground state unstable if there exists a $\mathbb{N} \ni m \neq N$ such that $E(t^m) \leqslant E(t^n)$ for all $0 \leqslant n \leqslant N$. Otherwise the ground state is stable. The time t^m is called turning time and will be further denoted by $\check{t}$.

The behaviour of $E(t^n)$ is illustrated by two numerical experiments. First we set $R_s = 5 < R_E$ and consider

$$S(x, z, t = 0) = \epsilon \cos(x) \,, \qquad \epsilon = 5 \cdot 10^{-4} \,, \qquad (4.8)$$

for $(x, z) \in \Omega_H^L$. The functional (4.7) is plotted as a crossed solid line in Figure 4 (left). Indeed, for this choice the energy norm decreases with time and the system remains stable. Next we set $R_s = 35 > R_L$ and again (4.8). For this unstable regime the functional is plotted as solid curves in Figure 4. The saturation profile and velocity field for this experiment are depicted in Figure 6. As to be expected, the initial periodic perturbation triggers growing instabilities in the saturation profile. At later times the influence of the lower boundary becomes noticeable (middle and bottom figures) and a steady state is reached. The corresponding saturation profile has the original eight fingers. This is due to the particular choice $a = 1$. Note that the energy $E(t^n)$ has a relative high value, Figure 4 (right).

As explained in *Pieters* [2001] this observation leads to an alternative method to analyse stability of the system: given wavenumbers a and system Rayleigh numbers R_s within relevant ranges, we can determine the turning times $\check{t}$, yielding a set of triples $(a, R_s, \check{t})$. Treating $\check{t}$ as parameter, we can construct stability curves similar to the ones in Section 3. The result is shown in Figure 5 and agrees with the curves obtained by the method of linearised stability (Figure 3).

This is due to the fact that the initial perturbations are small.

4.3. Numerical Experiments

We investigate the development of instabilities for more general initial perturbations. We take $R_s = 35$ and consider the following cases:

a. Stochastic perturbation, with

$$S(x, z, t = 0) = \epsilon \mu(x, z), \quad \text{for } (x, z) \in \Omega_H^L, \quad (4.9)$$

where μ is uniformly distributed in $[0, 1]$.

The computational results are shown in Figure 7. The system clearly selects a preferential wavenumber a^* for small and intermediate times. From the top figure in Figure 7 we deduce $a^* = 1.86$ approximately. At later times a steady state is reached which now has a saturation profile with twelve fingers and a relative low energy.

Taking $R_s = 35$ and considering the curve connecting the minima in Figure 3, we expect to find from linear stability $a = \bar{a} = 2.08$. Similarly, we obtain from the energy method, Figure 2, $a = \underline{a} = 1.82$. We observe that $\underline{a} \approx a^* < \bar{a}$. This is in agreement with the experimental results of Wooding et al. [1997a, b], see also Figure 10 in Section 5.

b. Non-periodic perturbation, with

$$S(x, z, t = 0) = \begin{cases} \epsilon & \text{for } -\frac{L}{3} < x < \frac{L}{3}, \ 0 < z < H, \\ 0 & \text{elsewhere}, \end{cases}$$

$$(4.10)$$

The results for flow and saturation are shown in Figure 8. Now, at small and intermediate times, no preferential wave number can be detected. At large times the resulting steady state shows ten fingers and an intermediate energy $E(t^n)$. We stopped the computations at $t = 15$. It could be possible that the steady state is not yet reached because the corresponding energy is still slightly increasing in time.

c. Combination of periodic perturbations, with

$$S(x, z, t = 0) = \epsilon \{ \cos(0.25x) + \cos(x) + \cos(2x) \}, \quad (4.11)$$

for $(x, z) \in \Omega_H^L$.

We observe that the modes with $a = 0.25$, 2 decay, while the one with $a = 1$ grows. The decay of $a = 0.25$ is in agreement with the stability analysis (Figure 3). However, Figures 2 and 3 seem to indicate competition between $a = 1$

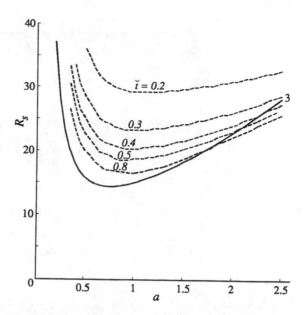

Figure 5. Stability curves for equilibrium boundary layer corresponding to the numerical method as described in *Pieters* [2001]. Dashed curves show system Rayleigh number R_s versus wavenumber a prior to equilibrium, treating $\check{t}$ as parameter. Solid curve 3 is taken from Figure 1.

and $a = 2$. Apparently, nonlinear effects cause decay of $a = 2$ and growth of $a = 1$. Note that the large time behaviour is identical to Figure 6, with the same energy.

5. DISCUSSION AND CONCLUSIONS

We have formulated a stability problem involving a porous medium saturated with saline water flowing vertically upwards through a horizontal surface. The upflowing water is assumed to evaporate completely at the surface. Salt saturation is established quickly and is sustained there, with excess salt precipitated on the surface. Below the surface, a saline boundary layer grows by diffusion in the counter direction to the upflow. If this layer remains stable under gravity, an equilibrium state is reached where the salinity (or density) profile is exponential, decreasing downwards towards the ambient upflow value.

Since the surface salinity and upflow rate are both taken constant, the layer is stable provided it is sufficiently thin; it is initially stable, but will tend to become less stable monotonically as the thickness increases by diffusion/dispersion. The system is least stable when the boundary layer has attained maximum thickness, which occurs at equilibrium. The equilibrium boundary layer thickness provides a length scale for the Rayleigh instability problem. If the porous medium has a lower boundary, it is assumed to be at a distance large relative to that scale.

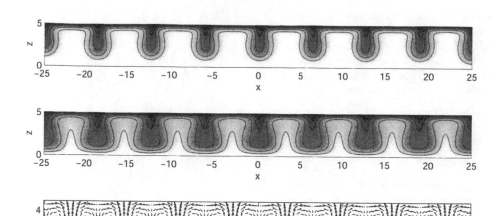

Figure 6. Periodic initial perturbation (4.8): contour plot of saturation at $t = 3$ (top) and $t = 15$ (middle), velocity field (bottom) at $t = 15$.

To study the stability of the boundary layer we have used two energy methods and the method of linearised stability. In terms of the system Rayleigh number R_s (2.12), the first give upperbounds for stability, the latter a lower bound for unstable behaviour. These bounds do not coincide (see Figure 1) and leave the possibility for decay of infinitesimal small perturbations and growth of large perturbations.

In the first energy method we follow *Homsy & Sherwood* [1976] and use (3.11) as constraint for perturbations. Assuming horizontal periodicity in the usual way, this con-

straint leads to a second order eigenvalue problem. Homsy and Sherwood constructed a numerical solution for the case of a porous layer of finite thickness with a (thermal) boundary layer at equilibrium. We explain their asymptotic result for large thickness in terms of Bessel functions. In particular we find that their stability bound corresponds to the square of the first root of the Bessel function J_0.

In the second energy method we use (3.5) as differential constraint and we consider the time dependent behaviour (growth) of the boundary layer. This leads to a sixth or-

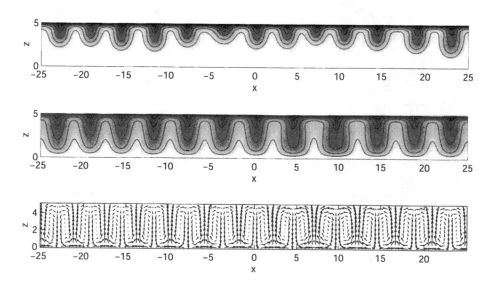

Figure 7. Stochastic initial perturbation (4.9): contour plot of saturation at $t = 3.5$ (top) and $t = 15$ (middle), velocity field (bottom) at $t = 15$.

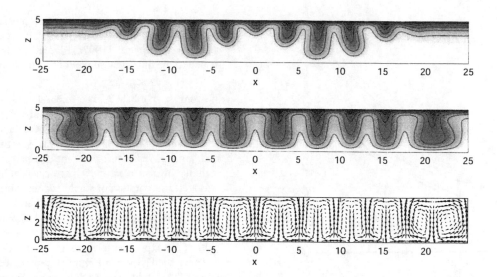

Figure 8. Non-periodic initial perturbation (4.10): contour plot of saturation at $t = 3.5$ (top) and $t = 15$ (middle), velocity field (bottom) at $t = 15$.

der eigenvalue problem which we solved by means of the Jacobi–Davidson method. Figure 2 shows the behaviour of the smallest positive eigenvalue versus the wavenumber a, with time as parameter. Figure 1 compares the two methods for the equilibrium case ($t = \infty$) and shows superior behaviour when using (3.5) instead of (3.11). This is explained in Appendix A. Given a R_s-value, we are now in a position to estimate the time during which the boundary layer grows in a stable manner. This is explained in Section 3.2.2, see (3.44).

The method of linearised stability estimates the onset of instabilities. We used the frozen profile approach which

allows us to incorporate the growth of the boundary layer in the analysis. As a result we arrive at a fourth order eigenvalue problem which we solved again by the Jacobi–Davidson method. The corresponding stability curves are shown in Figure 3. Now we are in the position to estimate an elapsed time beyond which the boundary layer becomes unstable. This is explained in Section 3.3, see (3.55).

In Section 4 we considered two-dimensional flow and study the growth of instabilities by means of the finite element method (with the stream function as flow variable). Introducing the energy functional (4.8) we are now in a position to estimate the elapsed time numerically for a given

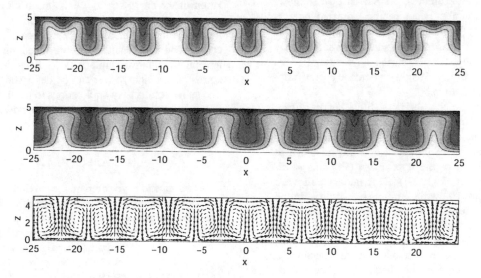

Figure 9. Combined periodic initial perturbation (4.10): contour plot of saturation at $t = 3$ (top) and $t = 15$ (middle), velocity field (bottom) at $t = 15$.

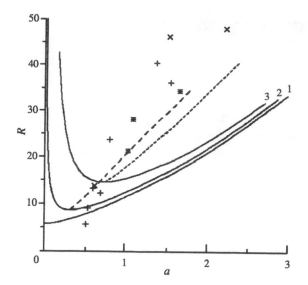

Figure 10. Comparison of theory (this paper) with experimental results (*Wooding et al.* [1997a, b]). Solid curves 1–3 give eigenvalues R versus wavenumber a for the equilibrium boundary layer (Figure 1). Curves of minima of R with respect to a for $t > 0$ increasing to equilibrium: by energy method (Figure 2, dashes), by linearised theory (Figure 3, short dashes). Symbols for experimental results are identified in the text.

initial perturbation. In this way we could qualitatively reproduce Figure 3 (linearised stability) for small periodic perturbations. The response of the system to other initial perturbations was investigated as well. The results are shown in Figure 7 (stochastic), Figure 8 (non-periodic) and Figure 9 (combination of periodic modes). The stochastic case shows a preferential wavenumber which can be estimated in terms of the curves tracing the minima in Figures 2 and 3. It is also worthwhile to note that different initial perturbations may lead to different steady states (occuring at large times). This follows clearly from Figure 4 (right), where the energy functional is plotted versus time.

Figure 10 repeats the equilibrium stability curves of Figure 1 and includes experimental measurements obtained using a tilted Hele–Shaw cell to simulate two-dimensional flow in a porous medium, with inflow of a saline solution and evaporation along part of the upper edge (*Wooding et al.* [1997a, b], *Simmons et al.* [1999]). Experimental points are represented in Figure 10 by the symbols $+$, $\times$ and $*$. In the experiments, the large scale Rayleigh number R_s based on finite "aquifer" depth was greater than 10^2 times the boundary layer R-value. Although the large scale flow in the experimental work differed from a simple vertical upflow, a uniform evaporation rate was modelled and a saline boundary layer of uniform thickness was observed to develop. Wavenumbers of initial instabilities, scaled to the equilib-

rium boundary layer thickness, were measured for a wide range of R-values. Previously, these observations were plotted by *Wooding et al.* [1997a, Figure 7] using wavenumbers scaled to the diffusion thickness and therefore equivalent to a/R in the present case.

From the published experimental data, stable boundary layers were observed for R-values of 5.8, 5.6 (two experiments), and smaller R. Unstable boundary layers resulted for R-values of 5.6 (one experiment), 8.9 (two experiments), and larger R. Except for the unexplained appearance of instability in one experiment performed at $R = 5.6$, there was a clear separation of stable and unstable layers into two ranges. If the single unstable result at $R = 5.6$ is not included, the theoretical lower bound of 8.590 obtained using the alternative energy method is in agreement with the results of the experimental studies.

The dashed curves in Figure 10 provide traces of the minima of the stability curves defined by the energy method in Figure 2 and by linearised stability analysis in Figure 3. For the data obtained by experimental simulation, either curve might be considered as an upper bound to the wavenumber of an instability which first appears. This is on the assumption that growth rate is zero at a critical point for stability, and a growing perturbation becomes significant when the boundary layer thickness scale has increased significantly. Clearly, however, the instabilities plotted in Figure 10 have been initiated by perturbations of small but finite amplitude, and the energy method with differential constraint provides the appropriate estimate. Three experimental points at the low-R end appear to be exceptional. These occur in a range where accurate observation becomes more difficult, and an inadvertent change of background conditions could have altered the wavenumber.

In general, we may conclude that the alternative formulation of the energy method has improved the quantitative and qualitative estimate of a lower bound to absolute stability, and is in agreement with experimental modelling. The comparison with results from linearised analysis yields interesting qualitative similarities, and stability properties of a growing boundary layer can be described in some detail. The above results have applications to the theory of stability of salt lakes and the salinization of groundwater.

APPENDIX A

In this appendix we compare the maximum problem (3.12) for the admissible perturbations $\mathbf{H}$ and $\widetilde{\mathbf{H}}$. In particular we show that $\widetilde{\mathbf{H}}$ can be identified with a proper subspace of $\mathbf{H}$. This explains why the differential constraint yields larger Rayleigh numbers than the integral constraint.

Let $(s, w) \in \widetilde{\mathbf{H}}$. For this given s we have the unique decomposition (*Temam* [1984])

$$se_z = v + \operatorname{grad} \varphi \quad (v, \varphi \text{ are } x, y\text{-periodic}) , \quad \text{(A1)}$$

where div $v = 0$ and $v \cdot n = 0$ on $\partial \mathcal{V}$. Here n denotes the unit normal at the boundary $\partial \mathcal{V}$. As in (3.5) we find

$$\Delta v_3 = \Delta_\perp s \quad \text{in } \mathcal{V},$$

where v_3 is the vertical component of v. This implies

$$\Delta(v_3 - w) = 0 \quad \text{in } \mathcal{V},$$

and the boundary conditions on $\partial \mathcal{V}$ give $v_3 = w$ in $\mathcal{V}$. Thus given $(s, w) \in \tilde{\mathbf{H}}$ we have obtained the pair (s, v) with div $v = 0$ and $v_3 = w$ in $\mathcal{V}$. Multiplying (A1) by v and integrating the result over $\mathcal{V}$ gives

$$\int_\mathcal{V} |v|^2 = \int_\mathcal{V} sw,$$

in other words, $(s, v) \in \mathbf{H}$.

The converse is not true. Given $(s, u) \in \mathbf{H}$ and using (A1) we obtain the vector field v satisfying $\Delta v_3 = \Delta_\perp s$ in $\mathcal{V}$. So $(s, v_3) \in \tilde{\mathbf{H}}$, but in general $v = u + \operatorname{curl} \Phi$ for a smooth vector field Φ which vanishes on $\partial \mathcal{V}$.

REFERENCES

Abramowitz, M., and I.A. Stegun, *Handbook of Mathematical Functions*, Dover, 1972.

Bear, J., *Dynamics of Fluids in Porous Media*, Elsevier, New York, 1972.

Caltagirone, J.-P., Stability of a saturated porous media layer subject to a sudden rise in surface temperature: comparison between the linear and energy methods, *Quart. J. Mech. Appl. Math.*, *33*, 47–58, 1980.

Davis, S. H., On the possibility of supercritical instabilities, in *Instability of Continuous Systems*, *Proc. IUTAM Symp.*, pp. 222–227, Springer–Verlag, Berlin, 1971.

De Josselin de Jong, G., Singularity distributions for the analysis of multiple fluid flow in porous media, *J. Geothermal Res.*, *65*, pp. 3739–3758, 1960.

Fokkema, D. R., G. L. G. Sleijpen and H. A. van der Vorst, Jacobi–Davidson style QR and QZ algorithms for the reduction of matrix pencils, *SIAM J. Sci. Comput.*, *20*, 94–125, 1999.

Gilman, A and J. Bear, The influence of free convection on soil salinization in arid regions, *Transport in Porous Media*, *24*, 275–301, 1996.

Homsy, G. M., Global stability of time-dependent flows: impulsively heated or cooled fluid layers, *J. Fluid Mech.*, *60*, 129–139, 1973.

Homsy, G. M., A. E. Sherwood, Convective instabilities in porous media with throughflow, *Lawrence Livermore Lab. Rep.*, UCRL-76539, 1975.

Homsy, G. M., A. E. Sherwood, Convective instabilities in porous media with throughflow, *Amer. Inst. Chem. Engrs. J.*, *22*, 168–174, 1976.

Jones, M. C. and J. M. Persichetti, Convective instability in packed beds with throughflow, *Amer. Inst. Chem. Engrs. J.*, *32*, 1555–1557, 1986.

Lapwood, E. R., Convection of a fluid in a porous medium, *Proc. Cambridge Phil. Soc.*, *44*, 508–521, 1948.

Nield, D. A., Convective instability in porous media with throughflow, *Amer. Inst. Chem. Engrs. J.*, *33*, 1222-1224, 1987.

Nield, D. A., A. Bejan, *Convection in Porous Media*, 2nd ed., Springer–Verlag, New York, 1992.

Pieters, G. J. M., Stability analysis for a saline boundary layer formed by uniform upflow using finite elements, *RANA Report 01-07*, Eindhoven University of Technology, 2001.

Simmons, C. T., K. A. Narayan and R. A. Wooding, On a test case for density-dependent groundwater flow and solute transport models: The salt lake problem, *Water Resources Res.*, *35*(12), 3607–3620, 1999.

Sleijpen, G. L. G., H. A. van der Vorst, A Jacobi–Davidson iteration method for linear eigenvalue problems, *SIAM J. Matrix Anal. Appl.*, *17*(2), 401–425, 1996.

Straughan, B., *The Energy Method, Stability and Nonlinear Convection*, vol. 91 of *Applied Mathematical Sciences*, Springer–Verlag, New-York, 1992.

Temam, R., *Navier–Stokes Equations, Theory and Numerical Analysis*, vol. 2 of *Studies in Mathematics and its Applications*, 3rd ed., Elsevier Science Publishers, Amsterdam, 1984.

Van Duijn, C. J., R. A. Wooding, G.J.M. Pieters, and A. van der Ploeg, Stability criteria for the boundary layer formed by throughflow at a horizontal surface of a porous medium, *RANA Report 01-12*, Eindhoven University of Technology, 2001a.

Van Duijn, C. J., R. A. Wooding, and A. van der Ploeg, Stability criteria for the boundary layer formed by throughflow at a horizontal surface of a porous medium: extensive version, *RANA Report 01-05*, Eindhoven University of Technology, 2001b.

Wooding, R. A., Rayleigh instability of a thermal boundary layer in flow through a porous medium, *J. Fluid Mech.*, *9*, 182–192, 1960.

Wooding, R. A., S. W. Tyler, and I. White, Convection in groundwater below and evaporating salt lake: 1. Onset of instability, *Water Resour. Res.*, *33*(6), 1199–1217, 1997a.

Wooding, R. A., S. W. Tyler, I. White, and P. A. Anderson, Convection in groundwater below and evaporating salt lake: 2. Evolution of fingers or plumes, *Water Resour. Res.*, *33*(6), 1219–1228, 1997b.

C. J. van Duijn, Department of Mathematics and Computer Science, Eindhoven University of Technology, P.O. Box 513, 5600 MB Eindhoven.

G. J. M. Pieters, Department of Mathematics and Computer Science, Eindhoven University of Technology, P.O. Box 513, 5600 MB Eindhoven.

A. van der Ploeg, MARIN, P.O. Box 28, 6700 AA Wageningen, The Netherlands.

R. A. Wooding, CSIRO Land and Water, G.P.O. Box 1666, Canberra, ACT 2601, Australia.

Injection of Dilute Brine and Crude Oil/Brine/Rock Interactions

Guoqing Tang and Norman R. Morrow

Department of Chemical and Petroleum Engineering, University of Wyoming, Laramie, Wyoming

Sensitivity of oil recovery to injection brine composition has been reported for a variety of circumstances including trends of increased recovery of crude oil with decrease in salinity. Absolute permeabilities of sandstones to synthetic reservoir brines and dilutions of these brines show little sensitivity to salinity when the initial brine and injected brine are of the same composition. With reservoir brine as the initial brine and injection of dilute brine, the pH of the outflow brine increased and absolute permeability to brine decreased, but never to less than 50% of its original value. Such changes, if any, were much less for rocks with low clay content. During the course of recovery of crude oil, interfacial tensions of crude oil and dilute effluent brine were reduced by about 25% relative to values for crude oil and reservoir brine. Effluent brine pH increased after injection of low salinity brine, but showed no response in the absence of an initial water saturation. Changes in brine composition resulting from flow through Berea sandstone were small. Fines production and permeability reduction resulting from injection of dilute brine was greatly reduced by the presence of crude oil.

1. INTRODUCTION

A series of studies has shown that the recovery of crude oil from sandstones containing clay can be markedly affected by brine composition. In particular, a trend of increase in oil recovery with decrease in salinity has been observed for a variety of crude oil and rock combinations. Necessary conditions for increased recovery are the presence of: 1) polar components in the oil phase, 2) clays and 3) an initial water saturation [*Tang and Morrow*, 1999].

It is well known that with injection of fresh water there is potential for formation damage; this presents a possible deterrent to injection of dilute brine for increased oil recovery. Extensive studies of brine/rock interactions have been made in the context of formation damage [*Jones*, 1964; *Khilar et al.*, 1983; *Vaidya et al.*, 1990; *Souto and Bazin*, 1993; *Miranda et al.*, 1993; and *Rahman et al.*, 1994].

Environmental Mechanics: Water, Mass and Energy Transfer in the Biosphere
Geophysical Monograph 129
Copyright 2002 by the American Geophysical Union
10.1029/129GM16

Migration of fine particles is widely recognized as the main cause of formation damage. Mueke, 1979; Sarker and Sharma, 1990; and Tang and Morrow, 1999, reported observations on fines migration in the presence of crude oil.

In study of crude oil/brine/rock interactions, an important guide to possible oil recovery mechanisms is provided by examining brine/rock interactions in both the absence and presence of an oil phase. Results presented in this paper focus on the effect of injecting dilute brine on rock and other properties. They include information relevant to oil recovery on absolute and relative permeabilities, effluent brine pH, interfacial tensions, dissolution of minerals, and initial water saturation.

2. EXPERIMENTAL

2.1 Crude Oil/Brine/Rock Systems

The aqueous phases were synthetic reservoir brines designated PB-RB, CS-RB, or DG-RB and dilutions of these brines and $CaCl_2$ brines with salinity ranging from 0.01 to 2%. Six kinds of rock, Berea sandstone, Bentheim

Table 1. Core properties.

Core	ϕ (%)	K_g (md)	Oil content (%)	Type
Berea sandstone	22.9-23.2	800-1100	0	outcrop
Bentheim sandstone	23.1-23.4	1800-2400	0	outcrop
DG sandstone	21.6-21.8	400-600	>50	reservoir
Clashach sandstone	17.1-19.0	1350-1688	0	outcrop
F/A Berea sandstone	23.2-23.3	750-850	0	outcrop

Table 2. Synthetic reservoir brine compositions.

	Concentration (ppm)							
Brine	Na^+	K^+	Ca^{2+}	Mg^{2+}	Cl^-	HCO_3^-	SO_4^{2-}	TDS
DG-RB*	4,267	7,237*	218	32	13,414	-	-	24,166
PB-RB	8,374	52	110	24	13,100	-	-	21,660
CS-RB	5,626	56	58	24	8,249	1,119	16	15,150

* high K^+ concentration was due to the addition of KCl to the injection water to prevent clay swelling in the target formation.

Table 3. Viscosity and density of cs crude oil.

Oil	Temp ($^{\circ}$C)	Viscosity (cP)	Density (g/cm^3)	Asphaltene (%)	Wax (%)
CS	22	70.5	0.891		
	50	23.6	0.860	0.78	12.6
	75	11.6	0.835		

Acid number 0.33±0.03; Base number 1.16±0.18 (*Buckley* et al. 1998)

sandstone, CS Reservoir Sandstone, DG reservoir sandstone, Clashach sandstone, and fired/acidized Berea sandstone, were tested. All core samples were 3.8 cm in diameter and 7.6-7.8 cm in length. In experiments that involved oil, CS crude oil was used. The properties of the rocks, brines and crude oil are listed in Tables 1, 2, and 3, respectively.

2.2 Procedures

All the core plugs were saturated with test brine and equilibrated for ten days. None of the plugs were reused. The brine permeability at 100% saturation of the test brine was used as the base brine permeability. For most tests, the cores were first equilibrated with a relatively saline brine and then flooded with various dilute brines to determine if injection of low-salinity brine caused changes in brine permeability (single-phase flow tests). All of these tests were run at room temperature.

Some of the cores were aged at 55°C with CS crude oil for an aging time of 10 days to induce a mixed-wet condition before injection of low salinity brines (two-phase

flow tests). Pressure drop at constant flow rate and effluent brine pH were measured vs. injected brine volume for both single and two-phase flow tests. In the investigation of dissolution of rock minerals, effluent brine was re-circulated through the core. Pressure drop versus oil recovery data was interpreted to obtain unsteady state relative permeabilities.

3. RESULTS AND DISCUSSION

3.1 Brine/Rock Interactions

3.1.1 Injection brine and initial brine of same composition. In the first series of tests, the injection brine composition was the same as the initial brine. The reservoir brines (CS-RB, PB-RB, and DG-RB) were diluted with distilled water by up to 100 times. The CaCl$_2$ brine salinity was varied from 0.01 to 2% (wt.%). The results presented in Figure 1 are absolute permeability to brine versus salinity. Permeability tended to decrease slightly with decrease in salinity. However, even when the salinities of synthetic brines were as low as 155 ppm (0.01 CS-RB), 217 ppm (0.01 PB-RB), or 242 ppm (0.01 DG-RB), the brine

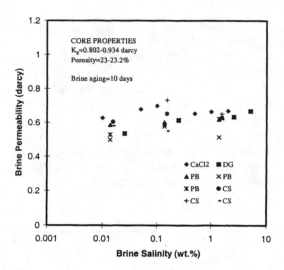

Figure 1. Effect of Salinity on Absolute Brine Permeability of Berea Sandstone When There is No Difference in Composition Between Initial Brine and Injected Brine

permeabilities were over 90% of the permeability for the corresponding reservoir brine. For $CaCl_2$ brine, the permeability was nearly constant even when the salinity was decreased from 2% to 0.01%. Thus little permeability reduction with dilution was observed when there was no difference in composition between the brine initially in place and the injected brine.

3.1.2 Injection of dilute brine. Cores in this test series were saturated with reservoir brine (RB) and aged for 10 days at room temperature. The flow rate was 18 ft/d. In these tests, the initial ionic equilibrium between aqueous phase and solid phase is disturbed by injection of dilute brine. Decrease in brine salinity results in cation exchange between the aqueous phase and the clay minerals in the rock. Exchange of metal ions from the clays with hydronium ions from the injected brine is the likely cause of increase in pH of the effluent brine. Rock-fluid interactions may also cause mobilization of fines that originally coated grain surfaces. The released fines can block the pore restrictions and thus reduce permeability. One mechanism of clay release is through reduction in van der Waals forces that accompany expansion of electrical double layers when the ionic strength is decreased.

The presence of potentially mobile fines is often cited as the main cause of formation damage resulting from injection of fresh water. However, it has also been concluded that the presence of mobile fines plays a key role in increased oil recovery that has been observed with injection of dilute brine [*Tang and Morrow*, 1997; and 1999]. It is therefore of interest to study the consequences of brine/rock interactions with respect to both improved recovery and formation damage that can result from injection of dilute brine.

DG Sandstone. DG reservoir sandstone was cleaned using toluene and methanol. The core was flooded with 6 PV DG-RB, followed by 0.01 DG-RB The results are presented in Figure 2. Switching the injection brine from DG-RB to 0.01 DG-RB resulted in increase in effluent brine pH, closely followed by reduction of brine permeability. This slight delay in permeability reduction may be related to the stabilizing effect of potassium ions and the possible presence of some fraction of organic material that was not removed by the toluene/methanol cleaning procedure. The maximum permeability reduction was about 32%. The effluent brine pH increased from 7.7 to 10.1. Production of fines was observed after 0.3-0.5 PV of 0.01 DG-RB had been injected. Fines production gradually decreased and ceased after about 3-5 PV of 0.01 DG-RB had been injected.

Comparison of the results shown in Figs. 1 and 2 indicates that the difference of ionic strength between initial and injection brines is a key factor in permeability reduction. This observation is qualitatively consistent with previous studies related to formation damage [*Jones*, 1964].

Berea Sandstone. Berea sandstone, a model rock widely adopted for laboratory study, contains about 8% of clay, mainly illite and kaolinite [*Ma and Morrow*, 1994]. The Berea sandstone used in this study had a gas permeability of 900 to 1000 md and porosity of 23%. After pre-equilibration with CS-RB, the core was flooded at 18 ft/d with CS-RB, followed by 0.01 CS-RB.

The results shown in Figure 3 are consistent with those shown in Figure 2. Switching injection brine from CS-RB to 0.01 CS-RB resulted in increase in effluent brine pH followed almost immediately by reduction in brine permeability. The effluent pH increased from 7.1 to a maximum of 10.1 and after falling to about 9 gradually

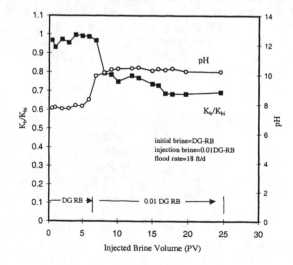

Figure 2. Change in K_b and Effluent pH with Injection of Diluted DG-RB into DG Reservoir Sandstone

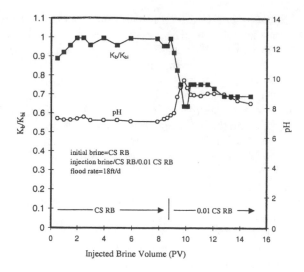

Figure 3. Change in K_b and Effluent pH with Injection of 0.01 CS-RB into Berea Sandstone

decreased to 8.3 after injection of 15 PV. The brine permeability decreased by about 35% of the absolute permeability to CS-RB. After switching the injection brine to 0.01 CS-RB, fines were observed in the effluent brine for the first 3 PV of injection but not thereafter. The produced fines were observed to settle after production.

Bentheim Sandstone. Bentheim is described as a clean sandstone because it has very low clay content. For these cores, porosity was about 23-24% and air permeability around 1500-2000 md. The test results are presented in Figure 4. The response to injection of dilute brine was evident but the changes in permeability and pH were much smaller than for Berea sandstone and DG reservoir sandstone. The brine permeability never fell below 83% of the value to CS-RB throughout injection of 14 PV of 0.01 CS-RB. The effluent pH increased after switching the injection brine to 0.01 CS-RB, but never rose above 9.0 for this relatively clay free sandstone. Slight production of fines was observed after injection of 0.01 CS-RB.

Clashach Sandstone. Results obtained for the Clashach sandstone (Figure 5), which has extremely low clay content, were similar to those for the Bentheim cores. Although the effluent pH increased slightly after the injection brine was switched from CS-RB to 0.01 CS-RB, there was no reduction of permeability or production of fines.

Fired and Acidized (F/A) Berea Sandstone. Berea sandstone cores were fired at 800°C, and then acidized with 2N HCl to remove metal oxides. Thereafter, the cores were flooded with CS-RB until the effluent pH was neutral. Based on the X-ray examination of fired Berea by Ma and Morrow, 1994,

water sensitive clays were stabilized for Berea sandstone fired at 800°C.

The results obtained for the fired and acidized Berea sandstone demonstrate the effect of removal of potentially mobile clay particles (compare Figs. 3 and 6). The brine permeability remained essentially constant, even after the core had been flooded with 18 PV of 0.01 CS-RB. Injection of dilute brine resulted in only slight increase in pH of the effluent brine. These observations support the conclusion that the increase in effluent pH observed for DG and Berea sandstone is related to cation exchange.

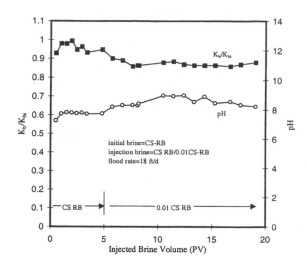

Figure 4. Change in K_b and Effluent pH with Injection of 0.01 CS-RB into Bentheim Sandstone

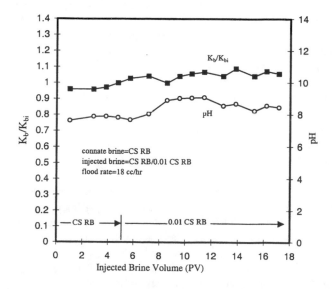

Figure 5. Change in K_b and Effluent pH with Injection of 0.01 CS-RB into Clashach Sandstone

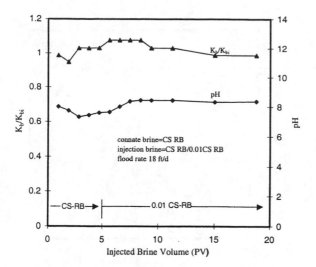

Figure 6. Change in K_b and Effluent pH with Injection of 0.01 CS-RB into Fired and Acidized Berea Sandstone

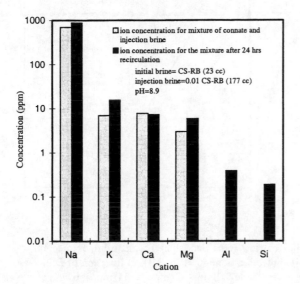

Figure 7. Change in Effluent Brine Composition With Recirculation 0.01 CS-RB through Berea Sandstone

3.1.3 Dissolution of minerals. Brine/rock interactions that result in dissolution of minerals at pore surfaces are a possible cause of change in wettability and oil recovery [*Tang and Morrow*, 1997 and 1999; and *Morrow et al.,* 1998). Preliminary studies were made of the change in composition of the effluent brine resulting from circulation of CS-RB through the core samples (Berea and Bentheim sandstones). The core initially 100% saturated with CS-RB (PV=23 cc) was flooded with 0.01 CS-RB for 177 cc. Thereafter, the combined brine (initial and injected brine) was recycled through the core for 24 hours at room temperature. The brine was then filtered to remove fines and

analyzed for K^+ by atomic adsorption spectroscopy and for other cations by inductively coupled plasma - optical emission spectroscopy. The test results are presented in Figure 7. Na^+, K^+, and Mg^{2+} increased slightly for both Berea and Bentheim sandstone. Traces of Al and Si were also found in the effluent. These results suggest that the dissolution of minerals from solid surfaces is slight. However, because wetting of a rock surface is dominated by the outermost layer of molecules, the contribution of dissolution of minerals to observed changes in wetting cannot be ruled out.

3.2 Brine/Oil Interactions

Lowering of the interfacial tension between water and oil phases underlies the mechanism of improved oil recovery by surfactant and alkaline flooding processes. Decrease in IFT results in a proportional increase in the capillary number, the ratio of viscous to capillary forces.

Figure 8 shows the measured IFT values between effluent brine and CS crude oil. For these tests, the CS-RB was the initial brine. Relative to injection of CS-RB, the results show that decrease in injection brine salinity resulted in decrease in IFT. When the brine salinity was decreased from CS-RB to 0.01 CS-RB, the IFT decreased by about one third. For all dilute brines except 0.01 CS-RB, the IFT exhibited a distinct minimum value after injection of about 3 PV of dilute brine. The minimum in IFT matched a maximum in pH of the effluent brine for both the pH and IFT behavior. Results for 0.01 CS-RB also showed decrease in IFT but only slight increase above the minimum value with continued injection.

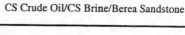

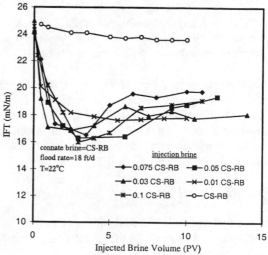

Figure 8. CS Crude Oil/Effluent Brine Interfacial Tensions vs. PV Injected Brine

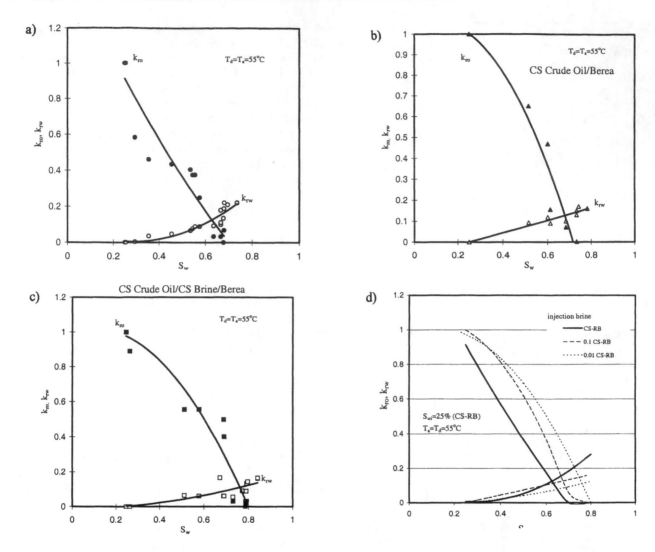

Figure 9. Relative Permeabilities ($S_{wi} \sim 25\%$); a) CS Crude Oil/CS Brine/Berea Sandstone (connate brine=injection brine=CS-RB); b) Injection of Dilute Brine (connate brine=CS-RB, injection brine=0.1 CS-RB); c) Injection of Dilute Brine (connate brine=CS-RB, injection brine=0.01 CS-RB); d) Injection Brine Salinity - Comparison (from a, b, and c).

From observations of the values of capillary numbers required for water-wet and mixed-wet conditions [*Zhou, et al.*, 1995], it is unlikely that the increase in oil recovery that accompanies reduction in salinity of the injection brine could be ascribed directly to the increase in capillary number resulting from decrease in IFT.

3.3 Oil/Rock/Brine Interactions

3.3.1 Relative permeability. Donaldson et al. (1969) pointed out that wettability can strongly affect relative permeability behavior because it is a major factor in control of the location, flow, and distribution of fluids in a porous medium. For the relative permeability tests, all cores were aged with CS crude oil and CS-RB at reservoir temperature (T_a=55°C) for 8 days (t_a) before the waterflood tests. The

initial CS-RB brine saturation was close to 25% for all tests. Relative permeabilities were measured by the unsteady-state JBN method [*Johnson, et al.*, 1959]. A high flood rate, 6 ft/d, was used to stabilize the flow and minimize end effects. The effect of brine salinity on unsteady-state relative permeability with change in injection brine concentration (0.01 CS-RB, 0.1 CS-RB and CS-RB) was measured at T_d=T_a=55°C.

The calculated relative permeability curves for three kinds of injection brines (CS-RB, 0.01 CS-RB, and 0.01-CS-RB) are presented in Figs. 9a, b, and c, respectively. Fluctuations in pressure caused scatter in the data which was analyzed without pre-smoothing. The results show the trend in relative permeabilities resulting from change in brine salinity. The smoothed curves are compared in Figure 9d. According to Craig's rule of thumb, 1971, for assessment of

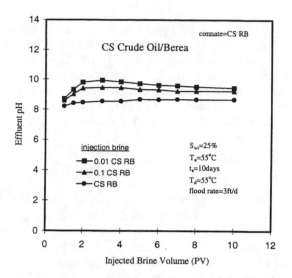

Figure 10. Effluent Brine pH for Three Injection Brine Salinities

wettability, all three relative permeability curves for the three brines corresponded to water-wet conditions.

Four features of the effect of brine salinity on the relative permeability were: (1) the saturation at the cross-over point between oil and water relative permeability curves tended to increase with decrease in the brine concentration. The cross-over saturation was 0.62 for CS-RB, 0.67 for 0.1 CS-RB, and 0.67 for 0.01 CS-RB; (2) residual oil saturation decreased with decrease in brine concentration; 0.31 for CS-RB, 0.28 for 0.1 CS-RB, and 0.2 for 0.01 CS-RB; (3) the relative permeability to water at maximum water saturation decreased with decrease in brine concentration; 0.23 for CS-RB, 0.17 for CS-RB, and 0.15 for 0.01 CS-RB; (4) oil relative permeability increased significantly with decrease in injection brine salinity. The relative permeability behavior indicates, according to traditional interpretations, that decrease in brine concentration resulted in transitions toward increased water-wetness for CS crude oil/CS brine/Berea.

3.3.2 Connate water saturation and effluent pH.
Berea sandstone, initially saturated with CS-RB, was flooded with CS crude oil to establish initial water saturation, S_{wi}. The cores were then aged with CS crude oil for 10 days at 55°C, and flooded with CS-RB, 0.1-CS-RB, or 0.01 CS-RB, respectively. For these three floods, values of the effluent pH increased with decrease in injection brine salinity from CS-RB to 0.01 CS-RB (Figure 10). Comparison of these results with those obtained for a 100% CS-RB saturated core shows that presence of crude oil did not prevent cation exchange between the injection brine and the rock surfaces. There are two possible reasons for this: (1) the smaller pores were still occupied by water (S_{wi}=25%); (2) in the larger pores, although occupied mainly by crude oil, water is

retained as thin films and in association with micropores at rock surfaces as bulk water. When pore spaces are invaded, the injection brine mixes with the initial water, and then interacts with the rock surfaces. Possible reasons for increase in effluent pH are: (1) cation exchange between the aqueous and solid phases; (2) dissolution of carbonate.

A question of more practical concern is how the injection brine salinity affects brine/rock interactions in the presence of crude oil. In a test of the role of an initial water saturation, a Berea sandstone core was first dried at 105°C to remove all adsorbed water and then 100% saturated with CS crude oil. The core was then aged for 10 days at 55°C and then flooded with 0.01 CS-RB. The effluent pH did not change with injection of dilute brine (Figure 11) and no fines production was observed. When the initial water saturation was zero, the otherwise water sensitive particles were initially coated with crude oil and showed no response to injection of dilute brine.

3.3.3 Oil recovery with different initial water saturation (CS-RB).
Decrease in injection brine salinity can result in an increased oil recovery at S_{wi}=25% [*Tang and Morrow*, 1997; and 1999]. Further study has been made of change in oil recovery with initial water saturation, S_{wi} (CS-RB).

The cores, originally saturated with CS-RB, were saturated with CS crude oil for S_{wi} ranging from 0 to 27% and aged at T_a=55°C for t_a=10 days. The aged cores were flushed with fresh CS crude oil before being flooded with 0.01 CS-RB.

The exploratory results shown in Figure 12 indicate that the initial brine saturation has a significant effect on breakthrough oil recovery and final recovery. For S_{wi}=0, the

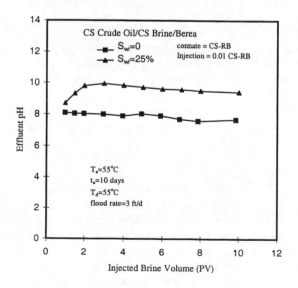

Figure 11. Effect of Initial Water Saturation on Effluent pH for Displacement of Oil by Dilute Brine

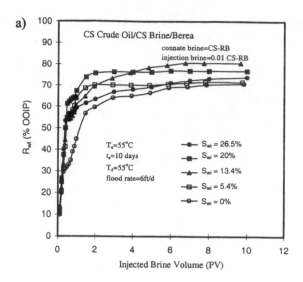

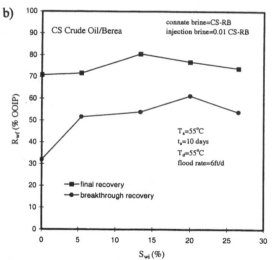

Figure 12. Effect of Initial Brine Saturation on Oil Recovery for Injection of Dilute Reservoir Brine; a) Waterflood Recoveries; b) Recovery at Breakthrough and Final Oil Recovery.

breakthrough recovery was only 32% but increased to 68% after injection of 10 PV. In the presence of an initial water saturation, the breakthrough and final oil recoveries were higher. For $S_{wi}>5.4\%$, the breakthrough recovery was more than 50% and the final oil recovery was about 70%. This result indicates that the presence of initial water saturation is necessary for improved oil recovery by injection of dilute brine. For this test series, a maximum in oil recovery of over 80% was obtained for an initial water saturation of 13.4%. However, the recovery curves exhibited crossover and no clear trend between initial water saturation and oil recovery.

For all values of initial water saturation, the production of fines for the oil/brine/Berea systems with injection of dilute brine (0.01 CS-RB) was much less than that for the brine/Berea systems. A small amount of fines was produced

at water breakthrough. Fines production ceased before 0.2 PV of production as compared to about 3 PV of production in the absence of crude oil (see 3.2). However, a light brown suspension of about 1-2 mm in depth formed at the oil/water interface of the produced liquids. The nature and location of suspended fines and/or droplets with respect to the interface and the neighboring phases deserves detailed investigation.

4. CONCLUSIONS

1. Injection brine salinity can have distinct interaction with rock surface mineralogy. Increase in effluent pH with injection of dilute brine is ascribed to cation exchange. Overall changes in brine composition resulting from flow of dilute brine through sandstone were very small.

2. The extent of formation damage on rock permeability resulting from injection of dilute brine is mainly dependent upon how the injection brine salinity is decreased. When injection brine and initial brines have the same composition, the decrease in brine permeability is small.

3. The presence of potentially mobile fines such as kaolinite is a key factor in reduction of brine permeability resulting from injection of dilute brine. If the active clay is destroyed or if the rock has very low clay content, little or no effect of injection brine on brine permeability is observed.

4. The presence of crude oil tends to inhibit release of fines from the rock surfaces. For the crude oil/brine/rock combinations studied to date, injection of fresh water caused only slight production of fines and these were concentrated at the oil-water interface.

5. Injection of dilute brine causes interfacial tensions of crude oil and reservoir brine to fall by about one third as determined for effluent brine and oil. This decrease in itself, is not expected to cause the increase in oil recovery observed for injection of dilute brine.

6. When the injection brine is more dilute than the connate brine, change of fluid distribution is indicated by relative permeability behavior. Residual oil saturation decreases with decrease in salinity.

Acknowledgements. Support for this work was provided by ARCO, BP/Amoco (U.K./U.S.A.), Chevron, ELF/Total/Gas de France/Institut Francais du Petrole (France), Exxon, JNOC (Japan), Marathon, Phillips, Shell (The Netherlands), Statoil (Norway), the Enhanced Oil Recovery Institute of the University of Wyoming, and the National Petroleum Technology Office Partnership through the U.S. Department of Energy.

REFERENCES

Buckley, J.S., Liu, Y., and Monsterleet, S., Mechanisms of wetting alteration by crude oils, *SPE Journal*, 3, 54-61, Mar, 1998.

Craig, F.C., Jr., *The Reservoir Engineering Aspects of Waterflooding*, SPE Monograph, Dallas, 141pp.. 1971.

Donaldson, E.C., Thomas, R.D. and Lorenz, P.B., Wettability Determination and Its Effect on Recovery Efficiency, *SPE Journal.*, 9, 13, 1969.

Johnson, E.F., Bossler, D.P. and Naumann, V.O., Calculation of relative permeability from displacement experiments, *Trans., AIME*, 192, 135-140, 1959.

Jones, F.O., Influence of chemical composition of water on clay blocking of permeability, *J. Pet. Tech*, April, 441, 1964.

Khilar,K.C., Fogler, H.S. and Ahluwalia, J.S., Sandstone water sensitivity: Existence of a critical rate of salinity decrease for particle capture, *Chemical Engineering Science*, 38, 5, 789-800, 1983.

Ma, S. and Morrow, N.R., Effect of firing on petrophysical properties of Berea sandstone, *SPE Form. Eval.*, Sept., 213-218, 1994.

Miranda, R.M. and Underdown, D.R., Laboratory measurement of critical rate: A novel approach for quantifying fines migration problems, *SPE 25432*, presented at the Production Operations Symposium held in Oklahoma City, OK, U.S.A., March 21-23, 1993.

Morrow, N.R., Tang, G., Valat, M. and Xie, X., Prospects of improved oil recovery related to Wettability and Brine Composition, *J. Pet. Sci. Engr.* 20, June, 267-276, 1998.

Muecke, T.W., Formation fines and factors controlling their movement in porous media, *J. Pet. Tech.*, Feb., 144-150, 1979.

Rahman, M.S., Rahman, S.S., and Arshad, A., "Control of Fines Migration: A Key Problem in Petroleum Production Industry," *SPE* 27362, presented at the SPE Intl Symposium on Formation Damage Control, Lafayette, Louisiana, 7-10 Feb., 1994.

Sarker, A.K. and Sharma, M.M., Fines migration in two-phase flow," *J Pet. Tech.*, May, 646-652, 1990.

Souto, E. and Bazin, B., Ion exchange between hydrogen and homoionic brines related to permeability reduction, *SPE 25203*, presented at the SPE International Symposium on Oilfield Chemistry held in New Orleans, LA, U.S.A., March 2-5, 1993.

Tang, G.Q. and Morrow, N.R., Salinity, temperature, oil composition and oil recovery by waterflooding, *SPE Reservoir Engineering*, Nov., 269-276, 1997.

Tang, G. and Morrow N.R., Influence of brine composition and fine particles on crude oil/brine/rock interactions and oil recovery," *J. Pet. Sci, Eng.*, 24, 2-4, 99-111 Dec. 1999.

Vaidya, R.N. and Fogler, H.S., Fines migration and formation damage: Influence of pH and ion exchange, *SPE Reservoir Engineering* , Nov., 325-330, 1992.

Zhou, X., Torsæter, O., Xie, X., and Morrow, N.R., The effect of crude oil aging time and temperature on the rate of water imbibition and long term recovery by imbibition, *SPE Formation Evaluation*, December, 10,4, 259-265, 1995.

Norman R. Morrow, Chemical and Petroleum Engineering, P.O. Box 3295, Laramie, WY 82071.

Gouqing Tang, Petroleum Engineering, Green Science Building, 367 Panama St., Room 077A, Stanford, CA, 94305-2220.

Multidimensional Flow of Water in Unsaturated Soils

Peter A. C. Raats

Wageningen University and Research Centre, Wageningen, The Netherlands

In his studies of flow in porous media, John Philip favored the macroscopic (Darcian) scale, although he also regularly paid attention to processes at the underlying pore scale. At both of these scales, he recognized the connections between general concepts in continuum mechanics and their particular forms in soil physics. This paper specifically explores concepts from continuum mechanics that can be used to interpret multi-dimensional flow patterns at the Darcian scale. The standard model for flow of water in saturated and unsaturated soils is presented. Techniques for solving multi-dimensional flow problems for specific classes of unsaturated soils are indicated briefly. Some concepts of the kinematics of continuous media are introduced, with emphasis on the velocity vector field. Euler's general solution of the mass balance equation for a compressible continuum is adapted to soil water. Numerous special cases of this general solution are derived, thus unifying the scattered literature dealing with specific spatial and/or temporal simplifications. The implications of Darcy's law for the velocity and volumetric flux vector fields are explored. The rotation vector fields of the volumetric flux and velocity vector fields are analyzed by considering their decompositions in terms of the Frenet trihedron. The expressions for these rotation vectors are found to be surprisingly simple, involving explicitly the soil physical properties that are known to govern multi-dimensional flows. Finally, some implications of Darcy's law for the nature of flow patterns of soil water are related to properties of general lamellar and complex lamellar vector fields.

1. INTRODUCTION

About 70 years ago, *Richards* [1931] consolidated the efforts of previous generations of soil physicists – notably Franklin H. King, Charles S. Slichter, Lyman J. Briggs, Edgar Buckingham, Willard Gardner and W.B. Haines – by

Environmental Mechanics: Water, Mass and Energy Transfer in the Biosphere
Geophysical Monograph 129
Copyright 2002 by the American Geophysical Union
10.1029/129GM17

formulating a general, macroscopic theory for movement of water in rigid, unsaturated soils (see *Philip* [1974]; *Raats et al.,* [this volume]). Richards theory fits experience in many branches of continuum mechanics: it combines the simplest possible balances of mass, expressed in the equation of continuity, and of momentum, expressed in Darcy's law. *Philip* [1973] recognized the continuum mechanical nature of this theory and its extensions with the words:

"This Darcy-scale attack on problems of flow in porous media evidently has some affinities with continuum mechanics, but there are profound differences: primarily differences associated with the scale of discourse, but also differences in philosophy. One is tempted, nevertheless, to remark that many of us workers in porous medium physics are rather like M. Jourdain (Molière, 1671) who spoke prose for forty years without knowing it: we have practiced something akin to continuum mechanics for about as long and never knew it."

Now again 28 years later, soil physicists still by and large are practicing continuum mechanics. Moreover, the thermo-mechanical continuum theory of mixtures is now the widely recognized framework for organizing and extending existing theories. This means that the reservation with regard to "differences associated with the scale of discourse, but also differences in philosophy" is now less relevant.

The theory of Richards can be formulated within the framework of the mechanical continuum theory of mixtures, provided that one recognizes from the outset the existence of the separate solid, liquid, and gaseous phases (for reviews see *Raats* [1984a, 1998]). Richards theory can also be justified on the basis of the continuum mechanical principles of surface tension and viscous flow at the pore scale (see e.g. *Miller & Miller* [1956]; *Whitaker* [1986]).

In this paper, the focus is on the macroscopic (Darcian) scale, specifically on the connections between general concepts in continuum mechanics and their particular forms in soil physics. A specific aim is to bring together scattered ideas that can be used to interpret multi-dimensional flow patterns.

The organization of this paper is as follows. In section 2, the standard model for flow of water in saturated and unsaturated soils is presented. Techniques for solving multi-dimensional flow problems are indicated briefly. In Section 3, some concepts of the kinematics of continuous media are introduced, with emphasis on the velocity vector field. In section 4, Euler's general solution of the mass balance equation for a compressible continuum is adapted to soil water. Numerous special cases of this general solution are derived, thus unifying the scattered literature dealing with specific spatial and/or temporal simplifications. In section 5, the implications of Darcy's law for the nature of flow patterns is discussed, with emphasis on the rotationality of the flow and the existence of scalar potentials and distance functions. In section 6, some concluding remarks are presented.

2. THE STANDARD MODEL FOR FLOW OF WATER IN SATURATED AND UNSATURATED SOILS

2.1. Balance of Mass

Assuming the density of the soil water to be constant, the balance of mass can be expressed as a volumetric balance equation:

$$\frac{\partial \theta}{\partial t} = -\nabla \cdot (\theta \mathbf{v}) - \lambda, \qquad (2.1)$$

where t denotes the time, θ the volumetric water content, $\mathbf{v}$ the velocity of the water, and λ the volumetric rate of uptake of water by plant roots. Further $\nabla = \mathbf{i}_x \partial/\partial x + \mathbf{i}_y \partial/\partial y + \mathbf{i}_z \partial/\partial z$ denotes the vector differential operator, where $\mathbf{i}_x, \mathbf{i}_y, \mathbf{i}_z$ are unit vectors in the orthogonal x, y, z directions. If the flow is steady, (2.1) reduces to

$$\nabla \cdot (\theta \mathbf{v}) = -\lambda, \qquad (2.2)$$

giving

$$\nabla \cdot (\theta \mathbf{v}) = 0 \qquad (2.3)$$

in the absence of plant uptake. A vector field, such as $\theta \mathbf{v}$, satisfying an equation of the form (2.3) is called solenoidal or divergence free. Some properties of solenoidal vector fields are discussed in sub-subsection 4.1.5.

In a saturated soil, the volumetric water content θ is independent of time and, assuming λ vanishes, equation (2.3) will apply, whether the flow is steady or unsteady. If the porous medium is homogeneous, then the volumetric water content will be independent of position $\mathbf{x}$, so that (2.3) further reduces to

$$\nabla \cdot \mathbf{v} = 0. \qquad (2.4)$$

So in this case both the volumetric flux vector field $\theta \mathbf{v}$ and the velocity vector field $\mathbf{v}$ are solenoidal. In unsaturated soils, (2.4) will strictly only apply if the soil is homogeneous and $\theta \mathbf{v}$ is purely gravitational. Nevertheless, in steady multi-dimensional flows the water content may be nearly uniform and (2.4) may apply approximately (see *Philip* [1984a, b]).

2.2. Darcy-Buckingham Equation

Given the constancy of the density of the soil water, Darcy's law can be written as:

$$\theta \mathbf{v} = -k \nabla h + k \nabla z, \qquad (2.5)$$

where k is the hydraulic conductivity, z the vertical coordinate taken positive downward, and the capillary pressure head h is defined by:

$$h = \frac{p - p_a}{\gamma g} = -\frac{p_c}{\gamma g}. \qquad (2.6)$$

Here p and p_a are the pressures of the aqueous and gaseous phases, p_c is the capillary pressure, and g is the gravitational constant. The capillary pressure head and the hydraulic conductivity are nonlinear functions of the volumetric water content θ. Moreover, the relationship $h(\theta)$ is hysteretic.

In discussions of flow patterns and of convective transport of solutes, it is often useful to consider the velocity vector field $\mathbf{v}$, rather than the volumetric flux vector field $\theta\mathbf{v}$. Dividing both sides of (2.5) by θ gives:

$$\mathbf{v} = -\frac{k}{\theta}\nabla h + \frac{k}{\theta}\nabla z. \qquad (2.7)$$

The hydraulic mobility k/θ represents the speed of the water when the driving force is unity [Raats and Gardner, 1974].

For future reference, three alternative forms of Darcy's law are also recorded:

$$\theta\mathbf{v} = -k\nabla H, \qquad (2.8)$$

$$\theta\mathbf{v} = -D\nabla\theta + k\nabla z, \qquad (2.9)$$

$$\theta\mathbf{v} = -\nabla\phi + k\nabla z, \qquad (2.10)$$

where $H = h - z$ denotes the total head, $D = k\,dh/d\theta = k/C$ the diffusivity (with $C = d\theta/dh$ the water capacity), and ϕ the matric flux potential defined by:

$$\phi - \phi_0 = \int_{h_0}^{h} k\,dh = \int_{\theta_0}^{\theta} D\,d\theta, \qquad (2.11)$$

where h_0 and θ_0 denote reference values and $\theta_0 = \theta(h_0)$. The volumetric flux $\theta\mathbf{v}$ is the sum of a matric component $-k\nabla h = -D\nabla\theta = -\nabla\phi$ and a gravitational component $k\nabla z$. The matric component of the volumetric flux is given by minus the gradient of ϕ, explaining why we call ϕ the matric flux potential. A transformation of the type introduced in (2.11) was already given around 1880 by Kirchhoff [1894] in his lectures on heat conduction. For this reason the transformation is often referred to as the Kirchhoff transformation and the potential ϕ is sometimes called the Kirchhoff potential (e.g., Philip [1989]).

Corresponding expressions for the velocity vector field $\mathbf{v}$ are obtained by dividing both sides of Equations (2.8)-(2.10) by θ. In addition, in analogy with (2.10) and (2.11) one can introduce the form

$$\mathbf{v} = -\nabla\phi_v + (k/\theta)\nabla z, \qquad (2.12)$$

where the matric velocity potential ϕ_v is defined by

$$\phi_v - \phi_{v0} = \int_{h_0}^{h} (k/\theta)\,dh = \int_{\theta_0}^{\theta} (D/\theta)\,d\theta. \qquad (2.13)$$

2.3. Soil Hydraulic Properties and Multi-dimensional Flows

In the context of Richards equation, the relationships among the water content θ, pressure head h, and hydraulic conductivity k specify the hydraulic properties of a soil. One can distinguish two groups of parametric expressions describing these hydraulic properties:

(1) A group yielding flow equations that can be solved analytically, in most cases as a result of linearization following one or more transformations.

(2) A group that is favored in numerical studies and to a large extent shares flexibility with a rather sound basis in Poiseuillean flow in networks of capillaries.

2.3.1. Group leading to analytical solutions. The parametric expressions belonging to the first group have been widely used for one-dimensional flow problems (see Raats [2001]; Raats et al., this volume). Examples include the delta function diffusivity leading to solutions with sharp fronts, the Brooks and Corey power functions leading to similarity solutions, and various expressions leading to the soluble Burgers equation. At least three parametric expressions belonging to the first group have also led to analytical solutions for multi-dimensional problems, all with key contributions from John Philip.

The class of linear soils with diffusivity D constant and hydraulic conductivity k linear in θ leads to a linear Fokker-Planck equation, which can be solved relatively easily (see Philip [1969, 1990]). It yields useful results for the integral aspects of the water balance, but is rather unreliable with respect to details of the distribution of the water content.

For the class of Brooks & Corey [1964, 1966] power function soils, Philip & Knight [1991] and Philip [1992] developed similarity solutions for redistribution of finite slugs of soil water applied from instantaneous 1-dimensional plane, 2-dimensional line, and 3-dimensional point sources near the soil surface. Until that time the ideas used in these works lay hidden under a cloud of rather abstract mathematics, to which, among others, a group of mathematicians at Leiden and Delft under the leadership of L.A. Peletier and C.J. van Duijn made important contributions. In his characteristic fashion, John Philip saw the opportunity to apply the technique to movement of soil water and, with the help of John Knight, worked out the details, including illuminating pictures and examples of practical calculations. Enough details are given so that irrigationists can apply their results to other situations. The authors admit that their results apply to a restricted class of soils. However, one might expect that, if and when this work stimulates numericists to consider the general case, the results in these papers would be a benchmark for the accuracy of the calculations and for qualitative interpretation of the results.

Of much wider use has been the class of Gardner soils with exponential dependence of the hydraulic conductivity upon the pressure head (see *Gardner* [1958]):

$$k = k_0 \exp \alpha (h - h_0),\qquad (2.14)$$

where α is an inverse characteristic length of the soil and the subscript 0 denotes reference values. For the Gardner class of soils, Darcy's law reduces to the linear form:

$$\theta \mathbf{v} = -\nabla \phi + \alpha \phi \nabla z.\qquad (2.15)$$

Introducing (2.15) in (2.3) gives:

$$\nabla^2 \phi = \alpha \phi \partial \phi / \partial z.\qquad (2.16)$$

The linearity of (2.15) and (2.16) in the matric flux potential ϕ opens the way to analytical solutions of steady multi-dimensional flow problems (see *Raats* [1970] for the early history of this linearization). *Philip* [1968] initiated the mathematical analysis of multi-dimensional flow problems on the basis of (2.16) with a paper on buried point sources. In the same year *Wooding* [1968] published an analysis of infiltration from shallow circular ponds, which later became the basis for the development of disc permeametry. Wooding's solution generalizes the solution for an electrified disk, which was the basis of the analysis of stomatal resistance. Later, Philip made numerous other contributions to this very fruitful line of research, often in co-operation with John Knight (for reviews see *Philip* [1988, 1989, 1998]; *Pullan* [1990]; *Raats* [1988]).

A subclass of the class of Gardner soils, of interest in the context of this paper, is the class with exponential dependence of the hydraulic mobility upon the pressure head:

$$k/\theta = k_0/\theta_0 \exp \alpha_v (h - h_0),\qquad (2.17)$$

where α_v is an inverse characteristic length of the soil. For this class of soils the velocity $\mathbf{v}$ is given by

$$\mathbf{v} = -\nabla \phi_v + \alpha_v \phi_v \nabla z.\qquad (2.18)$$

The inverse characteristic lengths α and α_v are related by

$$\alpha_v = \alpha - \frac{1}{\theta}\frac{d\theta}{dh} = \alpha - C/\theta.\qquad (2.19)$$

For a soil of the subclass of Gardner soils characterized by (2.17), not only α, but also C/θ is constant.

2.3.2. Group used in numerical solutions. In the second group of parametric expressions for the hydraulic proper-

ties, empirical expressions for the water retention characteristic are used to infer the pore size distribution, which in turn is used in expressions based on certain assumptions concerning the geometry of the pore system and the Poiseuille equation to calculate the hydraulic conductivity characteristic (see *Raats* [1992] for a review). The procedure in essence links physico-mathematical models at the Darcy and Navier-Stokes scales. In soil science the most commonly used models for the hydraulic properties are those of *Van Genuchten* [1980] and *Mualem* [1976]. The relationship of *Van Genuchten* [1980] reads:

$$S(h) = \frac{\theta(h) - \theta_r}{\theta_s - \theta_r} = (1 + |\alpha h|^n)^{-m},\qquad (2.20)$$

where S is the effective saturation ($0 \leq S \leq 1$), θ_r is the residual water content, θ_s is the water content at saturation, and α, n (> 1) and m are shape parameters. The steepness of $\theta(h)$ is determined by n, whereas m determines the value for S when $h = -1/\alpha$. Assuming $m = 1 - 1/n$ in (2.20), the corresponding $k(\theta)$ relationship according to the assumptions about the geometry of the pore system of *Mualem* [1976] is:

$$k(S) = k_s S^\lambda \left(1 - \left(1 - S^{1-1/m}\right)^m\right),\qquad (2.21)$$

where k_s is the hydraulic conductivity at saturation, and λ is the curve shape parameter representing a pore-size distribution index, sometimes set equal to 1/2 as suggested by Mualem.

Already in the early work of *Klute* [1952] and *Philip* [1955, 1957] it was clear that most flow problems in the unsaturated zone require, at least in part, numerical solutions. In a comprehensive review by *Breaster et al.* [1971], most numerical studies discussed concerned one-dimensional flow, including complications arizing from hysteresis, ponding and moving water tables. *Rubin* [1968] was the first to analyze a 2-dimensional flow, using a alternating direction implicit method. About a dozen studies of 2-dimensional problems followed in the next three years, all using various finite difference methods. The only 3-dimensional model found by *Breaster et al.* [1971] was the finite difference model of *Freeze* [1971].

Progress with multi-dimensional problems was hampered not only by the low speed of the available computers, but also by the finite difference methods used. Finite difference methods are awkward for handling curved boundaries and coping with anisotropic media for which the principal axis do not coincide with the coordinate axis. The first 2-dimensional numerical model overcoming these limitations was UNSAT2, a finite element model developed by Neuman (see *Neuman* [1973]; *Neuman et al.* [1975]; *Fed-*

des et al. [1975]). The model SWM II and the now popular HYDRUS-2D software package evolved from UN-SAT2 (see *Vogel* [1987] and *Simunek et al.* [1996]. Other more powerful 2- and 3-dimensional models that have been developed later include the integrated finite difference model TRUST [*Narasimhan*, 1976, 1978a, 1978b; *Reisenauer et al.*, 1982) and its descendants TOUGH [*Pruess,* 1987] and TOUGH2 [*Pruess,* 1991], the control volume method FUSSIM2 [*Heinen*, 1997; *Heinen and De Willigen,* 1998], and the finite (control) volume and finite element heat and mass computer code FEHM [*Zyvoloski et al.*, 1995]. The numerical methods for solution of the Richards equation now seem to have evolved to a satisfactory state and attention of model builders has shifted to post-Richards factors such as swelling and shrinkage [*Garnier et al.*, 1997], local non-equilibrium [*Selim and Ma*, 1998; *Vogel et al.*, 2000], and water uptake by plant roots [*Feddes and Van Dam*, 1999; *Heinen*, 1997; *Heinen and De Willigen*, 1998] and to linkage with other processes such as solute transport, aeration, chemical and biochemical reactions, and activity of plant roots.

The now available methods and faster computers have made it possible to efficiently solve transient flow problems involving complications such as (i) multi-dimensional regions that are partly and variably saturated, (ii) spatially variable soil physical properties, (iii) hysteresis of water retention, and (iv) uptake by plant roots. The HYDRUS-2D package was used by *De Vos et al.* [2000] to analyze a large field data set, yielding detailed information about the profile distribution of the soil physical properties and the time course and composition of the tile drain discharge. The FUSSIM2 model, including implementation of the modified dependent-domain theory of hysteresis of *Mualem* [1984], was used to simulate the 2-dimensional flow of water from surface drip sources and to bottom drains in closed, recirculating cropping systems in glasshouse horticulture [*Heinen and Raats*, 1999].

It is clear that in recent years analytical solutions, numerical solutions, and laboratory and field observations all have contributed to the rapid progress of our understanding of multi-dimensional flows. This paper is intended to review and further develop concepts that can be used to interpret such flows.

3. KINEMATICS OF THE SOIL WATER

3.1. Analysis of Deformation and Motion

Two approaches can be used to describe deformation and motion of the soil water (cf. *Raats* [1987a, b]). The spatial approach describes what happens in the course of time t to certain parcels $\mathbf{X}$ of the soil water. The material approach gives for any parcel $\mathbf{X}$ of the soil water the positions $\mathbf{x}$ occupied in the course of time t:

$$\mathbf{x} = \mathbf{x}\left(\mathbf{X}, t\right). \qquad (3.1)$$

As labels for parcels of the soil water, one can use locations $\mathbf{x}_\kappa = \mathbf{X}$ in the reference configuration κ at some reference time t_κ. Differentiation of the functional relationship (3.1) gives the two key concepts for describing the deformation and motion of the soil water, namely the deformation gradient tensor $\mathbf{F}$, and the velocity vector $\mathbf{v}$:

$$\mathbf{F} = \left(\partial \mathbf{x}/\partial \mathbf{X}\right)_t, \ \mathbf{v} = \left(\partial \mathbf{x}/\partial t\right)_\mathbf{X}. \qquad (3.2)$$

The velocity vector $\mathbf{v}$ defined by (3.2) can be used to relate the spatial time derivative $\left(\partial/\partial t\right)_\mathbf{x}$ and the material time derivative $\left(\partial/\partial t\right)_\mathbf{X}$:

$$\left(\partial/\partial t\right)_\mathbf{X} = \left(\partial/\partial t\right)_\mathbf{x} + \mathbf{v} \cdot \left(\partial/\partial \mathbf{x}\right)_t. \qquad (3.3)$$

The spatial coordinates $\mathbf{x}$ and material coordinates $\mathbf{X}$ are often referred to as, respectively, Eulerian and Lagrangian coordinates. However *Truesdell* [1954b, footnote 2 on p. 30-31] has shown that historically this cannot be justified (see also *Aris* [1989]). Commonly used notations and names for the material derivative $\left(\partial/\partial t\right)_\mathbf{X}$ are, respectively, D/Dt and convective or convected derivative.

The local properties of the deformation from a reference configuration at time t_κ to a configuration at time t are described by $\mathbf{F}$. Numerous concepts describing various aspects of the deformation can be derived from $\mathbf{F}$. If, from any configuration actually occupied, a continuous motion can reach the reference configuration, then $J = \det\mathbf{F}$ and then the polar decomposition theorem gives two unique, multiplicative decompositions of $\mathbf{F}$:

$$\mathbf{F} = \mathbf{RU} = \mathbf{VR}, \qquad (3.4)$$

where the rotation tensor $\mathbf{R}$ is orthogonal ($\mathbf{RR}^T = \mathbf{I}$, with the superscript T denoting the transpose and $\mathbf{I}$ the identity tensor), and the right and left hand stretch tensors $\mathbf{U}$ and $\mathbf{V}$ are symmetric ($\mathbf{U} = \mathbf{U}^T$, $\mathbf{V} = \mathbf{V}^T$). The geometric interpretations of the multiplicative decompositions (3.4) are very straightforward: the deformation corresponding locally to $\mathbf{F}$ may be regarded as resulting from pure stretches along three suitable, mutually orthogonal directions, followed by a rigid rotation of those directions, or from the same rotation, followed by the same stretches along the appropriate directions.

The velocity gradient tensor $\partial\mathbf{v}/\partial\mathbf{x}$ compares the current velocities of neighboring parcels of the soil water. The tensor $\partial\mathbf{v}/\partial\mathbf{x}$ can be additively decomposed in a symmetric stretching tensor $\mathbf{D}$ and a skew-symmetric vorticity or spin tensor $\mathbf{W}$

$$\partial \mathbf{v}/\partial \mathbf{x} = \mathbf{D} + \mathbf{W}, \qquad (3.5)$$

where

$$\mathbf{D} = 1/2 \left(\partial \mathbf{v}/\partial \mathbf{x} + \partial \mathbf{v}/\partial \mathbf{x}^T \right), \qquad (3.6)$$

$$\mathbf{W} = 1/2 \left(\partial \mathbf{v}/\partial \mathbf{x} - \partial \mathbf{v}/\partial \mathbf{x}^T \right). \qquad (3.7)$$

The geometric interpretation of the additive decomposition (3.5) is again very straightforward: the symmetric stretching tensor $\mathbf{D}$ describes the rate of stretch, the skew-symmetric spin tensor $\mathbf{W}$ describes the rate of rotation of a parcel of water. The spin tensor $\mathbf{W}$ is related to the vorticity $\nabla \times \mathbf{v}$ by:

$$\nabla \times \mathbf{v} = 2 \left(W_{yz}\mathbf{i}_x + W_{zx}\mathbf{i}_y + W_{xy}\mathbf{i}_z \right), \qquad (3.8)$$

where W_{yz}, W_{zx}, and W_{xy} are the components of W. The vorticity vector $\nabla \times \mathbf{v}$ represents the angular rotation with speed $1/2|\nabla \times \mathbf{v}|$ in a plane perpendicular to $\nabla \times \mathbf{v}$. The vorticity of soil water will be discussed in detail in section 5.

Differentiating the first of (3.2) with respect to t for fixed $\mathbf{X}$ or the second of (3.2) with respect to $\mathbf{X}$ for fixed t, it can be shown that the tensors $\partial \mathbf{v}/\partial \mathbf{x}$ and $\mathbf{F}$ are related by

$$(\partial \mathbf{F}/\partial t)_{\mathbf{X}} = (\partial \mathbf{v}/\partial \mathbf{x}) \mathbf{F}. \qquad (3.9)$$

According to (3.9), the material derivative of $\mathbf{F}$ is equal to the product of $\partial \mathbf{v}/\partial \mathbf{x}$ and $\mathbf{F}$ itself. Generally, solving this material, tensorial partial differential equation to obtain the deformation gradient tensor from one configuration to another will be difficult, since the components of $\partial \mathbf{v}/\partial \mathbf{x}$ at a particular parcel will generally be a function of time. However, of particular interest is the material, scalar partial differential equation resulting from taking the trace of (3.9):

$$J^{-1} (\partial J/\partial t)_{\mathbf{X}} = \nabla \cdot \mathbf{v}, \qquad (3.10)$$

where $J = \det \mathbf{F}$. Integration of (3.10) from some reference time t_κ to the current time t gives:

$$J = J_\kappa \exp \int_{t_\kappa}^{t} \nabla \cdot \mathbf{v} \mathrm{d}t. \qquad (3.11)$$

The determinant J of the deformation gradient tensor $\mathbf{F}$ is a measure of the volume of a parcel of soil water, and (3.11) describes the volumetric growth of such a parcel.

Equation (2.1) is the spatial form of the volumetric balance equation of the soil water. Adding $\mathbf{v} \cdot \nabla \theta$ to both sides of (2.1) and using (3.3) gives the material form:

$$(\partial \theta/\partial t)_{\mathbf{X}} = -\theta \nabla \cdot \mathbf{v} - \lambda. \qquad (3.12)$$

Substitution of (3.10) in (3.12) and rearranging gives:

$$(\theta J)^{-1} (\partial \theta J/\partial t)_{\mathbf{X}} = -\lambda/\theta. \qquad (3.13)$$

In (3.13), J transforms the volumetric water content θ to the configuration at the reference time t_κ. According to (3.13) the relative rate of change of θJ is equal to the source strength per unit volume of soil water $(-\lambda/\theta)$. Integration of (3.13) gives:

$$\theta = \theta_\kappa J^{-1} \exp \int_{t_\kappa}^{t} (-\lambda/\theta) \, \mathrm{d}t. \qquad (3.14)$$

Note that, since the configuration at time t_κ is the reference configuration, $J_\kappa = 1$. It is important to recall that (3.13) and (3.14) apply for a parcel $\mathbf{X}$ of soil water, not at fixed points $\mathbf{x}$ in space. According to (3.14) the water content at a parcel $\mathbf{X}$ at the current time t is the product of the initial water content θ_κ, a factor J^{-1} describing the effect of the deformation, and a factor describing the effect of the source strength. If the source strength vanishes, then (3.14) reduces to:

$$\theta = \theta_\kappa J^{-1}. \qquad (3.15)$$

3.2. Vector Fields, Vector Lines, and Vector Tubes

For any vector field $\mathbf{f}$:

$$\mathbf{f} = f\mathbf{s}, \qquad (3.16)$$

where $\mathbf{s}$ is the unit tangent vector field and the scalar field f is the magnitude of the vector field $\mathbf{f}$. Taking the divergence of $\mathbf{f}$ gives:

$$\nabla \cdot \mathbf{f} = \left(\nabla \cdot \mathbf{s} + \frac{1}{f} \frac{\delta f}{\delta s} \right) f, \qquad (3.17)$$

where the scalar field $\nabla \cdot \mathbf{s}$ is the divergence of the unit tangent vector field $\mathbf{s}$ associated with $\mathbf{f}$ and where $\delta/\delta s$ is the directional derivative along the vector line. Consider an infinitesimal vector tube with cross section $\Delta\sigma$. One can show that the divergence of the unit tangent vector field $\mathbf{s}$ is equal to the ratio of the rate of change of $\Delta\sigma$ along a vector line and $\Delta\sigma$ itself, i.e. is equal to the relative rate of change of $\Delta\sigma$ (see *Raats* [1974] and references given there):

$$\nabla \cdot \mathbf{s} = \frac{1}{\Delta\sigma} \frac{\delta \Delta\sigma}{\delta s}. \qquad (3.18)$$

Integration of (3.18) along a vector line gives:

$$\Delta\sigma = (\Delta\sigma)_0 \exp \int_{s_0}^{s} \nabla \cdot \mathbf{s} \mathrm{d}s, \qquad (3.19)$$

where $(\Delta\sigma)_0$ is the cross section at s_0. According to equation (3.19) the cross-sectional area $\Delta\sigma$ at some point s is proportional to the area $(\Delta\sigma)_0$ at the reference point s_0 and a function of the distribution of the divergence $\nabla \cdot \mathbf{s}$ between s_0 and s. As its name suggests, the divergence $\nabla \cdot \mathbf{s}$ measures the divergence of the infinitesimal vector tube. It involves only the vector line pattern, nothing about the magnitude of the underlying vector field $\mathbf{f}$.

3.3. Velocity and Volumetric Flux Vector Fields

The velocity vector field $\mathbf{v}$ can be written as the product of the unit tangent vector field $\mathbf{s}$ and the speed v:

$$\mathbf{v} = v(\mathbf{x}, t)\, \mathbf{s}(\mathbf{x}, t). \tag{3.20}$$

The interpretation of the vector field $\mathbf{s}$ as the unit tangent vector field of the velocity field $\mathbf{v}$ is not unique: the vector field $\mathbf{s}$ is also the unit tangent vector field of the volumetric flux vector field $\theta\mathbf{v}$ and, in view of (2.8), the vector field $-\nabla \cdot H$. The volumetric flux vector field $\theta\mathbf{v}$ of the water can be expressed as the product of the scalar field θ, the scalar field v, and the vector field $\mathbf{s}$:

$$\theta\mathbf{v} = \theta(\mathbf{x}, t)\, v(\mathbf{x}, t)\, \mathbf{s}(\mathbf{x}, t). \tag{3.21}$$

The unit tangent vector field defines the flow pattern, independent of the scalar magnitude fields $v(\mathbf{x}, t)$, $\theta(\mathbf{x}, t) v(\mathbf{x}, t)$, or $|\nabla H|(\mathbf{x}, t)$. Generally the flow pattern varies in the course of time. If the flow pattern is time invariant, then (3.21) reduces to

$$\theta\mathbf{v} = \theta(\mathbf{x}, t)\, v(\mathbf{x}, t)\, \mathbf{s}(\mathbf{x}). \tag{3.22}$$

Time invariant flow patterns are of particular interest since they allow the derivation of lumped models relating, for a family of stream lines, the input and the output [*De Valk and Raats*, 1995]. If the flow is steady, then (3.22) further reduces to:

$$\theta\mathbf{v} = \theta(\mathbf{x})\, v(\mathbf{x})\, \mathbf{s}(\mathbf{x}). \tag{3.23}$$

For steady flows, (2.1) reduces to (2.3) and $\theta\mathbf{v}$ is solenoidal (see also sub-subsection 4.1.5). Sometimes the further assumption that $\theta(\mathbf{x})$ can be replaced by an average value $\bar{\theta}$ is used [*Raats*, 1975; *Philip*, 1984a, 1984b], so that (3.23) reduces to:

$$\theta\mathbf{v} = \bar{\theta}v(\mathbf{x})\, \mathbf{s}(\mathbf{x}). \tag{3.24}$$

In this case, (2.1) reduces to (2.4) and $\mathbf{v}$ is solenoidal.

3.4. Balance of Mass Along a Stream Tube

Introducing the decomposition (3.22) of the volumetric flux vector $\theta\mathbf{v}$ in the spatial form (2.1) of the mass balance, and using (3.18) gives:

$$\frac{\delta\theta v}{\delta s} = -\frac{\partial\theta}{\partial t} - \theta v\nabla \cdot \mathbf{s} - \lambda =$$
$$-\frac{\partial\theta}{\partial t} - \frac{\theta v}{\Delta\sigma}\frac{\delta\Delta\sigma}{\delta s} - \lambda. \tag{3.25}$$

According to (3.25), positive contributions to $\delta(\theta v)/\delta s$ result from (i) a decrease of the water content with time, (ii) convergence of the streamlines, and (iii) a sink of water. If the flow is steady and the sink term λ vanishes, then (3.25) reduces to:

$$\frac{1}{\theta v}\frac{\delta\theta v}{\delta s} = -\nabla \cdot \mathbf{s} = -\frac{1}{\Delta\sigma}\frac{\delta\Delta\sigma}{\delta s}. \tag{3.26}$$

Integration of (3.26) along a stream line and using (3.19) gives:

$$\theta v = \theta_0 v_0 \exp - \left(\int_{s_0}^{s} \nabla \cdot \mathbf{s}\, \mathrm{d}s\right) = \theta_0 v_0 \frac{(\Delta\sigma)_0}{\Delta\sigma}. \tag{3.27}$$

According to (3.27), in a steady flow without sources or sinks of water the magnitude of the volumetric flux θv at some point s along a stream line can be calculated from its value $\theta_0 v_0$ at some other point s_0 and the distribution of $\nabla \cdot \mathbf{v}$ between s_0 and s. Also according to (3.27), the volumetric flux and the cross sectional area are inversely proportional to each other.

4. SOLUTION OF THE VOLUMETRIC MASS BALANCE

4.1. Three-dimensional Flows

4.1.1. Transient 3-dimensional flows. Without the source term, the volumetric mass balance equation (2.1) is analogous to the mass balance for a compressible fluid, the volumetric water content θ corresponding to the mass density of the fluid. The general solution of such equations was discovered by *Euler* [1770, sections 44-49; see also the commentary by *Truesdell* [1955]] in the context of compressible fluids. In modern times it was given, among others, by *Yih* [1957, equation (25)], *Krzywoblocki* [1958, case (5)] and *Truesdell and Toupin* [1960, equation (164.6)]. *Nelson* [1964, Appendix] introduced Euler's solution in porous media hydrodynamics. Written in the form of a 4-vector, the solution of the volumetric mass balance for transient, 3-dimensional flow of soil water is:

$$\theta\mathbf{1}, \theta v_x\mathbf{i}_x, \theta v_y\mathbf{i}_y, \theta v_z\mathbf{i}_z =$$

$$\begin{pmatrix} 1 & \mathbf{i}_x & \mathbf{i}_y & \mathbf{i}_z \\ \dfrac{\partial F}{\partial t} & \dfrac{\partial F}{\partial x} & \dfrac{\partial F}{\partial y} & \dfrac{\partial F}{\partial z} \\ \dfrac{\partial G}{\partial t} & \dfrac{\partial G}{\partial x} & \dfrac{\partial G}{\partial y} & \dfrac{\partial G}{\partial z} \\ \dfrac{\partial H}{\partial t} & \dfrac{\partial H}{\partial x} & \dfrac{\partial H}{\partial y} & \dfrac{\partial H}{\partial z} \end{pmatrix}. \tag{4.1}$$

where $\mathbf{1}$, $\mathbf{i}_x$, $\mathbf{i}_y$, $\mathbf{i}_z$ are the unit vectors, respectively, in the t, x, y, z directions, and where $F\left(t, x, y, z\right)$, $G\left(t, x, y, z\right)$, $H\left(t, x, y, z\right)$ are three families of surfaces partitioning the 4-dimensional t, x, y, z-space in cells or, interpreted alternatively, three families of moving surfaces partitioning the physical 3-dimensional x, y, z-space. Expansion of (4.1) gives

$$\theta\mathbf{1}, \theta v_x\mathbf{i}_x, \theta v_y\mathbf{i}_y, \theta v_z\mathbf{i}_z = \frac{\partial\left(F, G, H\right)}{\partial\left(\left(x, y, z\right)\right)}\mathbf{1},$$
$$-\frac{\partial\left(F, G, H\right)}{\partial\left(t, y, z\right)}\mathbf{i}_x, \frac{\partial\left(F, G, H\right)}{\partial\left(t, x, z\right)}\mathbf{i}_y, \frac{\partial\left(F, G, H\right)}{\partial\left(t, x, y\right)}\mathbf{i}_z. \quad (4.2)$$

It is easily shown that the 4-vector $\theta\mathbf{1}$, $\theta v_x\mathbf{i}_x$, $\theta v_y\mathbf{i}_y$, $\theta v_z\mathbf{i}_z$ satisfies the 3-D volumetric mass balance (2.1) without the source term. As often occurs with very general results, Euler himself seems to have noticed only a few of the implications of his general solution in special cases. These special cases were mostly discovered independently. In the remainder of this subsection, I will further interpret (4.1) and (4.2), and discuss the special cases of 3-dimensional flows with time invariant flow patterns, and of 3-dimensional steady flows. In the next two subsections, I will consider the special cases of, respectively, 2- and 1-dimensional flows. The overall result will be a unified presentation of scattered results in the literature. Truly 3-dimensional flows have rarely been analyzed thus far (see however *Larabi & de Smedt* [1994]).

4.1.2. Connection with Pfaffian differential equations. The families $F\left(t, x, y, z\right)$, $G\left(t, x, y, z\right)$, and $H\left(t, x, y, z\right)$ of surfaces in the 4-dimensional t, x, y, z space can also be seen as the solution of the Pfaffian differential equations for the path lines $\mathbf{x} = \mathbf{x}\left(\mathbf{X}, t\right)$, with $\mathbf{X}$ fixed and $-\infty < t < \infty$, of the parcels of water [*Pfaff*, 1818; *Sneddon*, 1957; *Truesdell and Toupin*, 1960, section 70]

$$dt = \frac{dx}{v_x} = \frac{dy}{v_y} = \frac{dz}{v_z}. \quad (4.3)$$

The solutions of these equations can be written as

$$F\left(t, x, y, z\right) = a, \quad G\left(t, x, y, z\right) = b, \quad H\left(t, x, y, z\right) = c. \quad (4.4)$$

Yih [1957] points out that, since the path lines are intersections of the 4-dimensional surfaces described by equations (4.4), they must lie in these surfaces, which are therefore material surfaces.

At a fixed instant $t = t_c$, equations (4.3) reduce to [*Goursat*, 1959, section 31; *Truesdell and Toupin*, 1960, section 70]:

$$\frac{dx}{v_x} = \frac{dy}{v_y} = \frac{dz}{v_z}. \quad (4.5)$$

The solutions of these equations can be written as

$$F\left(t_c, x, y, z\right) = a_c, \quad G\left(t_c, x, y, z\right) = b_c. \quad (4.6)$$

The two families of 3-dimensional surfaces described by equations (4.6) are instantaneous stream surfaces and their intersections are instantaneous streamlines at $t = t_c$.

4.1.3. Integral mass balance. Note that, with the help of (3.15), the expression for θ implied by (4.2) can be related to the volumetric water content in the reference configuration and the deformation gradient tensor $\mathbf{F}$ by:

$$\theta = \frac{\partial\left(F, G, H\right)}{\partial\left(x, y, z\right)} = \frac{\theta_\kappa}{J} = \theta_\kappa\det\frac{\partial\mathbf{X}}{\partial\mathbf{x}} = \theta_\kappa\det\mathbf{F}^{-1}. \quad (4.7)$$

This equation shows that the material coordinates $\mathbf{X}$ and the three families of surfaces $F\left(t, x, y, z\right)$, $G\left(t, x, y, z\right)$, and $H\left(t, x, y, z\right)$ are closely related.

At the current time t, consider the volume of water V_θ contained in the space V bounded by the six surfaces (cf., [*Yih*, 1957, equation (26)]; [*Truesdell and Toupin*, 1960, equations (155.4), (156.1), (156.2, material), (156.5, spatial), and (164.7)]):

$$F\left(t, x, y, z\right) = F_1, \quad \text{and} \quad F\left(t, x, y, z\right) = F_2, \quad (4.8)$$

$$G\left(t, x, y, z\right) = F_1, \quad \text{and} \quad G\left(t, x, y, z\right) = F_2, \quad (4.9)$$

$$H\left(t, x, y, z\right) = F_1, \quad \text{and} \quad H\left(t, x, y, z\right) = F_2. \quad (4.10)$$

This volume of water V_θ can be determined (i) by integrating the elements of volume of water dv_θ directly, or (ii) by integrating at the reference time $t = t_\kappa$ the reference volumetric water content θ_κ over the reference volume V_κ, or (iii) by integrating at the current time t the volumetric water content θ over the volume V, corresponding to the volume V_κ:

$$V_\theta = \int_{V_\theta} dv_\theta = \int_{V_\kappa} \theta_\kappa dv_\kappa = \int_V \theta dv =$$
$$\int\int\int_V \frac{\partial\left(F, G, H\right)}{\partial\left(x, y, z\right)}dx dy dz, \quad (4.11)$$

which can also be written as

$$V_\theta = \int_{H_1}^{H_2}\int_{G_1}^{G_2}\int_{F_1}^{F_2} dF dG dH =$$
$$\left(F_2 - F_1\right)\left(G_2 - G_1\right)\left(H_2 - H_1\right). \quad (4.12)$$

4.1.4. Time invariant, 3-dimensional flow patterns.

In subsection 4.1.2 we saw that the surfaces $F\left(t_c, x, y, z\right) = a_c$ and $G\left(t_c, x, y, z\right) = b_c$ in 3-dimensional x, y, z-space are the instantaneous stream surfaces at the fixed instant t_c, We now consider the possibility that these stream surfaces are steady (time-invariant). Then the flow pattern is steady, although the speed may still vary in time.

Assuming that the time-invariant stream surfaces already exist at the reference time t_κ, we set

$$F\left(t, x, y, z\right) = F\left(t_\kappa, x, y, z\right) = F_\kappa\left(x, y, z\right), \quad (4.13)$$

$$G\left(t, x, y, z\right) = G\left(t_\kappa, x, y, z\right) = G_\kappa\left(x, y, z\right). \quad (4.14)$$

Introducing (4.13) and (4.14) in (4.2) and expanding the determinants, noting $\partial F_\kappa/\partial t = 0$ and $\partial F_\kappa/\partial t = 0$, gives

$$
\begin{aligned}
&\theta\mathbf{1}, \theta v_x\mathbf{i}_x, \theta v_y\mathbf{i}_y, \theta v_z\mathbf{i}_z = \theta\mathbf{1}, \\
&\left(\frac{\partial F_\kappa}{\partial z}\frac{\partial G_\kappa}{\partial y} - \frac{\partial F_\kappa}{\partial y}\frac{\partial G_\kappa}{\partial z}\right)\frac{\partial H}{\partial t}\mathbf{i}_x, \\
&\left(\frac{\partial F_\kappa}{\partial x}\frac{\partial G_\kappa}{\partial z} - \frac{\partial F_\kappa}{\partial z}\frac{\partial G_\kappa}{\partial x}\right)\frac{\partial H}{\partial t}\mathbf{i}_y, \\
&\left(\frac{\partial F_\kappa}{\partial y}\frac{\partial G_\kappa}{\partial x} - \frac{\partial F_\kappa}{\partial x}\frac{\partial G_\kappa}{\partial y}\right)\frac{\partial H}{\partial t}\mathbf{i}_z. \quad (4.15)
\end{aligned}
$$

Note that (4.15) implies that the volumetric flux vector $\theta\mathbf{v}$ is given by:

$$
\begin{aligned}
\theta\mathbf{v} &= \theta v_x\mathbf{i}_x + \theta v_y\mathbf{i}_y + \theta v_z\mathbf{i}_z = \\
&\quad \left(\nabla F_\alpha \times \nabla G_\alpha\right)\frac{\partial H}{\partial t}. \quad (4.16)
\end{aligned}
$$

Some aspects of the class of time invariant flow patterns were discussed by *Truesdell* [1954b, section 28, p57; 1977, p. 110-111]. For this class of flows (i) the streamlines and the path lines coincide, and (ii) the velocity or volumetric flux vector fields change in the course of time only in magnitude, not in direction. A necessary and sufficient condition for such fields is that:

$$\mathbf{v} \times \left(\partial\mathbf{v}/\partial t\right)_\mathbf{x} = \mathbf{0}. \quad (4.17)$$

This means that the velocity vector $\mathbf{v}$ and the local acceleration vector $\left(\partial\mathbf{v}/\partial t\right)_\mathbf{x}$ are collinear. Time invariance of flow patterns facilitates the characterization of transport of solutes by means of transfer functions [*Raats*, 1978a, b; *De Valk and Raats*, 1995].

4.1.5. Steady, 3-dimensional flows.

The solution (4.2) of the mass balance for time-invariant 3-dimensional flow patterns reduces to the solution for steady 3-dimensional flow by setting:

$$
\begin{aligned}
H\left(t, x, y, z\right) &= H\left(t_\kappa, x, y, z\right) + \left(t - t_\kappa\right) = \\
&\quad H_\kappa\left(x, y, z\right) + \left(t - t_\kappa\right). \quad (4.18)
\end{aligned}
$$

Introducing (4.18) in (4.15) and (4.16) gives:

$$
\begin{aligned}
&\theta\mathbf{1}, \theta v_x\mathbf{i}_x, \theta v_y\mathbf{i}_y, \theta v_z\mathbf{i}_z = \theta\mathbf{1}, \\
&\left(\frac{\partial F_\kappa}{\partial z}\frac{\partial G_\kappa}{\partial y} - \frac{\partial F_\kappa}{\partial y}\frac{\partial G_\kappa}{\partial z}\right)\mathbf{i}_x, \\
&\left(\frac{\partial F_\kappa}{\partial x}\frac{\partial G_\kappa}{\partial z} - \frac{\partial F_\kappa}{\partial z}\frac{\partial G_\kappa}{\partial x}\right)\mathbf{i}_y, \\
&\left(\frac{\partial F_\kappa}{\partial y}\frac{\partial G_\kappa}{\partial x} - \frac{\partial F_\kappa}{\partial x}\frac{\partial G_\kappa}{\partial y}\right)\mathbf{i}_z, \quad (4.19)
\end{aligned}
$$

and

$$\theta\mathbf{v} = \theta v_x\mathbf{i}_x + \theta v_y\mathbf{i}_y + \theta v_z\mathbf{i}_z = \left(\nabla F_\alpha \times \nabla G_\alpha\right). \quad (4.20)$$

Steady 3-dimensional flows were discussed by *Truesdell* [1955, equation (203); pXV, line 2-3], *Yih* [1957, section VI, equation (20)], *Krzywoblocki* [1958, case (4)] *Truesdell and Toupin* [1960, section 163, equation (163.2)], *Nelson* [1964, equation (26)]. For steady 3-dimensional flows the volumetric mass balance equation (2.1) reduces to (2.3), expressing the solenoidal nature of the volumetric flux vector $\theta\mathbf{v}$ in this case. A characterization by the cross product of the gradients of two scalar fields, as in (4.20), is a general property of solenoidal vector fields (see *Raats* [1967]). It is nice to see that this characterization can be obtained as a special case from Euler's general solution (4.1).

The function $H\left(t, x, y, z\right) = H_\kappa\left(x, y, z\right) + \left(t - t_\kappa\right)$ in (4.18) describes the evolution of material surfaces. The function $H_\kappa\left(x, y, z\right) = H\left(t_\kappa, x, y, z\right)$ corresponds to the isochrones in discussions of purely convective transport of solutes (see *Raats* [1978a, b] and *De Valk & Raats* [1995] for details).

4.2. Two-dimensional Plane Flows

4.2.1. Transient 2-dimensional plane flows.

The solution of the mass balance for transient 3-dimensional flow reduces to the solution for transient 2-dimensional flow by setting:

$$F\left(t, x, y, z\right) = F\left(t, x, y_0, z\right), \quad (4.21)$$

$$G\left(t, x, y, z\right) = y, \quad (4.22)$$

$$H\left(t, x, y, z\right) = H\left(t, x, y_0, z\right). \quad (4.23)$$

Introducing (4.21), (4.22), and (4.23) in (4.2) gives:

$$\theta \mathbf{1}, \theta v_x \mathbf{i}_x, \theta v_y \mathbf{i}_y, \theta v_z \mathbf{i}_z = \frac{\partial (F, H)}{\partial (x, z)} \mathbf{1}, \frac{\partial (F, H)}{\partial (t, z)} \mathbf{i}_x,$$
$$(0) \mathbf{i}_y, \frac{\partial (F, H)}{\partial (t, x)} \mathbf{i}_z. \quad (4.24)$$

This corresponds to case 4 of *Krzywoblocki* [1958]. Expanding the determinants gives:

$$\theta \mathbf{1}, \theta v_x \mathbf{i}_x, \theta v_y \mathbf{i}_y, \theta v_z \mathbf{i}_z =$$
$$\left(\frac{\partial F}{\partial x} \frac{\partial H}{\partial z} - \frac{\partial F}{\partial z} \frac{\partial H}{\partial x} \right) \mathbf{1},$$
$$\left(\frac{\partial F}{\partial z} \frac{\partial H}{\partial t} - \frac{\partial F}{\partial t} \frac{\partial H}{\partial z} \right) \mathbf{i}_x,$$
$$(0) \mathbf{i}_y, \left(\frac{\partial F}{\partial t} \frac{\partial H}{\partial x} - \frac{\partial F}{\partial x} \frac{\partial H}{\partial t} \right) \mathbf{i}_z. \quad (4.25)$$

One can write (4.25) in the form

$$\theta \mathbf{1}, \theta v_x \mathbf{i}_x, \theta v_y \mathbf{i}_y, \theta v_z \mathbf{i}_z = \nabla_{txz} F \times \nabla_{txz} H, (0) \mathbf{i}_y. \quad (4.26)$$

Note that mathematically this case is analogous to steady 3-dimensional flow: the role of the t-coordinate in 2-dimensional transient flow corresponds to the role of the y-coordinate in steady 3-dimensional flow (cf. equation (64)).

Transient, 2-dimensional flows can be calculated in detail with current numerical models discussed in subsection 2.3. For example, using Hydrus-2D *Simunek et al.* [1996], *De Vos et al.* [2000] calculated the instantaneous spatial distribution of the volumetric flux vector $\theta \mathbf{v}$. The instantaneous streamlines $F_{txz} (t_c, x, y) = a_c$ are the tangent curves of such instantaneous volumetric flux vector fields.

4.2.2. Connection with Pfaff formulation. The differential equations for the path lines are:

$$dt = \frac{dx}{v_x} = \frac{dz}{v_z}. \quad (4.27)$$

The solutions of these equations can be written as

$$F (t, x, z) = a, \quad H (t, x, z) = c. \quad (4.28)$$

Since the path lines are intersections of the 3-dimensional surfaces described by equations (4.28), they must lie in these surfaces, which are therefore material surfaces.

At an instant $t = t_c$

$$\frac{dx}{v_x} = \frac{dz}{v_z}. \quad (4.29)$$

The solutions of this equation can be written as

$$F (t_c, x, z) = a_c. \quad (4.30)$$

The lines described by (4.30) are the instantaneous streamlines.

4.2.3. Integral mass balance. At the current time t, the mass contained in the space V of thickness Δy and bounded by the two pairs of curves:

$$F (t, x, z) = F_1, \quad \text{and} \quad F (t, x, z) = F_2, \quad (4.31)$$

$$H (t, x, z) = H_1, \quad \text{and} \quad H (t, x, z) = H_2, \quad (4.32)$$

is given by

$$V_\theta = \int_{V_\theta} dv_\theta = \int_{V_\kappa} \theta_\kappa dv_\kappa = \int_V \theta dv =$$
$$\Delta y \int \int_A \frac{\partial (F, H)}{\partial (x, z)} dx dz, \quad (4.33)$$

which can also be written as

$$V_\theta = \Delta y \int_{H_1}^{H_2} \int_{F_1}^{F_2} dF dH =$$
$$\Delta y (F_2 - F_1)(H_2 - H_1). \quad (4.34)$$

4.2.4. Time invariant, 2-dimensional plane flow patterns. The solution of the mass balance for transient 2-dimensional flow reduces to the solution for transient 2-dimensional flow with a time-invariant flow pattern by setting:

$$F (t, x, y, z) = F (t_\kappa, x, y_0, z) = F_\kappa (x, y_0, z), \quad (4.35)$$

Introducing (4.35) in (4.2) gives:

$$\theta \mathbf{1}, \theta v_x \mathbf{i}_x, \theta v_y \mathbf{i}_y, \theta v_z \mathbf{i}_z = \left(\frac{\partial F_\kappa}{\partial x} \frac{\partial H}{\partial z} - \frac{\partial F_\kappa}{\partial z} \frac{\partial H}{\partial x} \right) \mathbf{1},$$
$$\left(\frac{\partial F_\kappa}{\partial z} \frac{\partial H}{\partial t} \right) \mathbf{i}_x, (0) \mathbf{i}_y, \left(\frac{\partial F_\kappa}{\partial x} \frac{\partial H}{\partial t} \right) \mathbf{i}_z. \quad (4.36)$$

One can write (4.36) in the form:

$$\theta \mathbf{1}, \theta v_x \mathbf{i}_x, \theta v_y \mathbf{i}_y, \theta v_z \mathbf{i}_z = \nabla_{xz} F_\kappa \times \nabla_{txz} H, (0) \mathbf{i}_y. \quad (4.37)$$

The surfaces H=constant are material surfaces. For these flows with fixed streamlines, one can derive a parcel function for the infinitesimal stream tubes [*Wilson and Gelhar*, 1974, 1981; *Raats*, 1982].

4.2.5. Steady, 2-dimensional plane flows. The solution of the mass balance for transient 2-dimen-sional flow with time-invariant flow pattern reduces to the solution of steady 2-dimensional flow by setting

$$H\left(t,x,y,z\right)=H\left(t_\kappa,x,y_0,z\right)+\left(t-t_\kappa\right)=$$

$$H_\kappa\left(x,y_0,z\right)+\left(t-t_\kappa\right).\quad(4.38)$$

Introducing (4.38) in (4.36) gives

$$\theta\mathbf{1},\theta v_x\mathbf{i}_x,\theta v_y\mathbf{i}_y,\theta v_z\mathbf{i}_z=\left(\frac{\partial F_\kappa}{\partial x}\frac{\partial H_\kappa}{\partial z}-\frac{\partial F_\kappa}{\partial z}\frac{\partial H_\kappa}{\partial x}\right)\mathbf{1},$$

$$\left(\frac{\partial F_\kappa}{\partial z}\right)\mathbf{i}_x,(0)\,\mathbf{i}_y,\left(\frac{\partial F_\kappa}{\partial x}\right)\mathbf{i}_z,\quad(4.39)$$

or

$$\theta\mathbf{1},\theta v_x\mathbf{i}_x,\theta v_y\mathbf{i}_y,\theta v_z\mathbf{i}_z=\nabla_{xz}F_\kappa\times\nabla_{xz}H_\kappa,(0)\,\mathbf{i}_y.\quad(4.40)$$

The function F_κ is the D'Alembert stream function [*D' Alembert*, 1761]. According to *Truesdell and Toupin* [1960, section 161], in 1757 Euler introduced a more general type of stream function for pseudo-plane motions, but he did not note the especially simple properties of F_κ. The D'Alembert stream function was introduced by *Forchheimer* [1886] for what is now known as horizontal Dupuit-Forchheimer groundwater flow and by *Slichter* [1897] for vertical 2-D flow in saturated soils. Originally, I introduced this stream function for any steady flow in unsaturated soils [*Raats, 1967*]. Later, I showed that, for the Gardner class of soils defined by (2.14), the function F_κ satisfies a partial differential of the same form as the equation (2.16) for the matric flux potential [*Raats*, 1970]:

$$\nabla^2 F_\kappa=\alpha\frac{\partial F_\kappa}{\partial z}.\quad(4.41)$$

In (5.5) the function $H\left(t,x,y,z\right)=H_\kappa\left(x,y_0,z\right)+\left(t-t_\kappa\right)$ describes the evolution of material surfaces. The function $H_\kappa\left(x,z\right)$ describes plane isochrones. Following some pioneering work by *Batu and Gardner* [1978], *Philip* [1984a] calculated isochrones for a periodic distribution of surface source strength, both for flow to infinite depth and for flow to a water table at finite depth. In his analysis, Philip used the flow equation (2.16) for multidimensional steady flow and the assumption, expressed in (3.24), that $\theta\left(\mathbf{x}\right)$ can be replaced by an average value $\bar\theta$.

A similar derivation can be given for the Stokes' stream function for steady axisymmetric flows [*Raats*, 1971]. Corresponding isochrones were calculated by *Philip* [1984b] for surface and buried point sources. For the surface point source, *Clothier* [1984] evaluated the validity replacing $\theta\left(\mathbf{x}\right)$ by an average value $\bar\theta$.

4.3. One-dimensional Flows

4.3.1. Transient, 1-dimensional flow. The solution of the mass balance for transient 3-dimensional flow reduces to the solution for transient 1-dimensional flow by setting:

$$F\left(t,x,y,z\right)=x,\quad(4.42)$$

$$G\left(t,x,y,z\right)=y,\quad(4.43)$$

$$H\left(t,x,y,z\right)=H\left(t,x_0,y_0,z\right).\quad(4.44)$$

Introducing (4.42), (4.43), and (4.44) in (4.2) gives:

$$\theta\mathbf{1},\theta v_x\mathbf{i}_x,\theta v_y\mathbf{i}_y,\theta v_z\mathbf{i}_z=$$
$$\frac{\partial H}{\partial z}\mathbf{1},(0)\,\mathbf{i}_x,(0)\,\mathbf{i}_y,\frac{\partial H}{\partial t}\mathbf{i}_z.\quad(4.45)$$

The parcel function H for transient, one-dimensional flow was first introduced by *Euler* [1757, sections 48-49; see also the commentary *Truesdell* [1954a, equations 116 and 118]], and discussed later by *Kirchhoff* [1930], *Krzywoblocki* [1958, case 1], and *Truesdell and Toupin* [1960, section 161, equation 161.22]. The parcel function H for transient one-dimensional flow is analogous with the stream function F_κ for 2-dimensional plane flow derived in subsection 4.2.5. I first introduced the parcel function in a seminar at the University of Wisconsin in the fall of 1972 [*Raats*, 1972] and a year later at the Annual meeting of the American Society of Agronomy. The parcel function H_κ has been widely used in last two decades [*Smiles et al.*, 1981; *Wilson and Gelhar*, 1974, 1981; *Raats*, 1982, 1984b, 1987a, b; *Smiles*, 2000, this volume]. In an analysis of brine transport, *Van Duijn and Schotting* [1998; see also *Schotting*, 1998] use a transformation from t, z-coordinates to H, z-coordinates, referring to it as a variant of the Von Mises transformation in fluid mechanics.

4.3.2. Steady 1-dimensional flow. The solution of the mass balance for transient 1-dimensional flow reduces to the solution for steady 1-dimensional flow by setting:

$$H\left(t,x,y,z\right)=H\left(t_\kappa,x_0,y_0,z\right)-\theta_\kappa v_\kappa\left(t-t_\kappa\right)=$$
$$H_\kappa\left(x_0,y_0,z\right)-\theta_\kappa v_\kappa\left(t-t_\kappa\right).\quad(4.46)$$

Introducing (4.46) in (4.2) gives:

$$\theta\mathbf{1},\theta v_x\mathbf{i}_x,\theta v_y\mathbf{i}_y,\theta v_z\mathbf{i}_z=$$
$$\frac{\partial H_\kappa}{\partial z}\mathbf{1},(0)\,\mathbf{i}_x,(0)\,\mathbf{i}_y,-\theta_\kappa v_\kappa\mathbf{i}_z.\quad(4.47)$$

The idea of keeping track of parcels of water can also easily be generalized to steady flows in the presence of uptake of water by plant roots [*Raats*, 1975]

5. IMPLICATIONS OF DARCY'S LAW FOR THE NATURE OF THE FLOW PATTERNS

5.1. Rotationality of Flow of Soil Water

5.1.1. The rotationality of the flux $\theta\mathbf{v}$ and the velocity $\mathbf{v}$. Taking the curl of the Darcian expressions for the volumetric flux $\theta\mathbf{v}$ and the velocity $\mathbf{v}$ vector fields gives:

$$\nabla \times (\theta\mathbf{v}) = -$$
$$\nabla k \times \nabla H = -\nabla k \times \nabla h - \nabla k \times \nabla z, \quad (5.1)$$

$$\nabla \times \mathbf{v} = -\nabla (k/\theta) \times \nabla H =$$
$$- \nabla (k/\theta) \times \nabla h - \nabla (k/\theta) \times \nabla z. \quad (5.2)$$

Assuming that the hydraulic conductivity k and the hydraulic mobility k/θ are functions of the pressure head h, both ∇k and $\nabla (k/\theta)$ are collinear with ∇h, and hence the first terms on the right hand sides of (5.1) and (5.2) vanish:

$$\nabla \times (\theta\mathbf{v}) = -\nabla k \times \nabla z, \quad (5.3)$$
$$\nabla \times \mathbf{v} = -\nabla (k/\theta) \times \nabla z. \quad (5.4)$$

According to (5.3), the rotation vector $\nabla \times (\theta\mathbf{v})$ is perpendicular to the plane spanned by ∇k and $-\nabla z$ and according to (5.4) the vorticity vector $\nabla \times \mathbf{v}$ is perpendicular to the plane spanned by $\nabla (k/\theta)$ and $-\nabla z$. This means that both the rotation vector $\nabla \times (\theta\mathbf{v})$ and the vorticity vector $\nabla \times \mathbf{v}$ are always horizontal. It should be emphasized that the simple expressions (5.3) and (5.4) apply only if the soil is homogeneous and isotropic.

Again making use of the dependence of the hydraulic conductivity k and the hydraulic mobility k/θ upon the pressure head h, equations (5.3) and (5.4) can be rewritten as:

$$\nabla \times (\theta\mathbf{v}) = \frac{1}{k}\frac{dk}{dh}(-k\nabla h) \times \nabla z, \quad (5.5)$$

$$\nabla \times \mathbf{v} = \frac{1}{k/\theta}\frac{dk/\theta}{dh}(-(k/\theta)\nabla h) \times \nabla z. \quad (5.6)$$

Since $\nabla z \times \nabla z = 0$, equations of the form (5.5) and (5.6) also apply with the gradient of the pressure head ∇h replaced by the gradient of the total head $\nabla H = \nabla (h - z)$:

$$\nabla \times (\theta\mathbf{v}) = \frac{1}{k}\frac{dk}{dh}(-k\nabla H) \times \nabla z, \quad (5.7)$$

$$\nabla \times \mathbf{v} = \frac{1}{k/\theta}\frac{dk/\theta}{dh}(-(k/\theta)\nabla H) \times \nabla z. \quad (5.8)$$

Using (2.8) in (5.7) and (5.8), and making use of the fact that the cross product of the vertical component of $\mathbf{v}$ and $\nabla \times z$ is a zero vector, gives:

$$\nabla \times (\theta\mathbf{v}) = \frac{1}{k}\frac{dk}{dh}\theta\mathbf{v} \times \nabla z = \frac{1}{k}\frac{dk}{dh}\theta\mathbf{v}_{hor} \times \nabla z, \quad (5.9)$$

$$\nabla \times \mathbf{v} = \frac{1}{k/\theta}\frac{dk/\theta}{dh}\mathbf{v} \times \nabla z = \frac{1}{k/\theta}\frac{dk/\theta}{dh}\mathbf{v}_{hor} \times \nabla z. \quad (5.10)$$

According to (5.9) and (5.10), the direction of both the rotation vector $\nabla \times (\theta\mathbf{v})$ and the vorticity vector $\nabla \times \mathbf{v}$ is horizontal and perpendicular to $\mathbf{v}_{hor}$. According to (5.9), the magnitude of the rotation vector $\nabla \times (\theta\mathbf{v})$ is equal to the product of the magnitude of the horizontal component of the volumetric flux vector $\theta\mathbf{v}$ and $d\ln k/dh$. According to (5.10), the magnitude of the vorticity vector $\nabla \times \mathbf{v}$ is equal to the product of the magnitude of the horizontal component of the velocity $\mathbf{v}$ and $d\ln (k/\theta)/dh$.

Recall that for the Gardner class of soils, defined by (2.14), $d\ln k/dh = \alpha = $ constant, and for its variant, defined by (2.17), $d\ln (k/\theta)/dh = \alpha_v = $ constant. Hence, for these two classes of soils, respectively, equations (5.9) and (5.10) reduce to:

$$\nabla \times (\theta\mathbf{v}) = \alpha\theta\mathbf{v} \times \nabla z = \alpha\theta\mathbf{v}_{hor} \times \nabla z, \quad (5.11)$$

$$\nabla \times \mathbf{v} = \alpha_v\mathbf{v} \times \nabla z = \alpha_v\mathbf{v}_{hor} \times \nabla z. \quad (5.12)$$

Introducing (3.2), the definition of the velocity vector, in (5.12) and integrating with respect to time, shows that the rotation over the time interval $t - t_0$ is the product of α_v and the horizontal displacement.

In an early discussion of the rotation vector $\nabla \times (\theta\mathbf{v})$ of the volumetric flux vector field $\theta\mathbf{v}$, I derived (5.3) and referred to the rotation vector $\nabla \times (\theta\mathbf{v})$ as the vorticity [*Raats*, 1967], but this name should be reserved for the rotation vector $\nabla \times \mathbf{v}$ of the velocity vector field $\mathbf{v}$. The expression (5.10) for the vorticity vector $\nabla \times \mathbf{v}$ was also given earlier, but with the factor $\times \nabla z$ missing [*Raats*, 1982].

5.1.2. Intrinsic representation of rotation vectors. Some intrinsic geometric properties of vector fields become most transparent if various quantities are expressed in terms of the trihedron of Frenet, i.e., the unit tangent vector field $\mathbf{s}$, the unit principal normal vector field $\mathbf{n}$, and the unit binormal vector field $\mathbf{b}$. Subsections 3.2-3.4 already dealt with such intrinsic representations of features involving $\mathbf{s}$, but not $\mathbf{n}$

and **b**. The intrinsic representation of the curl of the vector field **f** is [*Truesdell*, 1954b]:

$$\nabla \times \mathbf{f} = \left(A\mathbf{s} + \frac{1}{f}\frac{\delta f}{\delta b}\mathbf{n} + \left(\kappa - \frac{1}{f}\frac{\delta f}{\delta n} \right) \mathbf{b} \right) f, \quad (5.13)$$

where the scalar field A is the abnormality, and the scalar field κ is the curvature of the system of vector lines. The curvature κ measures, as one moves along a streamline, the rate at which the unit tangent vector turns to the unit normal vector or, equivalently, the rate at which the normal plane turns about the binormal vector. Taking the inner product of **f** given by (3.16) and $\nabla \times \mathbf{f}$ given by (5.13) gives

$$\mathbf{f} \cdot \nabla \times \mathbf{f} = Af^2. \quad (5.14)$$

From (5.14) it follows that the abnormality A is defined by:

$$A = \frac{\mathbf{f} \cdot \nabla \times \mathbf{f}}{f^2} = \mathbf{s} \cdot \nabla \mathbf{s}. \quad (5.15)$$

From (5.13) it can be inferred that the expressions of the rotation vector fields of the volumetric flux vector field $\theta\mathbf{v}$ and of the velocity vector field **v**, expressed in terms of the Frenet trihedron $(\mathbf{s}, \mathbf{n}, \mathbf{b})$, are:

$$\nabla \times \theta\mathbf{v} = \theta v A\mathbf{s} + \frac{d\theta v}{db}\mathbf{n} + \left(\theta v\kappa - \frac{d\theta v}{db} \right)\mathbf{b}, \quad (5.16)$$

$$\nabla \times \mathbf{v} = vA\mathbf{s} + \frac{dv}{db}\mathbf{n} + \left(v\kappa - \frac{dv}{db} \right)\mathbf{b}. \quad (5.17)$$

Setting $\mathbf{f} = \mathbf{v}$ in (5.15), introducing (2.7) and (5.17), and making use of the facts that in the triple scalar product the dot and cross products commute and the cross product of a vector with itself is the zero vector, gives:

$$A = \frac{\mathbf{v} \cdot \nabla \times \mathbf{v}}{\mathbf{v} \cdot \mathbf{v}} =$$
$$- v^2 \frac{1}{k/\theta}\frac{dk/\theta}{dh}((\nabla h + \nabla z) \cdot (\nabla h \times \nabla z)) =$$
$$- v^2 \frac{1}{k/\theta}\frac{dk/\theta}{dh}((\nabla h \cdot \nabla h \times \nabla z) \cdot (\nabla h \cdot \nabla z \times \nabla z))$$
$$= 0. \quad (5.18)$$

Therefore, the first terms on the right hand sides of (5.16) and (5.17) vanish, implying that both the rotation vector $\nabla \times (\theta\mathbf{v})$ and the vorticity vector $\nabla \times \mathbf{v}$ lie in the **n**, **b** -plane

The name abnormality expresses that it is a measure of the departure of the unit tangent vector field **s** from the property of having a normal congruence of curves. It has been described also as 'the torsion of the curve system' and as 'the torsion of neighboring vector lines' [*Truesdell*, 1954b]. In

analogy with the term 'helicity' used in particle physics for the product of momentum and spin of a particle, in the modern literature of turbulence and magneto-hydrodynamics the quantity $\mathbf{f} \cdot \nabla \times \mathbf{f} = Af^2$ is called the helicity density, i.e. the helicity per unit volume [*Moffatt*, 1969; *Moffatt and Tsinobar*, 1992]. *Finnigan* [1990] discussed the intrinsic description in the context of fluid mechanics, indicating the importance of the abnormality for the onset of chaotic advection. The abnormality being zero for flow of water described by the Richards equation, evidently implies that for such flows chaotic flow patterns are not to be expected. As one reviewer pointed out, this is interesting and should be explored further.

5.2. Scalar Potentials and Distance Functions

5.2.1. For arbitrary vector fields. A vector field $\mathbf{f} = \mathbf{f}_l$ is called lamellar or conservative, if there exists a scalar potential field ϕ_1 such that [*Truesdell*, 1954b]:

$$\mathbf{f} = \mathbf{f}_l = \nabla \phi_1. \quad (5.19)$$

Mathematicians usually call such vector fields conservative. A lamellar vector field $\mathbf{f} = \mathbf{f}_l$ is everywhere normal to the equipotential surfaces $\phi_1 = $ constant constant. A vector field $\mathbf{f} = \mathbf{f}_l$ is lamellar if and only if its curl vanishes, i.e. if

$$\nabla \times \mathbf{f} = \nabla \times \mathbf{f}_l = 0. \quad (5.20)$$

For this reason lamellar fields are called also irrotational fields.

A vector field $\mathbf{f} = \mathbf{f}_{cl}$ is called complex-lamellar if there exist scalar fields ϕ_2 and ϕ_3 such that [*Truesdell*, 1954b]

$$\mathbf{f} = \mathbf{f}_{cl} = \phi_3 \nabla \phi_2. \quad (5.21)$$

A complex-lamellar vector field $\mathbf{f} = \mathbf{f}_{cl}$ is everywhere normal to the equipotential surfaces $\phi_2 = $ constant. A vector field $\mathbf{f} = \mathbf{f}_{cl}$ is complex-lamellar if, and only if, it is normal to its curl, i.e.

$$\mathbf{f} \cdot \nabla \times \mathbf{f} = \mathbf{f}_{cl} \cdot \nabla \times \mathbf{f}_{cl} = 0. \quad (5.22)$$

Any vector field **f** may be represented as the sum of a lamellar vector field $\mathbf{f}_l = \nabla \phi_1$ and a complex-lamellar vector field $\mathbf{f}_{cl} = \phi_2 \nabla \phi_3$ [*Truesdell*, 1954b]:

$$\mathbf{f} = \mathbf{f}_l + \mathbf{f}_{cl} = \nabla \phi_1 + \phi_3 \nabla \phi_2. \quad (5.23)$$

Truesdell [1954b] calls the scalars ϕ_1, ϕ_2, and ϕ_3 Monge potentials.

5.2.2. For velocity and flux vector fields in soil physics. Comparison of equations (2.8) and (5.19) shows that, if the soil is saturated and homogeneous with saturated hydraulic conductivity k_s and volumetric water content θ_s, then the velocity vector field $\mathbf{v}$ is a lamellar field. The scalar field $(k_s/\theta_s) H$ is then the scalar potential of the velocity vector field $\mathbf{v}$. The vector field $\mathbf{v}$ being lamellar implies that it is irrotational, i.e., that its curl vanishes. Of course, the viscous flow in individual pores will always be rotational, even if the macroscopic field $\mathbf{v}$ is irrotational [*Raats*, 1967].

Comparison of (2.8) and (5.21) shows that, if the soil is unsaturated and homogeneous, in general the volumetric flux vector field $\theta\mathbf{v}$ and the velocity vector field $\mathbf{v}$ are complex-lamellar fields (cf., *Nelson*, 1966; *Raats*, 1967). A significant implication is that the vector fields $\theta\mathbf{v}$ and $\mathbf{v}$ are normal to the surfaces of equal total head H. This makes it possible to infer the flow direction at any point from measurements of the total head H, or equivalently the pressure head h and gravitational head z, at a sufficient number of points. *Bouwer & Little* [1959] constructed lines of equal total head for steady plane flow in saturated and unsaturated soils from resistance network measurements and then, at the suggestion of E.C. Childs, drew the stream lines as orthogonals.

The expression (2.15) for the volumetric flux of the soil water is precisely of the form (5.23), with the matric component of the flux $\nabla(-\phi)$ being the lamellar part and the gravitational component of the flux $\alpha\phi\nabla z = k\nabla z$ being the complex-lamellar part. With this interpretation of (2.15), the matric flux potential ϕ, the hydraulic conductivity $\alpha\phi = k$ and the gravitational head z are the three Monge potentials.

6. CONCLUDING REMARKS

The brief and selective review in section 2 shows the key role of John Philip in analysis of multi-dimensional flow in rigid, unsaturated soils. The papers by him and his collaborators show that progress mostly depended on ideas from outside the narrow confines of soil physics and porous media hydrodynamics, particularly from the mathematical specialty of partial differential equations. In this paper, I have tried to demonstrate that continuum mechanics and differential geometry are also among the disciplines providing guidance.

Acknowledgements. I like to acknowledge the reviewers for their constructive criticism, especially the suggestions leading to the improved notation and to a more detailed review of numerical solutions of multi-dimensional problems. I like to thank Dr. Kees Rappoldt for introducing me to OzTEX, a Macinthosh implementation of TEX.

REFERENCES

Aris, R, *Vectors, Tensors, and the Basic Equations of Fluid Mechanics*, Dover, New York, 1989.

Batu, V., and W.R. Gardner, Steady-state solute convection in two dimensions with nonuniform infiltration, *Soil Sci. Soc. Amer. J.*, *42*, 18-22, 1978.

Breaster, C., G. Dagan, S. Neuman, and D. Zaslavsky, *A Survey of the Equations and Solutions of Unsaturated Flow in Porous Media*, First Annual Report (Part 1), Project No. A10-SWC-77, Grant No. FG-Is-287 made by USDA under PL480, Technion Israel Institute of Technology and Research and Development Foundation LTD, Hydrodynamics and Hydraulic Eng. Lab., 1971.

Brooks, R.H., and A.T. Corey, *Hydraulic Properties of Porous Media Affecting Fluid Flow*, Hydrology Paper 3, 27 pp., Colorado State University, Fort Collins, Colorado, 1964.

Brooks, R.H., and A.T. Corey, Properties of porous media affecting fluid flow, *J. Irrig. Drain. Div., Proc ASCE, 92*, 61-68, 1966.

Bouwer, H., and W.C. Little, A unifying numerical solution for two-dimensional steady flow problems in porous media with an electrical resistance network, *Soil Sci. Soc. Amer. Proc., 23*, 91-96, 1959.

Clothier, B.E., Solute travel times during trickle irrigation, *Water Resour. Res., 20*, 1848-1852, 1984.

D'Alembert, J.L., Remarques sur les lois du mouvement des fluides, *Opuscules Mathmatiques, 1*, 1761.

De Valk, C.F., and P.A.C. Raats, Lumped models of convective solute transport in heterogeneous porous media 1. One-dimensional media, *Water Resour. Res., 31*, 883-892, 1995.

De Vos, J.A., D.L.R. Hesterberg, and P.A.C. Raats, Water flow and nitrate leaching in a layered silt loam, *Soil Sci. Soc. of Amer. J., 64*, 517-527, 2000.

Euler, L., Principes généraux du mouvement des fluides. *Mém. Acad. Sci. Berlin, 11* (1755), 274-315, 1757. (=Leonhardi Euleri Opera Omnia (2) 12, 92-132, 1955.)

Euler, L., Sectio secunda de principiis motus fluidorum. *Novi Commm. Acad. Sci. Petrop., 14*, 270-386, 1770. (= Leonhardi Euleri Opera Omnia (2) 13 , 73-153, 1955.)

Feddes, R.A., and J.C. van Dam, Effects of plants on the upper boundary condition, in *Modelling of Transport Processes in Soils a Various Scales in Space and Time*, edited by J. Feyen, and K. Wiyo, pp. 391-405, Wageningen Pers, Wageningen, The Netherlands, 1999.

Feddes, R.A., S.P. Neuman, and E. Bresler, Finite element analysis of two-dimensional flow in soils considering water uptake by roots: II. Field applications, *Soil Sci. Soc. Amer. Proc., 39*, 231-242, 1975.

Finnigan, J.J., Streamline coordinates, moving frames, chaos and integrability in fluid flow, in *Topological Fluid Mechanics, Proceedings of the IUTAM Symposium, held 13-18 August 1989, at Cambridge, UK.*, edited by H.K. Moffatt, and A. Tsiboner, pp. 64-74, Cambridge University Press, Cambridge, UK., 1990.

Forchheimer, Ph., Ueber die Ergiebigkeit von Brunnen-Anlagen und Sickerschlitzen, *Zeitschrift des Architekten- und Ingenieur-Vereins in Hannover*, *32*, 540-563, 1886.

Freeze, R.A., Three-dimensional, transient, saturated-unsaturated flow in a groundwater basin, *Water Resour. Res.*, *7*, 347-366, 1971.

Gardner, W.R., Some steady-state solutions of the unsaturated moisture flow equation with application to evaporation from a water table, *Soil Sci.*, *85*, 228-232, 1958.

Garnier, P., E. Perrier, R. Angulo Jaramillo, and P. Baveye, Numerical model of 3-dimensional anisotropic deformation and 1-dimensional water flow in swelling soils, *Soil Sci.*, *162*, 410-420, 1997.

Goursat, E., *A Course in Mathematical Analysis, Vol. II, Part 2: Differential Equations*, Dover, New York, 1959.

Heinen, M., *Dynamics of Water and Nutrients in Closed, Recirculating Cropping Systems in Glasshouse Horticulture, with Special Attention to Lettuce Grown in Irrigated Sand Beds*, PhD Thesis, 270 pp., Wageningen Agricultural University, The Netherlands, 1997.

Heinen, M., and P.A.C. Raats, Hysteretic hydraulic properties of a coarse sand horticultural substrate, in *Characterization and Measurement of the Hydraulic Properties of Unsaturated Porous Media*, edited by M.Th. Van Genuchten, F.J. Leij, and L. Wu, Proceedings of an International Workshop organized by the U.S. Salinity Laboratory, USDA-ARS, and the Department of Soil and Environmental Sciences of the University of California, both at Riverside, CA, USA, and held 22-24 October, 1997 at Riverside, CA, USA, Part I, 467-476, University of California, Riverside, CA, USA, 1999.

Heinen, M., and P. De Willigen, *FUSSIM2: A Two-Dimensional Simulation Model for Water Flow, Solute Transport, and Root Uptake of Water and Nutrients in Partly Unsaturated Porous Media*, Quantitative Approaches in Systems Analysis No 20, 140 pp., DLO Research Institute for Agrobiology and Soil Fertility and the C.T. de Wit Graduate School for Production Ecology, Wageningen, The Netherlands, 1998.

Kirchhoff, G., *Vorlesungen über die Theorie der Wärme, Herausgegeben von M. Planck*, Teubner, Leipzig, 1894.

Kirchhoff, W., Reduktion simultaner partieller Differentialgleichungen bei hydrodynamischen Problemen, *J. Reine angew. Math.*, *164*, 183-195, 1930.

Klute, A., A numerical method for solving the flow equation for water in unsaturated material, *Soil Sci.*, *73*, 105-116, 1952.

Krzywoblocki, M.Z.v., On the stream functions in nonsteady three-dimensional flow, *J. Aeronaut. Sci.*, *25*, 67, 1958.

Larabi, A., and F. de Smedt, Solving three-dimensional hexahedral finite element groundwater models by preconditioned conjugate gradient methods, *Water Resour. Res.*, *30*, 509-521, 1994.

Miller, E.E., and R.D. Miller, Physical theory of capillary flow phenomena, *J. Appl. Phys.*, *27*, 324-332, 1956.

Moffatt, H.K., The degree of knottedness of tangled vortex lines, *J. Fluid Mech.*, *36*, 117-129, 1969.

Moffatt, H.K., and A. Tsinober, Helicity in laminar and turbulent flow, *Annu. Rev. Fluid Mech.*, *24*, 281-312, 1992.

Mualem, Y., A new model for predicting the hydraulic conductivity of unsaturated porous media, *Water Resour. Res. 12*, 513-522, 1976.

Mualem, Y., A modified dependent-domain theory of hysteresis, *Soil Sci.*, *137*, 283-291, 1984.

Narasimhan, T.N., and P.A. Whitherspoon, Numerical model for saturated-unsaturated flow in deformable porous media, 1. Theory, *Water Resour. Res.*, *12*, 657-664, 1976.

Narasimhan, T.N., P.A. Whitherspoon, and A.L. Edwards, Numerical model for saturated-unsaturated flow in deformable porous media, 2. The algorithm, *Water Resour. Res.*, *14*, 255-264, 1978a.

Narasimhan, T.N., and P.A. Whitherspoon, Numerical model for saturated-unsaturated flow in deformable porous media, 3. Applications, *Water Resour. Res.*, *14*, 1017-1034, 1978b.

Nelson, R.W., Stream functions for three-dimensional flow in heterogeneous porous media, *Proceedings of the 1963 General Assembly of Berkeley, International Association of Scientific Hydrology, Publication No. 64*, 290-301, 1964.

Nelson, R.W., Flow in heterogeneous porous mediums. 1. Darcian type description of two-phase systems, *Water Resour. Res.*, *2*, 487-495, 1966.

Neuman, S.P., Saturated - Unsaturated Seepage by Finite Elements, *J. of the Hydraulics Div.*, *99*, 2233-2250, 1973.

Neuman, S.P., R.A. Feddes, and E. Bresler, Finite element analysis of two-dimensional flow in soils considering water uptake by roots: I. Theory, *Soil Sci. Soc. Amer. Proc.*, *39*, 224-230, 1975.

Pfaff, J.F., Methodus generalis, aequationes differentiarum partialium, nec non aequationes differentiales vulgares, utrasque primi ordinis, inter quotcunque variabiles, complete integrandi, *Abhandlungen der Preussischen Akademie der Wissenschaften zu Berlin 1814-1815*, 76-136 [15 May 1815], 1818. Translated into German and annotated by G. Kowalewski as *Allgemeine Methode partielle Differentialgleichungen zu integrieren, Oswald's Klassiker der exakten Wissenschaften, Nr. 129*, Leipzig, 1902.

Philip, J.R., Numerical solution of equations of the diffusion type with diffusivity concentration dependent, *Trans. Faraday Soc.*, *51*, 885-892, 1955.

Philip, J.R., Numerical solution of equations of the diffusion type with diffusivity concentration dependent: 2. *Australian J. Phys.*, *10*, 29-42, 1957.

Philip, J.R., Steady infiltration from buried point sources and spherical cavities, *Water Resour. Res.*, *4*, 1039-1047, 1968.

Philip, J.R., Theory of infiltration, *Adv. Hydrosci.*, *5*, 215-296, 1969.

Philip, J.R., Flow in porous media, in *Proceedings of the 13th International Congress of Theoretical and Applied Mechanics, Moscow*, edited by E. Becker and G.K. Mikhailov, Springer, Heidelberg, pp. 279-294, 1973.

Philip, J.R., Fifty years progress in soil physics, *Geoderma*, *12*, 265-280, 1974.

Philip, J.R., Nonuniform leaching from nonuniform steady infiltration. *Soil Sci. Soc. Amer. J.*, *48*, 740-749, 1984a.

Philip, J.R., Travel times from buried and surface infiltration point sources, *Water Resour. Res.*, *20*, 990-994, 1984b.

Philip, J.R., Quasianalytic and analytic approaches to unsaturated flow, in *Flow and Transport in the Natural Environment: Advances and Applications*, edited by W.L. Steffen and O.T. Denmead, *Proceedings of the International Symposium on Flow and Transport in the Natural Environment, held in September 1987 at Canberra, Australia*, pp. 30-47 Springer-Verlag, Berlin, 1988.

Philip. J.R., The scattering analog for infiltration in porous media, *Rev. Geophys.*, *27*, 431-448, 1989.

Philip, J.R., How to avoid free boundary problems, in *Free Boundary Problems: Theory and Applications, Research Notes in Mathematics 185*, edited by K.H. Hoffman, and J. Sprekels, pp. 193-207, Longman, London, 1990.

Philip, J.R., Exact solutions for redistribution by nonlinear convection-diffusion, *J. Aust. Math. Soc. Ser.*, *B 33*, 363-383, 1992

Philip, J.R., Infiltration, in *Encyclopedia of Hydrology and Water Resources*, edited by R. Herschy, pp. 418-426, Chapman and Hall, London, 1998.

Philip, J.R., and J. H. Knight, Redistribution from plane, line, and point sources, *Irrigation Sci.*, *12*, 169-180, 1991.

Pruess, K., *TOUGH User's Guide, NUREG/CR-4645, SAND 86-7104. LBL-20700*, Sandia National Laboratories, Albuquerque, New Mexico., 1987.

Pruess, K., *TOUGH2 - A General-Purpose Numerical Simulator for Multiphase Fluid and Heat Flow, LBL-29400, UC-251*, 103pp., Lawrence Berkeley Laboratory, University of California Earth Science Division, 1991.

Pullan, A.J., The quasilinear approximation for unsaturated porous media flow, *Water Resour. Res.*, *26*, 1219-1234, 1990.

Raats, P.A.C., The kinematics of soil water, in *Isotope and Radiation Techniques in Soil Physics and Irrigation Studies*, pp. 191-201, International Atomic Energy Agency, Vienna, 1967.

Raats, P.A.C., Steady infiltration from line sources and furrows, *Soil Sci. Soc. Amer. Proc.*, *34*, 709-714, 1970.

Raats, P.A.C., Steady infiltration from point sources, cavities, and basins, *Soil Sci. Soc. Amer. Proc.*, *35*, 689-694, 1971.

Raats, P.A.C., Convection of solutes in soils, *Lecture notes for Soil Physics Seminar at the Soils Department, University of Wisconsin*, 9 pp. + 3 Figs., presented on October 3, 1972.

Raats, P.A.C., Steady flow patterns in saturated and unsaturated isotropic soils, J. Hydrol., 21, 357-369, 1974.

Raats, P.A.C., Distribution of salts in the root zone, *J. Hydrol.*, *27*, 237-248, 1975.

Raats, P.A.C., Convective transport of solutes by steady flow: 1. General theory, *Agric. Water Management*, *1*, 201-218, 1978a

Raats, P.A.C., Convective transport of solutes by steady flow: 2. Specific flow problems, *Agric. Water Management*, *1*, 219-232, 1978b.

Raats, P.A.C., Convective transport of ideal tracers in unsaturated soils. in *Proc. Symposium on Unsaturated flow and transport modelling, sponsored by Office of Nuclear Material Safety and Safeguards, U.S. Nuclear Regulatory Commission, held 32-24 March 1982 at Seattle. Washington, USA*, edited by E.M. Arnold, G.W. Gee, and R.W. Nelson, *NUREG/CP-0030, PNL-SA-10325*, pp. 249-265, Pacific Northwest Laboratory, Richland, Washington, USA, 1982.

Raats, P.A.C., Applications of the theory of mixtures in soil science. Appendix 5D, p. 326-343, in C. Truesdell, *Rational Thermodynamics, with an appendix by C.-C. Wang, Second Edition, corrected and enlarged, to which are adjoined appendices by 23 authors*, pp.326-343, Springer Verlag, New York, 1984a.

Raats, P.A.C., Tracing parcels of water and solutes in unsaturated zones, in *Pollutants in Porous Media: The Unsaturated Zone Between Soil Surface and Groundwater*, edited by B.Yaron, G. Dagan, and J. Goldshmid, *Proc. of International Workshop, held March 1983, at the Institute of Soils and Water of the Agricultural Research Organization in Bet Dagan, Israel*, pp. 4-16, Springer Verlag, Berlin, 1984b.

Raats, P.A.C., Applications of material coordinates in the soil and plant sciences, *Neth. J. Agric. Sci.*, *35*, 361-370, 1987a.

Raats, P.A.C., Applications of the theory of mixtures in soil science, *Math. Modelling*, *9*, 849-856, 1987b.

Raats, P.A.C., Quasianalytic and analytic approaches to unsaturated flow: commentary, in *Flow and Transport in the Natural Environment: Advances and Applications*, edited by W.L. Steffen, and O.T. Denmead, *Proceedings of the International Symposium on Flow and Transport in the Natural Environment, held in September 1987 at Canberra, Australia*, pp. 48-58, Springer-Verlag, Berlin, 1988.

Raats, P.A.C., A superclass of soils, in *Indirect methods for estimating the hydraulic properties of unsaturated soils*, edited by M.Th. van Genuchten, F.J. Leij, and L.J. Lund, *Proceedings of an International Workshop organized by the U.S. Salinity Laboratory, USDA-ARS, and the Department of Soil and Environmental Sciences of the University of California, both at Riverside, CA, USA, and held 11-13 Oct. 1989 at Riverside, CA, USA*, pp. 45-51, University of California, Riverside, CA, USA, 1992.

Raats, P.A.C., Spatial and material description of some processes in rigid and non-rigid saturated and unsaturated soils, in *Poromechanics. A tribute to Maurice A. Biot*, edited by J.-F. Thimus, Y. Abousleiman, A.H.-D. Cheng, O. Coussy, and E. Detournay, *Proceedings of the Biot Conference on Poromechanics, held September 14-16, 1998 at Louvain-la-Neuve, Belgium*, pp. 135-140, Balkema, Rotterdam, The Netherlands, 1998.

Raats, P.A.C., Developments in soil-water physics since the mid 1960s, *Geoderma*, *100*, 355-387, 2001.

Raats, P.A.C., and W.R. Gardner, Movement of water in the unsaturated zone near a water table, in *Drainage for Agriculture*, edited by J. van Schilfgaarde, *Agronomy Monograph 17*, p 311-357 and 401-405, American Society of Agronomy, Madison, Wisconsin, USA, 1974.

Reisenauer, A.E., K.T. Key, T.N. Narasimhan, and R.W. Nelson, *TRUST: A Computer Program for Variably Saturated Flow in Multidimensional Porous Media*, *NUREG/CR-2360*, U.S. Nuclear Regulatory Commission, Washington, D.C., 1982.

Richards, L.A., Capillary conduction of liquids through porous mediums, *Physics*, *1*, 318-333, 1931.

Rubin, J., Theoretical analysis of two-dimensional, transient flow of water in unsaturated and partly unsaturated soils, *Soil Sci. Soc. Amer. Proc.*, *32*, 607-615, 1968.

Schotting, R.J., *Mathematical Aspects of Salt Transport in Porous Media*, PhD thesis Delft University of Technology, 187p, 1998.

Van Duijn, C.J., and R.J. Schotting, Brine transport in porous media: On the use of Von Mises and similarity transformations, *Computational Geosciences*, *2*, 125-149, 1998.

Selim, H.M., and L. Ma (Eds.), *Physical Nonequilibrium in Soils, Modelling and Application*, Ann Arbor Press, Chelsia, Michigan, 1998.

Simunek, J., M. Senja, and M.Th. Van Genuchten, *The HYDRUS-2D Software Package for Simulating Water Flow and Solute Transport in Two-Dimensional Variably Saturated Media. Version 1.0*, Research Report, U.S. Salinity Laboratory, Riverside, CA., 1996.

Slichter, C.S., Theoretical investigation of the motion of ground water, *U.S. Geol. Survey, 19th Ann. Rep., Part 2*, 295-384, 1897.

Smiles, D.E., Material coordinates and solute movement in consolidating clay, *Chem. Engng. Sci.*, *55*, 773-781, 2000.

Smiles, D.E., K.M. Perroux, S.J. Zegelin, and P.A.C. Raats, Hydrodynamic dispersion during constant rate absorption of water by soil, *Soil Sci. Soc. Amer. J.*, *45*, 453-458, 1981.

Sneddon, I.N., *Elements of Partial Differential Equations*, McGraw-Hill, New York, 1957.

Truesdell, C., Editor's Introduction to Euler (1757), Part XIIA Contents of the 'General principles of the motion of fluids (1755)' and XIIB Comments on the paper summarized above, *Leonhardi Euleri Opera Omnia, (2) 12, LXXXIV-XCI*, 1954a.

Truesdell, C., *The Kinematics of Vorticity*, Indiana University Press, Bloomington, 1954b.

Truesdell, C., Editor's Introduction to Euler (1770), Part I. The first three sections of Euler's treatise on fluid mechanics (1766), *Leonhardi Euleri Opera Omnia (2) 13, X-XVIII*, 1955.

Truesdell, C., and R.A. Toupin, The classical field theories, in *Encyclopedia of Physics*, edited by S. Flügge, Vol III/1, pp. 226-793, Springer-Verlag, Berlin, 1960.

Truesdell, C.A., *A First Course in Rational Continuum Mechanics*, Academic Press, New York, 1977.

Van Genuchten, M.Th., A closed form equation for predicting the hydraulic conductivity of unsaturated soils, *Soil Sci. Soc. Am. J.*, *44*, 892-898, 1980.

Vogel, T., *SWM II - Numerical Model of Two-Dimensional Flow in a Variably Saturated Porous Medium*, Research Report no. 87, Dept.of Hydraul. and Catchment Hydrol., Agricult. Univ., Wageningen, The Netherlands, 1987.

Vogel, T., H.H. Gerke, R. Zhang, and M.Th. van Genuchten, Modeling flow and transport in a two-dimensional dual-permeability system with spatially variable hydraulic properties, *J. of Hydrology*, *238*, 78-89, 2000.

Wilson, J.L., and L.W. Gelhar, *Dispersive Mixing in a Partially Saturated Porous Medium*, Technical Report 191, Ralph M. Parsons Lab. for Water Resources and Hydrodynamics, Massachusetts Institute of Technology, Cambridge, Mass, 1974.

Wilson, J.L., and L.W. Gelhar, Analysis of longitudinal dispersion in unsaturated flow: I. The analytical method, *Water Resour. Res.*, *17*, 122-130, 1981.

Whitaker, S., Flow in porous media II: The governing equations for immiscible, two-phase flow, *Transport in Porous Media*, *1*, 105-125, 1986.

Wooding, R.A., Steady infiltration from a shallow circular pond, *Water Resour. Res.*, *4*, 1259-1273, 1968.

Yih, C.S., Stream functions in three-dimensional flows, *La Houille Blanche*, 445-450, 1957.

Zyvoloski, G, A., B. Robinson, Z. Dash, and L. Trease, *Models and Methods Summary for the FEHMN Application (POSTSCRIPT-FILE)*, Report LA-UR-94-3787, Los Alamos National Laboratory, Los Alamos, NM, USA, 1995.

Peter A.C. Raats, Paaskamp 16, 9301 KL Roden, The Netherlands (email: pac.raats@home.nl).

Effect of Temperature on Capillary Pressure

Steven A. Grant

U.S. Army Cold Regions Research & Engineering Laboratory, Hanover, New Hampshire

Jörg Bachmann

Institute of Soil Science, University of Hannover, Germany.

The effect of temperature on capillary pressure is one of several fascinating problems unearthed by J.R. Philip during his long career. In his classic paper written with Daniel de Vries, he assumed reasonably, but incorrectly, that the relative change in capillary pressure with temperature was equal to that of the surface tension of water. In fact the change for capillary pressure is roughly four times as large. Four mechanisms may be proposed to explain this discrepancy: expansion of water, expansion of entrapped air, solute effects on the surface tension of water, and temperature-sensitive contact angles. None of these explanations describes all of the pertinent data. A definitive explanation appears to be as elusive today as it has been at any time.

1. INTRODUCTION

John Philip's intellect, creativity, and productivity were so protean that his body of work can be viewed as a scientific bulldozer that created a vast terrain of important results but simultaneously left numerous lesser, but fascinating, problems for his grateful successors to examine in detail. An illustrative example of this simile is the subject of this chapter: the effect of temperature on capillary pressure, which was a minor point in a short (yet, nonetheless, very influential) paper that Philip published with Daniel de Vries in 1957.

Three aspects of *Philip and de Vries* [1957] are notable. First, given the problem addressed, the paper is astonishingly short, a scant ten undersize pages. (This chapter, one of many that have dwelt on a single equa-

tion in the paper, is three times as long.) Second, the paper has been extremely influential both in geophysics but no less so in engineering. A recent computerized inquiry for citations of the paper found 473 citations–a statistic that undoubtedly underestimated the influence of the paper. Given the paper's age, many modern authors have no doubt cited intermediate works without citing (or knowing) the original, seminal work. The third and most fascinating aspect is that both authors largely abandoned the topic with this paper. After leaving CSIRO, de Vries became a physics professor in his native Netherlands. While his professional obligations compelled him to work on other problems, de Vries retained an affinity and interest in the subject. (After his retirement he wrote a review paper revisiting the topic in which he quoted touchingly the French, *On revient toujours à ses premières amours* [de Vries, 1987].) Aside from a paper written during a second visit by de Vries nearly thirty year later, Philip appears not to have considered the problem further [*de Vries and Philip*, 1987].

In *Philip and de Vries* [1957] the authors developed what proved to be a seminal model describing the simul-

Environmental Mechanics: Water, Mass and Energy Transfer in the Biosphere
Geophysical Monograph 129
Copyright 2002 by the American Geophysical Union
10.1029/129GM18

taneous transfer of energy, liquid water, and water vapor in an unsaturated, nonisothermal porous medium. To describe these processes, they needed to estimate the change in capillary pressure with temperature. *Philip and de Vries* [1957] turned naturally to the so-called Young-Laplace equation:

$$p_c = \frac{2\gamma^{lg} \cos \Theta}{r} \tag{1}$$

where p_c is capillary pressure (in pascals); Θ, the contact angle between the solid and the liquid-gas interface (in degrees); γ^{lg}, the liquid-gas interfacial tension (in newtons per meter); and r, the apparent pore radius (in meters). At that time it was well established that the water film thickness for agricultural soils above the permanent wilting point was equivalent to many molecular thicknesses [*Taylor*, 1958]. The conventional wisdom of the day held that the contact angle of a liquid-gas interface at a solid wetted with two or more molecular layers of water was zero. As will be discussed in Section 3.4.1, it has only recently become apparent that this conventional wisdom was fallacious. Accordingly, it was not merely expeditious, but reasonable also, to assume a zero contact angle at a porous solid so wetted and, accordingly, $\cos \Theta = 1$. Equation (1) then became:

$$p_c = \frac{2\gamma^{lg}}{r}. \tag{2}$$

Taking the total derivative of equation (2) with respect to temperature and dividing through by p_c, they obtained:

$$\frac{1}{p_c}\frac{dp_c}{dT} = \frac{1}{\gamma^{lg}}\frac{d\gamma^{lg}}{dT} \tag{3}$$

where T is temperature (in kelvins).

As the old chestnut says, "Complex problems have simple, easy-to-understand wrong answers," and equation (3) appears to be wrong, on average, by a factor of 4. For soils studied $(1/p_c)(dp_c/dT)_{T=298\text{ K}}$ ranges from −0.00172 to -0.02829 K^{-1} with an average value of −0.00844 K^{-1}, whereas $(1/\gamma^{lg})(d\gamma^{lg}/dT)_{T=298\text{ K}}$ for pure liquid water is -0.002135 K^{-1} [*Grant*, in press]. The difference is small but much too large and too consistently observed to be attributable to experimental error.

Interestingly, why equation (3) is wrong has not been resolved after almost thirty years of detailed study by some of the most talented minds in geophysics and petroleum engineering. The problem was a popular topic assigned to doctoral students in these disciplines [*Wilkinson*, 1960; *Meeuwig*, 1964; *Haridasan*, 1970; *Jury*, 1973; *Okandan*, 1974; *Miller*, 1983; *Nimmo*, 1983; *Hopmans*, 1985; *Salehzadeh*, 1990; *She*, 1997]. There have been numerous explanations for this disparity, all of

which were based on reasonable suppositions about the nature of wetting liquids in porous media and all of which failed to give a completely satisfactory description of the phenomenon. Even though the phenomenon has not been resolved, much has been learned about porous media behavior. A great deal of this knowledge was due to failed attempts to understand the inadequacies of equation (3). In this Philip was wrong, but in being wrong, Philip opened up rich areas of study that has compelled geophysicists and engineers to explore the fundamentals of their understanding of natural phenomena.

The balance of this paper will present a selective survey of the experimental data describing the effect of temperature on capillary pressure. The paper will then review the explanations proposed to explain the phenomenon and suggestions for future research.

2. THE PHENOMENON

As far as we can determine, the effect of temperature on capillary pressure was first observed with a recording tensiometer by *Richards and Neal* [1937]. They acquired capillary pressure continuously in the field with a simple circular chart recorder, which showed that capillary pressures declined in the morning, as the soil warmed. The decrease in capillary pressure was most pronounced for tensiometer cups near the surface. While some of the phenomenon could have been due to the thermal expansion of liquids in the apparatus, their results indicated that capillary pressure decreased with increasing soil temperature.

Subsequently, the effect of temperature on capillary pressure was carefully studied in several laboratory studies. These studies generally, but not exclusively, consisted of determinations of capillary pressure saturation relations determined at more than one temperature. Figure 1 presents an example of data collected in these studies, which indicated that temperature was having an effect, though because the capillary pressure saturation relations were themselves so complex, it was difficult to grasp the nature of the effect.

In comparison, *Gardner* [1955] plotted capillary pressures of a soil sample maintained at a constant degree of saturation as it was heated and cooled. This plot is presented here as Figure 2. Due to its simplicity, Figure 2 demonstrates clearly and convincingly the nature of the temperature effect, which was obscured by the complexity of Figure 1, that p_c decreases linearly with temperature. The data of *Gardner* [1955] suggest that capillary pressure at a particular temperature is not precisely reproducible after cycles of heating and cooling. *Faybishenko* [1983] conducted a study similar to *Gardner* [1955], but, as presented in Figure 3, found that capillary pressures at his observational tempera-

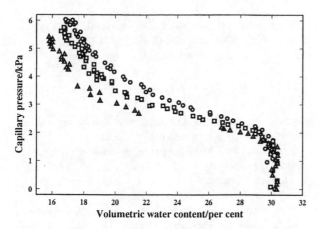

Figure 1. Capillary pressure saturation relations of a Plainfield sandy loam soil measured at 19.1 (circles), 34.1 (squares), and 49.2 °C (triangles) as reported by *Nimmo and Miller* [1986].

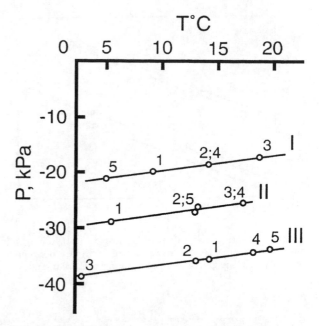

Figure 3. Capillary pressure measured by *Faybishenko* [1983] on a loam soil at three constant water contents, but at a range of temperatures. Roman numerals refer to water contents (I: 0.347 $m^3 \cdot m^{-3}$, II: 0.326 $m^3 \cdot m^{-3}$, III: 0.3 $m^3 \cdot m^{-3}$). Arabic numbers refer to steps in heating-and-cooling cycles.

tures were reproducible on cycles of heating and cooling and appeared to be linear functions of temperature. The results of *Gardner* [1955] and *Faybishenko* [1983] suggested that the effect of temperature on capillary pressure could be described by:

$$p_c = a_{p_c} + b_{p_c} T \qquad (4)$$

where a_{p_c} and b_{p_c} are empirical constants (in pascals and pascals per kelvin, respectively).

As with capillary pressure, all known liquid-gas interfacial tensions are well described as linearly decreasing

functions of temperature. Figure 4 presents the liquid-gas interfacial tensions of selected liquids from the comprehensive compilation of *Jasper* [1972]. We may write, therefore, the liquid-gas interfacial tensions as a similar linear function of temperature:

$$\gamma^{\mathrm{lg}} = a_{\gamma^{\mathrm{lg}}} + b_{\gamma^{\mathrm{lg}}} T \qquad (5)$$

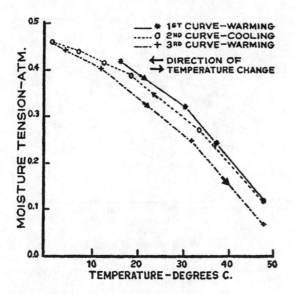

Figure 2. Capillary pressures presented by *Gardner* [1955] of a coarse sand at 2.2 % water content subjected to heating and cooling cycles.

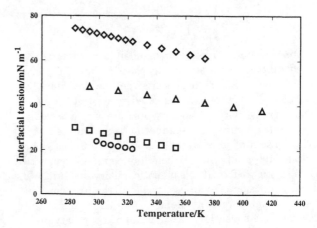

Figure 4. Liquid-gas interfacial tensions of selected liquids at a range of temperatures: acetone (circles), benzene (squares), ethylene glycol (triangles), and water (diamonds).

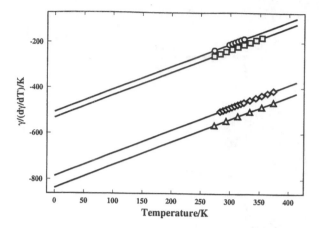

Figure 5. Normalized values of liquid-gas interfacial tensions as functions of temperature: acetone (circles), benzene (squares), ethylene glycol (triangles), and water (diamonds).

where $a_{\gamma^{lg}}$ and $b_{\gamma^{lg}}$ are empirical constants (in newtons per meter and newtons per meter-kelvin, respectively). To facilitate the comparison of liquid-gas interfacial tensions of disparate liquids, equation (5) can be normalized by dividing through by its temperature derivative:

$$\frac{\gamma^{lg}}{\left(\frac{d\gamma^{lg}}{dT}\right)} = \frac{a_{\gamma^{lg}}}{b_{\gamma^{lg}}} + T. \qquad (6)$$

We note here, and will return to, the fact that $a_{\gamma^{lg}}/b_{\gamma^{lg}}$ for pure water was equal to -766.45 K. Figure 5 replots the liquid-gas interfacial tension data presented in Figure 4 in terms of equation (6). Each line in Figure 5 had a slope of 1 and an intercept equal to $a_{\gamma^{lg}}/b_{\gamma^{lg}}$. Following this example, equation (4) can normalized by:

$$\frac{p_c}{\left(\frac{dp_c}{dT}\right)} = \frac{a_{p_c}}{b_{p_c}} + T. \qquad (7)$$

Figure 6 presents a plot for capillary pressures of selected porous media similar to that in Figure 5. The dashed line in Figure 6 presents the line described by equation (6), that is, the expected relation if the temperature sensitivity of capillary pressure were due exclusively to the temperature-induced changes in the interfacial tension of pure water. As in Figure 5, the slope is each line is unity, and their intercepts are equal to a_{p_c}/b_{p_c}. Figure 6 made clear that capillary pressure was a linear function of temperature, that the phenomenon differed from porous medium to porous medium, and that for most soils, the relative change in capillary pressure with temperature was very different from that of the liquid-gas interfacial tension.

In a wholly different theoretical treatment, *Grant and Salehzadeh* [1996] assigned the variable name β_0 to the ratio a_{p_c}/b_{p_c}. Equation (7) can be integrated to yield:

$$p_{c(T=T_f)} = p_{c(T=T_r)}\left(\frac{\beta_0 + T_f}{\beta_0 + T_r}\right). \qquad (8)$$

where T_r and T_f were the reference and observational temperatures (both in kelvins), respectively.

Subsequently, *Grant* [in press] estimated β_0 by nonlinear regression analysis for virtually all available water-air capillary pressure saturation relations measured at more than one temperature. His results were summarized in Table 1. He found that β_0 has values between -800 to -330 K, though generally far from -766 K. *Bachmann et al.* [in review] found that β_0 could be estimated well also from equation (7) and from transient flow experiments. *She and Sleep* [1998] found that equation (8) described well the capillary pressure saturation relations behavior of both water-air and water-tetrachloroethylene systems.

The preponderance of the studies conducted thus far have indicated that β_0 was largely unaffected by the degree of saturation. *Grant and Salehzadeh* [1996] found that the residuals of their nonlinear fits did not show a pronounced trend at the upper or lower extremes of the soils they studied. Similarly, *Bachmann et al.* [in review], who determined β_0 by calculating equation (7), found that β_0 was a weak linear function of water content.

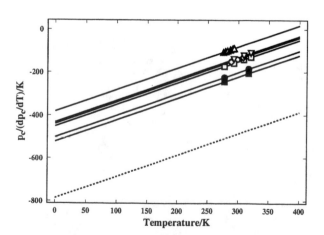

Figure 6. Normalized values of capillary pressure of selected porous media measured at different temperatures: Dubbs silt loam (circles), glass beads (squares), loam (upright triangles), Norfolk sandy loam (diamonds), Plainfield sand (inverted triangles), sand (filled circles), and silt (filled squares). The dashed line is the expected normalized value for the surface tension of pure water.

Table 1. Values of β_0 estimated by nonlinear regression analysis of capillary pressure saturation relations in the published literature and the corresponding relative values of the temperature derivative of capillary pressure and this temperature derivative to that of water's surface tension.

Matrix	D/I	$\beta_0\pm$SE	$(\frac{1}{p_c})(\frac{dp_c}{dT})$	$\dfrac{(\frac{1}{p_c})(\frac{dp_c}{dT})}{(\frac{1}{\gamma^{lg}_{H_2O}})(\frac{d\gamma^{lg}_{H_2O}}{dT})}$	Citation
Sand	D	$8.76\times10^{139}\pm0$		0.00	*Bachmann* [1998]
Hydrophobicized sand	D	$--1.51\times10^{139}\pm0$		0.00	*Bachmann* [1998]
Silt	D	-507.4 ± 26.4	-0.00478	2.24	*Bachmann* [1998]
Hydrophobicized silt	D	-603.4 ± 57.1	-0.00328	1.53	*Bachmann* [1998]
Soil	D	-674.3 ± 56.8	-0.00266	1.25	*Bachmann* [1998]
Hydrophobicized soil	D	-346.6 ± 12	-0.02064	9.67	*Bachmann* [1998]
Oakley Sand	D	-413.4 ± 15.3	-0.00868	4.06	*Constantz* [1982]
Oakley Sand (dynamic)	D	-419.7 ± 13	-0.00823	3.85	*Constantz* [1982]
Hanford Sandy Loam	D	-441.2 ± 36	-0.00699	3.27	*Constantz* [1982]
Hanford Sandy Loam (dynamic)	D	-448.8 ± 16.8	-0.00664	3.11	*Constantz* [1982]
Tipperary Sand	D	-498.2 ± 27.7	-0.00500	2.34	*Constantz* [1983]
Tipperary Sand	I	-440.4 ± 15.9	-0.00703	3.29	*Constantz* [1983]
Nonwelded tuff	D	-441 ± 17.2	-0.00700	3.28	*Constantz* [1991]
Nonwelded tuff	I	-598.1 ± 89.5	-0.00333	1.56	*Constantz* [1991]
Oakley sand	D	-436.2 ± 10.8	-0.00724	3.39	*Constantz* [1991]
Oakley sand	I	-391.8 ± 4.4	-0.01068	5.00	*Constantz* [1991]
Quartz sand	D	-783.9 ± 48	-0.00206	0.96	*Crausse* [1983]
Mixed sand	D	-384.2 ± 10.7	-0.01162	5.44	*Davis* [1994]
Mixed sand	I	-812.3 ± 612	-0.00194	0.91	*Davis* [1994]
Standard sand	D	-386.1 ± 6.7	-0.01137	5.33	*Davis* [1994]
Standard sand	I	$--4\times10^{12}\pm0$		0.00	*Davis* [1994]
Loam	I	-376.6 ± 12	-0.01275	5.97	*Faybishenko* [1983]
Loam	I	-381.9 ± 10.7	-0.01194	5.59	*Faybishenko* [1983]
silt loam	D	-437.8 ± 21.2	-0.00716	3.35	*Haridasan and Jensen* [1972]
Dundee silt loam	D	-566 ± 62.8	-0.00373	1.75	*Haridasan and Jensen* [1972]
Norfolk sandy loam	D	-439.3 ± 29.7	-0.00708	3.32	*Hopmans and Dane* [1986b]
Norfolk sandy loam	I	-370.6 ± 13.2	-0.01380	6.46	*Hopmans and Dane* [1986b]
Subalpine clay loam	D	-385.4 ± 13.6	-0.01146	5.37	*Meeuwig* [1964]
Mountain brush zone clay loam	D	-355.4 ± 14.1	-0.01747	8.18	*Meeuwig* [1964]
Millville silt loam	D	-522.7 ± 45.5	-0.00445	2.09	*Meeuwig* [1964]
Glass beads	D	-450.8 ± 1.5	-0.00655	3.07	*Nimmo and Miller* [1986]
Glass beads	I	-403.9 ± 1	-0.00946	4.43	*Nimmo and Miller* [1986]
Plainfield sand	D	-431.5 ± 5	-0.00750	3.51	*Nimmo and Miller* [1986]
Plainfield sand	I	-414.5 ± 3.8	-0.00859	4.03	*Nimmo and Miller* [1986]
Plano silt loam	D	-395.8 ± 3.5	-0.01024	4.80	*Nimmo and Miller* [1986]
Plano silt loam	I	-333.5 ± 1.6	-0.02829	13.25	*Nimmo and Miller* [1986]
Granular glass	D	-388.8 ± 9.7	-0.01103	5.17	*Novák* [1975]
Glass beads	?	-878.9 ± 18.4	-0.00172	0.81	*Salehzadeh* [1990]
Plano silt loam	D	-380.4 ± 2	-0.01216	5.69	*Salehzadeh* [1990]

Table 1. (continued)

Matrix	D/I	$\beta_0 \pm$ SE	$(\frac{1}{p_c})(\frac{dp_c}{dT})$	$\dfrac{(\frac{1}{p_c})(\frac{dp_c}{dT})}{(\frac{1}{\gamma_{H_2O}^{lg}})(\frac{d\gamma_{H_2O}^{lg}}{dT})}$	Citation
Plano silt loam	I	-356 ± 2	-0.01729	8.10	*Salehzadeh* [1990]
Elkmound sandy loam	D	$-2 \times 10^{15}\pm0$		0.00	*Salehzadeh* [1990]
Elkmound sandy loam	I	-398.8 ± 5.9	-0.00994	4.65	*Salehzadeh* [1990]
Sand	D	-468.8 ± 5	-0.00586	2.74	*She and Sleep* [1998]
Sand	I	-617.7 ± 59.9	-0.00313	1.47	*She and Sleep* [1998]
104-149 μm sand	D	-670 ± 30.5	-0.00269	1.26	*Wilkinson and Klute* [1962]
53-74 μm sand	D	-501.4 ± 8.1	-0.00492	2.30	*Wilkinson and Klute* [1962]
13.0-18.5 μm silt	D	-522.2 ± 9.4	-0.00446	2.09	*Wilkinson and Klute* [1962]

There is an old joke that goes something like this:

First person: My uncle thinks he is a chicken.
Second person: Why don't you take him to a psychiatrist?
First person: Because we need the eggs.

Equation (8) describes well the effect of temperature on capillary pressure, but the studies thus far have been unable to reconcile clearly equation (8) with any physical insight about capillary pressure in porous media. The following section reviews some of these attempts.

3. EXPLANATIONS OF THE PHENOMENON AND THEIR FAILURES

Three models have been developed to describe the enhanced sensitivity of capillary pressure to temperature, entrapped air, solutes, and temperature-sensitive contact angles. While the three will be discussed below, we begin with a digression to discuss the notion that the thermal expansion of water as a mechanism. It is likely that this mechanism was considered first, but never published because of its limitations. But it is a useful stage to begin this discussion since it illustrates the limitations of approaches based on the thermal expansion of the wetting liquid.

3.1. Water Expansion

While we know of no published works speculating on it, the first natural suggested mechanism to explain the discrepancy between the relative sensitivities of capillary pressure and surface tension of water would be due to the thermal expansion of water. If it is assumed that capillary pressure is solely a function of volumetric wa-

ter content and surface tension, the total derivative of capillary pressure with respect to temperature becomes:

$$\frac{dp_c}{dT} = \frac{\partial p_c}{\partial \theta}\frac{\partial \theta}{\partial T} + \frac{\partial p_c}{\partial \gamma^{ls}}\frac{\partial \gamma^{ls}}{\partial T}. \tag{9}$$

The $\partial p_c/\partial \theta$ term in equation (9) reveals the greatest difficulty with models of capillary pressure temperature sensitivity based on thermal expansion of the liquid or gas entrapped in it. $\partial p_c/\partial \theta$ is a highly nonlinear function of θ. (Or, if it is inverted, $\partial \theta/\partial p_c$ is a highly nonlinear function of p_c.) Figure 7 presents $dp_c/d\theta$ as a function of capillary pressure for the sand studied by *She and Sleep* [1998]. It is unrealistic to assume that any function multiplied by (or divided by) $dp_c/d\theta$ will be linear or nearly linear. Accordingly, realistic models based on thermal expansion are unlikely.

Unfortunately for the proponents of these mechanisms, the preponderance of available evidence has indicated that the temperature sensitivity of capillary pressure is, at best, a weak function of water content.

Equation (9) could be divided through by p_c to yield, after rearrangement:

$$\frac{1}{p_c}\frac{dp_c}{dT} = \frac{1}{p_c}\frac{\theta}{1}\frac{1}{\frac{\partial \theta}{\partial p_c}}\frac{1}{\theta}\frac{\partial \theta}{\partial T} + \frac{1}{\gamma^{ls}}\frac{\partial \gamma^{ls}}{\partial T}. \tag{10}$$

If it is assumed further that the relation between capillary pressure and water content is described by the equation of *van Genuchten* [1980]

$$\theta = \theta_r + (\theta_s - \theta_r)\left\{\frac{1}{[\alpha p_{c(T=T_r)}]^n + 1}\right\}^{\frac{n-1}{n}} \tag{11}$$

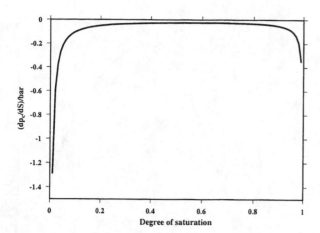

Figure 7. Partial derivative of capillary pressure with respect to degree of saturation for drainage by the sand sample studied by *She and Sleep* [1998].

where α (in reciprocal pascals) and n (dimensionless) are empirical parameters; θ is volumetric water content (in cubic meters of water per cubic meter of soil); and θ_s and θ_r are the saturated and residual volumetric water contents (in cubic meters of water per cubic meter of soil), the three elements of equation (10) become:

$$\frac{1}{p_c}\frac{dp_c}{dT} = \frac{1}{\beta_0 + T} \tag{12}$$

$$\frac{1}{\gamma^{\lg}}\frac{\partial \gamma^{\lg}}{\partial T} = \frac{1}{\frac{a_{\gamma^{\lg}}}{b_{\gamma^{\lg}}} + T}, \tag{13}$$

and

$$\frac{1}{p_c}\frac{\theta}{1}\frac{1}{\frac{\partial \theta}{\partial p_c}}\frac{1}{\theta}\frac{\partial \theta}{\partial T} =$$

$$\frac{[1 + (p_c\alpha)^n]\,\alpha_V\left\{\left[\left(\frac{1}{1+(\alpha p_c)^n} - 1\right)^{\frac{1}{n}-1}\right]\theta_r + \theta_s\right\}}{(n-1)\,(\alpha p_c)^n\,(\theta_r - \theta_s)} \tag{14}$$

where α_V (in reciprocal kelvins) is the volumetric coefficient of thermal expansion of water. At 298 K α_V has a value of 0.00026078 K^{-1}. Figure 8 presents these three elements of equation (10) for imbibition of an Elkmound sandy loam. The dark horizontal line approximates the relative effect of temperature on capillary pressure for this soil. The stippled horizontal line shows the relative effect of temperature on the surface tension of pure water. The dashed line shows the value of equation (14) for this soil. For this soil, the volumetric model was able to explain part of the observed discrepancy for some of the curves. The predictions of this model are not credible at the lower and higher extremes of capillary pressure.

3.2. Trapped Air Bubbles

A.J. Peck, a colleague of Philip's, suggested that trapped air was responsible for the enhanced sensitivity to temperature of capillary pressure [*Peck*, 1960].

This model and its successors [e.g., Chahal, 1964, 1995] suffer from the limitation of all models based on thermal expansion. Further, the central assumption of these models, that the volume of entrapped air increased with temperature, have not been supported by experiment.

Peck [1960] defined an apparent water volume as the sum of a "true" water volume and air bubbles trapped in the water that have no route to the external atmosphere:

$$\theta_{H_2O(l),app} = \theta_{H_2O(l)} + \theta_{g,bub} \tag{15}$$

where $\theta_{H_2O(l),app}$ was the apparent volumetric liquid water content (in cubic meters per cubic meter); $\theta_{H_2O(l)}$, volumetric liquid water content (in cubic meters per cubic meter); and $\theta_{g,bub}$, volumetric gas content in bubbles (in cubic meters per cubic meter). *Peck* [1960] assumed that the total differential of capillary pressure with respect to temperature was due to interfacial and volumetric effects

$$\frac{dp_c}{dT} = \frac{\partial p_c}{\partial \theta_{H_2O(l),app}}\frac{\partial \theta_{H_2O(l),app}}{\partial T}$$
$$+ \frac{p_c}{\gamma^{\lg}}\frac{\partial \gamma^{\lg}}{\partial T}. \tag{16}$$

Hopmans and Dane [1986b] carefully measured the volume of entrapped air in unsaturated soil columns at two temperatures and found that its volume actually decreased with temperature. A plausible explanation for this behavior was that it was due to the aqueous solubilities of gases, which decreased with increasing temperature [*Fogg and Gerrard*, 1990].

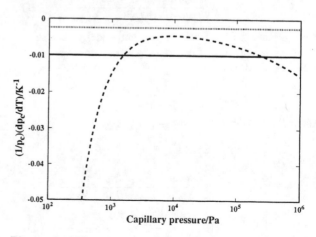

Figure 8. Plot of three elements of equation (10). Solid line: $(1/p_c)(dp_c/dT)$; dashed line: $(1/p_c)(\partial p_c/\partial \theta)(\partial \theta/\partial T)$; stippled line: $(1/\gamma^{\rm ls})(d\gamma^{\rm ls}/dT)$.

To make the problem more tractable, *Peck* [1960] assumed that the trapped air is composed of spherical bubbles with uniform radii. The pressure experienced by water in the porous matrix can be calculated by capillary pressure or by the radius of of the bubbles:

$$p_{g,bub} - \frac{2\gamma^{lg}}{r_{bub}} = p_g - p_c \qquad (17)$$

where $p_{g,bub}$ was pressure in gas bubbles (in pascals); p_g, pressure in the external gas phase (in pascals); and r_{bub}, radius of trapped bubble (in meters). The volume of $\bar{N}_{bub}$ uniformly sized spherical bubbles in the unit volume of the porous matrix was

$$\theta_{g,bub} = \bar{N}_{bub} V_{g,bub} = \frac{4\pi r_{bub}^3 \bar{N}_{bub}}{3} \qquad (18)$$

where $\bar{N}_{bub}$ was number of bubbles per unit volume (in units per cubic meter) and V_{bub}, volume of a single bubble (in cubic meters). Since the pressure within a bubble and its volume were related by the gas law:

$$p_{g,bub} V_{g,bub} = n_{g,bub} RT \qquad (19)$$

where R is the universal gas constant (in joules per kelvin-mole) and $\bar{n}_{g,bub}$, amount of gas trapped in bubbles per unit area (moles per cubic meter). Combining equations (17) and (19) yields

$$V_{g,bub}\left[p_g - p_c + 2\gamma^{lg}\left(\frac{4\pi\bar{N}_{bub}}{3V_{g,bub}}\right)^{1/3}\right] = n_{g,bub}RT. \quad (20)$$

Peck [1960] arrived at the following approximation:

$$\begin{aligned}
\frac{dp_c}{dT} \quad &\approx \quad \frac{p_c}{\gamma^{lg}}\frac{\partial\gamma^{lg}}{\partial T} \\
&+ \frac{\alpha_V \theta_{H_2O(l)} T(p_g - p_c)}{T(p_g - p_c)\partial\theta/\partial p_c - V_{g,bub}} \\
&+ \frac{V_{g,bub}(p_g - p_c) - T\frac{p_c}{\gamma^{lg}}\frac{\partial\gamma^{lg}}{\partial T}}{T(p_g - p_c)\partial\theta/\partial p_c - V_{g,bub}}. \quad (21)
\end{aligned}$$

From equation (7), it was known empirically that temperature sensitivity of capillary pressure for most soils was described well by:

$$\frac{dp_c}{dT} = \frac{p_c}{\beta_0 + T}. \qquad (22)$$

Accordingly, the equation

$$\begin{aligned}
\frac{p_c}{\beta_0 + T} \quad &\approx \quad \frac{p_c}{\gamma^{lg}}\frac{\partial\gamma^{lg}}{\partial T} \\
&+ \frac{\alpha_V \theta_{H_2O(l)} T(p_g - p_c)}{T(p_g - p_c)\partial\theta/\partial p_c - V_{g,bub}} \\
&+ \frac{V_{g,bub}(p_g - p_c) - T\frac{p_c}{\gamma^{lg}}\frac{\partial\gamma^{lg}}{\partial T}}{T(p_g - p_c)\partial\theta/\partial p_c - V_{g,bub}} \quad (23)
\end{aligned}$$

could be solved for $V_{g,bub}$ as a function of capillary pressure. This was done for the soil parameters for drainage

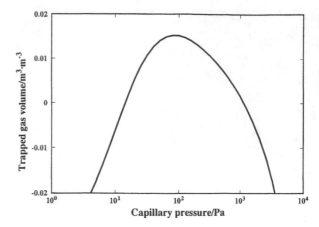

Figure 9. Predicted volumes of entrapped air of a Plano sandy loam as a function of capillary pressure for the model of *Peck* [1960] for the effect of temperature on capillary pressure to be consistent with the empirically determined value of β_0.

from a Plano silt loam (drainage) as estimated by *Grant and Salehzadeh* [1996]. Figure 9 presents the $V_{g,bub}$ so estimated.

The model of *Peck* [1960] must call upon widely varying and even negative values of entrapped air to describe the empirically described effect of temperature on capillary pressure. To be fair, these negative values of entrapped air may be due to the simplifying assumption of spherical bubbles. Negative bubbles may be calculated because the phenomenon is due to the expansion of entrapped air at pore throats, whose behavior is poorly represented by spherical bubbles. Subsequent researchers have found the model of *Peck* [1960] successful.

3.3. Solutes

Soil solutions are typically dilute (ca. 0.01 mol•kg^{-1}) aqueous solutions with a variety of inorganic and organic solutes. Since it is well known that solutes can have a pronounced effect on the thermophysical properties of the solution, it is reasonable to expect that the unexpectedly large effect of temperature capillary pressure may be due to the influence of solutes. Additionally, natural organic solutes sorb to mineral surfaces to form "conditioning films" with surface properties different from those of the pristine mineral surface [*Schneider*, 1996].

The solute-effect hypothesis proposes that the equation of *Philip and de Vries* [1957] is correct, but that the liquid-gas interfacial tension of pure water does not represent that of the pore solutions. Therefore, the equation of *Philip and de Vries* [1957] could be rewritten as:

$$\frac{1}{p_c}\frac{\mathrm{d}p_c}{\mathrm{d}T} = \frac{1}{\gamma_{\mathrm{sol}}^{\mathrm{lg}}}\frac{\mathrm{d}\gamma_{\mathrm{sol}}^{\mathrm{lg}}}{\mathrm{d}T} \qquad (24)$$

where $\gamma_{\mathrm{sol}}^{\mathrm{lg}}$ is the liquid-gas interfacial tension of the pore solution. Since $\frac{1}{p_c}\frac{\mathrm{d}p_c}{\mathrm{d}T}$ has been determined for most porous media, we can state, for the solute effect to be true:

$$\frac{1}{\gamma_{\mathrm{sol}}^{\mathrm{lg}}}\frac{\mathrm{d}\gamma_{\mathrm{sol}}^{\mathrm{lg}}}{\mathrm{d}T} \approx -0.008~\mathrm{K}^{-1} \qquad (25)$$

The experimental evidence reported by *Hopmans and Dane* [1986b] indicates that the condition presented in equation (25) is not met. They measured at several temperatures the liquid-gas interfacial tensions of solutions extracted from a glass bed sample and a soil sample. For solutions obtained from a porous matrix of glass beads $(1/\gamma_{\mathrm{sol}}^{\mathrm{lg}})(\mathrm{d}\gamma_{\mathrm{sol}}^{\mathrm{lg}}/\mathrm{d}T) = -0.0019~\mathrm{K}^{-1}$. For solutions obtained from a Norfolk sandy loam $(1/\gamma_{\mathrm{sol}}^{\mathrm{lg}})(\mathrm{d}\gamma_{\mathrm{sol}}^{\mathrm{lg}}/\mathrm{d}T) = -0.0020~\mathrm{K}^{-1}$.

Chemical thermodynamics presents the solute effect hypothesis with another difficulty. The temperature derivative of the interfacial tension $(\gamma^{\alpha\beta})$ is equal to the negative of the interfacial entropy per unit area $(s^{a,\alpha\beta}$, in joules per kelvin-meters squared):

$$\frac{\partial\gamma^{\alpha\beta}}{\partial T} = -s^{a,\alpha\beta}. \qquad (26)$$

Equation (25) becomes:

$$\frac{s^{a,\mathrm{lg_{sol}}}}{\gamma_{\mathrm{sol}}^{\mathrm{lg}}} \approx 0.008~\mathrm{K}^{-1} \qquad (27)$$

Equation (27) makes clear that, for the solute-effect hypothesis to be correct, the solutes must increase interfacial entropy, decrease interfacial tension, or some combination of the two.

Aqueous solutions are more ordered than than the pure solvent and it can be inferred that the entropies of aqueous solutions are lower than that of pure water. (Interestingly, the measurements of *Hopmans and Dane* [1986b] discussed above imply just such a reduction.) According to the development of *Cahn and Hilliard* [1958], the entropies of interfaces are identical to those of the bulk solution. Accordingly, the solute-effect hypothesis can only accepted tentatively if it markedly reduces interfacial tension.

For water to spread on a surface, the adhesive forces must exceed the cohesive forces within the bulk water. A contact angle of 90° arises when the solid surface tension is $\gamma^{\mathrm{lg}}/4$; a zero contact angle occurs when the surface tensions of solid and liquid are equal [*Letey et al.*, 2000]. If the interfacial tension of the solid is smaller

than the surface tension of pure liquid water, solutes in the soil solution may have a considerable impact on the contact angle. Increasing interfacial tension between the liquid and gas phases γ^{lg} may lead to greater contact angles, and a reduction of the liquid interfacial tension may decrease the contact angle. We are aware of only one study that found that the surface tension of soil solution was higher than for pure water. *Hartge* [1958] showed that liming his soils under study increased the concentration of inorganic cations in the solution, which increased the liquid surface tension to values around 76 mN•m^{-1} at 20 °C. More often, a reduction of the liquid interfacial tension was observed. The presence of both hydrophilic polar and hydrophobic structural units of natural organic compounds can be expected to promote accumulation at the liquid-gas interface, which thereby influences the solution surface tension [*Anderson et al.*, 1995]. Humic and fulvic acids, proteins, fatty acids, and other organic compounds of natural ecosystems possess both hydrophobic (aromatic rings and aliphatic hydrocarbons) and hydrophilic (oxygen-containing) functional groups. This suggests that, like synthetic surfactants, these compounds would exhibit significant surface activity. *Chen and Schnitzer* [1978] demonstrated that pyrolyzed fulvic acid, which had lost the functional groups, lowered the surface tension of water only slightly. The potential of humic and fulvic acids to lower the soil surface tension was shown by *Chen and Schnitzer* [1978]. For humic acid dissolved in water, a linear reduction to values of 43 mN•m^{-1} and for fulvic acid a hyperbolic decrease to values of 44 mN•m^{-1} were observed, whereas *Tschapek et al.* [1978] observed only a decrease in surface tension of diluted soil solutions of about 9 mN•m^{-1}. Temperature may have a significant effect on the solubility of surfactants in the soil solution. As reported by *Nimmo and Miller* [1986], the temperature effect on the surface tension of soil solutions is larger than for pure water. Unfortunately most of the measurements of the interfacial tension of soil solution were made without considering the temperature effect. It was shown that the solubility of fatty acids may increase with temperature by a factor of 2 to 3 for a temperature increase from 0 to 60 °C [*Singleton*, 1960, cited by *Nimmo and Miller*, 1986]. *Chen and Schnitzer* [1978] showed further that lipid-enriched leaf extracts of poplar and maple decreased the surface tension of water of 72 mN•m^{-1} effectively by 30 %. *Anderson et al.* [1995] demonstrated that humic acid-water solutions decrease linearly in surface tension with increasing concentration. The temperature had a substantial effect on the liquid interfacial tension. Surface tension reductions were linear with increasing temperature. Surface tension reductions per kelvin were found to be twice as high compared with pure water of –0.138 mN•m^{-1}•K^{-1}.

3.4. Contact angles

3.4.1. The nonzero contact angle phenomenon in soils.

The cosine of the contact angle at the intersection of the gas, liquid, and solid phases is related to the pertinent interfacial tensions by

$$\cos \Theta = \frac{\gamma^{ls} - \gamma^{gs}}{\gamma^{lg}} \qquad (28)$$

where γ^{ls} and γ^{gs} are the liquid-solid and gas-solid interfacial tensions (in newtons per meter), respectively [Rowlinson and Widom, 1982]. In order to discuss the wetting coefficient as a relevant factor affecting the temperature dependence of the capillary pressure, it is essential to realize that, in general, porous media exhibit nonzero contact angles with respect to water or soil solution. This statement should not be considered trivial, because in the past most scholars assumed complete wetting, except the few who focussed on eye-catching "extremely water repellent" soils [Doerr et al., 2000]. Until now, only a few studies indicate that, besides water repellency, "all other soils" or packings of glass beads [Lu et al., 1994] were interacting with the soil solution or water with nonzero contact angles.

Langmuir reported in 1919 in a lecture to the Faraday Society that an adsorbed monolayer of some organic compound could radically change the frictional and wetting properties of the solid surface [Zisman, 1964]. Selective sorption of organic molecules, as it was observed for aliphatic alcohol from water to silica surfaces [Tschapek, 1984], or the loss of water through evaporation, allows the deposition of solutes on the surface of the mineral.

Surface tensions of hard solids, like metals and minerals, range from 5000 mN•m^{-1} (high-energy surfaces), depending on their hardness and melting point, down to 9 mN•m^{-1} for closed packed -CF$_3$ groups [Zisman, 1964]. Soft organic solids have much lower melting points, and the surface tension is generally less than 100 mN•m^{-1}. The few measurements of interfacial tension in soil [Miyamoto and Letey, 1971] indicate small values. These authors reported interfacial tensions for quartz sand of about 43 mN•m^{-1}, for water-repellent soil of 25 mN•m^{-1}, and for silane-treated soil around 10 mN•m^{-1}. It has also been argued that all the inorganic soil minerals were hydrophilic because their surfaces usually hold ions and polar groups [Tschapek, 1984]. The hydrophilicity of minerals increases together with the densities of their surface charges and surficial polar groups.

Generally, a hydrophobic surface in contact with water can remain hydrophobic, as long as the interaction between water and a hydrophobic surface takes place through dispersive forces, while the polar forces remain free. According to the comprehensive review article by Doerr et al. [2000], the breakdown of hydrophobicity can be caused by the migration of surface-active substances in contact with water. The combination of surface properties and topology of the porous media emphasizes that observations were, at best, apparent contact angles, which cannot be related directly to the contact angle at interfaces within the medium [Philip, 1971].

When in studies about the temperature dependence of the contact angle, apparent contact angles were determined through capillary ascent, the tendency was a decrease of the contact angle with increasing temperature [King, 1981]. The soils studied by King [1981] had contact angles between 75 and 99 ° and were rated from not water repellent to severely water repellent. It was also shown that the temperature dependence of all soils increased with increasing contact angle and that in all cases it was considerably larger than estimated for pure water. Over the temperature range 0 to 36 °C, a negative linear relationship between capillary ascent and temperature was obtained. The height of water rise of the reference medium (ignited soil) was not affected by temperature. Further evidence for nonzero contact angles was provided by Siebold et al. [1997] with the capillary rise technique. For silica powder (<123 μm, 99.5% SiO$_2$) and for limestone particles (>460 μm, 98 % CaCO$_3$), contact angles of 56° to 79° were measured. The above findings indicate that a wetting coefficient <1 (i.e., contact angles > 0°) was not restricted to a few hydrophobic soils. It seems that weathered mineral surfaces or coatings and hydrophobic particles in the pore space of wettable mineral particles reduced the wettability of the high-energy surfaces to values < 72 mN•m^{-1}. It was further interesting to note that according to King [1981], soils classified with the conventional Water Drop Penetration Time (WDPT) Test as soils with a low degree of water repellency (WDPT <60 s) have contact angles up to 86°.

Experiments conducted with wettable soils and their water-repellent counterparts having identical textures showed that the contact-angle decrease with temperature is between −0.03°•K^{-1} and −0.26°•K^{-1} [Bachmann et al., 2001, Table 1]. These values agreed with those cited by She and Sleep [1998]. The results of Bachmann et al. [in review] suggested further that equation (16), which predicts an increase of the contact angle with increasing temperature, did not match the observed tendency to lower contact angles with increasing temperature in a partly saturated porous medium. However, Figure 10 shows contact angles measured with the sessile drop method for dry soil treated with different amounts of dimethyldimethylsilane. In this case, an increase of the contact angle with increasing tempera-

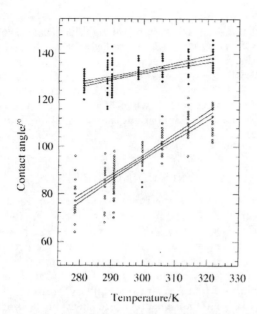

Figure 10.   Sessile drop contact angle measured on hydrophobic soil particles of the silt fraction. Soil was treated with 84 mL dimethyldichlorosilane per kg dry soil (closed symbols) and 21 mL silane per kg soil (open symbols). The method was reported by *Bachmann et al.* [2000].

ture was found. The largest temperature effects were observed for the soil with contact angles around 90° at 20 °C.

3.4.2. Interaction of capillary water and adsorbed water films.   The adsorption of water vapor leads to a decrease of the interfacial tension between the solid and the gas. Bangham and Razouk [cited in *Schrader*, 1993] stated that "...the adsorbed vapor phase and bulk liquid in contact with the solid surface must be regarded as distinct thermodynamic entities, separated in general by a discontinuity." Although not stated directly, the derivation of equation (32) did not include explicitly the physical nature of the thin water films either already adsorbed on the solid surface or adsorbed as droplets or menisci. Taking the case that the vapor phase was replaced by the liquid, then the decrease of the interfacial tension was $\gamma^{\mathrm{lg}}\cos\Theta$. The decrease of the interfacial tension taking place when water vapor was adsorbed was proportional to the temperature and the integral of the number of moles adsorbed per unit area at pressure p [*Schrader*, 1993]. An increase of water repellency with an increasing amount of water was phenomenologically observed for soils [see review paper *Doerr et al.*, 2000]. Most studies on the temperature effect on capillary pressure have been conducted for intermediate water contents. In this case it could be assumed that water films and menisci existed simultaneously. This

may have had important consequences for the contact angle of the solid-liquid interface. Under an initially dry condition, the wetting front proceeded like a jump behavior at the particle with the smallest diameter, while a very thin, unobservable water film may have existed on the surfaces. Under an initially wet condition, capillary rise occurred as a film thickening process [*Lu et al.*, 1994]. *Derjaguin and Churaev* [1986] suggested separating thin water films into two regions with different physical properties. It was assumed that thinner α-films are caused by structural forces and thicker β-films by electrostatic forces. The transition from an α- to a β-film is characterized by complete wetting (contact angle = 0°). It was found that the range of thickness of films varied between 3.0 and 27.0 nm. With an increasing contact angle, the film thickness decreased. This effect was observed on glass, quartz, and mica surfaces. *Derjaguin and Churaev* [1986] indicate also that an increasing temperature leads to thinner water films and increasing contact angles. The general behavior of a surface during adsorption of water vapor (drop formation or film formation) can be derived from water vapor adsorption isotherms [*Schrader*, 1993].

3.4.3. Temperature-sensitive contact angles.   *Grant and Salehzadeh* [1996] explored the notion that the phenomenon was due to temperature-induced changes in the contact angle. If a temperature-sensitive contact angle is accepted conditionally, then the temperature derivative of equation (1) becomes

$$\frac{1}{p_c}\frac{dp_c}{dT} = \frac{1}{\gamma^{\mathrm{lg}}}\frac{d\gamma^{\mathrm{lg}}}{dT} + \frac{1}{\cos\Theta}\frac{d\cos\Theta}{dT}. \quad (29)$$

As discussed in Section 2, for pure water

$$\frac{1}{\gamma^{\mathrm{lg}}}\frac{d\gamma^{\mathrm{lg}}}{dT} = \frac{1}{\frac{a}{b}+T}. \quad (30)$$

It remains therefore to derive an expression for

$$\frac{1}{\cos\Theta}\frac{d\cos\Theta}{dT}. \quad (31)$$

Equation (31) can be evaluated with a frequently cited, but rarely tested, expression first derived by *Harkins and Jura* [1944]:

$$-\Delta_{\mathrm{sg}}^{\mathrm{sl}}h^{\mathrm{s}} = \gamma^{\mathrm{lg}}\cos\Theta - T\frac{d\left(\gamma^{\mathrm{lg}}\cos\Theta\right)}{dT} \quad (32)$$

where $-\Delta_{\mathrm{sg}}^{\mathrm{sl}}h^{\mathrm{s}}$ is the enthalpy of immersion per unit area (in joules per square meter).

If it is assumed that

$$\frac{d\Delta_{\mathrm{sg}}^{\mathrm{sl}}h^{\mathrm{s}}}{dT} = 0$$

one derives immediately

$$\cos \Theta = \frac{-\Delta_{sg}^{sl} h^s + T\,\mathbf{C}_1}{a + bT} \qquad (33)$$

where $\mathbf{C}_1$ is a constant of integration. Equation (29) can then be evaluated directly:

$$\frac{1}{p_c}\frac{dp_c}{dT} = \frac{1}{\beta_0 + T} \qquad (34)$$

where

$$\beta_0 = \frac{-\Delta_{sg}^{sl} h^s}{\mathbf{C}_1}. \qquad (35)$$

Substituting equations (34) and (30) into equation (29) yields:

$$\frac{1}{\beta_0 + T} = \frac{1}{\frac{a}{b} + T} + \frac{1}{\cos\Theta}\frac{d\cos\Theta}{dT}. \qquad (36)$$

Clearly, if $d\cos\Theta/dT = 0$, as assumed implicitly by *Philip and de Vries* [1957], then $\beta_0 = a/b = -766.45$ K.

Equation (8) can be derived from first principles and describes virtually all available capillary pressure saturation relation data well. In spite of this, *She and Sleep* [1998] found a serious problem with equation (34). For consistency we are presenting their argument more in keeping with the treatment in this article. We hope they agree that this treatment reflects their ideas.

While there is a not inconsiderable uncertainty in the measurements, the available data indicate that the contact angle in soils is generally in a wide range between 0° and > 90° and that the contact angle is a decreasing function of temperature.

Rearrangement and simplification of equation (36) yields

$$\tan\Theta \frac{d\Theta}{dT} = \frac{180}{\pi}\left(\frac{1}{\frac{a}{b} + T} - \frac{1}{\beta_0 + T}\right). \qquad (37)$$

Equation (37) demonstrates the perils of formulating this wettability of porous media in terms of contact angles. The graph of the function $\tan\Theta$ has a discontinuity at 90°, $\lim \tan\Theta_{\Theta \to 90°}$ is $+\infty$ when approached from below 90° and $-\infty$ when approached from above. Since at 298 K

$$\frac{1}{\frac{a}{b} + T} = -0.002145$$

and at the same temperature the average value of β_0 for the soils studied thus far yields

$$\frac{1}{\beta_0 + T} = -0.00844,$$

it would be expected that

$$\tan\Theta \frac{d\Theta}{dT} \approx 0.36. \qquad (38)$$

For virtually all soils studied the right hand side of equation (38) is positive. This implies that $\tan\Theta$ and $\frac{d\Theta}{dT}$ must have the same sign. Assuming that $d\Theta/dT < 0$ and recalling that $\tan\Theta > 0$ for $0° > \Theta > 90°$ and $\tan\Theta < 0$ for $90° > \Theta > 180°$, equation (38) will hold only for $\Theta > 90°$, which is inconsistent with the "conventional wisdom" about the wettability of natural porous media.

4. CONCLUDING REMARKS

The effect of temperature on capillary pressure is a linearly decreasing function of temperature well described by equation (8). The parameter β_0 appears to be unaffected or weakly affected by water content. It is unlikely that a mechanism due to the thermal expansion of the soil solution or its constituents can describe the effect of temperature on capillary pressure. The most likely mechanisms are solute effects on the soil solution surface tension or temperature-induced changes in contact angles.

It is important to note that the general belief that soils exhibit nonzero contact angles may be invalid. Generally, the wetting coefficient may be one of three candidate mechanisms for the larger-than-expected temperature-dependence of the capillary pressure. In our opinion, wetting coefficients are temperature dependent. However, although the temperature effect is clearly observable for sessile drops on relatively dry soil particles, there are very few studies conducted to investigate the temperature effect on the contact angle of surfaces in contact with water. Generally, a larger temperature factor β_0 in equation (8) results in lower capillary forces at high temperatures because increasing temperature affects the capillary pressure in the same direction (to less negative values) as an increasing contact angle. This effect, however, cannot be quantified without additional measurements of the temperature dependence of the soil solution surface tension.

Further important research gaps can be identified. Soils are generally structured. However, no attempt has been made to investigate the wetting coefficient of outer and inner aggregate surfaces. Even when some experiments made on water/glass-bead systems indicate that solutes do not cause a temperature dependence, alternative methods of investigation (e.g., the calorimetric method) seem to confirm the assumption of a temperature dependence of the wetting coefficient according to equation (32).

Resolution of these issues will require novel experimental techniques and new physical insights. Forty-four years after its publication, the questions raised by *Philip and de Vries* [1957] continue to compel the geophysics community to the limits of understanding.

NOTATION

a_{p_c}	fitted parameter, Pa
$a_{\gamma^{ls}}$	fitted parameter, N/m
b_{p_c}	fitted parameter, Pa/K
$b_{\gamma^{ls}}$	fitted parameter, N/(m K)
n	van Genuchten equation parameter, dimension 1
$\bar{n}_{g,bub}$	amount of gas trapped in bubbles per unit area mol/m^3
$\bar{N}_{bub}$	number of bubbles per unit volume, m^{-3}
p_c	capillary pressure, Pa
p_g	pressure in the external gas phase, Pa
$p_{g,bub}$	pressure in gas bubbles, Pa
r	pore radius, m
r_{bub}	radius of trapped bubble, m
R	universal gas constant, J/(K mol)
$s^{s,\alpha\beta}$	interfacial entropy per unit area between α and β phases, J/(K m^2)
V_{bub}	volume of a single bubble, m^3
T	temperature, K
T_f	observational temperature, K
T_r	reference temperature, K
α	van Genuchten equation parameter, Pa^{-1}
α_V	the cubic expansion coefficient of water, K^{-1}
β_0	parameter, K
γ^{lg}	interfacial tension between the liquid and gas phases, N/m
γ^{ls}	interfacial tension between the liquid and solid phases, N/m
γ^{sg}	interfacial tension between the solid and gas phases, N/m
θ	volumetric soil-water content, m^3/m^3
$\theta_{g,bub}$	volumetric gas content in bubbles, m^3/m^3
$\theta_{H_2O(l)}$	volumetric liquid water content, m^3/m^3
$\theta_{H_2O(l),app}$	apparent volumetric liquid water content, m^3/m^3
θ_r	residual volumetric soil-water content, m^3/m^3
θ_s	saturated volumetric soil-water content, m^3/m^3
Θ	contact angle of the liquid-gas interface with the solid, $^\circ$

Acknowledgments. We thank Rienk R. van der Ploeg for constructive comments. This work was supported by Deutsche Forschungsgemeinschaft (DFG) project number Ba 1359/5-1 within the priority program SPP 1090 "Soils as source and sink for CO_2 - mechanisms and regulation of organic matter stabilization in soils" and the U.S. Army Engineer Research and Development Center work unit 61102/AT24/129/EE005 entitled "Chemistry of Frozen Ground."

REFERENCES

Anderson, M.A., A.Y.C. Hung, D. Mills, and M.S. Scott, Factors affecting the surface tension of soil solutions and solutions of humic acids, *Soil Sci.*, *160*, 111-116, 1995.

Bachmann, J., Measurement and simulation of nonisothermal moisture movement in water-repellent mineral soils, *Z. Pflanzenernähr Bodenk.*, *161*, 147-155, 1998 [In German].

Bachmann, J., R. Horton, S.A. Grant, and R.R. van der Ploeg, Temperature dependence of soil water retention curves of wettable and water repellent soils, *Soil Sci. Soc. Am. J.*, [in press].

Bachmann, J., R. Horton, R.R. van der Ploeg, and S.K. Woche, Modified sessile-drop method assessing initial soil water contact angle of sandy soil, *Soil Sci. Soc. Am. J.*, *64*, 564-567, 2000.

Chahal, R.S., Effect of temperature and trapped air on the energy status of water in porous media, *Soil Sci.*, *98*, 107-112, 1964.

Chahal, R.S., Effect of temperature and trapped air on matrix suction, *Soil Sci.*, *100*, 262-266, 1965.

Chen, Y., and M. Schnitzer, The surface tension of aqueous solutions of soil humic substances, *Soil Sci.*, *125*, 7-15, 1978.

Constantz, J., Temperature dependence of unsaturated hydraulic conductivity of two soils, *Soil Sci. Soc. Am. J.*, *46*, 466-470, 1982.

Constantz, J., Laboratory analysis of water retention in unsaturated zone materials at high temperature, pp. 147-164, in J. W. Mercer, P.S.C. Rao, and I. W. Marine, eds., *Role of the Unsaturated Zone in Radioactive and Hazardous Waste Disposal*, Ann Arbor, Michigan, Ann Arbor Press, 1983.

Constantz, J., Comparison of isothermal and isobaric water retention paths in non-swelling porous materials, *Water Resour. Res.*, *27*, 3165-3170, 1991.

Crausse, P., Etude fondamentale des transferts couplés de chaleur et d'humidité en milieu poreux non saturé, 207 pp., Thèse d'Etat, Institut National Polytechnique de Toulouse (France), 1983.

Doerr, S.H., R.A. Shakesby, and R.P.D. Walsh, Soil water repellency: its causes, characteristics and hydro-geomorphological significance, *Earth Sci. Rev.*, *51*, 33-65, 2000.

Davis, E.L., Effect of temperature and pore size on the hydraulic properties and flow of a hydrocarbon oil in the subsurface, *J. Contam. Hydrol.*, *16*, 55-86, 1994.

Derjaguin, B.V. and N.V. Churaev, Properties of water layers adjacent to interfaces, pp. 663-738, in, C.A. Croxton, ed., *Fluid Interfacial Phenomena*, New York, Wiley, 1986.

de Vries, D.A., The theory of heat and moisture transfer in porous media revisited, *Int. J. Heat Mass Transfer*, *30*, 1343-1350, 1987.

de Vries, D.A., and J.R. Philip, Soil heat flux, thermal conductivity, and the null-alignment method, *Soil Sci. Soc. Am. J.*, *50*, 12-17, 1987.

Faybishenko, B., Effect of temperature on moisture content, entropy, and water pressure in loam soils, *Pochvovedenie*, *12*, 43-48, 1983 [In Russian].

Fogg, P.G.T., and W. Gerrard, *Solubility of Gases in Liquids*, New York, Wiley, 1991.

Gardner, R., Relations of temperature to moisture tension of soil, *Soil Sci.*, *79*, 257-265, 1955.

Grant, S.A., Extension of a temperature-effects model for capillary-pressure saturation relations, *Water Resour. Res.*, *200* [in press].

Grant, S.A. and A. Salehzadeh, Calculation of temperature effects on wetting coefficients of porous solids and their capillary pressure functions, *Water Resour. Res.*, *32*, 261-270, 1996.

Haridasan, M., Effect of temperature on pressure head-water content relationship and conductivity of two soils, Ph.D. dissertation, Mississippi State Univ., Starkville, *Diss. Abstr. 32/07b:3740*, 1970.

Haridasan, M., and R.D. Jensen, Effect of temperature on pressure head-water content relationship and conductivity of two soils, *Soil Sci. Soc. Am. J.*, *36*, 703-708, 1972.

Harkins, W.D., and G. Jura, Surfaces of solids, 12., An absolute method for the determination of the area of a finely divided crystalline solid, *J. Am. Chem. Soc.*, *66*, 1362-1366, 1944.

Hartge, K.H., *Die Wirkung des Kalkes auf die Struktursta-bilität von Ackerböden*, Dissertation, Department of Horticulture, Technical University of Hannover, Germany. 1958 [in German].

Hopmans, W.J., Thermal effects on soil water transport, Ph.D. dissertation, Auburn Univ., Auburn, Alabama, *Diss. Abstr. 46/08b:2507*, 1985.

Hopmans, J.W., and J.H. Dane, Temperature dependence of soil hydraulic properties, *Soil Sci. Soc. Am. J.*, *50*, 4-9, 1986a.

Hopmans, W.J., and J.H. Dane, Temperature dependence of soil water retention curves, *Soil Sci. Soc. Amer. J.*, *50*, 562-567, 1986b.

Jasper, J.J., Surface tension of pure liquid compounds, *J. Phys. Chem. Ref. Data*, *1*, 841-1009, 1972.

Jury, W.A., Simultaneous transport of heat and moisture through a medium sand, Ph.D. dissertation, Univ. Wisconsin, Madison, Dissertation Abstracts Vol. 34/08b:3585, 1973.

King, P.M., Comparison of methods for measuring severity of water repellence of sandy soils and assessment of some factors that affect its measurement, *Aust. J. Soil Res.*, *19*, 275-285. 1981.

Letey, J., M.L.K. Carillo, and X.P. Pang, Approaches to characterize the degree of water repellency, *J. Hydrol.*, *231-232*, 61-65, 2000.

Lu, T.X., J. W. Biggar, and D.R. Nielsen, Water movement in glass beads porous media, 2., Experiments of infiltration and finger flow, *Water Resour. Res.*, *28*, 3283-3290, 1994.

Meeuwig, R.O., Effects of temperature on moisture conductivity in unsaturated soil, Ph.D. dissertation, Utah State Univ., Logan, *Diss. Abstr. 25/06:3180*, 1964.

Miller, M.A., Laboratory evaluation of in situ steam flushing for NAPL removal from soil, Ph.D. dissertation, Stanford Univ., Stanford, California, *Diss. Abstr. 44/05b:1568*, 1983.

Miyamoto, S. and J. Letey, Determination of solid-air surface tension of porous media, *Soil Sci. Soc. Amer. Proc.*, *35*, 856-859, 1971.

Novák, V., Non-isothermal flow of water in unsaturated soil, *J. Hydrol. Sci.*, *2*, 37-52, 1975.

Nimmo, J.R., The temperature dependence of soil-moisture characteristics, Ph.D. dissertation, Univ. Wisconsin, Madison, *Diss. Abstr. 44/06b:1858*, 1983.

Nimmo, J.R., and E.E. Miller, The temperature dependence of isothermal moisture vs. potential characteristics of soils, *Soil Sci. Soc. Am. J.*, *50*, 1105-1113, 1986.

Okandan, E., The effect of temperature and fluid composition on oil-water capillary pressure curves of limestone and sandstones and measurement of contact angle at elevated temperatures, Ph.D. dissertation, Stanford Univ., Stanford, California, *Diss. Abstr. 34/12b:6027*, 1974.

Peck, A.J., Change of moisture tension with temperature and air pressure, Theoretical, *Soil Sci.*, *89*, 303-310 , 1960.

Philip, J.R., Limitations on scaling by contact angle, *Soil Sci. Soc. Amer. Proc.*, *35*, 507-509. 1971.

Philip, J.R., and D.A. de Vries, Moisture movement in porous materials under temperature gradients, *Trans. Amer. Geophys. Union*, *38*, 222-232, 1957.

Richards, L.A., and O.R. Neal, Some field observations with tensiometers, *Soil Sci. Soc. Am. Proc.*, *1*, 71-91, 1937.

Rowlinson, J.S., and B. Widom, *Molecular Theory of Capillarity*, Clarendon Press, Oxford, 1982.

Salehzadeh, A., The temperature dependence of soil moisture characteristics of agricultural soils, Ph.D. dissertation, Univ. Wisconsin, Madison, *Diss. Abstr. 51/09b:4245*, 1990.

Schneider, R.P., Conditioning film–induced modification of substratum physicochemistry, Analysis by contact angles, *J. Colloid Interface Sci.*, *182*, 204-213, 1996.

Schrader, M.E., Sessile drops, Do they really stand on solid surfaces? pp. 109-121, in, Mittal, K.L., ed., *Contact Angle, Wettability and Adhesion*, Utrecht, The Netherlands, VSP, 1993.

She, H.Y., Laboratory evaluation of in situ steam flushing for NAPL removal from soil, Ph.D. dissertation, Univ. Toronto, *Diss. Abstr. 59/06b:2918*, 1997.

She, H.Y., and B. Sleep, The effect of temperature on capillary pressure-saturation relationships for air-water and tetrachloroethylene-water systems, *Water Resour. Res.*, *34*, 2587-2597, 1998.

Siebold, A., A. Walliser, M. Nardin, M. Opplinger, and J. Schultz, Capillary rise for thermodynamic characterization of solid particle surface, *J. Colloid Interface Sci.*, *186*, 60-70, 1997.

Singleton, W.S., Solution properties, p. 609-682 *in* K. Marley, ed., *Fatty acids*, New York, Interscience, 1960.

Taylor, S.A., The activity of water in soils, *Soil Sci.*, *86*, 83-90, 1958.

Tschapek, M., C.O. Scoppa, and C. Wasowski, The surface tension of soil water, *J. Soil Sci.*, *29*, 17-21, 1978.

Tschapek, M., Criteria for determining the hydrophilicity-hydrophobicity of soils, *Z. Pflanzenernähr Bodenk.*, *147*, 137-149, 1984.

van Genuchten, M.Th., A closed-form equation for predicting the hydraulic conductivity of unsaturated soils. *Soil Sci. Soc. Am. J.*, *44*, 892-898, 1980.

Wilkinson, G.E., The temperature effect on the equilibrium energy status of water held by porous media, Ph.D. dissertation, Univ. Illinois, Urbana-Champaign, *Diss. Abstr. 21/10:2916*, 1960.

Wilkinson, G.E. and A. Klute, The temperature effect on the equilibrium energy status of water held by porous media, *Soil Sci. Soc. Am. Proc.*, *26*, 326-329, 1962.

Zisman, W.A., Relation of equilibrium contact angle to liquid and solid construction, in, Gould, R.F. (ed.), *Contact angle, wettability and adhesion*, Advan. in Chem., Series 43, 1-51, Washington, D.C., Amer. Chem. Soc., 1964.

Steven A. Grant, U.S. Army Cold Regions Research & Engineering Laboratory, 72 Lyme Road, Hanover, NH 03755-1290. (e-mail: steven.a.grant@usace.army.mil)

Jörg Bachmann, Universitat Hannover, D-30419 Hannover, Germany. (e-mail: bachmann@ifbk.uni-hannover.de)

Soil Water Hysteresis Prediction Model Based on Theory and Geometric Scaling

Randel Haverkamp and Paolo Reggiani

Laboratoire d'Etude des Transferts en Hydrologie et Environnement, Grenoble, France

Peter J. Ross

CSIRO Land and Water, Indooroopilly, Australia

Jean-Yves Parlange

Cornell University, Ithaca, New York

Hysteresis in the water retention characteristic of a soil is important in prediction of soil hydraulic properties and in description of vadose zone flow and transport processes. Although a number of models have been proposed, some are entirely empirical and others are inconvenient to use. We apply a theoretical approach to derive parameters for a more suitable model based on the concept of rational extrapolation. Considering the water retention curve to be described by three parameters, the first defining the shape of the curve and the two others scaling the soil water pressure head and volumetric soil water content, three geometrical scaling conditions are derived. The first condition determines a shape parameter which is identical for all wetting and drying curves and independent of their scanning order; the second defines the relation between the pressure head scale and the water content scale specific to each curve in wetting or drying; and the third condition determines specific water content scale parameters according to the points of departure and arrival of each scanning curve. Equations necessary for the calculation of the different scale parameters are derived. Given the saturated water content, all main, primary and higher order scanning curves can be predicted from knowledge of only one curve, although in practice knowledge of a main or primary curve is desirable. Constraints such as the need for scanning curves to be closed and to lie inside curves of a lower scanning order are automatically satisfied. The method is illustrated using the van Genuchten water retention function but could also be applied to other functional forms. Results agree well with the existing data. It appears that knowledge

Environmental Mechanics: Water, Mass and Energy Transfer in
the Biosphere
Geophysical Monograph 129
Copyright 2002 by the American Geophysical Union
10.1029/129GM19

213

of water retention in the field is often hampered by uncertainty in the hysteresis history of the soil. The model eliminates the residual water content as a soil characteristic parameter.

1. INTRODUCTION

Knowledge of the soil water retention curve $h(\theta)$, relating soil water pressure head h [L] to volumetric soil water content θ [L^3/L^3], is important in field studies of water and solute transport in the unsaturated zone. A full description of the relationship requires at least three parameters: one shape parameter and a scale parameter for each of θ and h. While the shape parameter is strongly linked to soil texture, the water content and pressure head scale parameters are primarily related to soil structure [*Haverkamp et al.*, 1998]. The relationship is complicated by hysteresis where water content at a given pressure head is higher during drying than during wetting. Hysteresis is more pronounced for sands than for clay soils, particularly for sands with low initial water content profiles prior to wetting [e.g., *Vachaud and Thony*, 1971].

For convenience, hysteresis is usually ignored because its influence under field conditions is often masked by heterogeneities and spatial variability. However, many authors [e.g., *Nielsen et al.*, 1986; *Parker and Lenhard*, 1987; *Russo et al.*, 1989; *Heinen and Raats*, 1997; *Otten et al.*, 1997; *Whitmore and Heinen*, 1999; *Si and Kachanoski*, 2000] have shown it to be important in simulations of water transfer, solute transport, multiphase flow and/or microbial activities, and to disregard it leads to significant errors in predicted fluid distributions with concomitant effects on solute transport and contaminant concentrations [e.g., *Gilham et al.*, 1976; *Hoa et al.*, 1977; *Kool and Parker*, 1987; *Kaluarachchi and Parker*, 1987; *Mitchell and Mayer*, 1998].

Theoretically, it should be possible to predict hysteretis from first principles [e.g., *Hassanizadeh and Gray*, 1993]. However, this problem remains largely unsolved, and our rather sketchy understanding of soil structure suggests that only a few soil models will yield to this approach. Instead, the description of hysteresis in soils remains largely based on *Poulovassilis'* [1962] application of the independent domain theory to soils. Most hysteresis models presented in the literature [e.g., *Topp*, 1971; *Mualem*, 1974; *Mualem and Miller*, 1979] interpolate scanning water retention curves from both main drying and wetting curves. A simple linear interpolation method was proposed by *Hanks et al.* [1969] which assumed that the scanning curves can be replaced by straight lines, with slopes defined by the boundary curves. *Jaynes* [1985] concluded from a comparative study that "none of the interpolation methods was consistently better than the other" and suggested that because of its simplicity "the linear method appears to be the method of choice". Some empirical hysteresis models used geometric scaling [e.g., *Scott et al.*, 1983; *Kool and Parker*, 1987; *Parker and Lenhard*, 1987] with various methods to ensure closure of the scanning loops depending on possible air entrapment during rewetting [e.g., *Hopmans and Dane*, 1986].

As an alternative, *Parlange* [1976] presented a theoretical approach based on the concept of rational extrapolation. This theory is precise and robust. It requires only one boundary curve of the hysteresis envelope to predict the other boundary and all scanning curves in between. The basic equation, which estimates a drying curve starting at a given soil water pressure head h_{std} on the wetting curve, can be generalized to the differential equation:

$$\theta_d\left(h, h_{std}\right) = \theta_w\left(h\right) - \left[h - h_{std}\right]\frac{d\theta_w}{dh}, \qquad (1)$$

where the subscripts d and w refer to drying and wetting; the subscript *std* designates the pressure head h_{std} on the wetting branch from which a drying curve departs. As illustrated by Figure 1, Eq. (1) is only valid for $\theta_d \leq \theta_{std}$, where $\theta_{std} = \theta(h_{std})$. The integration of Eq. (1) gives an equation for the wetting curve starting at a pressure head $h_{stw}(\theta_{stw})$ on the drying curve [see *Parlange*, 1976]. Based on this concept, *Haverkamp and Parlange* [1986] simplified and reformulated the hysteresis model for the case of a *Brooks and Corey* [1964] water retention function. *Hogarth et al.* [1988] and *Liu et al.* [1995] generalized this explicit hysteresis model and demonstrated its accuracy against experimental observations.

An inconvenience of the *Parlange* [1976] model (Eq. (1)) is that it imposes a wetting curve without an inflection point. This is illustrated by Figure 1 taken from *Haverkamp and Parlange* [1986], where the main drying curve is given by a modified *Brooks and Corey* [1964] function. When considering other forms for the main drying water retention equation such as the *van Genuchten* [1980] function, the inconvenient condition imposed by Eq. (1) on the behavior of the main wetting curve remains. In general no analytical

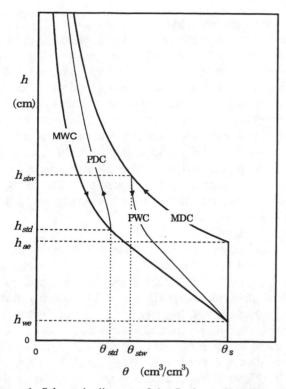

$$\theta^* = \frac{\theta - \theta_r}{\theta_S - \theta_r} = \left[1 + \left(\frac{h}{h_g}\right)^n\right]^{-m}, \qquad (2)$$

where θ^* is degree of saturation; θ is volumetric water content [L³/L³]; θ_S is the volumetric water content at natural saturation chosen as the water content scale parameter; θ_r is a parameter often referred to as the residual volumetric soil water content; h [L] is soil water pressure head, taken to be negative and expressed in cm of water; and h_g is the van Genuchten pressure head scale parameter (note that h_g is inversely proportional to the *Miller* [1980] characteristic length scale). The two dimensionless water retention shape parameters, m and n, are related by:

$$m = 1 - \frac{k_m}{n} \qquad \text{with} \qquad n > k_m, \qquad (3)$$

where k_m was initially introduced as an integer by *van Genuchten* [1980] to calculate closed-form analytical expressions for the hydraulic conductivity using the predictive conductivity models of *Burdine* [1953], when $k_m = 2$ and $n > 2$, or *Mualem* [1976a], when $k_m = 1$ and $n > 1$. For large negative pressure heads only the product mn is significant. *Haverkamp et al.* [1998] showed that its value is constant for a given soil as it remains independent of k_m, while m and n individually depend on k_m. This allows the shape parameter values calculated using the Burdine-mode to be converted into m and n values valid for the Mualem-mode of the van Genuchten equation and *vice-versa*. For the majority of soils (96 %) taken from the GRIZZLY soil database [*Haverkamp et al.*, 1998] mn lies in the interval $0 < mn < 1$. The product mn will be referred to as the *water retention shape indicator*.

The value of θ_r is generally estimated by fitting the water retention equation to measured $h(\theta)$ data points. This procedure reduces θ_r to an empirical fitting parameter restricted to the range of data points used. It gives doubtful results when applied beyond this range (e.g., in evaporation studies). The physical meaning of parameter θ_r is ambiguous. When wetting an oven dried soil sample, the θ_r-value of the main wetting curve should obviously be equal to zero. As the hysteresis loop should be closed, it follows that the θ_r-value of the main drying curve should theoretically also be equal to zero. Under field conditions a soil will rarely dry to zero water content. Small quantities of water are held by adsorptive forces and trapped in dead-end pores resulting in a non-zero θ_r-value. As this non-zero θ_r-value depends not only upon the pore geometry of the soil but also upon the initial conditions prior to wetting, the non-

Figure 1. Schematic diagram of the *Parlange* [1976] hysteresis model with the main wetting curve (MWC), the main drying curve (MDC), a primary drying curve (PDC) with starting point (θ_{std}, h_{std}) and a primary wetting curve (PWC) with starting point (θ_{stw}, h_{stw}).

solution can be found for the Parlange model applied to the van Genuchten water retention equation, although some solutions are available for special cases [*Braddock et al.*, 2001].

This study develops a simple analytical hysteresis prediction model using a convenient functional form such as that of *van Genuchten* [1980] for all wetting and drying curves. The *Parlange* [1976] model provides the theoretical basis, but instead of using the differential equation (1) directly we derive simple geometric scaling conditions. The analysis is presented in two parts: the first part deals with the theory of the model; the second tests it against data for various soils taken from the literature.

2. THEORY

The model uses the water retention relation proposed by *van Genuchten* [1980] but the method applies to other functions such as the *Brooks and Corey* [1964] relation.

The van Genuchten equation is given by:

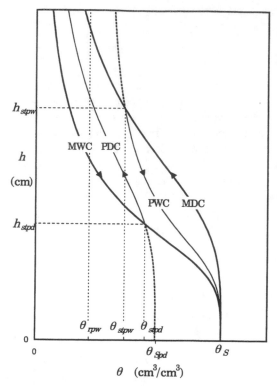

Figure 2. Schematic diagram of the hysteresis model with the main wetting curve (MWC), the main drying curve (MDC), a primary wetting curve (PWC) with starting point $(\theta_{stpw}, h_{stpw})$ and a primary drying curve (PDC) with starting point $(\theta_{stpd}, h_{stpd})$.

zero θ_r-value cannot be considered as a soil characteristic parameter. In our model the value of θ_r is related to the wetting and drying history prior to measurement of the $h(\theta)$ data points. Setting $\theta_r = 0$ for the main hysteresis loop, the scanning curves (e.g., a primary wetting curve) will have non-zero θ_r-values. The non-zero capillary θ_r-value is then attributed to a wetting or drying curve of a higher scanning order rather than to the main wetting or drying curve. This eliminates θ_r as a soil characteristic parameter (at least for soils with unimodal behavior). We note that even though a zero θ_r-value seems the obvious choice for the main drying/wetting loop, the hysteresis model described here can still be applied when a non-zero θ_r-value is preferred.

2.1. Hysteresis Model

The model is shown schematically in Figure 2. The boundary hysteresis loop consists of the main wetting curve (MWC) and main drying curve (MDC). If the wetting process is truncated and reversed to drying at a pressure head h_{stpd} on the main wetting curve, a primary drying curve (PDC) results. Similarly, a primary wetting curve (PWC)

departs from the main drying curve at a pressure head h_{stpw} on the main drying curve and finishes at saturation θ_S (Figure 2). This condition implies that changes in the volume of entrapped air during rewetting [*Hopmans and Dane*, 1986] are disregarded. In general, any wetting (or drying) curve is defined by its point of departure from and arrival at a drying (or wetting) curve of a scanning order one lower. For example, in Figure 2 a secondary wetting curve (SWC) would depart from the primary drying curve (PDC) at some pressure head h_{stsw} and rejoin the primary drying curve (PDC) at the point $(\theta_{stpd}, h_{stpd})$ where the PDC departed from its wetting parent (MWC). Further wetting would then continue along the MWC.

The basic hypothesis for our hysteresis model is that all wetting and drying curves, whatever their scanning order, have the form of the normalized van Genuchten water retention equation (Eq. (2)); that is, they have shape similarity, which does not necessarily mean shape identity. Each curve is characterized, *a-priori*, by its specific shape parameters (m and n), pressure head scale parameter (h_g) and water content scale parameters i.e., θ_r and/or θ_S depending upon the starting and end points of the curves. For example, the PDC (Figure 2) which departs from the MWC at a point $(\theta_{stpd}, h_{stpd})$, ends at $\theta_{rpd} = 0$, and can be prolonged beyond the hysteresis envelope to the θ-axis (dashed line). When $h = 0$ then $\theta = \theta_{Spd}$, which is fully determined by the starting point θ_{stpd} at which the primary drying curve departs. Although the physical meaning of the water content scaling parameter θ_{Spd} is slightly abstract, its value is mathematically well-defined. The residual water content values of the various scanning curves are mathematically defined in a similar way, i.e., when $h \to -\infty$ then $\theta \to \theta_r$. For example, for the PWC (Figure 2) $\theta_r = \theta_{rpw}$, even though its value lies outside the hysteresis envelope. Obviously, $\theta_r = 0$ for the main wetting and drying curves, as well as for the primary drying curves.

Table 1 gives the parameter notation we use for the various scanning curves.

This set of wetting and drying equations with their specific shape and scale parameters defines the basic framework of the hysteresis model. To minimize the number of parameters for prediction purposes, relations between the different parameters must be determined. As shown by the arrows in Table 1, there are two sets of relations: one links parameters of curves belonging to the same scanning order but different families (e.g., main wetting curve ⇔ main drying curve); the other links parameters of curves belonging to the same family but different scanning order (e.g., main wetting curve ⇔ primary wetting curve). Only then can the hysteresis model be used to predict the hysteretic behavior of a soil from a simple set of water reten-

Table 1. Table showing the specific parameter notations used for the main and scanning curves in wetting and drying. Parameter k designates the scanning order.

Scanning order	SHAPE and SCALE Parameters	
	Wetting	Drying
Main curves $(k=0)$	$m_{mw}, n_{mw}, \theta_S, h_{gmw}$	$m_{md}, n_{md}, \theta_S, h_{gmd}$
Primary curves $(k=1)$	$m_{pw}, n_{pw}, \theta_S, \theta_{rpw}, h_{gpw}$	$m_{pd}, n_{pd}, \theta_{Spd}, h_{gpd}$
Secondary curves $(k=2)$	$m_{sw}, n_{sw}, \theta_{Ssw}, \theta_{rsw}, h_{gsw}$	$m_{sd}, n_{sd}, \theta_{Ssd}, \theta_{rsd}, h_{gsd}$
Tertiary curves $(k=3)$	$m_{tw}, n_{tw}, \theta_{Stw}, \theta_{rtw}, h_{gtw}$	$m_{td}, n_{td}, \theta_{Std}, \theta_{rtd}, h_{gtd}$
Quarternary curves $(k=4)$	$m_{qw}, n_{qw}, \theta_{Sqw}, \theta_{rqw}, h_{gqw}$	$m_{qd}, n_{qd}, \theta_{Sqd}, \theta_{rqd}, h_{gqd}$

tion measurements. While the model of *Parlange* [1976] provides a suitable theoretical basis for deriving the relations between the shape and scale parameters of the main and scanning curves, the concept of geometric scaling allows the determination of the scale parameters as a function of the starting and arrival points. First, the relations between the shape and scale parameters of the main and scanning curves are addressed.

2.1.1. Shape and scale parameters

i) Routing between the main wetting and main drying curves

Starting with the main loop, the main wetting curve (MWC) expressed in the form of the van Genuchten function (Eq. (2)) is given by:

$$\theta_{mw}^* \equiv \frac{\theta_{mw}}{\theta_{Smw}} = \left[1 + \left(\frac{h}{h_{gmw}} \right)^{n_{mw}} \right]^{-m_{mw}} \quad (4)$$

and the main drying curve (MDC) by:

$$\theta_{md}^* \equiv \frac{\theta_{md}}{\theta_{Smd}} = \left[1 + \left(\frac{h}{h_{gmd}} \right)^{n_{md}} \right]^{-m_{md}}, \quad (5)$$

where the subscripts mw and md refer to the main wetting and main drying curves respectively. The loop for the main wetting and drying curves should obviously be closed with $\theta_{Smw} = \theta_{Smd} = \theta_S$ and $\theta_{rmw} = \theta_{rmd} = 0$ (Figure 2). While the relations between the specific wetting parameters m_{mw}, n_{mw}, h_{gmw} and drying parameters m_{md}, n_{md}, h_{gmd} are unknown a-priori, the MWC should not cross the MDC.

The differential equation (1) of *Parlange* [1976] does not permit direct calculation of a main wetting curve following the van Genuchten form. However, it does provides a theoretical basis to derive appropriate conditions for a van Genuchten main wetting curve so that it approximates the solution of the differential equation. The area under the water retention curves is chosen to provide these conditions since it is directly related to the work done in wetting or drying the soil. Integration of Eq. (1) gives:

$$\int \theta_{pd}(h)dh = \int \theta_{mw}(h)dh - \int \left(h - h_{stpd} \right) d\theta_{mw} + c, \quad (6)$$

where c is a constant of integration; and h_{stpd} is the soil water pressure head value on the main wetting curve at which the primary drying curve departs. The subscript pd denotes the primary drying curve. Evaluating the last integral by parts yields:

$$\int \theta_{pd}(h)dh = 2\int \theta_{mw}(h)dh - \left[h - h_{stpd} \right]\theta_{mw} + c. \quad (7)$$

When $h_{stpd} = 0$, then θ equals θ_S and the primary drying curve becomes the main drying curve. Applying Eq. (7) to the functions of van Genuchten describing the main wetting and main drying curves (Eqs. (4) and (5)), gives:

$$\theta_S \int_0^h \left[1 + \left(\frac{\bar{h}}{h_{gmd}} \right)^{n_{md}} \right]^{-m_{md}} d\bar{h} =$$

$$2\theta_S \int_0^h \left[1 + \left(\frac{\bar{h}}{h_{gmw}} \right)^{n_{mw}} \right]^{-m_{mw}} d\bar{h} \quad (8)$$

$$-\theta_S h \left[1 + \left(\frac{\bar{h}}{h_{gmw}} \right)^{n_{mw}} \right]^{-m_{mw}}.$$

For large h/h_{gmw} (or h/h_{gmd}) the van Genuchten function behaves as a power function in h with an exponent of $-m_{mw}n_{mw}$ (or $-m_{md}n_{md}$). On integrating an exponent $(1 - m_{mw}n_{mw})$ or $(1 - m_{md}n_{md})$ is obtained. Hence, unless the product $m_{mw}n_{mw} > 1$ (or $m_{md}n_{md} > 1$) the area under the curve becomes infinite [*Fuentes et al.*, 1991]. This is a consequence of choosing the van Genuchten function, but the same is true for the *Brooks and Corey* [1964] water retention function. In the case $0 < m_{mw} n_{mw} \leq 1$ (or $0 < m_{md} n_{md} \leq 1$), which covers the majority of field soils, it is therefore necessary either to modify the original van Genuchten equation (Eq. (2)) or to seek conditions that cancel the terms in h. As the former option is considered beyond the scope of this study, the solution for the original van Genuchten equation is analyzed.

When $0 < m_{mw} n_{mw}$ and/or $m_{md} n_{md} \leq 1$ each term of Eq. (8) is expanded about $1/h = 0$ as follows:

$$\left[1 + \left(\frac{h}{h_{gmd}} \right)^{n_{md}} \right]^{-m_{md}} = \left(\frac{h_{gmd}}{h} \right)^{m_{md} n_{md}} - m_{md} \left(\frac{h_{gmd}}{h} \right)^{(1+m_{md})n_{md}} + \dots . \quad (9)$$

Integrating and retaining only the leading terms, Eq. (8) becomes:

$$\frac{h_{gmd}}{1 - m_{md} n_{md}} \left(\frac{h_{gmd}}{h} \right)^{m_{md} n_{md} - 1} = \frac{2 h_{gmw}}{1 - m_{mw} n_{mw}} \left(\frac{h_{gmw}}{h} \right)^{m_{mw} n_{mw} - 1}$$

$$- h_{gmw} \left(\frac{h_{gmw}}{h} \right)^{m_{mw} n_{mw} - 1} + \dots \quad (10)$$

To cancel these terms we require:

$$m_{mw} n_{mw} = m_{md} n_{md} \quad (11)$$

and

$$h_{gmd} = h_{gmw} \left(1 + m_{mw} n_{mw} \right)^{1/m_{mw} n_{mw}} , \quad (12)$$

which are the conditions necessary to satisfy the Parlange hysteresis model (Eq. (1)). As k_m is considered to remain constant for wetting and drying, it follows from Eqs. (3) and (11) that:

$$m_{mw} = m_{md}$$

and

$$n_{mw} = n_{md} . \quad (13)$$

When $m_{mw} n_{mw}$ and/or $m_{md} n_{md} = 1$ then $h_{gmd} = 2 h_{gmw}$ (Eq. (12)). Similar results were reported by *Haines* [1930] and *Bouwer* [1966] who showed on field measurements that the drying pressure head scale parameter (h_{ae}) of the *Brooks and Corey* (1964) water retention equation is twice the wetting scale parameter (h_{we}) when associating h_{ae} and h_{we} with the air entry and water entry pressure heads, respectively (Figure 1). Likewise, *Kool and Parker* [1987] reported a ratio $h_{gmd}/h_{gmw} = 2.08$ with the pressure head scale parameters h_{gmd} and h_{gmw} calculated independently using a best fit to measured $h(\theta)$ data points of 8 soils ranging from clay loam to sand.

When $m_{mw} n_{mw}$ and/or $m_{md} n_{md} > 1$ Eq. (7) can be solved for the main hysteresis loop with $h_{stpd} = 0$ without using the conditions (11) and (12). The solution gives a relation between the ratio h_{gmd} / h_{gmw} and the shape parameters m_{mw}, n_{mw} and m_{md}, n_{md} [*Haverkamp et al.*, 1998]:

$$\frac{h_{gmd}}{h_{gmw}} = 2 \frac{n_{md}}{n_{mw}} \frac{\Gamma(m_{md})}{\Gamma(m_{mw})} \frac{\Gamma\left(\frac{1}{n_{mw}}\right) \Gamma\left(m_{mw} - \frac{1}{n_{mw}}\right)}{\Gamma\left(\frac{1}{n_{md}}\right) \Gamma\left(m_{md} - \frac{1}{n_{md}}\right)} , \quad (14)$$

where $\Gamma()$ refers to the classical gamma function. Obviously, Eq. (14) is rather laborious to use for routine purposes. On the other hand, the conditions (11) and (12) (used for $m_{mw} n_{mw}$ and/or $m_{md} n_{md} \leq 1$) are still valid as they serve to match areas under the curves for large negative pressure head values. Since for most field soils $m_{mw} n_{mw}$ and/or $m_{md} n_{md} < 1$ [*Haverkamp et al.*, 1998], and since results obtained with conditions (11) and (12) agree well with experimental data for $m_{mw} n_{mw}$ and/or $m_{md} n_{md} > 1$, condition (11) is adopted hereafter for both cases $0 < m_{mw} n_{mw} \leq 1$ and $m_{mw} n_{mw} > 1$. Combining Eqs. (13) and (14) gives then the extremely simple condition for the drying and wetting pressure head scale parameters:

$$h_{gmd} = 2 h_{gmw} \quad \text{for} \quad \begin{cases} m_{mw} n_{mw} \geq 1 \\ m_{md} n_{md} \geq 1 \end{cases} . \quad (15)$$

The behavior of the ratio h_{gmd}/h_{gmw} as a function of $m_{mw} n_{mw}$ is shown in Figure 3.

In short, to satisfy the theoretically based hysteresis model of Parlange (Eq. (1)) the van Genuchten functions describing the main wetting and main drying curves should have identical shapes (i.e., $m_{mw} = m_{md}$ and $n_{mw} = n_{md}$) and can be collapsed into one unique dimensionless function $h^*(\theta^*)$ where the dimensionless water pressure head h^* is defined by the ratio h/h_{gmd} or h/h_{gmw}. For the sake of con-

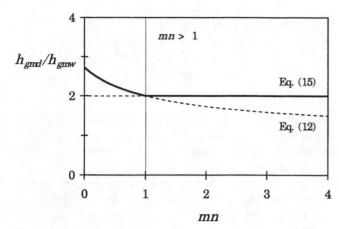

Figure 3. The ratio h_{gmd}/h_{gmw} as a function of the water retention shape indicator mn.

ciseness the main wetting and main drying shape parameters m_{mw}, m_{md}, n_{mw} and n_{md} are referred to, hereafter, as m and n.

The next step concerns the relation between curves of lower and higher scanning orders.

ii) Routing between main and primary scanning curves

Equation (1) used for the previous analysis is not only applicable to the main loop. The primary scanning curves can be analyzed in a similar way. Starting with the scanning curve emanated from the main wetting curve, a primary drying curve (PDC) expressed in the form of the van Genuchten function (Eq. (2)) is given by:

$$\theta_{pd}^* \equiv \frac{\theta_{pd}}{\theta_{Spd}} = \left[1 + \left(\frac{h}{h_{gpd}}\right)^{n_{pd}}\right]^{-m_{pd}} , \qquad (16)$$

where θ_{Spd} is the value of saturation specific for this primary drying curve; and m_{pd}, n_{pd}, and h_{gpd} are the specific shape and pressure head scale parameters. As mentioned before, the residual water content θ_{rpd} equals zero for primary drying curves (Figure 2).

Applying Eq. (16) together with the MWC equation (4) to Eq. (7), gives:

$$\theta_{Spd} \int_{h_{stpd}}^{h} \left[1 + \left(\frac{\bar{h}}{h_{gpd}}\right)^{n_{pd}}\right]^{-m_{pd}} d\bar{h} =$$

$$2\,\theta_S \int_{h_{stpd}}^{h} \left[1 + \left(\frac{\bar{h}}{h_{gmw}}\right)^{n_{mw}}\right]^{-m_{mw}} d\bar{h} \qquad (17)$$

$$-\,\theta_S \left[h - h_{stpd}\right] \left[1 + \left(\frac{\bar{h}}{h_{gmw}}\right)^{n_{mw}}\right]^{-m_{mw}} ,$$

which is different from Eq. (8) by its interval of integration. Evaluating the solution of Eq. (17) in a similar way to that of Eq. (8), we have for the shape parameters:

$$m_{pd} = m_{mw} = m$$

and

$$n_{pd} = n_{mw} = n , \qquad (18)$$

and for the scale parameters:

$$h_{gpd} = h_{gmw} \left[\frac{\theta_S}{\theta_{Spd}}\left(1 + mn\right)\right]^{1/mn} \qquad (19)$$

or

$$h_{gpd} = h_{gmd} \left[\frac{\theta_S}{\theta_{Spd}}\right]^{1/mn} . \qquad (20)$$

Equation (20) expresses the variation of the pressure head scale parameter as a function of the change in water content scale when routing from the main drying to the primary drying curve (i.e., different scanning orders but within the same family). Consequently, Eq. (20) is valid whether $0 < mn \le 1$ or $mn > 1$.

Hence, the shape identity between the primary drying curve and the main drying (or main wetting) curve is maintained (Eq. (18)). However, the primary drying pressure head scale parameter h_{gpd} changes as compared to the main drying scale parameter h_{gmd} (Eq. (20)). As shown by Figure 4, the ratio h_{gpd}/h_{gmd} increases non-linearly with decreasing water content θ_{Spd}, i.e., the smaller the value of θ_{Spd}, the steeper is the primary drying curve. This behavior becomes gradually more important for decreasing mn.

Next, the shape and scale parameters of the primary wetting curve (PWC) are explored. The PWC expressed in the form of the van Genuchten function (Eq. (2)), is given by:

$$\theta_{pw}^* \equiv \frac{\theta_{pw} - \theta_{rpw}}{\theta_S - \theta_{rpw}} = \left[1 + \left(\frac{h}{h_{gpw}}\right)^{n_{pw}}\right]^{-m_{pw}} , \qquad (21)$$

where θ_{rpw} is the residual water content value specific for the primary wetting curve; and m_{pw}, n_{pw}, and h_{gpw} are the specific shape and pressure head scale parameters. As men-

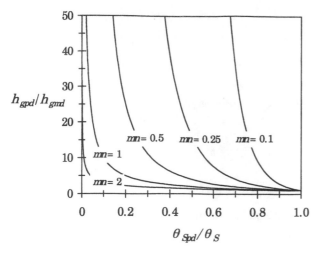

Figure 4. The ratio h_{gpd}/h_{gmd} as a function of θ_{Spd}/θ_S for different values of mn.

tioned before, the water content at natural saturation $\theta_{Spw} = \theta_S$ for the primary wetting curve (Figure 2). Thus, the system disregards possible air entrapment during rewetting.

Writing Eq. (7) for the case of the secondary drying curve (SDC), we have:

$$\int \theta_{sd}(h)\,dh = 2\int \theta_{pw}(h)\,dh - \left[h - h_{stsd}\right]\theta_{pw} + c \, , \quad (22)$$

where the subscript sd refers to the SDC; and h_{stsd} is the starting point on the primary wetting curve (PWC) from which the SDC emanates. When $h_{stsd} = 0$, then by definition, the secondary drying curve collapses into the main drying curve over the pressure head interval $h_{stpw} \le h \le 0$. So, the relation between the main drying and the primary wetting curve can be expressed by:

$$\int_0^h \theta_{md}(h)\,dh = 2\int_0^h \theta_{pw}(h)\,dh - h\,\theta_{pw} \, . \quad (23)$$

Applying Eqs. (5) and (21) to Eq. (23) and integrating, the analysis of the leading terms (evaluated similarly to those for Eqs (8) and (17)) gives for the shape parameters:

$$m_{pw} = m_{md} = m$$

and $$\quad (24)$$

$$n_{pw} = n_{md} = n \, ,$$

and for the scale parameters:

$$h_{gpw} = h_{gmd}\left[\frac{\theta_S - \theta_{rpw}}{\theta_S}\left(1 + mn\right)\right]^{-1/mn} \quad (25)$$

or combined with Eq. (12):

$$h_{gpw} = h_{gmw}\left[\frac{\theta_S}{\theta_S - \theta_{rpw}}\right]^{1/mn}, \quad (26)$$

where Eq. (26) is valid independent of the value of mn. When $\theta_{rpw} = 0$, then $h_{gpw} = h_{gmw}$ (Eq. (26)) and Eq. (25) is identical to Eq. (12) which is fully consistent as the PWC becomes the MWC.

Hence, the shape identity between the primary wetting curve and the main wetting (or main drying) curve is maintained (Eq. (24)). However, as for the previous case of the primary drying curve, the primary wetting pressure head scale parameter h_{gpw} changes as compared to the main wetting scale parameter h_{gmw} (Eq. (26)). The ratio h_{gpw}/h_{gmw} increases non-linearly with decreasing water content scale ($\theta_S - \theta_{rpw}$), i.e., the bigger the value of θ_{rpw}, the steeper is the primary drying curve.

The next step concerns the secondary scanning curves and the generalization of the routing equations.

iii) Secondary and higher order scanning curves

Starting with the secondary loop, a secondary wetting curve (SWC) expressed in the form of the van Genuchten function (Eq. (2)) is given by:

$$\theta_{sw}^* \equiv \frac{\theta_{sw} - \theta_{rsw}}{\theta_{Ssw} - \theta_{rsw}} = \left[1 + \left(\frac{h}{h_{gsw}}\right)^{n_{sw}}\right]^{-m_{sw}} \quad (27)$$

and a secondary drying curve (SDC) by:

$$\theta_{sd}^* \equiv \frac{\theta_{sd} - \theta_{rsd}}{\theta_{Ssd} - \theta_{rsd}} = \left[1 + \left(\frac{h}{h_{gsd}}\right)^{n_{sd}}\right]^{-m_{sd}}, \quad (28)$$

where the subscripts sw and sd refer to the secondary wetting and drying curves respectively. Unlike primary drying and/or wetting curves which use only one single water content scale parameter (i.e., θ_{Spd} for the PDC and θ_{rpw} for the PWC), the second and higher order scanning curves are defined as functions of two unknown water content scale parameters (e.g., θ_{Ssw} and θ_{rsw} for the SWC, Eq. (27)).

As shown before, Eq. (7) relates the parameters of two water retention equations of consecutive scanning order, but it also alternates the family. Hence, the series of scanning curves emanating from the main wetting curve slaloms down the rank of scanning orders following the sequence MWC, PDC, SWC, TDC, ... The series is shown sche-

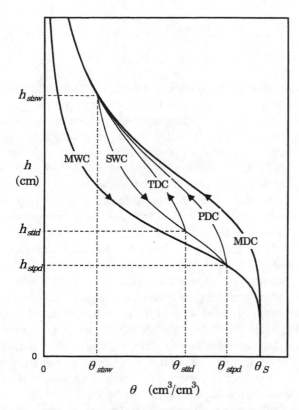

Figure 5. Schematic diagram of the hysteresis model with the main wetting curve (MWC), the main drying curve (MDC), a primary drying curve (PDC) with starting point (θ_{stpd}, h_{stpd}), a secondary wetting curve (SWC) with starting point (θ_{stsw}, h_{stsw}) and a tertiary drying curve (TDC) with starting point (θ_{sttd}, h_{sttd}).

matically in Figure 5. The alternation (represented by the shaded areas in Table 1) is illustrated here for the secondary wetting curve (SWC). When applying Eq. (7) to the secondary wetting curve, the shape and pressure head scale parameters can be related to those of either the primary drying curve (PDC) or the tertiary drying curve (TDC), depending upon the value of the starting point h_{sttd} on the secondary wetting curve where the tertiary drying curve departs. When $h_{sttd} < h_{stpd}$, then Eq. (7) relates the SWC to the TDC :

$$\int \theta_{td}(h)\,dh = 2\int \theta_{sw}(h)\,dh - \left[h - h_{sttd}\right]\theta_{sw} + c \ . \quad (29)$$

However, when h_{sttd} equals its limiting value h_{stpd} (the hysteresis loop should be closed) then the SWC relates to the PDC:

$$\int \theta_{pd}(h)\,dh = 2\int \theta_{sw}(h)\,dh - \left[h - h_{stpd}\right]\theta_{sw} + c \ . \quad (30)$$

Obviously, if wetting were to continue beyond the point (θ_{stpd}, h_{stpd}), then the MWC would be rejoined and the system would retain no memory of the secondary wetting curve.

First, we consider the case of the secondary wetting curve with $h_{sttd} = h_{stpd}$ (Eq. (30)). The integral analysis used for the main and primary loops, gives for the shape parameters:

$$\text{and} \quad \begin{aligned} m_{sw} &= m_{pd} = m \\ n_{sw} &= n_{pd} = n \ , \end{aligned} \quad (31)$$

and for the scale parameters:

$$h_{gsw} = h_{gpd}\left[\frac{\theta_{Ssw} - \theta_{rsw}}{\theta_{Spd}}\left(1 + mn\right)\right]^{-1/mn} . \quad (32)$$

Combination of Eqs. (19) and (32) gives then:

$$h_{gsw} = h_{gmw}\left[\frac{\theta_S}{\theta_{Ssw} - \theta_{rsw}}\right]^{1/mn} . \quad (33)$$

When the analysis is applied to Eq. (29) with $h_{sttd} < h_{stpd}$ (TDC) the shape and pressure head scale relations take the form:

$$\begin{aligned} m_{td} &= m_{sw} = m \\ n_{td} &= n_{sw} = n \ , \end{aligned} \quad (34)$$

and:

$$h_{gtd} = h_{gsw}\left[\frac{\theta_{Ssw} - \theta_{rsw}}{\theta_{Std} - \theta_{rtd}}\left(1 + mn\right)\right]^{1/mn} \quad (35)$$

or combined with Eqs. (20) and (33):

$$h_{gtd} = h_{gmd}\left[\frac{\theta_S}{\theta_{Std} - \theta_{rtd}}\right]^{1/mn} . \quad (36)$$

Obviously, the procedure presented so far for the SWC, and the TDC can be repeated for all other curves belonging to the sequences of scanning curves emanating from the main wetting curve (shaded areas in Table 1) and the main drying curve (non-shaded areas in Table 1). Instead of continuing the detailed description of the shape and scale parameter relations of each scanning loop, we will pass directly to the generalized equations.

As for the first three scanning loops, it can easily be shown that the shape parameters m and n remain identical for all wetting and drying curves whatever their scanning order:

and
$$m = m_{kw} = m_{kd}$$
$$n = n_{kw} = n_{kd} \ ,$$
(37)

where k designates the scanning order (n.b., the case $k = 0$ refers to the main loop). Hence, to satisfy the theoretically based hysteresis model (Eq. (1)) of *Parlange* [1976], the van Genuchten functions describing the wetting and drying curves should have identical shapes and collapse into one unique dimensionless function of the form:

$$\theta_{kd}^* = \frac{\theta_{kd} - \theta_{rkd}}{\theta_{Skd} - \theta_{rkd}} = \left[1 + \left(\frac{h}{h_{gkd}} \right)^n \right]^{-m} = \left[1 + \left(h^* \right)^n \right]^{-m} \quad (38)$$

for drying, and

$$\theta_{kw}^* = \frac{\theta_{kw} - \theta_{rkw}}{\theta_{Skw} - \theta_{rkw}} = \left[1 + \left(\frac{h}{h_{gkw}} \right)^n \right]^{-m} = \left[1 + \left(h^* \right)^n \right]^{-m} \quad (39)$$

for wetting. By contrast, the pressure head and water content scale parameters of each scanning curve change according to the scanning order and the family. The relations linking these specific scale parameters can be generalized for the two families (n.b., when staying within the same family the equations are valid whatever the interval of mn). For the drying family we have:

$$h_{gkd} = h_{gmd} \left[\frac{\theta_S}{\theta_{Skd} - \theta_{rkd}} \right]^{1/mn} , \quad (40)$$

and for the wetting family:

$$h_{gkw} = h_{gmw} \left[\frac{\theta_S}{\theta_{Skw} - \theta_{rkw}} \right]^{1/mn} . \quad (41)$$

The link between h_{gmw} and h_{gmd} is given by:

$$h_{gmd} = \alpha \, h_{gmw} \ , \quad (42)$$

where α is the pressure head scale ratio which depends upon the value of mn:

$$\alpha = \left(1 + mn \right)^{1/mn} \quad \text{for} \quad 0 < mn \leq 1$$
$$\alpha = 2 \quad \text{for} \quad mn > 1 \ . \quad (43)$$

Note that the shape identity condition (37) derived for the hysteresis model presented here (i.e., the shape parameters m and n are independent of the hysteresis history of a soil), is consistent with the fact that the water retention shape indicator mn is mainly texture dependent [*Haverkamp et al.*, 1998]. This condition (37) was chosen intuitively as the leading hypothesis for three empirical hysteresis models based on the use of the van Genuchten water retention equation [i.e., *Scott et al.*, 1983; *Kool and Parker*, 1987; *Parker and Lenhard*, 1987].

Equations (37) to (43) define the general framework of the hysteresis model. However, the scale equations (40) and (41) are not yet expressed in terms of the starting point of any particular wetting or drying scanning curve. For example, for the PDC ($k = 1$) with the main wetting curve being known, Eq. (40) defines the pressure head scale parameter (h_{gpd}) as a function of the water content scale parameter (θ_{Spd}). As there is an infinite number of primary drying curves depending on the starting point h_{stpd} on the MWC, the scale parameter equation must be defined in terms of h_{stpd}. Only then can the specific water content and pressure head scales be calculated for each particular primary drying curve. To do so, we use a geometric scaling approach which is described next.

2.1.2. Geometric scaling

i) Primary loop

Starting with the primary loop we first analyze the PDC. Equations (19) and/or (20) are expressed in terms of the two variables θ_{Spd} and h_{gpd} which depend upon the reversal point h_{stpd} where the PDC leaves the MWC. Evidently, h_{stpd} is part of both the MWC and PDC (Figure 2) and from Eqs. (4) and (16) for $h = h_{stpd}$ we have:

$$\left[1 + \left(\frac{h_{stpd}}{h_{gmw}} \right)^n \right] = \left(\frac{\theta_S}{\theta_{Spd}} \right)^{1/m} \left[1 + \left(\frac{h_{stpd}}{h_{gpd}} \right)^n \right] . \quad (44)$$

Combining Eqs. (20) and (44) and transforming, we have:

$$\theta_{Spd} = \theta_S \left[1 + \left(\frac{h_{stpd}}{h_{gmw}} \right)^n - \left(\frac{h_{stpd}}{h_{gmd}} \right)^n \right]^{-m} \quad (45)$$

and

$$h_{gpd} = h_{gmd} \left[1 + \left(\frac{h_{stpd}}{h_{gmw}} \right)^n - \left(\frac{h_{stpd}}{h_{gmd}} \right)^n \right]^{1/n} . \quad (46)$$

So, from the knowledge of one boundary curve of the hysteresis envelope and the starting point h_{stpd} on the main wetting curve, Eqs. (45) and (46) fully define the corresponding primary drying curve guaranteeing the closure of the scanning loop.

The procedure for the primary wetting curve (PWC) is similar. The two variables θ_{rpw} and h_{gpw} in the scale parameter equation (26) have to be expressed in terms of the starting point $(\theta_{stpw}, h_{stpw})$ on the MDC. Using Eqs. (5) and (21) we have:

$$\theta_S \left[1 + \left(\frac{h_{stpw}}{h_{gmd}} \right)^n \right]^{-m} =$$

$$\theta_{rpw} + (\theta_S - \theta_{rpw}) \left[1 + \left(\frac{h_{stpw}}{h_{gpw}} \right)^n \right]^{-m} . \quad (47)$$

Equation (47) fully defines the unknown θ_{rpw}. when the primary wetting scale parameter h_{gpw} is replaced by h_{gmd} using Eq. (26). However, the calculation of θ_{rpw} from Eqs. (26) and (47) is not altogether straightforward, because θ_{rpw} enters implicitly. For the higher order scanning loops two unknowns have to be determined (e.g., θ_{Ssw} and θ_{rsw} for the SWC) as a function of the starting and arrival points of the scanning curves. The procedure followed here is based on the principle of geometric scaling as conditioned by the unique dimensionless function for the wetting and drying curves (Eqs. (38) and (39)). The method guarantees closure of the scanning loops.

Before going into detail it is useful to revisit briefly two empirical scaling techniques used in the literature [i.e., *Kool and Parker*, 1987; and *Parker and Lenhard*, 1987]. Both methods rescale the drying scanning curves from the main drying curve and the wetting scanning curves from the main wetting curve. The scaling equations proposed by *Kool and Parker* [1987] are based on closure of the scanning loops at the point of departure only. As the arrival points of the primary loop are well-defined i.e., $\theta_{Spw} = \theta_S$ for the PWC and $\theta_{rpd} = 0$ for the PDC, the prediction of the primary loop remains closed. However, for the secondary and higher order scanning loops the scaling technique of *Kool and Parker* [1987] has closure problems which can cause mass balance errors [*Jaynes*, 1984]. The technique proposed by *Parker and Lenhard* [1987] forces closure of the scanning loops by

using the points of both departure and arrival. The authors used the van Genuchten expression for both the main wetting and drying curves with identical shape parameters and two different pressure head scale parameters (e.g., h_{gmw} and h_{gmd}). The consecutive scanning curves are calculated by rescaling the water content values only. Hence, the scanning curves are expressed by composed functions which are different from the van Genuchten expression used for the main curves. This approach is fundamentally different from that followed here where all wetting and drying curves are expressed by the van Genuchten equation (which therefore necessitates scaling of both pressure head and water content values).

As shown by Eqs. (38) and (39), all wetting and drying curves can be represented by one unique dimensionless function $\theta^*(h^*)$ independent of their scanning order. As the starting point $(\theta_{stpw}, h_{stpw})$ is part of both the MDC and PWC, Eq. (39) gives:

$$\theta_{rpw} = \frac{\left[\theta_{md}(h_{stpw}) - \theta_S \, \theta_{pw}^*(h_{stpw}) \right]}{\left[1 - \theta_{pw}^*(h_{stpw}) \right]} , \quad (48)$$

where θ_{Spw} is set equal to θ_S. To solve Eq. (48) in terms of the parent curve (MDC), $\theta_{pw}^*(h_{stpw})$ is expressed by the use of Eqs. (26) and (42):

$$\theta_{pw}^*(h_{stpw}) = \left[1 + \alpha^n \left(\frac{\theta_S - \theta_{rpw}}{\theta_S} \right)^{1/m} \left(\frac{h_{stpw}}{h_{gmd}} \right)^n \right]^{-m} , \quad (49)$$

where α is the pressure head scale ratio defined by Eq. (43). The value θ_{rpw} is then calculated by the simultaneous use of Eqs. (48) and (49) using a root finding technique. The algorithm presented in Appendix uses simple bisection, but more efficient methods are available [*Press et al.*, 1992].

When the influence of θ_{rpw} is considered to be negligible for the calculation of θ_{pw}^*, Eq. (49) gives $\theta_{pw}^*(h_{stpw}) = \theta_{mw}^*(h_{stpw})$ or $h_{gpw} = h_{gmw}$ yielding a first order explicit approximation of θ_{rpw} through Eq. (48). This first order approximation of $h_{gpw} = h_{gmw}$ corresponds to the scaling equation given by *Kool and Parker* [1987], when applied to the primary loop for which the scaling equation is still valid in terms of closure.

Equation (48) imposes a condition on the value of θ_{rpw}. As $\theta_{rpw} \geq 0$, the numerator of the right hand term of Eq. (48) should be positive:

$$\theta_{md}(h_{stpw}) - \theta_S \, \theta_{pw}^*(h_{stpw}) \geq 0 . \quad (50)$$

Combination of Eqs. (12), (49) and (50) leads to the condition:

$$\theta_{rpw} \leq \theta_S \frac{[mn]}{[1+mn]}, \qquad (51)$$

which implies $h_{gpw} \leq h_{gmd}$ (Eqs. (12) and (26)). Hence, the interval of possible θ_{rpw}-values decreases with mn, especially for small values of mn, such as are found for clay soils, where the range of θ_{rpw}-values is reduced (e.g., for $mn = 0.1$, $0 < \theta_{rpw} \leq 0.09 \, \theta_S \, \text{cm}^3/\text{cm}^3$).

The last step concerns the definition of the secondary and higher order scanning curves in terms of the parent curves of a scanning order one lower.

ii) Secondary and higher order scanning curves

The problem for the secondary and higher order scanning curves is slightly more complicated than that encountered for the primary curves, because two unknowns have to be calculated (e.g., θ_{Ssw} and θ_{rsw} for the SWC) instead of one. Starting with the secondary loop we first analyze the SWC. As shown by Figure 5, any SWC is defined by its point of departure (θ_{stsw}, h_{stsw}) from and arrival (θ_{stpd}, h_{stpd}) at a drying curve of a scanning order one lower (PDC), where the arrival point is the departure point of the lower order curve (PDC) from its own lower order curve (MWC). Hence, the SWC has two points in common with the PDC.

Applying the geometric scaling approach (as for the PWC) and substituting both points in the appropriate general equations (Eqs. (38) and (39)) shows that for any secondary wetting curve we must have:

$$[\theta_{Ssw} - \theta_{rsw}] = \frac{[\theta_{pd}(h_{stpd}) - \theta_{pd}(h_{stsw})]}{[\theta^*_{sw}(h_{stpd}) - \theta^*_{sw}(h_{stsw})]} \qquad (52)$$

and

$$\theta_{rsw} = \frac{[\theta_{pd}(h_{stsw})\theta^*_{sw}(h_{stpd}) - \theta_{pd}(h_{stpd})\theta^*_{sw}(h_{stsw})]}{[\theta^*_{sw}(h_{stpd}) - \theta^*_{sw}(h_{stsw})]}, \quad (53)$$

where $\theta_{pd}(h_{stsw}) = \theta_{sw}(h_{stsw})$ and $\theta_{pd}(h_{stpd}) = \theta_{sw}(h_{stpd})$. To solve Eqs. (52) and (53), $\theta^*_{sw}(h_{stpd})$ and $\theta^*_{sw}(h_{stsw})$ are expressed in terms of the parent curve (PDC) by the use of Eqs. (41) and (43):

$$\theta^*_{sw}(h_{stpd}) = \left[1 + \alpha^n \left(\frac{\theta_{Ssw} - \theta_{rsw}}{\theta_{Spd} - \theta_{rpd}}\right)^{1/m} \left(\frac{h_{stpd}}{h_{gpd}}\right)^n\right]^{-m} \qquad (54)$$

and

$$\theta^*_{sw}(h_{stsw}) = \left[1 + \alpha^n \left(\frac{\theta_{Ssw} - \theta_{rsw}}{\theta_{Spd} - \theta_{rpd}}\right)^{1/m} \left(\frac{h_{stsw}}{h_{gpd}}\right)^n\right]^{-m} . \quad (55)$$

The value of ($\theta_{Ssw} - \theta_{rsw}$) is then calculated by the simultaneous use of Eqs. (52), (54) and (55). Subsequently, the individual values of θ_{rsw} and θ_{Ssw} are determined from Eqs. (52) and (53) with $\theta^*_{sw}(h_{stpd})$ and $\theta^*_{sw}(h_{stsw})$ known from Eqs. (54) and (55). The bisection method used in Appendix begins by bracketing ($\theta_{Ssw} - \theta_{rsw}$) between 0 and θ_S, then replacing one of the end points by the midpoint to keep the solution bracketed, and so on. This method is guaranteed to converge, though more efficient methods are available [Press et al., 1992].

Obviously, once again a first order approximation can be written by setting $h_{gsw} = h_{gpw}$ which corresponds to $\theta^*_{sw}(h_{stpd}) = \theta^*_{pw}(h_{stpd})$ and $\theta^*_{sw}(h_{stsw}) = \theta^*_{pw}(h_{stsw})$. These first order approximations of $\theta^*_{sw}(h_{stpd})$ and $\theta^*_{sw}(h_{stsw})$ can be useful to provide a first estimate of θ_{rsw}.

The procedure for the calculation of the secondary drying curve (SDC) is very similar, so we will pass directly to the generalized equations.

The generalized expressions which describe the scanning curves as a function of their respective points of departure from and arrival at the scanning curve of an order one lower, are different for wetting and drying. If the departure and arrival points are referred to as (h_1, θ_1) and (h_2, θ_2) respectively, then substitution in the appropriate general water retention equation (Eq. (39)) shows that, for any wetting curve, we must have:

$$[\theta_{Skw} - \theta_{rkw}] = \frac{[\theta_{(k-1)d}(h_1) - \theta_{(k-1)d}(h_2)]}{[\theta^*_{kw}(h_1) - \theta^*_{kw}(h_2)]} \qquad (56)$$

and

$$\theta_{rkw} = \frac{[\theta_{(k-1)d}(h_2)\theta^*_{kw}(h_1) - \theta_{(k-1)d}(h_1)\theta^*_{kw}(h_2)]}{[\theta^*_{kw}(h_1) - \theta^*_{kw}(h_2)]}, \quad (57)$$

where k designates the scanning order. To solve Eqs. (56) and (57), $\theta^*_{kw}(h_1)$ and $\theta^*_{kw}(h_2)$ are expressed in terms of the parent (drying) curve of a scanning order one lower (k-1) using Eqs. (41) and (43):

$$\theta^*_{kw}(h_1) = \left[1 + \alpha^n \left(\frac{\theta_{Skw} - \theta_{rkw}}{\theta_{S(k-1)d} - \theta_{r(k-1)d}}\right)^{1/m} \left(\frac{h_1}{h_{g(k-1)d}}\right)^n\right]^{-m}$$

$$(58)$$

and

$$\theta_{kw}^*(h_2) = \left[1 + \alpha^n \left(\frac{\theta_{Skw} - \theta_{rkw}}{\theta_{S(k-1)d} - \theta_{r(k-1)d}}\right)^{1/m} \left(\frac{h_2}{h_{g(k-1)d}}\right)^n\right]^{-m}.$$

(59)

To calculate individual values of θ_{Skw} and θ_{rkw}, $(\theta_{Skw} - \theta_{rkw})$ is determined using Eqs. (56), (58) and (59) simultaneously. The iterative procedure is the same as for the SWC. Obviously, Eqs. (56) and (57) are only applicable for $k \geq 1$, because for $k=0$ (main wetting curve) $\theta_{S0w} = \theta_{Smw} = \theta_S$ and $\theta_{r0w} = \theta_{rmw} = 0$.

The generalized expressions for the drying scanning curves are derived in a similar way:

$$[\theta_{Skd} - \theta_{rkd}] = \frac{[\theta_{(k-1)w}(h_1) - \theta_{(k-1)w}(h_2)]}{[\theta_{kd}^*(h_1) - \theta_{kd}^*(h_2)]}$$

(60)

and

$$\theta_{rkd} = \frac{[\theta_{(k-1)w}(h_2)\theta_{kd}^*(h_1) - \theta_{(k-1)w}(h_1)\theta_{kd}^*(h_2)]}{[\theta_{kd}^*(h_1) - \theta_{kd}^*(h_2)]},$$

(61)

where $\theta_{kd}^*(h_1)$ and $\theta_{kd}^*(h_2)$ are given by:

$$\theta_{kd}^*(h_1) = \left[1 + \alpha^{-n} \left(\frac{\theta_{Skd} - \theta_{rkd}}{\theta_{S(k-1)w} - \theta_{r(k-1)w}}\right)^{1/m} \left(\frac{h_1}{h_{g(k-1)w}}\right)^n\right]^{-m}$$

(62)

and

$$\theta_{kd}^*(h_2) = \left[1 + \alpha^{-n} \left(\frac{\theta_{Skd} - \theta_{rkd}}{\theta_{S(k-1)w} - \theta_{r(k-1)w}}\right)^{1/m} \left(\frac{h_2}{h_{g(k-1)w}}\right)^n\right]^{-m}$$

(63)

To solve Eqs. (60) and (61) for the individual values of θ_{Skd} and θ_{rkd}, first the term $(\theta_{Skd} - \theta_{rkd})$ is determined iteratively using simultaneously Eqs. (60), (62) and (63).

With the last series of equations (Eqs. (56) through (63)) the theoretical framework of the hysteresis model is complete.

2.1.3. Summary of theory

Summarizing the model we showed that, when choosing the van Genuchten equation to describe the water retention

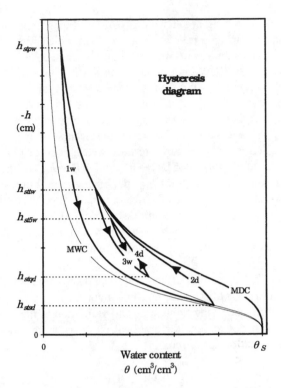

Figure 6. Schematic diagram of the hysteresis model showing hypothetical wetting and drying paths.

curve, all wetting and drying curves (whatever their scanning order) have identical shapes (the same mn). The curves can be collapsed into one unique dimensionless expression (Eq. (38) and/or Eq. (39)). The pressure head scale parameters (h_{gkw} or h_{gkd}) are specific for each curve depending on the family (i.e., wetting or drying family) and the scanning order (k). They are expressed as a function of the shape parameter (mn) and the specific water content scale parameters i.e., θ_{Skd} and θ_{rkd} for drying (Eq. (40)), or θ_{Skw} and θ_{rkw} for wetting (Eq. (41)). The water content scale parameters depend upon the family and the scanning order (hysteresis history); they are calculated as a function of the points of departure and arrival of the specific scanning curve (Eqs. (56) through (63)).

To clarify the behavior of the hysteresis model, we consider the prediction of $h(\theta)$ relations for a hypothetical case illustrated in Figure 6 for a path sequentially passing through reversal pressure heads h_{stpw}, h_{stsd}, h_{sttw}, h_{stqd} and h_{st5w}. The path follows scanning curves of increasing order represented by the alternating non-shaded areas given in Table 1. The calculations were carried out with the algorithm presented in Appendix which gives a short numerical code for calculating water content and updating the hysteresis state. Starting at saturation (θ_S), the drying path follows the MDC up to a reversal point h_{stpw} at which wetting begins. The subsequent wetting path *1w* lies on a primary wetting curve (PWC) defined by the starting point h_{stpw}. At a reversal point h_{stsd} the drying process starts again following path *2d* which is a segment of a secondary drying curve (SDC) defined by its starting and arrival points (h_{stsd} and h_{stpw}). At reversal point h_{sttw} rewetting occurs following path *3w* which is a segment of a tertiary wetting curve (TWC) of the closed loop $h_{stsd} \rightarrow SDC \rightarrow h_{sttw} \rightarrow TWC \rightarrow h_{stsd}$. If the soil resumes drying at a point h_{stqd} the drying path follows path *4d* which is a segment of a quaternary drying curve. However, if wetting were to continue beyond point h_{stqd} and subsequently point h_{stsd}, then the primary wetting curve would be resumed and the system would retain no memory of the loop $h_{stsd} \rightarrow SDC \rightarrow h_{sttw} \rightarrow TWC \rightarrow h_{stsd}$. Figure 6 clearly shows that hysteretic effects will usually become small beyond third- or fourth-order scanning curves.

Before discussing the results and validation of the model, it is most convenient for practical purposes to recapitulate the application procedure.

3. PROCEDURE

The following *step by step* procedure should be used to calculate a series of successive wetting and/or drying scanning curves from a given set of water retention data:

1) Identify the family and scanning order (k) of the experimental data used for the calculation (e.g., main or primary wetting data).

2) Calculate the function parameters (θ_{Skd}, θ_{rkd}, h_{gkd} and m) or (θ_{Skw}, θ_{rkw}, h_{gkw} and m) by fitting Eqs. (38) or (39) to the experimental data according to the family (i.e., drying or wetting).

3) Calculate the shape parameter n and the pressure head scale ratio α using Eqs. (3) and (43) respectively.

4) Decide upon the starting point $h_{st(k+1)d}$ or $h_{st(k+1)w}$ from which the higher order drying or wetting scanning curve should depart.

5) Calculate ($\theta_{S(k+1)d} - \theta_{r(k+1)d}$) using simultaneously Eqs. (60), (62) and (63), or ($\theta_{S(k+1)w} - \theta_{r(k+1)w}$) using Eqs. (56), (58) and (59) depending on the family.

6) Determine the residual water content value $\theta_{r(k+1)d}$ from Eq. (61) or $\theta_{r(k+1)w}$ from Eq. (57) using the results of step 5.

7) Calculate $\theta_{S(k+1)d}$ from Eq. (60) or $\theta_{S(k+1)w}$ from Eq. (56) using the values of $\theta_{r(k+1)d}$ or $\theta_{r(k+1)w}$ determined in step 6.

8) Finally, calculate the pressure head scale parameters $h_{g(k+1)d}$ by Eqs. (40) and (43) or $h_{g(k+1)w}$ by Eqs. (41) and (43), using parameter α calculated in step 3.

While this procedure describes the sequence of steps required to calculate higher order scanning curves, a similar sequence determines lower order parent curves.

4. MATERIAL and TEST CRITERIA

Twenty three different soils were chosen from the literature to validate the hysteresis model. Most of the experimental data (21 soils) was taken from the three independent soil databases of *Mualem* [1976b], UNSODA [*Leij et al.*, 1996] and GRIZZLY [*Haverkamp et al.*, 1998]. Using the 7th American soil classification system of the US Department of Agriculture, USDA, [1960] the soils cover 6 different texture classes, i.e., sand, sandy loam, sandy clay loam, silt loam, silty clay loam and clay loam. As was expected, most soils were sands (11) for which the effect of hysteresis is more important than for heavy soils such as clays. For 16 soils, the data came from laboratory experiments carried out on disturbed samples; for 6 other soils the data were obtained on soil samples combined with field measurements. Only for one soil (soil n° 23) were the data fully measured in the field.

The soil series, texture classes, authors and literature sources are given in Table 2.

While the hysteresis history of the laboratory soils was mostly well documented, the reported data for the field soils gave little or no information on the hysteresis history. For three soils (*Rubicon* sandy loam, *Caribou* silt loam and *Rideau* clay loam) reported by *Topp* [1969 and 1971], the main drying data clearly suggest a bi-modal behavior probably caused by dynamical effects [*Topp*, 1966].

The generalized wetting and drying water retention equations (Eqs. (38) and (39)) postulate exact functional relationships between the variables θ and h. The most elementary examination of water retention data, however, indicates that measurement points (θ_i, h_i) do not lie exactly on a smooth function such as given by Eqs. (38) and/or (39). A more realistic hypothesis is to consider the water retention curve fitted to a scatter of measurement data, as an estimate of the true curve $h(\theta)$. Taking the most common example where the drying data are measured after wetting of the soil sample, the estimate of the generalized drying curve is expressed by:

$$\tilde{\theta}_{kd}(h) = \tilde{\theta}_{rkd} + \left[\tilde{\theta}_{Skd} - \tilde{\theta}_{rkd}\right]\left[1 + \left(\frac{h}{\tilde{h}_{gkd}}\right)^{\tilde{n}}\right]^{-\tilde{m}} + \dot{e}, \quad (64)$$

where $\tilde{\theta}_{Skd}$, $\tilde{\theta}_{rkd}$, $\tilde{h}_{gkd}$, $\tilde{m}$ and $\tilde{n}$ are all estimates of the unknown parameters θ_{Sd}, θ_{rd}, h_{gd}, m and n; $\tilde{\theta}_{kd}$ is the water content value calculated by the drying water retention curve

Table 2. Soil names, texture classes, authors and literature sources for soils used for the validation of the hysteresis model. *NA* stands for *Not Applicable*. The principal literature sources are the soil databases of *Mualem* [1976 b], UNSODA [*Leij et al.*, 1996] and GRIZZLY [*Haverkamp et al.*, 1998].

No	Soil Series / Location	Texture class	Authors	Literature Source
			Laboratory Measurements	
1	Del Monte	Sand	*Liakopoulos*, 1966	Mualem, code 4108
2	*NA*	Sand	*Poulovassilis*, 1970	Mualem, code 4106
3	*NA*	Sand	*Poulovassilis*, 1970	Mualem, code 4107
4	Molonglo	Sand	*Talsma*, 1970	Mualem, code 4126
5	Grenoble	Sand	*Vauclin*, 1971	*NA*
6	Grenoble 1	Sand	*Vachaud and Thony*, 1971	GRIZZLY, code 8
7	Grenoble 3	Sand	*Elmaloglou*, 1980	GRIZZLY, code 10
8	Grenoble 5	Sand	*Touma et al.*, 1984	GRIZZLY, code 12
9	Las Cruces	Sand	*Dane and Hruska*, 1983	UNSODA, code 1310
10	Rubicon	Sandy loam	*Topp*, 1969	Mualem, code 3501
11	Glendale 1	Sandy clay loam	*Dane and Hruska*, 1983	UNSODA, code 1300
12	Ioa	Silt loam	*Green et al.*, 1964	Mualem, code 3305
13	Caribou	Silt loam	*Topp*, 1971	Mualem, code 3301
14	Manawatu	Silt loam	*Clothier and Smettem*, 1990	UNSODA, code 2140
15	Rideau	Clay loam	*Topp*, 1971	Mualem, code 3101
16	Glendale 2	Clay loam	*Dane and Hruska*, 1983	UNSODA, code 1301
			Combined Field and Laboratory Measurements	
17	Femic 1	Sand	*Bouten*	UNSODA, code 3340
18	Femic 2	Sand	*Bouten*	UNSODA, code 3341
19	Twyfordÿ (hor.1)	Sandy loam	*Clothier and Smettem*, 1990	UNSODA, code 2150
20	Twyfordÿ (hor.2)	Sandy loam	*Clothier and Smettem*, 1990	UNSODA, code 2151
21	Tantalus 1	Silty clay loam	*Ahuja and El-Swaify*, 1974	UNSODA, code 2020
22	Tantalus 2	Silty clay loam	*Ahuja and El-Swaify*, 1974	UNSODA, code 2021
23	Tomelloso	Silt loam	*Haverkamp et al.*, 1997	*NA*

for any given value of h; and e is a variable associated with the model (or equation) error which is supposed to follow a normal distribution (centered at zero) with a finite variance $\sigma^2(e)$.

The procedure for using the hysteresis model described in the previous section involves two steps. For the example of measured drying data, the first step consists of the calculation of the unknown parameters $\tilde{\theta}_{Skd}$, $\tilde{\theta}_{rkd}$, $\tilde{h}_{gkd}$, $\tilde{m}$ and $\tilde{n}$ by fitting the appropriate drying curve to the series of sample observations (θ_i, h_i). The values of $\tilde{\theta}_{Skd}$, $\tilde{\theta}_{rkd}$, $\tilde{h}_{gkd}$, $\tilde{m}$ and $\tilde{n}$ are chosen so as to make the sum of the squared residuals Σe^2 as small as possible and the goodness of fit is expressed by:

$$s(e) = \sqrt{\frac{\sum_{i=1}^{N_P} \left[\tilde{\theta}_{kd}(h_i) - \theta_i \right]^2}{N_P}} \quad , \qquad (65)$$

where $s(e)$ is the estimator of the standard deviation $\sigma(e)$ [cm^3/cm^3]; the subscript i refers to the sample measurements (θ_i, h_i) with $i = 1, \dots, N_P$; and the water content values $\tilde{\theta}_{kd}(h_i)$ are calculated by Eq. (38). The quantity $s(e)$ is non-negative and varies with the spread of the data points from the function (38). The second step consists of the prediction of the successive wetting scanning curve by the use of θ_{Skd}, θ_{rkd}, h_{gkd}, $\tilde{m}$ and $\tilde{n}$ optimized in the first step. When experimental wetting scanning data are available the goodness of prediction can be expressed, once again, by the use of Eq. (65) but with $\tilde{\theta}_{kd}(h_i)$ replaced by the wetting values $\tilde{\theta}_{kw}(h_i)$. The latter water content values $\tilde{\theta}_{kw}(h_i)$ are calculated by the appropriate wetting scanning curve equation (Eq. (39)).

However, the experimental data points (θ_i, h_i) are affected by measurement errors which are not taken into account by the definition of criterion (65). These errors tend to mask 'true' values of volumetric water content and soil water pressure head, so measured values $\tilde{\theta}_i$ and $\tilde{h}_i$ may be considered as estimates of the true values:

$$\begin{aligned} \tilde{\theta}_i &= \theta_i + u_i \\ \tilde{h}_i &= h_i + v_i \end{aligned} \quad , \qquad (66)$$

where u_i and v_i are measurement errors distributed with expected values of zero and standard deviations of $\sigma(u_i)$ and $\sigma(v_i)$, respectively. Since none of the literature sources used for this study estimated measurement errors, the water content error is chosen to be ± 0.01 cm^3/cm^3 for laboratory experiments and ± 0.02 cm^3/cm^3 for field experiments [*Sinclair and Williams*, 1979; *Haverkamp et al.*, 1984]. Similarly, the measurement error associated with soil water pressure head is taken to be ± 2 cm for laboratory experiments and ± 5 cm for field experiments. These measurement errors are expressed as two standard deviations $\sigma(u_i)$ and $\sigma(v_i)$. Hence, for each observation $\sigma(u_i) = 0.005$ cm^3/cm^3 and $\sigma(v_i) = 1$ cm for laboratory experiments, and $\sigma(u_i) = 0.01$ cm^3/cm^3 and $\sigma(v_i) = 2.5$ cm for field experiments.

While the effect of the measurement errors is superimposed on that of the equation error (Eq. (64)), it also affects the values of the optimized function parameters $\tilde{\theta}_{Skd}$, $\tilde{\theta}_{rkd}$, $\tilde{h}_{gkd}$, $\tilde{m}$ and $\tilde{n}$ (when using a drying curve). Therefore, the fitted parameters should be considered as biased estimators of θ_{Skd}, θ_{krd}, h_{gkd}, m and n. When subsequently using the fitted parameters for the prediction of a scanning curve of an order one higher, the inaccuracies in $\tilde{\theta}_{Skd}$, $\tilde{\theta}_{rkd}$, h_{gkd}, $\tilde{m}$ and $\tilde{n}$ are obviously reflected in the results of prediction. For the case of a linear regression, this effect can easily be quantified in terms of a confidence interval [e.g., *Johnston*, 1963; *Sinclair and Williams*, 1979; *Haverkamp et al.*, 1984], but for the case of a non-linear regression such as used for the water retention equation, the analysis based on the classical principle of least squares can not be applied. Recently, *Haverkamp et al.* [1998] described the uncertainty analysis applied to the hysteresis model presented in this study and the reader is referred to that study for more information. Here, only the results on the goodness of fit and goodness of prediction (Eq. (65)) are reported.

5. RESULTS AND DISCUSSION

For convenience, the laborious statistical notation applied in the previous section is disregarded hereafter. The procedure for using the model for interpreting measured data is dependent on the data available. The complete set of curves is defined by three unknowns mn, h_{gmd} (or h_{gmw}) and θ_S. If points on the main drying curve have been measured, Eq. (5) can be fitted to give mn, h_{gmd} and $\theta_{Smd} = \theta_S$. If any drying curve of scanning order k is known, Eq. (38) can be fitted to obtain mn, h_{gkd}, θ_{rkd} and θ_{Skd}. A similar procedure is followed by fitting Eq. (4) for the MWC or Eq. (39) for any wetting scanning curve when wetting data are available. The MWC will seldom be observed under field conditions as this curve applies to an initially oven dry sample. Therefore, wetting data should generally be attributed to wetting scanning curves. The model input parameters, such as m, h_{gmd} and θ_{Smd} for the MDC, were calculated using a least squares fit procedure minimizing $s(e)$ given by Eq. (65). As the shape parameter n is directly related to m by Eq. (3), it should be replaced in terms of m before fitting. The value of $k_m = 2$ corresponding to the Burdine mode was chosen in Eq. (3). However, note that the Mualem-mode of the water retention curve with $k_m = 1$ gives almost the same goodness of fit, because the shape indicator mn is practically independent of the value of k_m. Only the individual values of m and n change.

The fitted and predicted characteristic soil parameters are given in Tables 3a, b and c.

5.1. Quality of Experimental Soil Data

Even though the experimental water retention data used for this study are collected from the literature and, therefore, can be assumed to have been checked for quality, other features such as data scatter and number of data points influence the representativeness of the regression parameters (e.g., m, θ_S and h_{gmd} for the MDC) when calculated over the experimental water retention data. Therefore, it is appropriate to examine the quality of the experimental data of the 23 soils in terms of the standard deviations of the three regression parameters. As the drying water retention data are generally used for the best-fit procedure and the wetting data for the validation of the hysteresis model, the analysis is carried out for the drying data.

To estimate $s(m)$, $s(\theta_S)$ and $s(h_{gmd})$ we assume that the errors u_i and v_i introduced in Eq. (66) are random errors belonging to a Gaussian distribution of mean zero with standard deviations equal to $\sigma(u_i)$ and $\sigma(v_i)$. For given values of $\sigma(u_i)$ and $\sigma(v_i)$, repeated samples of u_i and v_i are generated using a Monte-Carlo technique. The sample size N_{MC} is chosen such that robust statistics are guaranteed, i.e., $N_{MC} = 1000$. Hence, the values of θ_i and h_i vary from sample to sample as a consequence of the drawings from the u and v distributions in each sample. Calculating m, θ_S and h_{gmd} for each set of sample observations, series of N_{MC} values of

Table 3. Characteristic soil parameters obtained by fitting to experimental main drying data (Table 3a) and primary wetting data (Table 3b), together with the system parameters used for the prediction of main and primary wetting curves and secondary drying curves. Three series of prediction parameters are presented for soil n° 23. Each series refers to a different soil depth, i.e., n° 23 a corresponds to the top layer ($0 < z < 50$ cm), n°23 b to the horizon at 70 cm depth and n° 23 c to the layer at 90 cm depth. The characteristic soil parameters of the bi-modal soils (Table 3c) are calculated by fitting to experimental primary wetting data.

Table 3a	Fitting Main Drying Curve (MDC)						Prediction Primary Wetting Curve (PWC)			
Soil no.	θ_S (cm^3/cm^3)	θ_S/ε	$-h_{gmd}$ (cm)	m	$m\,n$	$s(e)$ (cm^3/cm^3)	θ_{rpw} (cm^3/cm^3)	$-h_{gpw}$ (cm)	θ_{stpw} (cm^3/cm^3)	$s(e)$ (cm^3/cm^3)
1	0.2993	0.86	116.56	0.518	2.150	$6.9\ 10^{-3}$	0.0233	60.52	0.0302	$1.0\ 10^{-2}$
2	0.2724	0.81	24.37	0.553	2.478	$2.6\ 10^{-3}$	0.0750	13.88	0.0931	$3.0\ 10^{-3}$
3	0.2612	0.82	29.17	0.654	3.784	$2.6\ 10^{-3}$	0.0857	16.21	0.0939	$6.0\ 10^{-3}$
4	0.2710	0.80	10.99	0.334	1.003	$2.6\ 10^{-3}$	0.0423	6.51	0.0852	$1.2\ 10^{-2}$
5	0.3776	0.83	39.09	0.526	2.220	$1.4\ 10^{-2}$	0.1049	22.63	0.1364	$7.7\ 10^{-3}$
6	0.3553	0.83	48.86	0.551	2.451	$1.2\ 10^{-2}$	0.0995	27.93	0.1241	$1.7\ 10^{-2}$
7	0.2902	0.75	38.20	0.466	1.744	$1.8\ 10^{-2}$	0.0375	20.68	0.0537	$6.8\ 10^{-3}$
8	0.3120	0.84	38.20	0.481	1.857	$9.9\ 10^{-3}$	0.0609	21.47	0.0851	$1.7\ 10^{-2}$
11	0.3825	0.75	23.38	0.087	0.189	$9.7\ 10^{-3}$	0.0213	12.66	0.1339	$9.1\ 10^{-3}$
12	0.5375	0.99	45.35	0.133	0.308	$9.3\ 10^{-3}$	0.0313	23.06	0.1332	$3.2\ 10^{-2}$
14	0.3986	0.87	82.29	0.146	0.340	$3.4\ 10^{-3}$	0.0733	63.24	0.2967	$6.3\ 10^{-3}$
16	0.3748	0.76	25.15	0.061	0.130	$5.5\ 10^{-3}$	0.0005	9.93	0.0043	$1.7\ 10^{-2}$
17	0.3247	0.75	16.44	0.320	0.940	$1.5\ 10^{-2}$	0.0210	8.69	0.0413	$2.3\ 10^{-2}$
18	0.3970	0.80	24.35	0.149	0.350	$2.1\ 10^{-2}$	0.0428	14.31	0.1651	$2.6\ 10^{-2}$
19	0.3964	0.75	22.40	0.152	0.358	$8.9\ 10^{-3}$	0.0390	12.72	0.1479	$9.1\ 10^{-3}$
20	0.2934	0.78	36.64	0.444	1.599	$3.3\ 10^{-3}$	0.0662	21.50	0.1006	$8.8\ 10^{-3}$
21	0.6849	0.94	6.71	0.030	0.062	$2.8\ 10^{-3}$	0.0244	4.56	0.4164	$1.1\ 10^{-2}$
22	0.6480	0.97	10.20	0.018	0.037	$2.0\ 10^{-3}$	0.0161	7.47	0.4451	$5.4\ 10^{-3}$

Table 3b	Fitting Primary Wetting Curve (PWC)						Prediction Secondary Drying Curve (SDC)			
Soil no.	θ_S (cm^3/cm^3)	θ_{rpw} (cm^3/cm^3)	θ_S/ε	$-h_{gpw}$ (cm)	$m\,n$	$s(e)$ (cm^3/cm^3)	θ_{Ssd} (cm^3/cm^3)	θ_{rsd} (cm^3/cm^3)	$-h_{gsd}$ (cm)	$s(e)$ (cm^3/cm^3)
9	0.3863	0.0545	0.97	14.81	3.238	$9.8\ 10^{-3}$	0.1644	0.0054	37.18	$5.3\ 10^{-3}$
23 a	0.4457	0.0594	0.83	380.98	2.556	$1.5\ 10^{-2}$	0.4227	0.0002	735.71	$9.8\ 10^{-3}$
23 b	-	-	-	-	-	-	0.3657	0.0007	779.06	$5.4\ 10^{-3}$
23 c	-	-	-	-	-	-	0.2963	0.0016	847.17	$3.3\ 10^{-3}$

Table 3c	Fitting Primary Wetting Curve (PWC)						Prediction Main Drying Curve (MDC)			
Soil no.	θ_S (cm^3/cm^3)	θ_{rpw} (cm^3/cm^3)	$-h_{gpw}$ (cm)	$m\,n$		θ_{stpw} (cm^3/cm^3)	$s(e)$ (cm^3/cm^3)	θ_S (cm^3/cm^3)	θ_{rmd} (cm^3/cm^3)	$-h_{gmd}$ (cm)	$s(e)$ (cm^3/cm^3)
10	0.3805	0.1434	14.29	0.778	0.1737	$4.6\ 10^{-3}$	0.3805	0.1148	31.71	$5.8\ 10^{-2}$
13	0.4380	0.2859	15.22	0.517	0.3032	$9.9\ 10^{-4}$	0.4380	0.2769	30.52	$2.2\ 10^{-2}$
15	0.4168	0.2690	11.09	0.626	0.2790	$1.2\ 10^{-3}$	0.4168	0.2628	22.57	$1.4\ 10^{-2}$

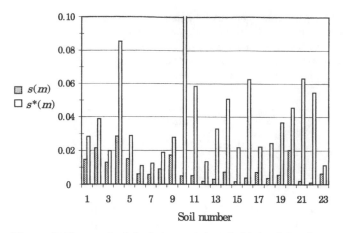

Figure 7. The standard deviations $s(m)$ and $s*(m)$ of the shape parameter m calculated as a function of the 23 soils used in this study.

m, θ_S and h_{gmd} are generated. The distributions of these values form the sampling distributions of m, θ_S and h_{gmd} for which the respective standard deviations $s(m)$, $s(\theta_S)$ and $s(h_{gmd})$ can be calculated. Their values are used as the quality indicators of the experimental data points used. As the values of the standard deviation $s(\theta_S)$ calculated for the 23 soils are all of the same order of magnitude, with an average of $\bar{s}(\theta_S) = 0.0031$ cm³/cm³, only the standard deviations of m and h_{gmd} are shown in Figures 7 and 8. The relative standard deviations $s*(m)$ and $s*(h_{gmd})$ are defined by $s*(m) = s(m)/m$ and $s*(h_{gmd}) = s(h_{gmd})/h_{gmd}$.

The results of Figure 7 show that the standard deviations $s(m)$ calculated for the group of sands (soils n° 1-9) are generally larger than those calculated for the other soils. On the contrary, the relative standard deviations $s*(m)$ of the sands are smaller. Hence, the uncertainty in parameter m increases with the value of m. However, two soils clearly deviate from this general trend, i.e., soil n° 4 with $s*(m) = 0.0853$ and soil n° 10 with $s*(m) = 0.1130$. While there is no apparent reason for soil n° 4 (a series of 10 experimental drying data regularly spread without any apparent scatter) to give such large uncertainty $s(m) = 0.0285$, the problem for soil n° 10 is more evident as this soil has a typical bi-modal behavior which affects the drying data in particular. The shape parameters m and n of these two soils must be considered with caution in the analysis presented hereafter.

The standard deviations of the pressure head scale parameter h_{gmd} are shown in Figure 8. Note that the values of $s*(h_{gmd})$ are multiplied by a constant factor 25. This allows presentation of the histograms of both $s(h_{gmd})$ and $s*(h_{gmd})$ in the same figure. Generally speaking, the standard devia-

tion $s(h_{gmd})$ increases with h_{gmd}. This explains that the values of $s(h_{gmd})$ calculated for the non-sand soils (i.e., soils n° 11-16 and n° 19-23) are slightly larger than those calculated for the sands (soils n° 1-9). Only soil n° 14, the *Manawatu* silt loam taken from *Clothier and Smettem* [1990], shows an unusual large uncertainty $s(h_{gmd}) = 9.97$ cm. Apart from the fact that only 6 experimental data points are available for this soil and that the water retention curve is steep and, therefore sensitive to the value of h_{gmd}, there is no apparent reason for the large uncertainty in h_{gmd}. Nevertheless, the results of this soil should be interpreted with great caution.

5.2. Scanning Order

The hysteresis history of soils n° 1-8 was fully described by the different authors, as the laboratory experiments were carried out expressly to demonstrate hysteresis. The drying data of these soils were measured after total wetting of the entire soil sample yielding MDC data. The drying process was then truncated at a given pressure head h_{stpw} on the main drying curve and reversed to yield wetting PWC data. For these soils, the MDC data are chosen for the fitting procedure and the PWC data are used for the validation of the prediction procedure. The results are given in Table 3a.

For soils n° 8-23, the description of the experimental conditions was often less well documented. This lack of precise information introduces an uncertainty in the choice of the appropriate scanning equation to be used for the parameter identification. The problem is illustrated for soil n° 9 (*Las Cruces* sand) taken from *Dane and Hruska* [1983]. The experiment was started by wetting an air dried soil column yielding PWC data. After saturation of the surface layer, the infiltration was stopped and the drying process

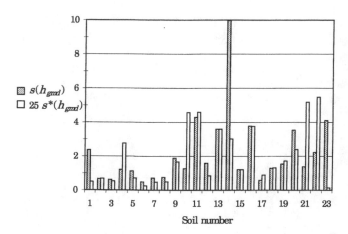

Figure 8. The standard deviations $s(h_{gmd})$ and $s*(h_{gmd})$ of the pressure head scale parameter h_{gmd} calculated as a function of the 23 soils used in this study.

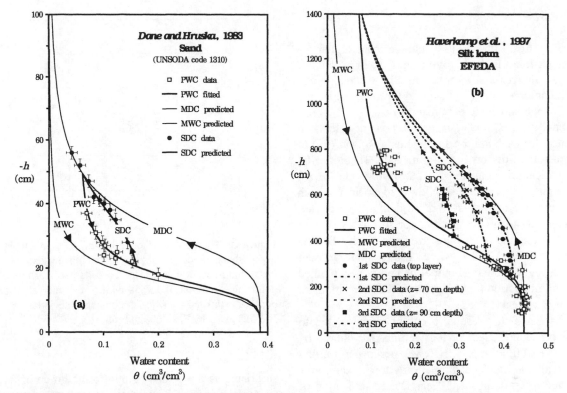

Figure 9. Primary wetting curves (PWC) fitted to measured primary wetting data (   ) for sand (9a) and silt loam (9b), soils n° 9 and 23 taken from *Dane and Hruska* [1983] and *Haverkamp et al.* [1997] respectively. The predicted main wetting curve (MWC) and secondary drying curves (SDC) are presented together with experimental secondary drying data ($\bullet$), ($\times$) and ($\blacksquare$). The horizontal and vertical bars correspond to the measurement errors in water content and soil water pressure head (i.e., ± 0.01 cm^3/cm^3 and ± 2 cm for soil n° 9 (9a), and ± 0.02 cm^3/cm^3 and ± 5 cm for soil n° 23 (9b)) associated with each observation (θ_i, h_i).

was measured. However, at the moment the infiltration was stopped only the top layer was saturated whereas the deeper soil layers were probably still not fully saturated. Hence, the drying data measured at some depth below the soil surface is more likely to belong to a secondary drying curve (SDC) than to the main drying curve (MDC) such as reported by *Dane and Hruska* [1983]. Fitting the main drying curve to such data obligatory leads to erroneous parameter values. As an example, the water content at natural saturation calculated by the main drying equation fitted to the drying data of soil n° 9 gives the value $\theta_S = 0.162$ cm^3/cm^3 which seems highly unrealistic as a porosity of $\varepsilon = 0.3962$ cm^3/cm^3 was reported by *Dane and Hruska* [1983]. The resulting ratio θ_S/ε takes a value of 0.41 which is considerably smaller than the interval $0.8 < \theta_S/\varepsilon < 1$ typically reported for field soils [*Rogowski*, 1971]. Therefore, the ratio θ_S/ε can be used as an indicator to decide upon the scanning order of the drying data. Hereafter, $\theta_S/\varepsilon = 0.75$ is chosen as the threshold value below which the drying data are considered to belong to a secondary drying scanning curve rather than to the main drying curve. As a result, for soils n° 9 and 23, the PWC data are used for the fitting procedure and the SDC data for the validation of prediction. The results are reported in Table 3b and illustrated in Figures 9a and b.

The scanning order being determined for the different soils, the prediction ability of the hysteresis model is tested. As stated before, the model is essentially based on three geometrical conditions: i) the first imposes the shape parameter (*mn*) to be identical for all wetting and drying curves independent of their scanning order; ii) the second condition determines the pressure head scale parameter specific for each drying and wetting curve; and iii) the third condition gives the specific water content scale parameters. The validity of these conditions is tested separately. The shape identity is verified first.

5.3. Shape Parameters

As shown by Eqs. (38) and (39), the main and higher order scanning curves in both drying and wetting collapse into one non-dimensional expression. Since experimental data are available in drying and wetting for all test soils, the two sets of shape and scale parameters (i.e., θ_{Skd}, θ_{rkd}, h_{gkd}, m_d and n_d; and θ_{Skw}, θ_{rkw}, h_{gkw}, m_w and n_w) can be calculated independently for each soil by fitting Eqs. (38) and (39) to the different sets of drying and wetting data. Theoretically, the shape indicators calculated for the drying data ($m_d n_d$) and the wetting data ($m_w n_w$) should be identical for each soil. When experimental data of more than one secondary drying curve were available, such as for soil n° 23 (Figure 8b), each scanning curve was analyzed separately. The results of soils n° 10, 13 and 15 were not taken into account as these soils show a clear bi-modal behavior as explained in more detail in Section 6. All together 23 combinations of $m_w n_w$ and $m_d n_d$ were analyzed.

Before discussing the results, a complication related to the best-fit procedure used for the wetting data should be noted. For several field soils, very few experimental wetting data were available. As an example, only three data points were reported for soils n° 14, 19 and 20. Identification of the full set of system parameters θ_S, θ_{rpw}, h_{gpw} and m of Eq. (39) on such limited series of data points is obviously not possible. To overcome this problem the best-fit procedure was carried out by imposing physical constraints on some of the parameters. For the case of soils n° 14, 19 and 20, only two independent parameters i.e., m and h_{gpw} were optimized while the values of θ_S and θ_{rpw} were taken from the analysis carried out over the main drying data. For the group of soils for which 8 or fewer experimental data points were available (i.e., soils n° 5, 21, 22, 23b and 23c) the three parameters m, θ_{rpw} and h_{gpw} were optimized. For the other soils the complete set of system parameters was calculated.

The comparison between $m_w n_w$ and $m_d n_d$ is given in Figure 10. A linear regression of the form $m_w n_w = a\, m_d n_d$ gives a coefficient a equal to 0.961, with a squared correlation coefficient of $r^2 = 0.995$, which shows only slight deviation from the 1:1 line. This result clearly validates the assumption that the wetting and drying shape indicators may be taken equal, i.e., $m_w n_w = m_d n_d = mn$. Since k_m is chosen constant in Eq. (3) for both wetting and drying, it follows that $m_w = m_d = m$ and $n_w = n_d = n$. Especially for the soils with $mn \leq 1$, which covers most of the field soils, the shape identity condition is extremely well verified.

Even though all soils lie within the 95% confidence interval of $\pm 2\, s(e)$ with $s(e) = 0.203$, soils n° 3, 4, 14 and 23c are at the limit. The results for soils n° 4 and 14 are not

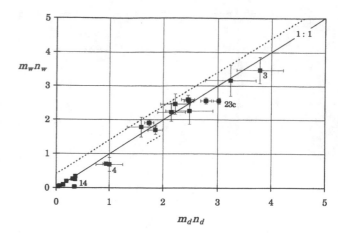

Figure 10. The wetting shape indicator $m_w n_w$ as a function of the drying shape indicator $m_d n_d$ calculated independently by fitting to the wetting and drying data of 20 soils used for this study. The dotted lines confine the 95% confidence interval and the error bars are estimated by twice the standard deviations $s(m_d n_d)$ and $s(m_w n_w)$ given by Eqs. (67) and (68) respectively.

surprising as these soils belong to the group of three problem soils shown to be affected by large uncertainties in m and h_{gmd} (see Section 5.1). The difficulty of soil n° 23c is illustrated by Figure 9b (3rd SDC). Only 7 experimental data points are available for the best-fit procedure. As all data points are concentrated around the inflection point (Figure 9b), the precision in the determination of $m_d n_d$ is questionable. Finally, soil n° 3 is the only soil which marginally satisfies the shape identity condition ($m_w n_w = m_d n_d = mn$) without any apparent reason.

The uncertainties in the values of $m_w n_w$ and $m_d n_d$ can be evaluated through the standard deviations $s(m_d n_d)$ and $s(m_w n_w)$. Using Eq. (3), the equations for $s(m_d n_d)$ and $s(m_w n_w)$ are given by:

$$s\left(m_d n_d\right) = \frac{\left[2 + m_d n_d\right]^2}{2}\, s(m) \qquad (67)$$

and

$$s\left(m_w n_w\right) = \frac{\left[2 + m_w n_w\right]^2}{2}\, s(m). \qquad (68)$$

Setting the errors of $m_w n_w$ and $m_d n_d$ respectively to plus or minus twice the standard deviations $s(m_w n_w)$ and $s(m_d n_d)$, the individual error bars can be calculated for each soil (Figure 10). The soils with small mn-values (e.g., $mn \leq 1$) are exposed to the smallest errors as their standard deviation $s(m)$ is generally small (Figure 7). The soils which show large error bars (e.g., soil n° 3) are those which are characterized by a combination of large mn and $s(m)$ values. Note that the uncertainties in $m_w n_w$ and $m_d n_d$ of soil n° 3, which only marginally satisfies the shape identity condition, are such that the error bars are still crossing the bisector.

As mentioned at the beginning of this chapter, we chose to apply the hysteresis model with the residual water content of the main loop set to zero. However, for the drying data of soils n° 1, 5, 6, 8 and 17, reported as main drying data, a non-zero θ_{rmd}-value would be preferable in terms of the goodness of fit criterion (Eq. (65)). Whether or not the experimental drying data belong in reality to a SDC curve with $\theta_{rsd} \neq 0$ rather than to the MDC, it is worthwhile noting that the shape identity condition ($m_w n_w = m_d n_d = mn$) is fully maintained when fitting a main drying equation with a non-zero θ_{rmd}-value. As an example, the results calculated for soil n° 6 taken from *Vachaud and Thony* [1971] are shown in Figure 11. When setting θ_r of the main loop *a-priori* equal to zero, the fit of Eq. (38) to the set of drying data yields a drying shape indicator $m_d n_d = 2.451$ with a goodness of fit $s(e) = 1.17\ 10^{-2}\ cm^3/cm^3$ (Table 3a). Applying the best-fit procedure to the experimental wetting data, reported as PWC data, with the physical constraint $\theta_{Spw} = \theta_S$, we obtain $m_w n_w = 2.596$ with $s(e) = 1.10\ 10^{-2}\ cm^3/cm^3$. When performing the same procedure for the case where θ_r of the main loop is not set equal to zero, then we have $m_d n_d = 3.513$ with $s(e) = 8.96\ 10^{-3}\ cm^3/cm^3$ for the MDC and $m_w n_w = 3.346$ with $s(e) = 9.10\ 10^{-3}\ cm^3/cm^3$ for the PWC. Apart from the fact that the goodness of fit improves with a non-zero θ_r, it is clear that the shape identity is well maintained even though the numerical value of mn increases. Obviously, the value of $m_d n_d$ increases when introducing a non-zero θ_r-value in the van Genuchten expression (MDC), because a squatter curve provides a better fit to the experimental data of soil n° 6 (Figure 11).

5.3.1. Summary shape parameters

Summarizing the results obtained in this Section 5.3, the shape identity condition (37) derived by the hysteresis model presented in this study has been verified for all soils tested.

5.4. Pressure Head Scale Parameters

The analysis of the pressure head scale parameter equations is less straightforward than that used to validate the shape parameter identity. As main drying data and primary wetting data are available for most soils (apart from soils n°

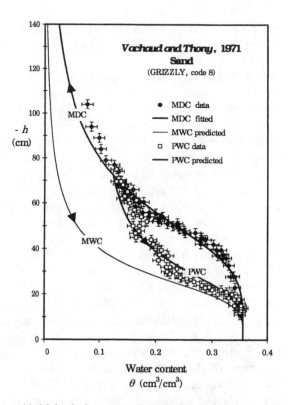

Figure 11. Main drying curve (MDC) fitted to measured drying data (●) for sand n° 6 taken from *Vachaud and Thony* [1971]. The predicted main wetting curve (MWC) and primary wetting curve (PWC) are presented together with experimental primary wetting data (   ). The horizontal and vertical bars correspond to the measurement errors in water content and soil water pressure head (i.e., ± 0.01 cm³/cm³ and ± 2 cm).

9 and 23 for which primary wetting and secondary drying data were reported), the pressure head scale parameters h_{gmd} and h_{gpw} can be calculated independently for each soil by fitting the appropriate water retention equation to the different sets of drying and wetting data. The values of h_{gmw} can then be predicted for the best fit values of h_{gmd} by the use of Eqs. (42) and (43) depending on the values of mn. However, the prediction of the primary wetting pressure head scale parameter h_{gpw} from the calculated value of h_{gmw} requires information on the water content scale parameter θ_{rpw}, which depends on the precision with which the starting point (θ_{stpw}, h_{stpw}) of the PWC is known. Consequently, when comparing the predicted values $(h_{gpw})_{pred}$ with those calculated by the best fit procedure $(h_{gpw})_{fit}$, possible amplifying or compensatory effects cannot be excluded. As no main drying data were available for soils n° 9 and 23, the comparison was carried out over the values $(h_{gpw})_{fit}$, and $(h_{gpw})_{pred}$ was calculated backward from the fitted secondary

wetting pressure head scale parameter $(h_{gsd})_{fit}$. Once again the data of the bi-modal soils n° 10, 13 and 15 are not taken into account.

The comparison between $(h_{gpw})_{pred}$ and $(h_{gpw})_{fit}$ is given in Figure 12. A linear regression of the form $(h_{gpw})_{pred} = b\,(h_{gpw})_{fit}$ gives a coefficient b equal to 1.005 with a squared correlation coefficient of $r^2 = 0.999$ which shows nearly no deviation from the 1:1 line. Obviously, this extremely high correlation is partially forced by soils n° 1 and 23 which have the largest values of h_{gpw} (Tables 3a and b). When soils n° 1 and 23 are excluded, the squared correlation coefficient falls down to $r^2 = 0.985$ with a regression coefficient $b = 1.071$. The foregoing results clearly validate the conditions (40) to (43), which relate the wetting and drying pressure head scale parameters. Apart from soil n° 14, *Manawatu* silt loam, all soils lie within the 95% confidence interval of $\pm\,2\,s(e)$ with $s(e) = 3.672$ cm. It was noted in Section 5.1 that soil n° 14 showed a large uncertainty in h_{gmd} calculated over the drying data. This uncertainty automatically affects the value of $(h_{gpw})_{pred}$ which is calculated from h_{gmd}. Moreover, the value of $(h_{gpw})_{fit}$ was calculated over a series of only 3 experimental data points in wetting making its value also very questionable. These reasons which are independent of the hysteresis model, explain the large discrepancy between $(h_{gpw})_{pred}$ and $(h_{gpw})_{fit}$ observed in Figure 12. Therefore, we prefer to ignore this soil for the calculation of $s(e)$ in order not to widen artificially the 95% confidence interval.

The calculation of the error bars used in Figure 12 is explained next. The horizontal error bars are associated with $(h_{gpw})_{fit}$ and are estimated by:

$$s\left(h_{gpw}\right)_{fit} = s\left(h_{gmd}\right) \quad , \tag{69}$$

where $s(h_{gmd})$ results from the uncertainty analysis given in Section 5.1. The estimation of the vertical error bars associated with $(h_{gpw})_{pred}$ is more complicated as two different steps are involved. First, the main wetting pressure head scale parameter is predicted from h_{gmd} using Eqs. (42) and (43). The uncertainty associated with $(h_{gmw})_{pred}$ is estimated by:

$$s\left(\left(h_{gmw}\right)_{pred}\right) = s\left(h_{gmd}\right) \quad \text{for} \quad mn > 1 \ , \tag{70}$$

and

$$s\left(\left(h_{gmw}\right)_{pred}\right) = \frac{s\left(h_{gmd}\right)}{\left[1 + mn\right]^{1/mn}} + \frac{h_{gmd}}{m\left[1 + mn\right]^{1/mn}}\left[\frac{\ln\left(1 + mn\right)}{2\,m} - \frac{1}{1 + m}\right]s\left(m\right) \tag{71}$$

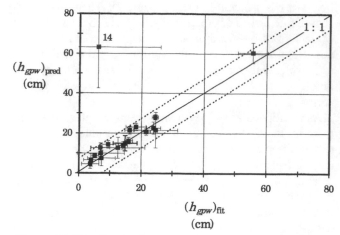

Figure 12. The fitted primary wetting pressure head scale parameter $(h_{gpw})_{fit}$ as a function of the predicted primary wetting pressure head scale parameter $(h_{gpw})_{pred}$. The dotted lines confine the 95% confidence interval. The error bars are estimated by twice the standard deviations $s((h_{gpw})_{fit})$ and $s((h_{gpw})_{pred})$ given by Eqs. (69) and (72) respectively.

for $0 < mn \leq 1$. The second step predicts h_{gpw} from the values of h_{gmw} using Eq. (41). This introduces a supplementary uncertainty associated with the precision in θ_S and θ_{rpw}. The equation for estimating the error $s((h_{gpw})_{pred})$ as a function of $s((h_{gmw})_{pred})$ and $s(m)$ is taken from *Haverkamp et al.* [1998]:

$$s\left(\left(h_{gpw}\right)_{pred}\right) = \left[\frac{\theta_S}{\theta_S - \theta_{rpw}}\right]^{1/mn} s\left(\left(h_{gmw}\right)_{pred}\right) + \frac{h_{gpw}}{2}\ln\left(\frac{\theta_S}{\theta_S - \theta_{rpw}}\right)\frac{\left[2 + mn\right]^2}{mn^2}\,s\left(m\right) \ , \tag{72}$$

where Eq. (72) is written so that it maximizes the error of $(h_{gpw})_{pred}$. As expected, the uncertainty in $(h_{gpw})_{pred}$ increases with increasing θ_{rpw}. Combining Eqs. (70), (71) and (72) we have an estimation of the standard deviation of $(h_{gpw})_{pred}$ in terms of $s(m)$ and $s(h_{gmd})$. Setting the error of $(h_{gpw})_{pred}$ to plus or minus twice the value of $s((h_{gpw})_{pred})$, the individual error bars can be calculated for each soil. As shown by Figure 12, the imprecision in the prediction of h_{gpw} is of the same order of magnitude as that observed for the best fit values of h_{gpw} validating once again the pressure head scale conditions (40) to (43) derived by the hysteresis model.

In addition to the previous test, the values of h_{gmd} and h_{gpw} calculated by individual fitting to the different sets of

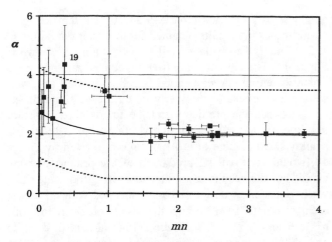

Figure 13. Parameter α as a function of the water retention shape indicator mn. The points (■) correspond to α_{fit} calculated for each soil and the continuous line represents the theoretically derived relation (43). The dotted lines confine the 95% confidence interval. The error bars are estimated by twice the standard deviations $s(mn)$ and $s(\alpha_{fit})$ given by Eqs. (67) and (73) respectively.

$$s\left(\alpha_{fit}\right) = \frac{\alpha_{fit}}{\left(h_{gmd}\right)_{fit}}\left[s\left(\left(h_{gmd}\right)_{fit}\right) + \alpha_{fit}\ s\left(\left(h_{gmw}\right)_{pred}\right)\right],\ (73)$$

where Eq. (73) is written so that it maximizes the error of α_{fit}. The values of $s((h_{gmd})_{fit})$ are given by the uncertainty analysis addressed in Section 5.1. As $(h_{gmw})_{pred}$ is calculated from $(h_{gpw})_{fit}$ using Eq. (41), the uncertainties associated with the precision in θ_S and θ_{rpw} have to be taken into account for the estimation of $s((h_{gmw})_{pred})$:

$$s\left(\left(h_{gmw}\right)_{pred}\right) = \left[\frac{\theta_S - \theta_{rpw}}{\theta_S}\right]^{1/mn} s\left(\left(h_{gpw}\right)_{fit}\right) -$$
$$\frac{\left(h_{gpw}\right)_{fit}}{mn^2}\left[\frac{\theta_S - \theta_{rpw}}{\theta_S}\right]^{1/mn}\ln\left(\frac{\theta_S - \theta_{rpw}}{\theta_S}\right)\ s(m)\ . \quad (74)$$

For convenience, $s((h_{gpw})_{fit})$ is chosen equal to $s((h_{gmd})_{fit})$. Combining Eqs. (73) and (74) we have an estimation of the standard deviation of α_{fit} in terms of $s(m)$ and $s(h_{gmd})$. Setting the error of α_{fit} to plus or minus twice the value of $s(\alpha_{fit})$, the individual error bars can be calculated for each soil.

Generally speaking, the uncertainty in α_{fit} increases with decreasing values of mn (Figure 13). This is mainly due to the fact that, according to Eq. (74), $s((h_{gmw})_{pred})$ increases with increasing pressure head scale values. Taking into account these uncertainties in α_{fit}, the behavior of Eq. (43) which slightly underestimates α for small mn (i.e., $0 < mn < 1$) is still satisfactory. Moreover, when using Eq. (43) to predict the wetting pressure head scale parameter, h_{gpw}, for soils with $0 < mn < 1$, the effect of the underestimation in α on the prediction of the wetting scanning curve is hardly noticeable due to the steepness of the water retention curve. As an example, Figure 14 shows the results obtained for soil n° 14 with $mn = 0.34$. As mentioned before (Section 5.1), the data for this soil are highly problematic with an uncertainty of 12% in the best fit value of h_{gmd}. Among the 23 soils tested in this study, it is the only soil to give erratic pressure head scale prediction (Figure 12). The predicted value of $(h_{gpw})_{pred}$ equals -63.24 cm, which is ten times larger than the best fit value $(h_{gpw})_{fit} = -6.04$ cm. In spite of this large discrepancy, the predictive results shown in Figure 14 still give reasonable agreement with the observed primary wetting data.

5.4.1. Summary pressure head scale parameters

Summarizing the results obtained in this Section 5.4, the pressure head scale relations (Eqs. (40) to (43)) derived by

drying and wetting data of each soil can be used to verify the relation between h_{gmd} and h_{gmw} in Eq. (42). Calculating backward the best-fit values of $(h_{gmw})_{pred}$ from $(h_{gpw})_{fit}$, the values of α_{fit} defined as the ratio $(h_{gmd})_{fit}$ over $(h_{gmw})_{pred}$ should theoretically follow the relation $\alpha(mn)$ given by Eq. (43). Obviously, this test is less dominated by the large values of h_{gmd} than the previous test (Figure 12). Hence, it gives a better picture of the validity of the pressure head scale relation in the range of small mn-values which covers most of the field soils. The results are shown in Figure 13.

When $mn \geq 1$, the best fit values of α are very close to those predicted by Eq. (43). However, when $mn < 1$, the prediction slightly underestimates the values of α_{fit}. Nevertheless, all soils lie within the 95% confidence interval of $\pm 2\ s(e)$ with $s(e) = 0.775$, with the exception of soils n° 14 and 19. When taking into account the uncertainties in α_{fit} indicated by the vertical errors bars, soil n° 19 still lies within the confidence interval. However, soil n° 14 is totally out of range with $\alpha_{fit} = 24.75$. Therefore, this value is ignored for the calculation of $s(e)$ in order not to widen the 95% confidence interval unrealistically.

The horizontal error bars used in Figure 13 are identical to those given by Eq. (67). The estimation of the vertical error bars is more complicated as two steps are involved. First, the uncertainty associated with α_{fit} is expressed as a function of the standard deviations of $(h_{gmd})_{fit}$ and $(h_{gmw})_{pred}$ using Eq. (42) (*Haverkamp et al.* [1998]):

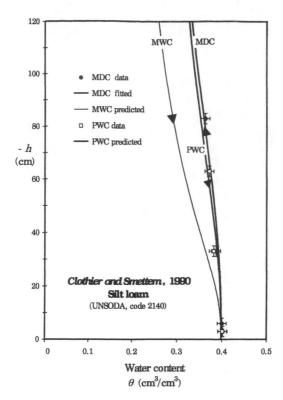

Figure 14. Main drying curve (MDC) fitted to measured drying data (•) for sand n° 14 taken from *Clothier and Smettem* [1990]. The predicted main wetting curve (MWC) and primary wetting curve (PWC) are presented together with experimental primary wetting data ( ). The horizontal and vertical bars correspond to the measurement errors in water content and soil water pressure head (i.e., ± 0.01 cm³/cm³ and ± 2 cm) associated with each observation (θ_i, h_i).

the hysteresis model presented in this study have been verified for all soils tested with the exception of soil n° 14 which showed large uncertainty in the experimental soil pressure head data.

5.5. Water Content Scale Parameters

The third stage of verification concerns the water content scale parameters. First, soils n° 1 - 8 with a well described hysteresis history are analyzed. The MDC data were chosen for the fitting procedure and the PWC data were used for the validation of the prediction procedure. As shown by Table 3a, the goodness of fit calculated by Eq. (65) over the measured MDC data points is of the same order of magnitude as the goodness of prediction calculated over the PWC data points. On average the goodness of fit for this group of soils ($\bar{s}(e) = 8.57 \ 10^{-3} \ \text{cm}^3/\text{cm}^3$) is slightly smaller than the

goodness of prediction ($\bar{s}(e) = 9.94 \ 10^{-3} \ \text{cm}^3/\text{cm}^3$). Figure 15 illustrates the results obtained for the laboratory soils n° 1, 3, 7 and 8; the results of soil n° 6 were already shown in Figure 11. The agreement between predicted and measured primary wetting curves is good for all soils. Soil n° 3 was chosen on purpose to show to what extent the slight difference between the predicted and the fitted shape indicator mn (see Figure 10) affects the overall results. Indeed, a slightly smaller value of $m_w n_w$, such as suggested by Figure 10, would improve the prediction of the primary wetting curve, but the results are still extremely good, especially when considering the error bars associated with the measurement points. Even though the results obtained for soils n° 1, 6 and 8 shown in Figures 15a, 11 and 15d could probably be improved by considering a non-zero θ_{rmd}-value in the original MDC expression, the predicted PWC passes through the scatter of measurement points. The results shown in Figures 11 and 15 support the validity of the geometric scaling approach used to calculate the water content scale parameters θ_S and θ_{rpw}.

The second group of soils includes soils n° 10 to 23 for which the hysteresis history is less well documented. The shape indicator mn is smaller than 1 for this group with the exception of soils n° 20 and 23 (Tables 3a and b). Hence, the hysteresis effect is less evident as the water retention curves are much steeper than those observed for sandy soils with $mn > 1$. The elongation of the water retention curve causes the inflection point to be less pronounced leading to less precision in the determination of the pressure head scale parameter h_{gmd}. This problem is reflected in the error $s(h_{gmd})$ detailed in Section 5.1 (Figure 8). Similar to the foregoing group of soils, the goodness of fit calculated by Eq. (65) over the measured MDC data points is of the same order of magnitude as the goodness of prediction calculated over the PWC data points (Table 3a). On average the goodness of prediction for this group of soils ($\bar{s}(e) = 1.43 \ 10^{-2} \ \text{cm}^3/\text{cm}^3$) is slightly inferior to the goodness of fit ($\bar{s}(e) = 8.72 \ 10^{-3} \ \text{cm}^3/\text{cm}^3$). As an example, Figure 16 shows the results obtained for soils n° 11, 12, 20 and 21 taken from the finer textured soil classes, i.e., sandy clay loam, silt loam, sandy loam and silty clay loam, respectively. The agreement between predicted and measured primary wetting curves is good for all soils. The examples show how easily the wetting and drying measurement points can be confused for soils which cannot be classified as sandy soils, especially when the hysteresis history is not well known, as is the case for most field measurements. Even though the difference between the wetting and drying pressure head scale parameters increases (Eq. (42)), the water content scales reduce to such a narrow band that the hysteresis effects become almost negligible, i.e., the steeper

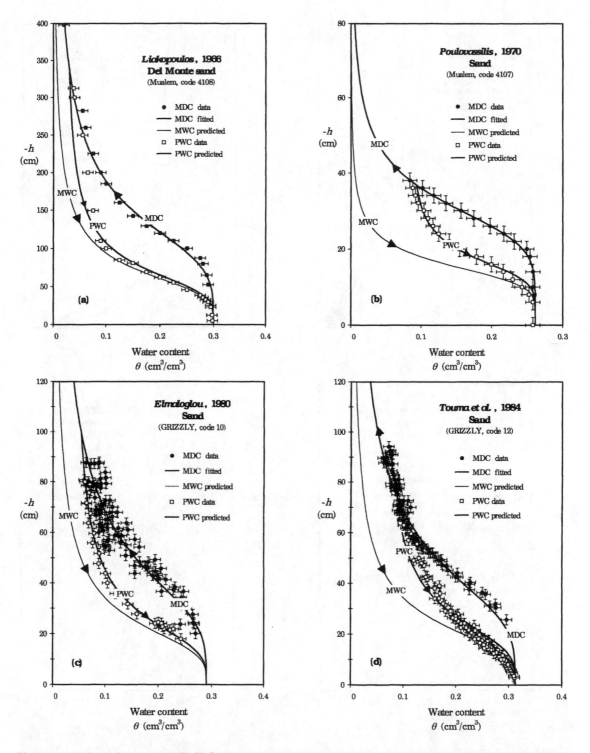

Figure 15. Main drying curve (MDC) fitted to measured drying data (•) for sands n° 1, 3, 7 and 8 taken from *Liakopoulos* [1966], *Poulovassilis* [1970], *Elmaloglou* [1980] and *Touma et al.* [1984] respectively. The predicted main wetting curve (MWC) and primary wetting curve (PWC) are presented together with experimental primary wetting data (   ). The horizontal and vertical bars correspond to the measurement errors in water content and pressure head (i.e., ± 0.01 cm³/cm³ and ± 2 cm) associated with each observation (θ_i, h_i).

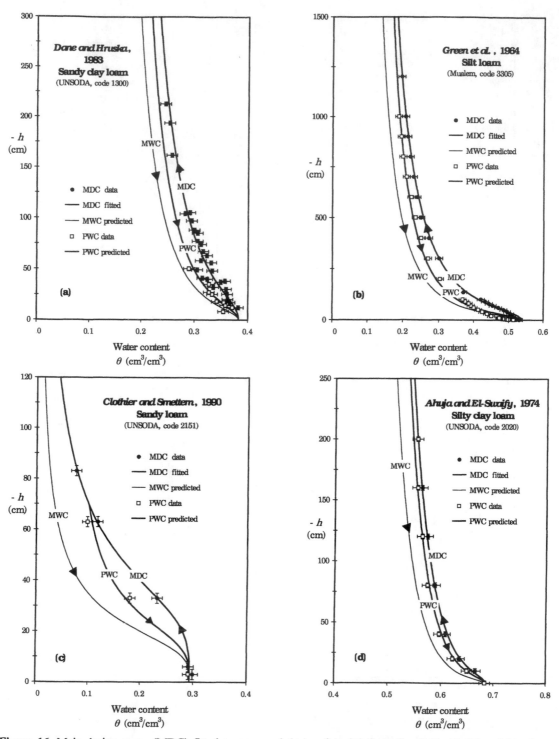

Figure 16. Main drying curve (MDC) fitted to measured drying data (●) for soils n° 11, 12, 20 and 21 taken from *Dane and Hruska* [1983], *Green et al.* [1964], *Clothier and Smettem* [1990] and *Ahuja and El-Swaify* [1974] respectively. The predicted main wetting curve (MWC) and primary wetting curve (PWC) are presented together with experimental primary wetting data (   ). The horizontal and vertical bars correspond to the measurement errors in water content and soil water pressure head (i.e., ± 0.01 cm³/cm³ and ± 2 cm) associated with each observation (θ_i, h_i).

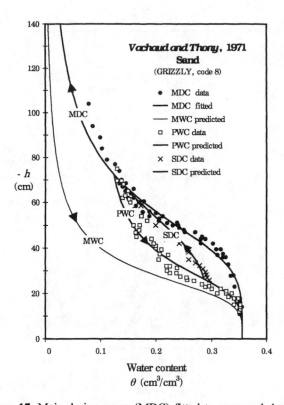

Figure 17. Main drying curve (MDC) fitted to measured drying data (•) for sand n° 6 taken from *Vachaud and Thony* [1971]. The predicted main wetting curve (MWC), primary wetting curve (PWC) and secondary drying curve (SDC) are presented together with experimental primary wetting data ( ) and secondary drying data (×).

goodness of fit calculated over the measured MDC data points ($s(e) = 1.2 \ 10^{-2}$ cm^3/cm^3).

5.5.1. Summary water content scale parameters

Summarizing the results obtained in this Section 5.5, the water content scale relations (Eqs. (56) to (63)) derived by the hysteresis model presented in this study have been verified for all soils tested.

6. LIMITATIONS ON THE USE OF THE MODEL

So far, the results obtained using the prediction model agree well with the available data on hysteresis taken from the literature. However, there are some limitations to its use.

The first limitation, already mentioned, concerns changes in the volume of entrapped air during rewetting, that is $\theta_{Spw} \neq \theta_S$. This effect may occur when temperature changes are involved [e.g., *Hopmans and Dane*, 1986]. Even though no rigorous theoretical solution is available in the literature, the effect could be incorporated in the model using the purely empirical relationship presented by *Aziz and Settari* [1979].

The second limitation concerns the θ_r-value of the main loop. The generalized pressure head scale relations (Eqs. (40) and (41)) and water content scale equations (Eqs. (56) to (63)) derived from the hysteresis model are based on the assumption of $\theta_{rmd} = \theta_{rmw} = 0$. A non-zero value of θ_{rmd} or θ_{rmw} does not fundamentally change these equations. While Eqs. (40) and (41) should be used with ($\theta_S - \theta_{rmd}$) instead of θ_S, Eqs. (56) to (63) remain as they stand. However, the introduction of a non-zero θ_r-value has a non-negligible effect on the value of mn which increases with a non-zero θ_r as shown before in the context of Figure 11.

The third limitation concerns bi-modal systems. Three exemplary soils (i.e., soils n° 10, 13 and 15 taken from *Topp* [1969 and 1971]) with a typical bi-modal behavior are included in the 23 test soils. As the bi-modal behavior principally affects the drying curve, the characteristic soil parameters were calculated by fitting to the wetting data (Table 3c). The data for drying were then used for comparison with the predicted MDC. Figures 18a and b show the results obtained for the soils n° 10 (Rubicon sandy loam) and n° 15 (Rideau clay loam). Note that it was necessary to apply the hysteresis model with a non-zero θ_{rmd}-value. Due to the bi-modal behavior of the MDC the prediction based on a unimodal PWC is obviously not satisfactory. Bi-modality is likely to be at least partially caused by the laboratory treatment of the soil during the packing of the column, e.g., unstable particle arrangements. Whether or not

the water retention curve, the less noticeable are the hysteresis effects.

The last test used to verify the validity of the geometric scaling approach focuses on prediction of a series of scanning curves with different scanning orders. Soil n° 6 [*Vachaud and Thony*, 1971] is used for this test as it is the only soil for which sufficient experimental data were reported. After calculating the model parameters from the main drying data, a sequence of primary wetting and secondary drying curves was predicted. The results are shown in Figure 17. For the sake of clarity, the error bars are not presented. Even though the agreement between predicted and measured primary wetting and secondary drying curves is good, the prediction would have been even better with a non-zero θ_r-value for the MDC (Section 5.3, Figure 11). The goodness of prediction calculated with Eq. (65) for the SDC ($s(e) = 8.1 \ 10^{-3}$ cm^3/cm^3) and for the PWC ($s(e) = 1.7 \ 10^{-2}$ cm^3/cm^3) is of the same order of magnitude as the

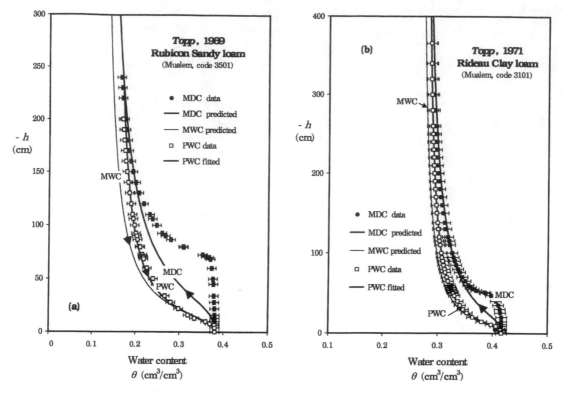

Figure 18. Primary wetting curves (PWC) fitted to measured primary wetting data ( ) for soils n° 10 (18a) and n° 15 (18b), taken from *Topp* [1969] and *Topp* [1971] respectively. The predicted main wetting curves (MWC) are presented together with experimental main drying data (●). The horizontal and vertical bars correspond to the measurement errors in water content and soil water pressure head (i.e., ± 0.01 cm³/cm³ and ± 2 cm) associated with each observation (θ_i, h_i).

the bi-modality is artificially generated, it is only effective for the range of small negative pressure head values. For example, for the Rideau clay loam (Figure 18b) the behavior of the water retention curve tends to the unimodal shape for $h < -60$ cm. This corresponds to the range where the predicted MDC merges into the measured drying data supporting at the same time the validity of the hysteresis model presented here. Similar results were observed for the other two bi-modal soils n° 10 (Figure 18a) and 13.

7. CONCLUSIONS

A model of soil hysteresis based on theory and geometric scaling has been presented, together with a short algorithm for calculating water content and updating the hysteresis state in numerical flow codes. The theoretical based model of *Parlange* [1976] was applied to determine the relation between the main wetting and drying curves using the *van Genuchten* [1980] equation to describe the water retention curve. We showed that all wetting and drying curves, what-

ever their scanning order, can be collapsed into one unique dimensionless expression defined as a function of three parameters, one determining the shape of the curve and the two others scaling the soil water pressure head and volumetric water content. The shape parameter mn is identical for all wetting and drying curves. The pressure head scale parameters h_{gkw} or h_{gkd} are specific for each curve depending on the family (i.e., wetting or drying family) and the scanning order k. They are expressed as a function of the shape parameter mn and the specific water content scale parameters given by θ_{Skd} and θ_{rkd} for drying and θ_{Skw} and θ_{rkw} for wetting. The water content scale parameters depend upon the family and the scanning order, i.e., the hysteresis history. They are calculated as a function of the points of departure and arrival of the specific scanning curve.

The model uses extrapolation and requires only the water retention shape parameter mn, together with the water content and pressure head scale parameters θ_S and h_{gmd}. Given these input parameters, full description of soil hysteresis, i.e., main, primary and higher order scanning curves, can be predicted. The shape and scale parameters mn, θ_S and h_{gmd} are best obtained from the main drying curve or a primary wetting curve, otherwise the hysteresis history must be known. The method is formulated to facilitate its use in numerical flow analyses.

The model was tested on 23 soils (taken from the literature) covering 6 textural classes. With the exception of one soil with a large uncertainty in the experimental soil pressure head data, the predicted and measured results agree extremely well for all soils, e.g., the goodness of prediction is of the same order of magnitude as the goodness of fit. The shape parameter condition as well as the pressure head and water content scale conditions have been verified for all soils supporting the validity of the hysteresis model presented here.

Even though the residual water content θ_r of the main loop was chosen equal to zero, a non-zero value can easily be introduced into the hysteresis model. The model does not allow prediction of soil hysteresis for bi-modal soils.

8. APPENDIX

An algorithm suitable for incorporating hysteresis in the soil water retention relation into a computer program must store and update the current hysteretic state as well as return the value of content θ corresponding to a given soil water pressure head h (here we regard θ as the dependent variable and h as the independent variable). The state is defined entirely by a list of points (h_i, θ_i) at which reversal from drying to wetting or wetting to drying has occurred, plus a 'previous point' (h_p, θ_p), which is the last known point attained and therefore a potential reversal point. The first point in the list will normally be $(0, \theta_S)$ corresponding to the start of the main drying curve (only for an oven dry soil would the state ever be on the main wetting curve). In addition to the state, it is necessary for practical calculation to store the pressure head scale parameter h_g associated with each scanning curve. Thus the burden of storage is three values per reversal, but when the pressure head goes outside the range of a scanning loop the six values for the loop are deleted. In addition, it is convenient to store the water content scale parameters for the current curve.

The following algorithm updates the hysteresis state and calculates $\theta(h)$. In the algorithm, $F(h,h_g)=[1+(h/h_g)^n]^{-m}$ while θ_{sr} and θ_r refer to the current curve (with $\theta_{sr}=\theta_S-\theta_r$). A simple bisection method is used to find θ_{sr}, but more efficient methods are possible. The points of reversal from wetting to drying and vice-versa are (h_i, θ_i) with $i=1,\ldots,j$. The initialization sets the first of these, together with the

associated pressure head scale parameter h_{g1}, then the 'previous point' (h_p, θ_p), then the parameters θ_r and θ_{sr} for the main drying curve. Values for θ_S, h_{gmd}, m and n are assumed known. The value of θ_r is taken as zero, although this is not necessary for the calculations. If the algorithm is used within an iterative method for finding h, then j and (h_p, θ_p) should not be changed permanently until a final value of h is found; this not allowed for here.

8.1. Hysteresis Algorithm

Initialise   **If** $mn<1$ **Then** $\alpha=(1+mn)^{1/mn}$ **Else** $\alpha=2$
$\Delta=0.000001$; $h_1=0$; $\theta_1=\theta_S$; $h_{g1}=h_{gmd}$; Reset

Function $\theta(h)$
  If $h\geq0$ **Then**
    Reset; $\theta=\theta_S$
  Else

    $j_S=j$
    If Reversal **Then** AddPoint
    Do While OutOfRange; $j=j-2$; **End Do**
    If $j\neq j_S$ **Then** GetParameters
    $\theta=\theta_r+\theta_{sr}F(h,h_{gj})$
    $h_p=h$; $\theta_p=\theta$
  End If

End Function

Reset: $j=1$; $h_p=h_1$; $\theta_p=\theta_1$; $\theta_r=0$; $\theta_{sr}=\theta_S-\theta_r$
Reversal: $(h-h_p)(h_p-h_j)<0$
AddPoint: $j=j+1$; $h_j=h_p$; $\theta_j=\theta_p$
OutOfRange: **If** $j=1$ **Then** *False* **Else** ($h\geq\max(h_{j-1},h_j)$ **Or**
                  $h\leq\min(h_{j-1},h_j)$)
GetParameters: **If** $j=1$ **Then**
                  $\theta_r=0$; $\theta_{sr}=\theta_S-\theta_r$
               Else

                 If $j>j_S$ **Then**
                   If $h_j>h_{j-1}$ **Then** $h_g=\alpha h_{gj-1}$ **Else**
                   $h_g=h_{gj-1}/\alpha$, $\theta_{srold}=\theta_{sr}$; $x_1=0$;
                   $x_2=\theta_S$
                   Do
                     $\theta_{sr}=0.5(x_1+x_2)$;
                     $h_{gj}=h_g(\theta_{srold}/\theta_{sr})^{1/mn}$;
                     $F_1=F(h_j,h_{gj})$; $F_2=F(h_{j-1},h_{gj})$;

                     $y=(\theta_j-\theta_{j-1})/(F_1-F_2)$
                     If $\theta_{sr}>y$ **Then** $x_2=\theta_{sr}$
                            Else $x_1=\theta_{sr}$
                   Until $|\theta_{sr}-y|<\Delta$
                 Else
                   $F_1=F(h_j,h_{gj})$; $F_2=F(h_{j-1},h_{gj})$;
                   $\theta_{sr}=(\theta_j-\theta_{j-1})/(F_1-F_2)$
               End If

$$\theta_r = (\theta_{j-1} F_1 - \theta_j F_2)/(F_1 - F_2)$$
End If

9. PARAMETER NOTATION

a	regression coefficient;
b	regression coefficient;
c	constant of integration;
e	deviation between measured and fitted water content values (L^3/L^3);
h	soil water pressure head taken to be negative (L);
h^*	non-dimensional soil water pressure head;
h_{ae}	drying pressure head scale parameter of the Brooks and Corey water retention equation (L);
h_{we}	wetting pressure head scale parameter of the Brooks and Corey water retention equation (L);
h_g	general notation for pressure head scale parameter of the van Genuchten water retention equation (L);
h_{gkd}	pressure head scale parameter of any drying curve of scanning order k (L);
$\tilde{h}_{gkd}$	estimate of the pressure head scale parameter h_{gkd} (L);
h_{gkw}	pressure head scale parameter of any wetting curve of scanning order k (L););
h_{gmd}	specific pressure head scale parameter of the main drying curve (L);
h_{gmw}	specific pressure head scale parameter of the main wetting curve (L);
h_{gpd}	specific pressure head scale parameter of the primary drying curve (L);
h_{gpw}	specific pressure head scale parameter of the primary wetting curve (L);
h_{gsd}	specific pressure head scale parameter of the secondary drying curve (L);
h_{gsw}	specific pressure head scale parameter of the secondary wetting curve (L);
h_{gtd}	specific pressure head scale parameter of the tertiary drying curve (L);
h_{gtw}	specific pressure head scale parameter of the tertiary wetting curve (L);
h_{stpd}	soil water pressure head value on the main wetting curve from which the primary drying curve departs (L);
h_{stpw}	soil water pressure head value on the main drying curve from which the primary wetting curve departs (L);
h_{stsd}	soil water pressure head value on the primary wetting curve from which the secondary drying curve departs (L);
h_{stsw}	soil water pressure head value on the primary drying curve from which the secondary wetting curve departs (L);
h_{sttd}	soil water pressure head value on the secondary wetting curve from which the tertiary drying curve departs (L);
h_{sttw}	soil water pressure head value on the secondary drying curve from which the tertiary wetting curve departs (L);
k	scanning order;
k_m	integer value relating shape parameters m_d and n_d as well as m_w and n_w ($k_m = 2$);
m	general notation for the shape parameter of the van Genuchten water retention equation;
$\tilde{m}$	estimate of the shape parameter m;
m_{md}	specific shape parameter of the main drying curve ($m_{md} = m$);
m_{mw}	specific shape parameter of the main wetting curve ($m_{mw} = m$);
m_{pd}	specific shape parameter of the primary drying curve ($m_{pd} = m$);
m_{pw}	specific shape parameter of the primary wetting curve ($m_{pw} = m$);
m_{sd}	specific shape parameter of the secondary drying curve ($m_{sd} = m$);
m_{sw}	specific shape parameter of the secondary wetting curve ($m_{sw} = m$);
m_{td}	specific shape parameter of the tertiary drying curve ($m_{td} = m$);
m_{tw}	specific shape parameter of the tertiary wetting curve ($m_{tw} = m$);
mn	water retention shape indicator of the van Genuchten water retention equation;
MDC	main drying curve;
MWC	main wetting curve;
n	general notation for the shape parameter of the van Genuchten water retention equation;
$\tilde{n}$	estimate of the shape parameter n;
n_{md}	specific shape parameter of the main drying curve ($n_{md} = n$);
n_{mw}	specific shape parameter of the main wetting curve ($n_{mw} = n$);
n_{pd}	specific shape parameter of the primary drying curve ($n_{pd} = n$);
n_{pw}	specific shape parameter of the primary wetting curve ($n_{pw} = n$);
n_{sd}	specific shape parameter of the secondary drying curve ($n_{sd} = n$);
n_{sw}	specific shape parameter of the secondary wetting curve ($n_{sw} = n$);

n_{td} specific shape parameter of the tertiary drying curve ($n_{td} = n$);

n_{tw} specific shape parameter of the tertiary wetting curve ($n_{tw} = n$);

N_P total number of water content measurements;

PDC   primary drying curve;

PWC   primary wetting curve;

r^2 squared correlation coefficient;

$s(e)$ goodness of fit expressed in terms of volumetric water content (L^3/L^3);

$s(h_{gmd})$ standard deviation of h_{gmd};

$s(h_{gmw})$ standard deviation of h_{gmw};

$s(h_{gpw})$ standard deviation of h_{gpw};

$s(m)$ standard deviation of m;

$s(m_d n_d)$ standard deviation of $m_d n_d$;

$s(m_w n_w)$ standard deviation of $m_w n_w$;

$s(\theta_S)$ standard deviation of θ_S;

SDC   secondary drying curve;

SWC   secondary wetting curve;

TDC   tertiary drying curve;

TWC   tertiary wetting curve;

u_i measurement error associated with measured water content value θ_i (L^3/L^3);

v_i measurement error associated with measured soil water pressure head value h_i (L);

α pressure head scale ratio;

ε soil porosity (L^3/L^3);

θ volumetric soil water content (L^3/L^3);

θ^* general notation for degree of saturation;

θ_{kd} volumetric soil water content of any drying curve of scanning order k (L^3/L^3);

$\tilde{\theta}_{kd}$ estimate of water content θ_{kd} (L^3/L^3);

θ_{kd}^* non-dimensional soil water content of any drying curve of scanning order k;

θ_{kw} volumetric soil water content of any wetting curve of scanning order k (L^3/L^3);

θ_{kw}^* non-dimensional soil water content of any wetting curve of scanning order k;

θ_{md} volumetric soil water content of the main drying curve (L^3/L^3);

θ_{md}^* non-dimensional soil water content of the main drying curve;

θ_{mw} volumetric soil water content of the main wetting curve (L^3/L^3);

θ_{mw}^* non-dimensional soil water content of the main wetting curve;

θ_{pd} volumetric soil water content of the primary drying curves;

θ_{pd}^* non-dimensional soil water content of the primary drying curves;

θ_{pw} volumetric soil water content of the primary wetting curves (L^3/L^3);

θ_{pw}^* non-dimensional soil water content of the primary wetting curves;

θ_r general notation for the residual volumetric soil water content (L^3/L^3);

θ_{rkd} residual water content of any drying curve of scanning order k (L^3/L^3);

$\tilde{\theta}_{rkd}$ estimate of residual water content θ_{rkd} (L^3/L^3);

θ_{rkw} residual water content of any wetting curve of scanning order k (L^3/L^3);

θ_{rpd} residual volumetric soil water content of the primary drying curves with $\theta_{rpd} = 0$;

θ_{rpw} residual volumetric soil water content of the primary wetting curves (L^3/L^3);

θ_{rsd} residual volumetric soil water content of the secondary drying curves (L^3/L^3);

θ_{rsw} residual volumetric soil water content of the secondary wetting curves (L^3/L^3);

θ_{rtd} residual volumetric soil water content of the tertiary drying curves (L^3/L^3);

θ_{rtw} residual volumetric soil water content of the tertiary wetting curves (L^3/L^3);

θ_{sd} volumetric soil water content of the secondary drying curves (L^3/L^3);

θ_{sd}^* non-dimensional soil water content of the secondary drying curves;

θ_{sw} volumetric soil water content of the secondary wetting curves (L^3/L^3);

θ_{sw}^* non-dimensional soil water content of the secondary wetting curves;

θ_{stpd} volumetric soil water content on the main wetting curve associated with the pressure head value (h_{stpd}) from which the primary drying curve departs (L^3/L^3);

θ_{stpw} volumetric soil water content on the main drying curve associated with the pressure head value (h_{stpw}) from which the primary wetting curve departs (L^3/L^3);

θ_{stsd} volumetric soil water content on the primary wetting curve associated with the pressure head value (h_{stsd}) from which the secondary drying curve departs (L^3/L^3);

θ_{stsw} volumetric soil water content on the primary drying curve associated with the pressure head value (h_{stsw}) from which the secondary wetting curve departs (L^3/L^3);

θ_{sttd} volumetric soil water content on the secondary wetting curve associated with the pressure head value (h_{sttd}) from which the tertiary drying curve departs (L^3/L^3);

θ_{sttw} volumetric soil water content on the secondary drying curve associated with the pressure head value (h_{sttw}) from which the tertiary wetting curve departs (L^3/L^3);

θ_S general notation for volumetric water content at natural saturation (L^3/L^3);

θ_{Skd} saturated water content of any drying curve of scanning order k (L^3/L^3);

$\tilde{\theta}_{Skd}$ estimate of saturated water content θ_{Skd} (L^3/L^3);

θ_{Skw} saturated water content of any wetting curve of scanning order k (L^3/L^3);

θ_{Spd} volumetric soil water content at saturation of the primary drying curves (L^3/L^3);

θ_{Spw} volumetric soil water content at saturation of the primary wetting curves with $\theta_{Spw} = \theta_S$ (L^3/L^3);

θ_{Ssd} volumetric soil water content at saturation of the secondary drying curves (L^3/L^3);

θ_{Ssw} volumetric soil water content at saturation of the secondary wetting curves (L^3/L^3);

θ_{Std} volumetric soil water content at saturation of the tertiary drying curves (L^3/L^3);

θ_{Stw} volumetric soil water content at saturation of the tertiary wetting curves (L^3/L^3);

θ_{td} volumetric soil water content of the tertiary drying curves (L^3/L^3);

θ_{td}^* non-dimensional soil water content of the tertiary drying curves;

θ_{tw} volumetric soil water content of the tertiary wetting curves (L^3/L^3);

θ_{tw}^* non-dimensional soil water content of the tertiary wetting curves;

$\sigma(u_i)$ standard deviation associated with measurement error u_i (L^3/L^3);

$\sigma(v_i)$ standard deviation associated with measurement error v_i (L);

Acknowledgements. This study was partially supported by the National Scientific Research Council (*Centre National de la Recherche Scientifique* – CNRS) through the funding of a visit of one of the authors to Grenoble.

REFERENCES

Ahuja, L. R., and El-Swaify, Hydrologic characteristics of benchmark soils of Hawaii's forest watersheds, *Report University of Hawaii*, 1974.

Aziz, K., and A. Settari, *Petroleum Reservoir Simulation*, Applied Science, Barking, England, 395-401, 1979.

Bouwer, H., Rapid field measurement of air entry value and hydraulic conductivity of soils as significant parameters in flow system analysis, *Water Resour. Res.*, 2, 729-738, 1966.

Brooks, R. H., and C. T. Corey, Hydraulic properties of porous media, *Hydrol. Paper 3*, Colorado State University, Fort Collins, 1964.

Burdine, N. T., Relative permeability calculations from pore-size distribution data, *Petr. Trans., Am. Inst. Mining Metall. Eng.*, 198, 71-78, 1953.

Clothier, B. E., and K. R. J. Smettem, Combining laboratory and field measurements to define the hydraulic properties of soil, *Soil Sci. Soc. Am. J.*, 54, 299-304, 1990.

Dane, J. H., and S. Hruska, In-situ determination of soil hydraulic properties during drainage, *Soil Sci. Soc. Am. J.*, 4, 619-624, 1983.

Elmaloglou, S., Effets des stratifications sur les transferts de matières dans les sols, *Thèse de Docteur-Ingénieur, Institut National Polytechnique de Grenoble*, France, 1980.

Gilham, R. W., A. Klute, and D. F. Heermann, Hydraulic properties of a porous medium; Measurements and empirical representation, *Soil Sci. Soc. Am. J.*, 40, 203-207, 1976.

Green, R. E., R. J. Hanks, and W. E. Larson, Estimates of field infiltration by numerical solution of the moisture flow equation, *Soil Sci. Soc. Am. Proc.*, 28, 15-19, 1964.

Fuentes, C., R. Haverkamp, J.-Y. Parlange, W. Brutsaert, K. Zayani, and G. Vachaud, Constraints on parameters in three soil-water capillary retention equations, *Transport in Porous Media*, 6, 445-449, 1991.

Haines, W. B., Studies in the physical properties of soil: V. The hysteresis effect in capillary properties, and the modes of moisture distribution associated therewith, *J. Agric. Sci.*, 20, 97-116, 1930.

Hanks, R. J., A. Klute, and E. Bresler, A numeric method for estimating infiltration, redistribution, drainage and evaporation of water from soil, *Water Resour. Res.*, 5, 1064-1069, 1969.

Hassanizadeh, S. M., and W. G. Gray, Thermodynamic basis of capillary pressure in porous media, *Water Resour. Res.*, 29, 3389-3405, 1993.

Haverkamp, R., M. Vauclin, and G. Vachaud, Error analysis in estimating soil water content from neutron probe measurements: I. Local standpoint, *Soil Sci.*, 137, 78-90, 1984.

Haverkamp, R., and J.-Y. Parlange, Predicting the water-retention curve from particle size distribution: 1. Sandy soils without organic matter, *Soil Sci.*, 142, 325-339, 1986.

Haverkamp, R., J. L. Arrue, and M. Soet, Soil physical properties within the root zone of the vine area of Tomelloso. Local and spatial standpoint, In *Final integrated report of EFEDA II (European Field Experiment in a Desertification Area) Spain*, Ed. J. F. Santa Olalla, CEE project n° CT920090, Brussels, chapter 3, 1997.

Haverkamp, R., C. Zammit, F. Bouraoui, K. Rajkai, J. L. Arrúe, and N. Heckmann, GRIZZLY, Grenoble Catalogue of Soils: Survey of soil field data and description of particle-size, soil water retention and hydraulic conductivity functions, Laboratoire d'Etude des Transferts en Hydrologie et Environnement (LTHE), Grenoble Cedex 9, France, 1998.

Heinen, M., and P. A. C. Raats, Hysteretic hydraulic properties of a coarse sand horticultural substrate, In *Characterization and Measurement of the Hydraulic Properties of Unsaturated Porous Media I*, Eds. M. Th. van Genuchten, F. J. Leij and L. Wu, Dept. of Environmental Sciences, University of California, Riverside, CA., 467-476, 1997.

Hoa, N. T., R. Gaudu, and C. Thirriot, Influence of hysteresis effects on transient flows in saturated-unsaturated porous media, *Water Resour. Res.*, 13, 992-996, 1977.

Hogarth, W. L., J. Hopmans, J.-Y. Parlange, and R. Haverkamp, Application of a simple soil water hysteresis model, *J. Hydrol.*, 98, 21-29, 1988.

Hopmans, J. W., and J. H. Dane, Temperature dependence of soil water retention curves, *Soil Sci. Soc. Am. J.*, 50, 562-567, 1986.

Jaynes, D. B., Comparison of soil-water hysteresis models, *J. Hydrol.*, 75, 287-299, 1985.

Johnston, J., *Econometric methods*, McGraw-Hill, New York, 1963.

Kaluarachchi, J. J., and J. C. Parker, Effects of hysteresis with air entrapment on water flow in the unsaturated, *Water Resour. Res.*, 23, 1967-1976, 1987.

Kool, J. B., and J. C. Parker, Development and evaluation of closed-form expressions for hysteretic soil hydraulic properties, *Water Resour. Res.*, 23, 105-114, 1987.

Leij, F. J., W. J. Alves, M. Th. van Genuchten, and J. R. Williams, The UNSODA - Unsaturated Soil Hydraulic Database. User's Manual Version 1.0. Report *EPA/600/R-96/095*. National Risk Management Research Laboratory, Office of Research and Development, U.S. Environmental Protection Agency, Cincinnati, Ohio 45268, pp. 1-103, 1996.

Lenhard, R. J., and J. C. Parker, A model for hysteretic constitutive relations governing multiphase flow: 2. Permeability-saturation relations, *Water Resour. Res.*, 23, 2197-2206, 1987.

Liakopoulos, A. C., Theoretical approach of the infiltration problem, *Bull. of IASH*, Vol. XII (1), 69-110, 1966.

Liu, Y., J.-Y. Parlange, and T. S. Steenhuis, A soil water hysteresis model for fingered flow data, *Water Resour. Res.*, 31, 2263-2266, 1995.

Miller, E. E., Similitude and scaling of soil-water phenomena. pp. 300-318. *In* D. Hillel (ed.) *Application of soil physics*. Academic Press, New York, U.S.A., 1980.

Mitchel, R. J., and A. S. Mayer, The significance of hysteresis in modeling solute transport in unsaturated porous media, *Soil Sci. Soc. Am. J.*, 62, 1506-1512, 1998.

Mualem, Y., A conceptual model of hysteresis, *Water Resour. Res.*, 10, 514-520, 1974.

Mualem, Y., A new model for predicting the hydraulic conductivity of unsaturated porous media, *Water Resour. Res.*, 12, 513-522, 1976a.

Mualem, Y., A catalogue of the hydraulic properties of unsaturated soils. Research Project n° 442, *Hydrodynamics and Hydraulic Laboratory, Technion, Israel Institute of Technology*, Haifa, Israel, pp. 1-100, 1976b.

Mualem, Y., and E. E. Miller, A hysteresis model based on an explicit domain-dependence function, *Soil Sci. Soc. Am. J.*, 43, 1067-1073, 1979.

Nielsen, D.R., M. T. van Genuchten, and J. W. Biggar, Water flow and solute transport processes in the unsaturated zone. *Water Resour. Res.*, 22, 89-108, 1986.

Otten, W, P. A. C. Raats, and P. Kabat, Hydraulic properties of root zone substrates used in greenhouse horticulture, In *Characterization and Measurement of the Hydraulic Properties of Unsaturated Porous Media I*, Eds. M. Th. van Genuchten, F. J. Leij and L. Wu, Dept. of Environmental Sciences, University of California, Riverside, CA., 477-488, 1997.

Parlange, J.-Y., Capillary hysteresis and relationship between drying and wetting curves, *Water Resour. Res.*, 12, 224-228, 1976.

Parker, J. C., and R. J. Lenhard, A model of hysteretic constitutive relations governing multiphase flow, 1. Saturation-pressure relations, *Water Resour. Res.*, 23, 2187-2196, 1987.

Poulovassilis, A., Hysteresis in pore water and application of the concept of independent domains, *Soil Sci.*, 93, 405-412, 1962.

Poulovassilis, A., Hysteresis in pore water in granular porous bodies. *Soil Sci.*, 109: 5-12, 1970.

Press, W. H., B. P. Flannery, S. A. Teukolsky, and W. T. Vetterling, *Numerical Recipes*, Cambridge University Press, New York, 1992.

Rogowski, A. S., Watershed physics: Model of soil moisture characteristics, *Water Resour. Res.*, 7, 1575-1582, 1971.

Russo, D., W. A. Jury, and G. L. Butters, Numerical analysis of solute transport during transient irrigation: 1. The effect of hysteresis and profile heterogeneity. *Water Resour. Res.*, 25, 2109-2118, 1989.

Scott, P. S., G. J. Farquhar, and N. Kouwen, Hysteretic effects on net infiltration, Advances in Infiltration, *Am. Soc. Agric. Eng.*, Publ. 11-83, 163-170, 1983.

Si, B. C., and R. G. Kachanoski, Unified solution for infiltration and drainage with hysteresis: Theory and field test. *Soil Sci. Soc. Am. J.*, 64, 30-36, 2000.

Sinclair, D. F., and J. Williams, Components of variance involved in estimating soil water content and water content change using a neutron moisture meter, *Aust. J. Soil Res.*, 17, 237-247, 1979.

Soil Classification System (7th Approximation), *Soil Survey and Conservation Service, U.S. Department of Agriculture*, Government Printing Office Washington 25, D.C., USA, 1960.

Talsma, T., Hysteresis in two sands and the independent domain models, *Water Resour. Res.*, 6, 964-970, 1970.

Topp, G. C., Surface tension and water contamination as related to the selection of flow system components, *Soil Sci. Soc. Am. Proc.*, 30, 128-129, 1966.

Topp, G. C., Soil-water hysteresis in a sandy loam compared with the hysteretic domain model, *Soil Sci. Soc. Am. J.*, 33, 645-651, 1969.

Topp, G. C., Soil-water hysteresis: the domain theory extended to pore interaction conditions, *Soil Sci. Soc. Am. J.*, 35, 219-225, 1971.

Touma, J., G. Vachaud, and J.-Y. Parlange, Air and water flow in a sealed, ponded vertical soil column: Experiment and model, *Soil Sci.*, 137, 4181-187, 1984.

Vachaud, G., and J.-L. Thony, Hysteresis during infiltration and redistribution in a soil column at different initial water contents, *Water Resour. Res.*, 7, 111-127, 1971.

van Genuchten, M. Th., A closed form equation for predicting the hydraulic conductivity of unsaturated soils, *Soil Sci. Soc. Am. J.*, 44, 892-898, 1980.

Vauclin, M., Effets dynamiques sur la relation succion-teneur en eau lors d'écoulements en milieu non saturé, *Thèse de Docteur-Ingénieur, Université Scientifique et Médicale de Grenoble*, France, 1971.

Whitmore, A. P., and M. Heinen, The effect of hysteresis on microbial activity in computer simulation models. *Soil Sci. Soc. Am. J.*, 63, 1101-1105, 1999.

R. Haverkamp, Laboratoire d'Etude des Transferts en Hydrologie et Environnement, LTHE (UMR 5564, CNRS, INPG, UJF, IRD), BP 53x, 38041, Grenoble, Cedex 9, France (email: randel.haverkamp@hmg.inpg.fr).

P. Reggiani, Laboratoire d'Etude des Transferts en Hydrologie et Environnement, LTHE (UMR 5564, CNRS, INPG, UJF, IRD), BP 53x, 38041, Grenoble, Cedex 9, France (email: paolo.reggiani @hmg.inpg.fr).

P. J. Ross, CSIRO Land and Water, Cunningham Laboratory, Indooroopilly, Qld. 4067, Australia (email: peter.ross@bne.clw.csiro.au).

J.-Y. Parlange, Cornell University, Department of Agricultural Engineering, Ithaca, 14853 N.Y., USA (email: jp58@cornell.edu).

How Useful are Small-Scale Soil Hydraulic Property Measurements for Large-Scale Vadose Zone Modeling?

Jan W. Hopmans and Don R. Nielsen

Hydrology, Department of Land, Air and Water Resources, University of California, Davis, California

Keith L. Bristow

CSIRO Land and Water, Townsville, Australia

A major challenge that recurs throughout the geophysical sciences is the downscaling (disaggregation) and upscaling (aggregation) of flow or transport processes and their measurement across a range of spatial or temporal scales. Such needs arise, for example, when field-scale behavior must be determined from soil hydraulic data collected from a limited number of in situ field measurements or analysis of small soil cores in the laboratory. The scaling problem cannot be solved by simple consideration of the differences in space or time scale, for several reasons. First, spatial and temporal variability in soil properties create uncertainties when changing between scales. Second, flow and transport processes in geophysics and vadose zone hydrology are highly nonlinear. We present a historical overview of the theory of scaling procedures, and demonstrate the application of various aggregation techniques, such as scaling and inverse modeling, to aggregate laboratory-scale soil hydraulic properties to larger scale effective soil hydraulic properties. Examples of application of these aggregation techniques from the pore scale to the watershed scale are demonstrated. We conclude that the development of new instrumentation to characterize soil properties and their variation across spatial scales is crucial. Moreover, the inherent complexity of flow in heterogeneous soils, or soil-like materials, and the need to integrate theory with experiment, requires innovative and multidisciplinary research efforts to overcome limitations imposed by current understanding of scale-dependent soil flow and transport processes.

INTRODUCTION

For the past few decades, soil scientists have applied soil hydraulic data to characterize flow and transport processes in large-scale heterogeneous vadose zones, using measurement scales that are typically much smaller. For example, prediction of soil-water dynamics at the field-scale is derived from the measurement of soil hydraulic properties from laboratory cores, collected from a limited number of sampling sites across large spatial extents, often using large sampling spacings. Typically, the measurement scale for soil hydraulic characterization is in the order of 10 cm, with a sample spacing of 100 m or larger.

This scale-transfer question is being asked more frequently than ever, mostly because of water quality issues resulting from chemical contamination of soil, groundwater, and surface water systems worldwide.

Environmental Mechanics: Water, Mass and Energy Transfer in the Biosphere
Geophysical Monograph 129
Copyright 2002 by the American Geophysical Union
10.1029/129GM20

Appropriate answers are expected, requiring the estimation of appropriate soil hydraulic parameters for use in describing the behavior of pollutant plumes at field or landscape scales. In 1991, the U.S. National Research Council [*NRC*, 1991] identified the scaling of dynamic nonlinear behavior of hydrologic processes as one of the priority research areas that offer the greatest expected contribution to a more complete understanding of hydrologic sciences. Simultaneously, with the increasing awareness of the crucial role of the soil and the vadose zone in the management of chemical loadings to groundwater and surface waters, soil hydraulic characterization is needed to predict transport and fate of agricultural and industrial chemicals at the regional or landscape scale. Routine measurement of soil hydraulic properties are usually conducted in the laboratory [*Dane and Hopmans*, 2002], using small-size soil cores, collected in-situ from a few representative or many random locations within the region of interest, or is done in the field on small field plots at the meter-scale [*Green et al.*, 1986] using either direct or inverse methods [*Hopmans et al.*, 2002]. Soil parameters obtained from cm-scale measurements (laboratory scale) are included in numerical models with a grid or element size ten times as large or larger, with the numerical results extrapolated to field-scale conditions. A critical analysis of the assumptions made when applying small-scale (subgrid) parameters to the model grid scale, was presented by *Beven* [1989].

Irrespective of scale, transient isothermal unsaturated water flow in non-swelling soils is described by the so-called Richards' equation,

$$C(h_m)\frac{\partial h_m}{\partial t} = \nabla \cdot \left[\mathbf{K}(h_m)\nabla \left(h_m - z \right) \right] \qquad (1)$$

which provides the soil water matric potential (h_m), water content (θ) and water flux density as a function of time and space, using one-, two-, or three-dimensional flow models (e.g.,*Šimůnek et al.*, 1995]. In (1), $\mathbf{K}$ is the unsaturated hydraulic conductivity tensor (L T^{-1}), and z denotes the gravitational head (L) to be included for the vertical flow component only. The function C(h_m) in (1) is the so-called soil water capacity, and represents the slope of the soil water retention curve. Both the soil water retention and unsaturated hydraulic conductivity functions are highly nonlinear, with both h_m and K varying many orders of magnitude over the water content range of significant water flow. These nonlinearities make the application of (1) across spatial scales inherently problematic. Specifically, the averaging of processes determined from discrete small-scale samples may not describe the true soil behavior involving larger spatial structures. Moreover, the dominant physical flow processes may vary between

spatial scales. For example, *Dooge* [1997] discussed the hypothesis that the mathematical model represented by Eq (1) may not be applicable to describe unsaturated water flow at the watershed scale. Because their measurement is time-consuming, the number of measured hydraulic data is usually limited, and is usually far less than statistically required to fully characterize soil heterogeneity. As a result, data assimilation techniques, such as linear regression analysis, pedotransfer functions and neural networks [*Pachepski et al.*, 1999, *Schaap et al.*, 1998] have been developed to derive soil hydraulic functions from other, easier-to-obtain soil properties. Considering that soil hydraulic measurements are typically conducted for small measurement volumes and that the natural variability of soils is enormous, the main question asked, is how small-scale measurements can provide information about large-scale flow and transport behavior [*Gelhar*, 1986].

Field experiments have confirmed that soil heterogeneity controls flow and transport, including preferential flow. Initial attempts of the prediction of large-scale flow problems used deterministic modeling. Although studies such as those of *Hills et al.* [1991] showed a qualitatively acceptable comparison between field-measured and predicted water contents using a deterministic approach, other studies have shown the need for either distributed physically-based modeling [*Loague and Kyriakidis*, 1997] or stochastic modeling [*Famiglietti and Wood*, 1994] at the watershed scale, mostly because it will require an enormous amount of data to accurately represent the multi-dimensional soil heterogeneity. Similarly, stochastic approaches have been developed to characterize field-scale soil water flow, e.g. by using scaling and Monte-Carlo analysis [*Hopmans et al.*, 1988; *Hopmans and Stricker*, 1989], stochastic modeling [*Mantoglou and Gelhar*, 1987] and geostatistical methods [*Yeh and Zhang*, 1996; *Rockhold*, 1999].

Alternatively, the conceptual characterization of the flow system may be simplified by modeling the key flow mechanisms for representative elementary areas (REA) only, and for which effective hydrological parameters can be defined [*Duffy*, 1996; *Famiglietti and Wood*, 1994], assuming that variability within a REA is statistically homogeneous. Although partially successful in surface hydrology, it has been determined that the size of the REA is event-dependent, controlled by initial conditions and rainfall intensity [*Blöschl et al.*, 1995; *Grayson et al.*, 1997]. Surface hydrological modeling at watershed scales has also demonstrated that the spatial distribution of hydrological processes is controlled by the spatial organization of key soil properties (*Blöschl and Sivapalan*, 1995; *Merz and Plate*, 1997]. The need to incorporate the spatial organization of these key soil properties, such as the soil hydraulic functions, is also recognized in soil science. Specifically, we refer to the treatise by *Roth et al.* [1999], outlining a conceptualization of the control of soil heterogeneity on soil flow and transport processes, using the so-called scaleway approach. In this approach, the soil is conceptualized by a hierarchical heterogeneous medium with discrete spatial scales, that each may require different effective process models with distinct effective material properties.

Assuming that soils are statistically heterogeneous, *Freeze* [1975] questioned the existence of a uniquely defined equivalent saturated hydraulic conductivity, and concluded that its value is likely a function of the boundary conditions and soil geometry. In his review of some basic issues of the consequences of natural soil variability for prediction of soil hydraulic properties, *Philip* [1980] extended the work of *Freeze* [1975] to unsaturated, scale-heterogeneous soils, to hypothesize that the field application of theory of heterogeneous soil systems might lead to "trans-science." Although his conclusion appears bleak at first reading, Philip ends this remarkably insightful paper with a plea to 'not abandon the task of seeking to understand as much about these systems as we possibly can.' It is in this spirit that many soil scientists have continued their analysis of theories, both deterministic and stochastic, and soils, both at the field and landscape scale, to increasingly elucidate the control of small-scale processes on larger-scale flow behavior.

The following was partly presented at the joint 'Soil Science Society of America-German Soil Science Society' meeting in Osnabrueck [*Hopmans and Bristow*, 2000], and specifically asks whether routine soil hydraulic measurements conducted in the laboratory can successfully be applied to larger-scale flow and transport problems.

We will make an attempt to answer this question, while adopting the relativist concept of *Baveye and Sposito* [1984], assuming that soil property values are dependent on the scale of the measurement that can vary between soil properties. We also agree with *Baveye and Sposito* [1984] and *Beven* [2001] that the physical laws of the model scale must be consistent with the measurement scale. Hence, laboratory-measured soil hydraulic properties are appropriate input for laboratory-scale columns studies and simulation models. When applied to larger spatial scales, we offer two alternatives. First, the laboratory-scale soil hydraulic properties may be spatially distributed across the larger spatial scale of interest, assuming that the integrated flow behavior can be determined from aggregation of many individual, soil column-like flow processes.

Second, laboratory-measured soil hydraulic functions can be used as initial estimates, and improved by using inverse modeling (IM), conditioned by scale-appropriate boundary conditions and flow measurements. Hence, we concur that prediction of hydraulic behavior of heterogeneous soils is likely impossible from the a priori knowledge of the homogeneous soil components that make up the heterogeneous soil, and that its estimation can only be accomplished using scale-appropriate measurements.

SCALE-DEPENDENCY OF SOIL PROPERTIES AND PROCESSES

Upscaling requires integration and aggregation of spatial information into larger spatial units, e.g., as in the estimation of an effective field soil water retention or conductivity curve from small-scale laboratory core measurements. As clearly pointed out by *Baveye and Boast* [1999], *Darcy*'s experiment [1856] can in effect be interpreted as yielding an upscaled, effective saturated hydraulic conductivity. In contrast, the downscaling requires the disaggregating of scale information to smaller scales, e.g., by discerning the contribution of different soil structures to the effective soil hydraulic conductivity function (*Kasteel et al.*, 2000). In a statistical sense, one may refer to scale as the spatial correlation length or integral scale of the measurement, property, or process [*Dagan*, 1986]. In his classic treatise, Dagan distinguishes between the laboratory core scale, local scale, and regional scale. He defines the laboratory scale as equal to the representative elementary volume (REV), for which the mean is a constant deterministic quantity, and the variance approaches zero [*Bear,* 1972]. It is this scale, for which the Darcy equation can be used as the equation of motion, as derived from the volume averaging of the Stokes equations [*Whitaker*, 1986]. At the next, larger scale, *Dagan* [1986] defines the local scale, where the soil is heterogeneous, but stationary in the mean and variance. We interpret the regional scale as the spatial dimension at which the relevant soil properties become nonstationary.

Nonstationary Soil Properties

When increasing spatial scales, soil properties typically become nonstationary [*Russo and Jury*, 1987a], as evidenced by the delineation of soil map units in a soil survey. Much of the early soil spatial analysis, specifically geostatistics, was based on the intrinsic hypothesis of stationarity (stationarity of spatial differences), however, it is questionable whether this stationary model is realistic [*Webster*, 2000].

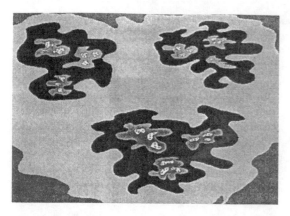

Figure 1. Conceptual model of evolving hetereogeneity [after Wheatcraft and Tyler, 1988].

Nevertheless, as was pointed out by *Kavvas* [1999], the averaging of hydrologic observations or aliasing at a larger observation scale may remove nonstationary trends at the smaller observation scale. Specifically, as one moves through a hierarchical sequence of increasing sampling scales, nonstationarities at smaller spatial scales may be eliminated. An ideal instrument, such as the mathematical tool by *Cushman* [1984], will work similarly and filter out the high frequency variability component of a spatial signal to yield a scale-specific measurement. The remaining nonstationary trends of the natural variability or process scales [*Blöschl and Sivapalan,* 1995] may be determined from the power spectrum, covariance analysis, and by wavelet analysis [*Lark and Webster,* 1999].

As an example, a model of evolving heterogeneity [*Sposito,* 1986], assuming fractal heterogeneous soil properties, was presented by *Wheatcraft and Tyler* [1986], showing a pattern of heterogeneity that is scale-independent across a large range of scales (Figure 1).

Natural patterns of soil variability may show embedded, organizational structures as in Figure 1, that are not necessarily fractal, but that lead to nonstationary soil properties or processes. However, as pointed out by *Cushman* [1990], spatial patterns of soil properties within and between scales (structural hierarchy) might be different from the organization of the soil hydrological processes (functional hierarchy) across spatial scales. As different flow processes may be dominant at each scale, different mathematical relationships may be required to describe the underpinning physical process at each scale [*Klemeš,* 1983]. As one moves towards a larger spatial scale, soil properties may change from deterministic to random, with the smaller-scale variations filtered out by the larger-scale process, thereby eliminating nonstationary trends at the smaller spatial scales.

The spatial organization and its evolvement across spatial scales can be defined in various ways. The example of Figure 1 shows a discrete and a continuous variation pattern between and within the main spatial units, respectively. As noted by *Gelhar* [1986], at each field-of-view, the large-scale variation, which causes the nonstationairity of the specific soil variable or process, can be regarded as deterministic, whereas the smaller scale variations within each main unit can be treated stochastically. Alternatively, one can define the types of variability as ordered (between main units) and disordered (within main units), or as macroscopic and microscopic variations. Most appropriately, the hierarchical heterogeneous soil medium can be described by the structural and textural definitions or the scaleway approach of *Roth et al.* [1999], with the structural elements describing the dominating soil patterns that affect the physical mechanisms operating at the a priori defined field-of-view. In contrast, the textural patterns within the structural units are merely perturbations of the main processes, and can be described statistically. Thus, when characterizing soil hydraulic variability for the prediction of soil hydrologic processes, it is assumed that the occurrence and location of these structural elements are dominating, and must be accurately determined. We conclude that stationarity of a soil hydrological process or parameter is dependent on the scale of observation.

Analysis of process scales

Scale-dependent, nonstationary processes exhibit statistical properties that are different than what is usually assumed in geostatistical analysis. Specifically, the REV [*Bear,* 1972] cannot be defined, as the soil property changes value, when increasing the scale of observation. Moreover, the spatial correlation structure of nonstationary fields will depend on the spatial extent or sampling area of the data, resulting in variograms with multiple sills (see Figure 2) that occur at correlation lengths of the multiple process scales [*Gelhar,* 1986]. In addition, *Rodriquez-Iturbe et al.* [1995] demonstrated that nonstationary properties will show a power law decay of the variance, resulting in a linear relation between variance and observation scale, when plotted on a log scale.

The power spectrum is determined from the Fourier transform of the autocovariance functions and represents the partitioning of the sample variance into spatial frequency components [*Greminger et al.,* 1985]. Process scales occur at spectral peaks, whereas spectral gaps represent spatial scales with minimum spectral variance. An example of a hypothetical power spectrum

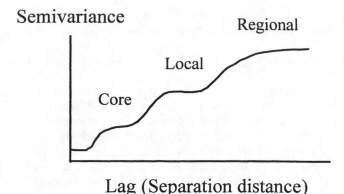

Figure 2. Hypothetical variogram for scale-dependent hydraulic conductivity. Adapted from *Gelhar*, [1986].

is demonstrated in Figure 3, with a small-scale (core scale) component, superimposed on two large-scale components (local and regional scale).

The Nyquist frequency of the power spectrum determines that the smallest process scale that can be examined is twice the sampling scale [*Cushman*, 1984]. In other words, if the sampling distance is d (spatial frequency is 1/d), then no fluctuations in processes with size smaller than 2d (or larger than 1/2d) can be observed. Using measurement scales larger than defined by the Nyquist frequency would merely show noise, rather than describing spatial trends. Analogously, *Russo and Jury* [1987b] demonstrated for stationary fields that the correlation length of the process scale could only be accurately estimated if the sampling distance is smaller than half of the range of the underlying process. This constraint is usually not an issue for most soil hydraulic measurements, as the soil sampling scale (about 10 cm) is usually smaller than the natural process scale.

According to the scaleway or nested approach of Roth et al. [1999], subsequent aggregation of information and the modeling of flow and transport at one specific scale, provides the required information at the next, larger scale level. However, rather than implying that this type of analysis is needed across many spatial scales, we argue that likely only two scale levels need to be considered within the spatial domain of interest. For example, if the scale of interest is an agricultural field, one defines the structural elements based on the dominant physical mechanism that causes the major differences in soil water regime between the structural units. Most recently, *Becker and Braun* [1999] defined these units as hydrotopes or hydrological response units, based on differences between vegetation types, shallow groundwater presence, soil type or hillslope. *Wood* [1995] and *Famiglietti and Wood* [1994] described the

aggregated watershed response by aggregation of total watershed runoff using area-weighted average runoff values of REA's with different topography indices. Likewise, in his review on scale issues in hydrological models, Beven [1995] introduced the simple patch model for scale-dependent modeling, with a patch defined as any area of the landscape that has broadly similar hydrological response in terms of the quantities of interest. In soil hydrological studies, soil map units may define the structural units across the landscape [*Ferguson and Hergert*, 1999] or may be indicative of geologic hydrofacies as identified using the transition probability geostatistical method [*Weizmann et al.*, 1999]. The smaller spatial scale level of the textural information within structural units is distributed either deterministically or stochastically, e.g. using scaling of soil hydraulic properties from laboratory soil cores [*Hopmans and Stricker*, 1989]. The upscaling from the textural to the structural scale level may result in effective, scale-appropriate soil hydraulic functions that may differ in form and parameter values between scales, but serve a similar function across scales. The subsequent distribution of the structural units is deterministic (distributed modeling) and their aggregation to the scale of interest may be possible by simple mass conservation principles, e.g by the fractional area approach (addition or averaging). It is important to realize that the spatial organization of structures might be caused by different soil processes at different spatial scales.

SCALING ACROSS SPATIAL SCALES

Whereas we have presented a general framework to measure and model flow and transport across spatial scales, various mathematical and analytical tools

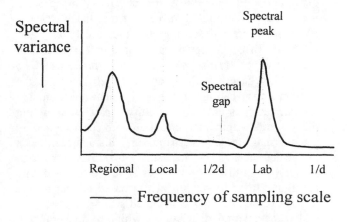

Figure 3. Schematic presentation of power spectrum, showing various process scales (d = sampling spacing).

are needed to aggregate soil hydraulic information across spatial scales. Specifically, we review scaling and inverse modeling.

Scaling and Monte Carlo Analysis

Most of the uncertainty in the assessment of water flow in unsaturated soils at the field scale can be attributed to soil spatial variability caused by soil heterogeneity. The knowledge of constitutive relationships for the unsaturated hydraulic conductivity, water saturation, and soil water matric potential are essential in using (1). The exact nature of the functional dependence of these flow variables with water content differs among soil types with different particle size compositions and pore size geometry within a heterogeneous field soil. The scaling approach has been extensively used to characterize soil hydraulic spatial variability and to develop a standard methodology to assess the variability of soil hydraulic functions and their parameters. The single objective of scaling is to coalesce a set of functional relationships into a single curve using scaling factors that describe the set as a whole (e.g. structural unit). The concept of this approach has been developed principally from the theory of microscopic geometric similitude as proposed by *Miller and Miller* [1956]. The procedure consists of using scaling factors to relate the hydraulic properties in a given location to the mean properties at an arbitrary reference point. *Philip* [1967] designated this type of variability scale-heterogeneity, emphasizing that the spatial variation of soil properties is fully embodied in the spatial variability of the scaling factor. Instead of using pore radius as the microscopic characteristic length, similarity of pore size distribution [*Kosugi and Hopmans*, 1998] was used to scale soil water retention curves for soils that exhibit a lognormal pore-size distribution. In this study, the physically based scale factors were computed directly from the physically based parameters describing the individual soil water retention [*Kosugi*, 1996] and unsaturated hydraulic conductivity [*Tuli et al.*, 2001] functions. The physically based scaling concept provides for the simultaneous scaling of the soil water retention and unsaturated hydraulic conductivity functions, assuming that all soils within a structural unit are characterized by a lognormal pore-size distribution. This approach leads to scaled-mean soil hydraulic functions for each structural unit that may serve as effective soil hydraulic functions. In addition, physically based scaling results in a set of lognormally distributed scaling factors, from which the textural distribution within a structural unit can be characterized. Using Monte Carlo analysis, stochastic soil water flow modeling can be conducted,

with scaling factors generated from a known probability density function [*Hopmans and Stricker*, 1989].

Inverse Modeling

The inverse method offers a powerful procedure to estimate flow properties across spatial and temporal scales. As numerical models have become increasingly sophisticated and powerful, inverse methods are applicable to laboratory and field data, no longer limited by the physical dimensions of the soil domain, or type of imposed boundary conditions. Inverse methods might be especially appropriate for estimating regional-scale effective soil hydraulic parameters, from boundary condition measurements. For example, *Eching et al.* [1994] estimated field-representative hydraulic functions using inverse modeling of Eq. (1) with field drainage flow rate serving as the lower boundary condition for the Richards' flow equation applied at the field-scale. The application of inverse modeling to estimate soil hydraulic functions for laboratory soil cores has been extensively reviewed by *Hopmans et al.* [2002].

The inverse modeling approach mandates the combination of experimentation with numerical modeling. Since the optimized hydraulic functions are needed as input to numerical flow and transport models for prediction purposes, it is an added advantage that the hydraulic parameters are estimated using similar numerical models as used for predictive forward modeling, with similar grid sizes so that the estimated effective hydraulic properties include the within grid integration of real soil variability. Although application of inverse methodology may suffer from non-uniqueness (e.g., *Beven*, 2001), the application of inverse methods in general to estimate soil hydraulic functions across spatial scales is very promising. This technique has demonstrated potential as an excellent new tool for a wide spectrum of transient laboratory and field experiments, yielding effective or lumped hydraulic properties that pertain to the scale of interest. We will demonstrate various applications in the following examples.

EXAMPLES OF SCALING APPLICATONS ACROSS SPATIAL SCALES

We demonstrate the application of the various aggregation techniques, to estimate effective large-scale soil hydraulic properties from small-scale laboratory measurements on soil cores. We start with the measurement of REV of porosity at the pore-scale, and present scaling applications at the soil core, field plot, field, and watershed scale, respectively. As the

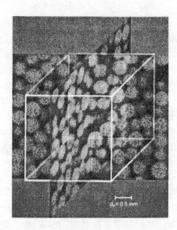

Figure 4a. Three-dimensional image of dry glass beads (light gray) and pore space (dark grey) [after *Clausnitzer and Hopmans*, 1999].

examples will show, the conceptualization of separating soil heterogeneity into textural and structural elements allows the integration of small-scale soil hydraulic properties to larger spatial scales, possibly resulting in scale-dependent soil hydraulic properties.

Pore Scale

Although (1) is not applicable at the pore scale, this example is shown to demonstrate the existence of a REV for porosity, for the first time as we know [*Clausnitzer and Hopmans*, 1999], using x-ray computed tomography (CT). Using the three-dimensional spatial distribution of x-ray attenuation as a proxy, porosity measurements for a glass bead medium were conducted for increasing measurement volumes. X-ray CT measurements were conducted in a random pack of uniform glass beads within a vertical Plexiglas cylinder of 4.76 mm inner diameter. The bead diameter, d_p, was 0.5 mm and the spatial resolution was 18.4 micrometer, resulting in $(18.4 \ \mu m)^3$ voxel volumes (see Fig. 4a). In this example, the single structural unit is represented by the glass beads pack, and textural variations are defined by porosity changes at a measurement scale larger than the REV. Starting from the original three-dimensional data set of attenuation values, increasingly larger volumes were extracted, all centered at the same location, beginning with 8x8x8 voxels and incrementing the cube side length, L, of the averaging volume by 4 voxel lengths (0.0736 mm) in each step. The sequence of porosity calculations with increasing volume size was conducted twice, first with the initial 8x8x8 averaging volume centered in the air phase, and subsequently with the averaging volume centered in the glass phase. The resulting curves are

presented in Fig. 4b, suggesting a REV of about 3 to 5 times the bead diameter. In an independent, modeling study, *Zhang et al.*, [2000] showed that the REV may depend on the quantity being represented, as suggested by *Baveye and Sposito* [1984]. Thus, the REV for porosity may be different than for the Darcy scale at which (1) may be applicable, and for which soil hydraulic properties can be defined.

Soil Core Scale

Both *Roth et al.* [1999] and *Kasteel et al.* [2000] have shown that the spatial structure of soil hydraulic properties at the core-scale must be known, to accurately predict solute transport through the soil core. Using soil bulk density, as measured by x-ray CT, to proxy for soil hydraulic properties, two distinct soils were characterized within a 16-cm diameter soil core. The resulting image of the dense (light) and less dense soil matrix (dark) is shown in Figure 5, delineating the high and low-conductive soil materials or structural elements within the core. In this example, no textural variations within the structural elements were assumed.

The hydraulic properties of the more-conductive soil material were determined using a network model from independently measured pore geometry. Assuming a

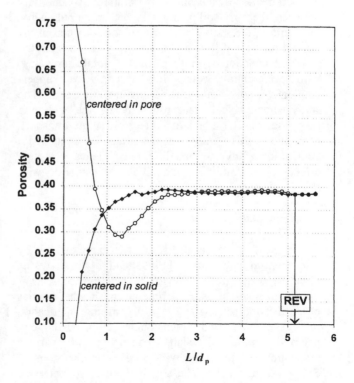

Figure 4b. Estimated porosity for a cubic domain with increasing size within the glass-bead pack [after *Clausnitzer and Hopmans*, 1999].

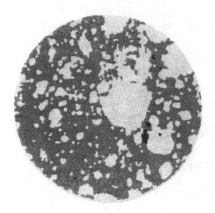

Figure 5. Illustration of two soil density classes in one specific cross-section of a 16-cm soil core [after *Kasteel et al.*, 2000].

trial value for the conductivity of the low-conductivity structure, an effective saturated conductivity for the whole soil core was estimated from a composite conductivity, with weighting factors determined by the volume fraction of each soil material. Subsequently, it was demonstrated using a three-dimensional flow and transport code, that the simulated breakthrough of a chloride solution in the unsaturated soil core could be matched reasonably well with breakthrough measurements, if the ratio of saturated conductivity between the two soil materials was optimized, while maintaining the spatial structure of soil variation in the three-dimensional flow and transport code. Specifically, different conductivity ratios affected solute spreading, with preferential water flow through the higher-conductive soil structure dominating transport. The study showed that nonstationarity of the hydraulic properties can have a large effect on solute transport. Moreover, this study demonstrated a successful application of inverse modeling, to estimate a core-scale effective hydraulic conductivity functions with structural contraints.

Field Plot and Field Scale

The next larger scale level was investigated by *Wildenschild and Jensen* [1999a and b] to study the effective water flow behavior in heterogeneous, two-dimensional soil slabs. Experiments consisted of a series of infiltration experiments with varying application rates, in a two-dimensional 100x100x3 cm, heterogeneous soil slab, consisting of various realizations of packings with 5 different sands, using 5x5x8 cm unit cells. The soil hydraulic properties for each sand type were measured in the laboratory first, and represented the structural units. The soil tank was

instrumented with strategically placed tensiometers and TDR probes, to estimate local soil hydraulic properties and their spatial distribution, from measurements during various steady state flow regimes, with water flux rates determined by a rain application device. After incorporating the distributed soil heterogeneity deterministically, using the laboratory-measured soil hydraulic functions for the individual sand types, a two-dimensional flow and transport model [*Simunek et al*, 1999] was able to predict the measured spatial variability of soil water flow in the soil slab. In addition, effective soil hydraulic properties for the whole slab were determined using simple statistical averages (geometric and arithmetic), as well as by inverse modeling using the measurements of water content and matric potential in the transient stages between the steady state experiments. In either case, effective hydraulic properties were able to describe the average transient soil water behavior for the heterogeneous soil system, as determined from two-dimensional transient water flow modeling.

In the field experiment by *de Vos et al.* [2000], nine soil horizons were classified into four different hydrologic zones, each determined by different soil hydraulic functions, representing the functional soil structural elements. The tile-drained field was 62.5 by 12 m with a center drain at about the 1 m soil depth. Soil water matric potential, groundwater level, piezometric heads, at various locations within the experimental field, and field discharge rate and nitrate concentrations were measured during the 1991-92 leaching period. Soil water retention, saturated and unsaturated hydraulic conductivity data for the 4 characteristic zones were measured from laboratory soil cores. The HYDRUS-2D model [*Simunek et al.*, 1999] was used to simulate the two-dimensional flow regime and nitrate transport in the field, and the drainage rate and nitrate concentration in the drain outlet. Field-effective soil water retention and hydraulic conductivity functions were estimated using an inverse modeling approach, by adjusting the hydraulic parameters that were measured from the laboratory soil cores. Regarding the calibration of the field-representative hydraulic conductivity function, the saturated hydraulic conductivity for each of the four functional soil layers was adjusted, so that the simulated groundwater level-drainage rate relationship matched the measured data, using the constraint that effective saturated hydraulic conductivity values were within their laboratory-measured ranges. Laboratory-measured soil water retention curves were adjusted to match simulated with measured groundwater level and drainage rate during the monitoring period.

Watershed scale

At the watershed scale, *Hopmans and Stricker* [1989] used a stochastic-deterministic model to simulate soil water flow in the spatially heterogeneous Hupsel watershed. Using various laboratory techniques, soil water retention and unsaturated hydraulic conductivity functions were measured for each of 3 hydrologically-distinct soil layers that were widely present in the 650 ha watershed. Simultaneous scaling was used to model the spatial variability of the soil hydraulic data, yielding reference curves and a set of scaling factors for each of the 3 identified soil layers. The objective of this experimental study was to quantify the impact of soil spatial variability on the water balance of the watershed.

The spatial organization of the various soil types was selected based on the starting depth of a clay layer, since it largely controlled spatial variations in groundwater level and drainage rate within the watershed. The influence of small-scale local variations in soil hydraulic properties on water flow within each structural unit was simulated stochastically, using Monte-Carlo simulations, from random generation of scale factors for each of the 3 distinct soil layers. Computer simulations with SWAP [*van Dam et al.*, 1997] were conducted for a dry (1976) and a wet year (1982), yielding mean and variance of evapotranspiration, groundwater level, and drain discharge for each structural unit. Subsequently, the same hydrological variables were either simply averaged or added, to yield watershed-representative values. Without any further calibration, independently-measured and simulated groundwater level and watershed discharge were close for both years. This last example shows how the proposed deterministic-stochastic approach using the structure-texture concept was successfully applied at the watershed scale, using laboratory-core soil hydraulic functions.

Using the laboratory-measured soil hydraulic functions as a starting point, *Feddes et al.* [1993a] subsequently demonstrated that almost equally good agreement was found, by using a single set of effective soil hydraulic functions, representative for the whole watershed. Their numerical exercise demonstrated that an area-average, effective parameterization of the soil hydraulic functions can be applied to (1). Results obtained by using the scaled reference hydraulic functions were almost as close as using an inverse approach, by optimization of the hydraulic functions via minimization of the residuals between measured and simulated soil hydrological variables. In a later study, *Kabat et al.* [1997] concluded that effective soil hydraulic properties could successfully describe area-

average evaporative and soil moisture fluxes at the 10-100 km^2 scale, provided that the averaged area contained a single soil type only. This was concluded with the understanding that the estimated effective properties are merely <u>calibration</u> parameters, which do not necessarily have the physical meaning implied by application of the Darcy flow equation.

CONCLUDING REMARKS

Although most of the presented examples show that some kind of fitting is needed along the way, the estimated soil hydraulic properties using small-scale laboratory soil cores can be effective in estimating large-scale, effective soil hydraulic properties. We are also convinced that significant progress in the understanding of fundamental flow processes in heterogeneous soils is possible only if scale-appropriate measurement technologies are available. Innovative examples of such instruments that are explored to characterize subsurface flows across spatial scales include the application of noninvasive techniques [*Hopmans et al.*, 1999], such as x-ray tomography, electromagnetic induction, electrical resistivity, seismic reflection, and microwave remote sensing [*Jackson et al.*, 1999; *Hollenbeck et al.*, 1996; *Mattikalli et al.*, 1998]. Present theory and applications of remote sensing may potentially help improve the understanding of large-scale hydrological processes such as runoff, infiltration and evapotranspiration, including their spatial distribution and scale-dependency. The monitoring of transient soil moisture changes by remote sensing may provide the essential information to estimate up-scaled soil hydraulic parameters such as the saturated hydraulic conductivity or unsaturated hydraulic parameters, using the inverse modeling approach. An excellent example of such an application was presented by *Feddes et al.* (1993b), who demonstrated that remote sensing of soil surface temperature and soil moisture combined may provide the essential information to estimate effective soil hydraulic parameters at the catchment scale. The work of *Ahuja et al.* (1993) support this potential application of remote sensing, and showed that spatial variations in surface soil moisture can be related to spatial variations in effective values of soil profile saturated hydraulic conductivity. In their review of scaling field soil-water behavior, *Nielsen et al.* [1998] suggested that increased efforts to measure field-based soil hydraulic data are needed to extend the application of (1) to the landscape-scale.

In his analysis, *Philip* [1980] used the analytical solution of a simple one-dimensional sorptivity experiment to determine whether a sample-mean sorptivity value could be predicted from sorptivity values of the individual soil components that made up a deterministic heterogeneous soil. His results indicated that the averaging of spatially-variable soil parameters does not necessarily result in an average soil water flow behavior. Even now, after a further two decades of dedicated research in soil physics and vadose hydrology, we must agree with *Philip*'s [1980] final statement, 'that our adventures into trans-science will be least likely to lead to disaster if we are as well informed as possible about stochastic heterogeneous systems.' Hence, we conclude that the development of new instrumentation to characterize soil properties and their variation across spatial scales is crucial. Moreover, the inherent complexity of flow in heterogeneous soils or soil-like materials and the need to integrate theory with experiment, requires innovative and multidisciplinary research efforts to break the deadlock, imposed by current understanding of scale-dependent soil flow and transport processes.

Acknowledgements. The first author was awarded a fellowship by the Land and Water Resources Research and Development Corporation (LWRRDC) and CSIRO Land and Water, Davies Laboratory in Townsville, Australia, to support his sabbatical leave in the Davies Laboratory. We thank the Soil Science Society of America for supporting the joint SSSA-DBG Conference in Osnabrueck, Germany, where part of this review was presented. Finally, the in-depth discussions with Gerrit Schoups (Hydrologic Sciences Graduate Group at University of California, Davis) helped in shaping the final version of the manuscript.

REFERENCES

Ahuja, L.R., O. Wendroth, and D.R. Nielsen, Relationship between initial drainage of surface soil and average profile saturated hydraulic conductivity, *Soil Sci. Soc. Am. J*, 57, 19-25, 1993.

Baveye, P., and G. Sposito, The operational significance of the continuum hypothesis in the theory of water movement through soils and aquifers, *Water Resour. Res.*, 20, 521-530, 1984.

Baveye, P, and C.W. Boast, Physical scales and spatial predictability of transport processes in the environment, in *Assessment of Non-Point Source Pollution in the Vadose Zone*, edited by D.L. Corwin, K. Loague, and T.R. Ellsworth, p. 261-280, Geological Monograph Series 108, American Geophysical Union,Washington, D.C., 1999.

Bear, J., *Dynamics of Fluids in Porous Media*, Dover Publications, Inc.,New York, 1972, 764 pp.

Becker, A., and P. Braun, Disaggregation, aggregation and spatial scaling in hydrological modeling, *J. of Hydrology*, 217, 239-252, 1999.

Beven, K., Changing ideas in hydrology – The case of physically-based models, *J. of Hydrology*, 105, 157-172, 1989.

Beven, K., Linking parameters across scales: subgrid parameterizations and scale dependent hydrological models, *Hydrological Processes*, 9, 507-525, 1995.

Beven, K., How far can we go in distributed hydrological modeling, *Hydrology and Earth System Sciences*, 5, 1-12, 2001.

Blöschl, G., and M. Sivapalan. Scale issues in hydrological modeling: A review, *Hydrological processes*, 9, 251-290, 1995.

Blöschl, G., R.B.Grayson, and M. Sivapalan. On the representative elementary area (REA) concept and its utility for distributed rainfall-runoff modeling, 1995, *Hydrological processes*, 9, 313-330, 1995.

Clausnitzer, V., and J.W. Hopmans, Determination of phase-volume fractions from tomographic measurements in two-phase systems, *Adv Water Resour.*, 22(6), 577-584, 1999.

Cushman, J.H., On unifying the concepts of scale, instrumentation, and stochastics in the development of multiphase transport theory, *Water Resour. Res.*, 20(11), 1668-1676, 1984.

Cushman, J.H., An introduction to hierarchical porous media, in *Dynamics of Fluids in Hierarchical Porous Media*, edited by J.H. Cushman, p 1-6, Academic Press, San Diego, CA, 1990.

Dagan, G., Statistical theory of groundwater flow and transport: pore to laboratory, laboratory to formation, and formation to regional scale, *Water Resour Res.*, 22(9), 120S-134S, 1986.

Dane, J.H., and J.W. Hopmans, Water Retention and Storage, Chapter 3.3.1 Introduction, in *Methods of Soil Analysis, Part 1, Physical Methods*, edited by J.H. Dane and G.C.Topp, Agronomy 9, 2001; Madison, WI, 2002.

Darcy, H., Les fontaines publiques de la ville de Dijon, Victor Valmont, Paris, France, 1856.

De Vos, J.A., D. Hesterberg, and P.A.C. Raats, Nitrate leaching in a tile-drained silt lam soil, *Soil Sci. Soc. Am. J.*, 64, 517-527, 2000.

Dooge, J.C.I., Scale issues in Hydrology, In *Reflections on Hydrology, Science and Practice*, Edited by N. Buras, , pp 85-143, American Geophysical Union, Washington D.C., 1997.

Duffy. C.J., A two-state integral-balance model for soil moisture and groundwater dynamics in complex terrain, *Water Resour. Res.*, 32, 2421-2434, 1996.

Eching, S.O., J.W. Hopmans, and W.W. Wallender. Estimation of in situ unsaturated soil hydraulic functions from scaled cumulative drainage data. *Water Resour. Res.*, 30, 2387-2394, 1994a.

Famiglietti, J.S. and E.F. Wood, Multiscale modeling of spatially variable water and energy balance processes, *Water Resour. Res.*, 30, 3061-3078, 1994.

Feddes, R.A, G.H.de Rooij, G.H., J.C. van Dam, P. Kabat, P., Droogers, and J.N.M. Stricker, Estimation of regional effective soil hydraulic parameters by inverse modeling, In, D. Russo an G. Dagan (Editors), *Water Flow and Solute Transport in Soils*, Advanced Series in Agricultural Sciences, Springer-Verlag, NY, pp. 211-231, 1993a.

Feddes, R.A., M. Menenti, P. Kabat and W.G.M. Bastiaanssen, Is large scale inverse modeling of unsaturated flow with areal average evaporation and surface soil moisture as estimated from remote sensing feasible, *J. of Hydrology*, 143, 125-152, 1993b.

Ferguson, R.B., and G.W. Hergert, Sampling and spatial analysis techniques for quantifying soil map unit composition, in *Assessment of Non-Point Source Pollution in the Vadose Zone*, edited by D.L. Corwin, K. Loague, and T.R. Ellsworth, p. 79-91, Geological Monograph Series 108, American Geophysical Union,Washington, D.C., 1999.

Freeze, R.A., Stochastic Conceptual Analysis of One-Dimensional Groundwater Flow in Nonuniform Homogeneous Media, *Water Resour. Res.*, 11, 725-741, 1975.

Gelhar, L.W. , Stochastic subsurface hydrology from theory to applications, *Water Resour Res.* , 22, 135S-145S, 1986.

Grayson,R.B., A.W. Western, and F.H.S. Chiew. Preferred states in spatial soil moisture patterns: Local and nonlocal controls. *Water Resour. Res.* 33:2897-2908, 1997.

Green, R.E., L.R. Ahuja, and S.K. Chong, Hydraulic conductivity, diffusivity, and sorptivity of unsaturated Soils: Field Methods, In *Methods of Soil Analysis, Part 1, Physical and Mineralogical Methods,* Second Edition, edited by A. Klute, pp. 771-798, American Society of Agronomy, Inc, Madison, WI, 1986.

Greminger, P.J., Y.K. Sud, and D.R. Nielsen, Spatial variability of field-measured soil-water characteristics, *Soil Sci. Soc. Am. J.*, 49(5), 1075-1082, 1985.

Hills, R.G., P.J. Wierenga., D.B. Hudson, and M.R. Kirkland, The Second Las Cruces Trench Experiment: Experimental Rresults and Three-Dimensional Flow Predictions, *Water Resour. Res.* 27, 2707-2718, 1991.

Hollcnbcck, K.J., T.J. Schmuggc, G.M. Hornberger, and J.R. Wang, Identifying soil hydraulic heterogeneity by detection of relative change in passive microwave remote sensing observations, *Water Resour. Res.* 32, 139-148, 1996.

Hopmans, J.W., J.M.H. Hendrickxs, and J.S. Selker, Emerging Measurement Techniques for Vadose Zone Characterization, In, V*adose zone hydrology: cutting across disciplines*, Edited by M.B. Parlange and J.W. Hopmans, pp. 279-316, Oxford University Press, New York, 1999.

Hopmans, J.W., J. Simunek, N. Romano, and W. Durner, Chapter 3.6. Simultaneous determination of water transmission and retention properties, In *Methods of Soil Analysis, Part 1, Physical Methods*, Edited by J.H. Dane and G.C. Topp, Agronomy 9, Madison, WI, 2002.

Hopmans, J.W., and K.L. Bristow, Soil hydraulic properties at different scales – How helpful are small-scale measurements for large-scale modeling?, Proceedings of the First Joint Congress of the Soil Science Society of America and the German Soil Science Society, Sept. 18-22, Osnabrueck, Germany, 2000.

Hopmans, J.W., and J.N.M. Stricker, Stochastic analysis of soil water regime in a watershed, *J. Hydrology* , 105, 57-84, 1989

Hopmans, J.W., H. Schukking, and P.J.J. F. Torfs, Two-dimensional steady state unsaturated water flow in heterogeneous soils with autocorrelated soil hydraulic properties. Water Resour. Res. 24:2005-2017, 1988.

Jackson, T.J., E.T. Engman, and T.J. Schmugge, Microwave observations of soil hydrology, In *Vadose Zone Hydrology – Cutting Across Disciplines*, Edited by M.B. Parlange and J.W. Hopmans, pp.317-333, Oxford University Press Inc, New York, 1999.

Jury, W.A., G. Sposito, and R.E. White, A transfer function model of solute movement through soil 1. Fundamental concepts. *Water Resour. Res.* 22:243-247, 1986.

Kabat, P., R.W.A. Hutjes, and R.A. Feddes, The scaling characteristics of soil parameters: from plot scale heterogeneity to subgrid parameterization. *J. Hydrology*, 190, 363-396, 1997.

Kasteel, R., H.-J Vogel, and K. Roth, From local hydraulic properties to effective transport in soil, *European J. of Soil Sci.*, 51, 81-91, 2000.

Kavvas, M.L., On the coarse-graining of hydrological processes with incrasing scales, *J. of Hydrology*, 217, 191-202, 1999.

Klemeš, V., Conceptualization and scale in hydrology, *J. of Hydrology,* 65, 1-23, 1983.

Kosugi, K., Lognormal distribution model for unsaturated soil hydraulic properties, Water Resour. Res., 32, 2697-2703, 1996.

Kosugi, K., and J.W. Hopmans, Scaling water retention curves for soils with lognormal pore size distribution. Soil Sci. Soc. Amer. J., 62, 1496-1505, 1998.

Lark, R.M., and R. Webster, Analysis and elucidation of soil variation using wavelets, *European J. of Soil science*, 50, 185-206, 1999.

Loague, L., and P.C. Kyriakidis, Spatial and temporal variability in the R-5 infiltration data set: Déjà vu and rainfall-runoff simulations, *Water Resour. Res.*,33, 2883-2895, 1997.

Mantoglou, A., and L.W. Gelhar, Stochastic modeling of large-scale transient unsaturated flow systems, *Water Resour. Res*, 23, 37-46, 1987.

Mattikalli, N.M, E.T. Engman, T.J. Jackson and L.R. Ahuja, Microwave remote sensing of temporal variations of brightness temperature and near-surface soil water content during a watershed-scale field experiment, and its application to the estimation of soil physical properties, *Water Resour. Res.*, 34, 2289-2299.

Merz B., and E.J. Plate, An analysis of the effects of spatial variability of soil and soil moisture on runoff, *Water Resour. Res.* , 33, 2909-2922, 1997.

Miller, E.E., and R.D. Miller, Physical theory for capillary flow phenomena, *J. Appl. Physics*, 27, 324-332, 1956.

National Research Council, *Opportunities in the Hydrologic Sciences*, National Academy Press, Washington, D.C.,1991.

Nielsen, D.R., J.W. Hopmans, and K. Reichardt, An emerging technology for scaling field soil-water behavior, In *Scale Dependence and Scale Invariance in Hydrology*, Edited by G. Sposito, pp. 136-166, Oxford Univ. Press. , 1998.

Pachepsky, Y.A., W.J. Rawls, and D.J. Timlin, The current status of pedotrnasfer functions: Their accuracy, reliability, and utility infield- and regional scale modeling, in *Assessment of·Non-Point Source Pollution in the Vadose Zone*, edited by D.L. Corwin, K. Loague, and T.R. Ellsworth, p. 223-234, Geological Monograph Series 108, American Geophysical Union,Washington, D.C., 1999.

Philip, J.R., Sorption and infiltration in heterogeneous media, *Aust. J. Soil Res.*, 5, 1-10, 1967.

Philip., J.R., Field Heterogeneity: Some Basic Issues, *Water Resour Res.*, 16(2), 443-448, 1980.

Rockhold, M.L., Parameterizing flow and transport models for field-scale applications in heterogeneous, unsaturated soils, in *Assessment of Non-Point Source Pollution in the Vadose Zone*, edited by D.L. Corwin, K. Loague, and T.R. Ellsworth, p. 243-260, Geological Monograph Series 108, American Geophysical Union,Washington, D.C., 1999.

Rodriguez-Iturbe, I., G.K. Vogel, R. Rigon, D. Entekhabi, F. Castelli, and A. Rinaldo, On the spatial organization of soil moisture fields. *Geophysical Review Letters*, 22, 2757-2760, 1995.

Roth, K., H.-J. Vogel, and R. Kasteel, The Scaleway: A Conceptual Framework for Upscaling Soil Properties, in *Modelling of Transport Porcesses in Soils at Various Scales in Time and Space*, edited by J. Feyen and K. Wiyo, pp. 477-490, Intern. Workshop of EurAgEng's Field of Interest on Soil and Water, 24-26 November, 1999, Leuven, Wageningen Pers, Wageningen, 1999.

Russo, D., and W.A. Jury, A theoretical study of the estimation of the correlation scale in spatially variable fields, 2. Nonstationary fields, *Water Resour. Res.*, 23, 1269-1279, 1987a.

Russo, D., and W.A. Jury, A theoretical study of the estimation of the correlation scale in spatially variable fields, 1. Stationary fields, *Water Resour. Res.*, 23, 1257-1268, 1987b.

Schaap, M.G., F.J. Leij, and M.Th. van Genuchten, Neural Network Analysis for Hierarchical Prediction of Soil Water Retention and Saturated Hydraulic Conductivity. *Soil Sci. Soc. Amer. J.* 62:847-855, 1998.

Šimůnek, J., M. Šejna, and M. Th. van Genuchten, The HYDRUS-2D software package for simulating two-dimensional movement of water, heat, and multiple solutes in variably saturated media. Version 2.0, *IGWMC - TPS - 53*, International Ground Water Modeling Center, Colorado School of Mines, Golden, Colorado, 251pp., 1999.

Šimůnek, J., K. Huang, and M. Th. van Genuchten, The SWMS_3D code for simulating water flow and solute transport in three-dimensional variably saturated media. Version 1.0, *Research Report No. 139*, U.S. Salinity Laboratory, USDA, ARS, Riverside, California, 155pp., 1995.

Sposito, G., W.A. Jury, and V.K. Gupta,,Fundamental problems in the stochastic convection-dispersion model of solute transport in aquifers and field soils, *Water Resour. Res.*, 22, 77-88, 1986.

Toride, N, and F.J. Leij, Convective-dispersive stream tube model for field-scale solute transport: II. Examples and calibration. *Soil Sci. Soc. Am. J.* , 60, 352-361, 1996.

Tuli, A., K. Kosugi, and J.W. Hopmans, Simultaneous scaling of soil water retention and unsaturated hydraulic conductivity functions assuming lognormal pore size distribution, *Advances in Water Resour.*, In Press., 2001.

Van Dam, J.C., J. Huygen, J.G. Wesseling, R.A. Fedes, P. Kabat, R.E.V. van Walsum, P. Groenendijk, and C.A. van Diepen, *Theory of SWAP Version 2.0, SC-DLO*, Wageningen Agricultural University, Report 71, Department of Water Resources, Wageningen, the Netherlands, 1997.

Vogel, H.-J, and K. Roth, Quantitative morphology and network representation of soil pore structure. *Adv in Water Resour.*, 24, 233-242, 2001.

Webster, R., Is soil variation random?, *Geoderma*, 97, 149-163, 2000.

Weissmann, G.S., S.F. Carle, and G. E. Fogg, Three-dimensional hydrofacies modeling based on soil surveys and transition probability geostatistics, *Water Resourc. Res.*, 35, 1761-1770, 1999.

Wheatcraft, S.W., and S.W. Tyler, An explanation of scale-dependent dispersivity in heterogeneous aquifers using concepts of fractal geometry, *Water Resour. Res*, 24, 566-578, 1988.

Whitaker, S., Flow in Porous Media I: A theoretical Derivation of Darcy's Law, *Transport in Porous Media*, 1, 3-25, 1986.

Wildenschild, D., and K.H. Jensen, Laboratory investigations of effective flow behavior in unsaturated hetereogeneous sands, Water Resour. Res, 35, 17-27, 1999.

Wildenschild, D., and K.H. Jensen, Numerical modeling of observed effective flow behavior in unsaturated hetereogeneous sands, *Water Resour. Res*, 35, 29-42, 1999b.

Wood, E.F., Scaling behavior of hydrological fluxes and variables: empirical studies using a hydrological model and remote sensing data, *Hydrol. Process.*, 9, 331-346, 1995.

Yeh, T.-C., and J. Zhang, A geostatistical inverse method for variably saturated flow in the vadose zone, *Water Resour. Res.*, 32, 2757-2766, 1996.

Zhang, D., R. Zhang, S. Chen, and W.E. Soll, Pore scale study of flow in porous media: Scale dependency, REV, and statistical REV, *Geophys. Rev. Lett.*, 27, 1195-1198, 2000.

Keith L. Bristow, CSIRO Land and Water and CRC Sugar, Davies Laboratory, PMB Aitkenvale, Townsville, Qld 4814, Australia. Phone:+61-7-4753-8596, fax:+61-7-4753-8600,email:Keith.Bristow@clw.csiro.au.

Jan W. Hopmans, Hydrology, Department of Land, Air and Water Resources, 123 Veihmeyer Hall, University of California, Davis, CA 95616. Phone: 530-752-3060, fax:530-752-5262, e-mail:jwhopmans@ucdavis.edu.

Don R. Nielsen, Hydrology, Department of Land, Air and Water Resources, 123 Veihmeyer Hall, University of California, Davis, CA 95616. Phone: 530-753-5730, fax:530-752-5262,email:drnielsen@ucdavis.edu.

The Role of Estimation Error in Probability Density Function of Soil Hydraulic Parameters: Pedotop Scale

Miroslav Kutílek

Czech Technical University, Prague, Czech Republic, Professor Emeritus,

Miroslav Krejca

School of Technological Education, Písek, Czech Republic

Jana Kupcová-Vlašimská

Konstruktiva, Prague, Czech Republic

For modeling of transport processes and for prognosis of their results, the knowledge of hydrodynamic parameters is required. Soil hydrodynamic parameters are determined in the field by methods based upon certain approximations and the procedure of inverse solution is applied. The estimate of a parameter P includes therefore an error e. Hydrodynamic parameters are variable to a different degree like other soil properties even over the region of one pedotaxon and the knowledge of their probability density function PDF is frequently required. We have used five approximate infiltration equations for the estimation of sorptivity S and saturated hydraulic conductivity K. Distribution of both parameters was determined with regard to the type of applied infiltration equation. PDF of parameters was not identical when we compared the parameter estimates derived by various infiltration equations. As it follows from this comparative study, the estimation error e deforms PDF of the parameter estimates.

1. INTRODUCTION

John R. Philip deserves the merit for shifting soil physics from empirical studies to theoretical analysis of processes. He was the first to present theoretical solutions of soil hydrological problems, among them infiltration of water in soil [*Kutílek and Rieu*, 1998]. His infiltration equations [*Philip*, 1957a] and equations of soil physicists inspired by

his approach are applied here to the study on probability density function PDF of soil hydraulic parameters.

Soils are spatially variable natural bodies due to their genetic development, actual vegetation cover, the recent type of land use and methods of soil cultivation. All soil properties vary at relatively short distance with various coefficients of variation CV ranging from less than 15% to over 100% [*Wilding*, 1985, *Jury*, 1989] on one soil taxon within a landscape unit of a few ha. Soil physical properties are grouped into two classes: 1. Capacity parameters which have a static character with CV usually small. 2. Transport characteristics including hydrodynamic parameters and fluxes which are dynamic, their CV is relatively high [*Kutílek and Nielsen*, 1994]. For modeling of transport processes and for prognosis on their results, the knowledge of

Environmental Mechanics: Water, Mass and Energy Transfer in the Biosphere
Geophysical Monograph 129
Copyright 2002 by the American Geophysical Union
10.1029/129GM21

hydrodynamic parameters is required. Using the term transport processes, we understand by it e.g. fluxes of water in soil-plant-atmosphere system, recharging of groundwater, fluxes of solutes in soil profile and consequent potential pollution of groundwater. Owing to the spatial variability of hydrodynamic parameters their statistical evaluation is applied and for a proper understanding of their variation, the probability density function PDF is estimated.

For the determination of soil hydrodynamic parameters in the field we use methods which are based upon certain approximations [Angulo-Jaramillo et al., 2000] and the procedure of inverse solution is applied. We are searching the value of a certain soil parameter P, however our observation is equal to $(P + e)$, where e is the estimation error related to the applied method.

The aim of this research is to study the role of the estimation error e upon PDF of soil hydrodynamic parameters. Since we do not know a priory the value of P, we are restricted in this study to the relative comparative procedure. Thus we have evaluated a population of infiltration tests by various methods and the PDF of resulting physical parameters have been determined.

2. MATERIALS AND METHODS

2.1. Site Description

Experiments have been performed on two sites:

1. Arenic Chernozem of the carbonate variety (FAO) in Central Bohemia (Czech republic). The soil has been developed on a Quaternary fluvial terrace of gravel sand overlaid by sandy-loamy topsoil of variable thickness. Fluvial origin of the parent material is the main factor causing greater spatial variability of soil physical properties than in soils of site 2. The soil is typical by its low structural stability in A horizon. We assume therefore that the pore size distribution is not constant during wetting and in the time span of infiltration. Infiltration tests were performed on a rectangular net with spacing of 7.5 m. The net consisted of 10 rows and 7 columns, total number of infiltration tests was 70.

2. Deep Ferralsol in Nigeria, with loamy texture and highly stable aggregates. Infiltration tests were performed on four transects. Two parallel transects (A, B) were at the mutual distance 38 m and the distance of two parallel transects (C, D) located perpendicularly to (A, B) was 8 m. Spacing of infiltration tests in transects was 2 m. Total number of infiltration tests was 128.

2.2. Infiltration Tests

Double ring infiltrometers were used with ponding depth (= positive pressure head on the soil surface) in ranges 2 to 2.5 cm, the diameter of the inner ring was 37.5 cm, of the outer ring 60 cm. We have assumed that the experimental conditions were very close to the conditions of one-dimensional infiltration with Dirichlet boundary conditions on the top boundary. The mean initial soil water content of the topsoil on site 1 was $\theta_i = 0.222$ at the depth 10-15 cm. θ_i was not determined on site 2. Since there was no rain during the time of infiltration measurement and the tests were performed after a dry rainless period, we have assumed that θ_i of the topsoil did not change practically within the time of the whole set of measurements. In the tests, we have measured cumulative infiltration I [L] in time t up to 120 min. The time t_c of the quasi steady infiltration rate, $q_c = dI/dt = $ const. was reached in less than 120 min., q[LT^{-1}]. This was due to limited accuracy in measuring technique and q was oscillating at $t > t_c$ around the mean q_c which was taken as quasi steady infiltration rate.

2.3. Evaluation of Infiltration Tests

The measured data $I(t)$ were fitted to the following approximate infiltration equations derived from the theoretical solution of the process:

Green-Ampt's [1911] equation which is exact for Dirac delta soil [Philip, 1957b]

$$I(t) = Kt + \lambda \ln(1 + \frac{I(t)}{\lambda}) \quad (1)$$

where K is saturated hydraulic conductivity [LT^{-1}], and the parameter $\lambda = (\theta_S - \theta_i)(h_0 - h_f)$, with h_0 the pressure head on the soil surface [L], and h_f the pressure head on the wetting front.

Philip's [1957b] two terms algebraic equation was derived from the infinite series solution [Philip, 1957a]

$$I(t) = St^{1/2} + At \quad (2)$$

with S sorptivity [LT$^{-1/2}$], which is an estimate of the real sorptivity of infinite series solution [Kutílek et al., 1988] and parameter A [LT^{-1}] which is related to K by $A = mK$. Theoretically based and most frequently used m is close to $m = 0.361$ [Philip, 1987].

The three parameters equation derived from the Philip's [1957a] infinite series solution [Kutílek and Krejca, 1987, quotation and details see in Kutílek and Nielsen, 1994] is

$$I(t) = c_1 t^{1/2} + c_2 t + c_3 t^{3/2} \quad (3)$$

with parameters c_1 [LT$^{-1/2}$], c_2 [LT^{-1}] and c_3 [LT$^{-3/2}$] related to the Philip's series solution. The estimate of sorptivity is $S = c_1$, and approximation of $K = (3c_1c_3)^{1/2} + c_2$.

Table 1. First two best fitting distributions of S [LT$^{-1/2}$], sorptivity, and of q_1 [LT^{-1}], infiltration rate after the first minute of infiltration

Equation	Site 1				Site 2					
	Short set		Full set		Transects A, B		Transects C, D		Transects A, B, C, D	
	S	q_1	S	q_1	S	q_1	S	q_1	S	q_1
-	-	-	E	-	B	-	E	-	E	N
			G		W		G		G	G
(2)	LG	-	-	-	G	-	N	-	N	-
	E				E		W		G	
(3)	LG	-	E	-	G	-	N	-	E	-
	G		G		E		W		G	
(4)	LG	-	E	-	G	-	E	-	N	-
	E		G		LG		G		G	
(5)	LG	-	-	-	LG	-	G	-	G	-
	G				E		E		E	

Type of probability density function: N - normal, LG - lognormal, W - Weibull, G - gamma, B - beta, E – Erlang. Number of infiltration tests in short set: 49, in full set: 70, in transects A, B: 64, in transects C, D: 64, and in all transects A, B, C, D: 128.

Swartzendruber's [1987] three parameters equation is

$$I(t) = \frac{S}{c_4}\left[1 - \exp(-c_4 t^{1/2})\right] + Kt \quad (4)$$

where S is the approximation of sorptivity, and c_4 [T$^{-1/2}$] is a parameter.

Brutsaert's [1977] equation is

$$I(t) = Kt + \frac{S^2}{BK}\left\{1 - \frac{1}{1+(BKt^{1/2})/S}\right\} \quad (5)$$

where B is an empirical dimensionless coefficient.

S and K in the above equations are estimates of sorptivity and saturated hydraulic conductivity. During the fitting procedure, they behave as fitting parameters [*Kutílek et al.,* 1988]. Their physical interpretation corresponds to our earlier statement that the results of our evaluation procedures are equal to $(P + e)$, where P is the physical parameter, in our case either S or K, and e is the estimation error related to the applied method.

We have used the chi-square goodness-of-fit test and Kolmogorov-Smirnov test on the fit [*Statgraphics,* 1985] for the determination of the probability density function of sorptivity and of saturated hydraulic conductivity estimates obtained by equations (1), (2), (3), (4), (5). Further on, for comparative reasons, PDF of the directly measured experimental data q_1 and q_{120} were evaluated, too.

3.   RESULTS AND DISCUSSION

In many of instances, the evaluation of PDF according to Kolmogorov-Smirnov test and by the chi-square goodness-of-fit test have offered the same results. The second procedure was sensitive to the number of classes and due to this requirement the PDF differed in some instances, as it could be expected theoretically. In Tables 1 and 2, there are results obtained by Kolmogorov-Smirnov test.

On experimental site 1 with Arenic Chernozems, the infiltration data were not applicable to the fitting procedure according to equation (1) of Green and Ampt. Therefore, this equation was excluded from PDF testing on site 1. Philip's two parameters equation offered non realistic values of hydraulic conductivity K with either $K = 0$ or even $K < 0$ in about 30% of tests. Brutsaerts equation (5) led to unrealistic values of parameter $B \to 0$ or $B = 0$ in 16 % of instances. They were identical with tests belonging to 30% of non-realistic output in K according to eq. (2). We have therefore formed a shortened set of 49 infiltration tests, where we have omitted the infiltration experiments not applicable for fitting according to eq. (2). We have determined PDF of parameters obtained from equations (2), (3), (4), and (5) in this short set. Full set with all 70 infiltration experiments was kept for PDF determination of parameters obtained by eq. (3) and (4).

On site 2 with Ferralsols all equations were applicable and no fitting problems occurred there. In parallel transects A, B, the number of infiltration tests was 64 and the same number of tests was performed in parallel transects C, D.

3.1. PDF of Sorptivity

Evaluation of distribution functions of sorptivity S estimates is in Table 1. The PDF of S are supplemented by evaluation of infiltration rate after the first minute, q_1, which is theoretically closely related to sorptivity. However, PDF of q_1 differ in the majority of instances from PDF of S. The

Table 2. First two best fitting distributions of K [LT^{-1}], saturated hydraulic conductivity, and of q_{120} [LT^{-1}], infiltration rate after 120 minutes of infiltration

Equation	Site 1				Site 2				Site 2	
	Short set		Full set		Transects A, B		Transects C, D		Transects A, B, C, D	
	K	q_{120}	K	q_{120}	K	q_{120}	K	q_{120}	K	q_{120}
-	-	B	-	W	-	LG	-	G	-	LG
		E		B		E		B		G
(1)	-	-	-	-	LG	-	G	-	E	-
					E		E		G	
(2)	LG	-	-	-	LG	-	E	-	LG	-
	G				E		G		G	
(3)	W	-	B	-	LG	-	E	-	LG	-
	N		W		G		G		G	
(4)	B	-	W	-	LG	-	E	-	LG	-
	W		E		G		B		E	
(5)	W	-	-	-	LG	-	LG	-	LG	-
	B				E		N		G	

Type of probability density function: N - normal, LG - lognormal, W - Weibull, G - gamma, B - beta, E – Erlang. Number of infiltration tests in short set: 49, in full set: 70, in transects A, B: 64, in transects C, D: 64, and in all transects A, B, C, D: 128.

only exception is in site 2, all transects A, B, C, D together, where S obtained by equations (2) and (4) show the same PDF as q_1. There is distinct curtosis and skewness of S and q_1 distributions indicating that in the majority of instances S is not normally distributed. The exception is in 4 cases, where normal PDF was evaluated. Lognormal PDF was equally frequent as gamma PDF and both were more frequent than other tested PDF. For q_1 we have found normal PDF in one case, while Erlang PDF was prevailing in remaining sets. We are concluding that the PDF of S estimates depends not only upon distribution of theoretical S which is unknown to us, but upon the nature of approximation in the applied equation. A not unique type of PDF in directly measured q_1 is probably due to a lack of accuracy even in the measuring technique with double ring infiltrometers and with no strictly constant pressure head on the surface. In addition to it, the variation in soil heterogeneity on vertical may cause variation of divergence of flow paths and a non-unique deviation from theoretically assumed 1-dimensional flow. Low population in short sets may play a role in discussed differences, too, especially due to soil heterogeneity. For population of 128 experiments (all transects A, B, C, D) there was an increased identity in PDF between q_1 and S estimates.

3.2. PDF of Saturated Hydraulic Conductivity

In Table 2 there are the results of distribution evaluation of estimates of saturated hydraulic conductivity K together with evaluation of quasi steady infiltration rate after 120 minutes, q_{120}. The PDF of q_{120} and PDF of K estimates was

more frequently identical than was the case of S and q_1. However, a complete identity did not exist. A distinct curtosis and skewness of K and q_{120} distributions was found indicating that there is not normal distribution in all instances. The type of prevailing PDF was different on site 1 and on site 2. On site 1 was the Weibull PDF more frequent than other distributions. On site 2, lognormal PDF was distinctly prevailing and the next most frequent was Erlang PDF. This was dominant in transects C, D. There is a tendency showing that the increase of population in the studied sets leads to a better agreement between PDF of directly measured q_{120} and PDF of the K estimates. A greater variation of PDF types in site 1, when compared to site 2 is probably due to the instability of soil structure on site 1. Owing to it K was probably time dependent.

4. CONCLUSIONS

The comparison of PDF for the estimates of parameters S and K shows that their PDF is deformed by the error e, i.e. we are determining PDF of an estimate $(P + e)$, where P is the studied soil parameter. The error e is a summation of two components: 1. Error due to the approximate character of the equation applied for fitting the experimental data. 2. Error due to the difference between the field reality and the physical soil conditions assumed in the theoretical development of equations, i.e. the assumed soil physical homogeneity in depth and time. The instability of soil structure and thus induced non-constant pore size distribution in time increases the role of the second component of the estimation error e. This is a probable reason for a greater variation of

PDF of the studied parameters in the soil with a low stability of aggregates.

Acknowledgement. The authors thank Dr. O.A. Folorunso for offering us his infiltration data.

REFERENCES

Angulo-Jaramillo, R., J. P. Vandarvaere, S. Roulier, J. L. Thony, J. P. Gaudet, and M. Vauclin, Field measurement of soil surface hydraulic properties by disc and ring infiltrometers: A review and recent developments, *Soil Tillage Res., 55,* 1-29, 2000.

Brutsaert, W., Vertical infiltration in dry soil, *Water Resour. Res,, 13,* 363-368, 1977.

Green, W.H., and G. A. Ampt, Studies on soil physics: I. Flow of air and water through soils, *J. Agric. Sci., 4,* 1-24, 1911.

Jury, W.A., Spatial variability of soil properties, in: *Vadose Zone Modelling of Organic Pollutants,* edited by S. C. Hern, and S. M. Melancon, pp. 245-269, Lewis Publishers, 1989.

Kutílek, M., M. Krejca, R. Haverkamp, L. L. Rendón, and J. Y. Parlange, On extrapolation of algebraic infiltration equations, *Soil Technol., 1,* 47-61, 1988.

Kutílek, M., and D. R. Nielsen, *Soil Hydrology,* Catena Verlag, 370 pp., 1994.

Kutilek, M., and M. Rieu, Introduction to the Symposium 1: New concepts and theories in soil physics, *Soil Tillage Res., 47,* 1-4, 1998.

Philip, J.R., The theory of infiltration: 1. The infiltration equation and its solution, *Soil Sci., 83,* 345-357, 1957a.

Philip, J.R., The theory of infiltration: 4. Sorptivity and algebraic infiltration equations, *Soil Sci., 84,* 329-339, 1957b.

Philip, J.R., The infiltration joining problem, *Water Resour. Res., 23,* 2239-2245, 1987.

Statgraphics, The Braegen Group Inc., Toronto, Ontario, 1985.

Swartzendruber, D., A quasi-solution of Richards´ equation for the downward infiltration of water into soil, *Water Resour. Res., 23,* 809-817, 1987.

Wilding, L.P., Spatial variability: Its documentation, accomodation and implication to soil surveys, in: *Soil Spatial Variability,* edited by D.R. Nielsen, and J. Bouma, Pudoc, Wageningen, pp.125-178, 1985.

Corresponding author: M. Kutílek, Czech Technical University, Nad Patankou 34, CZ 160 00 Prague 6, Czech Republic, Tel./Fax (420) 2 3333 6338, *E-mail address:* kutilek@ecn.cz.

M. Krejca, School of Technological Education, Komenskeho 86, 397 11 Pisek, Czech Republic.

J. Kupcova – Vlašimská, Konstruktiva, Limuzska 8, 110 00 Prague 10, Czech Republic.

Searching Below Thresholds: Tracing the Origins of Preferential Flow Within Undisturbed Soil Samples

Milena Císlerová, Tomáš Vogel, Jana Votrubová and Alice Robovská

CTU Prague, Faculty of Civil Engineering, Prague, Czech Republic

Infiltration outflow experiments, performed on undisturbed soil samples of coarse sandy loam taken in Korkusova Hut (KH) revealed the presence of preferential flow and "steady state" infiltration rate instability, compared to results of the same experiment performed on an undisturbed sample of fine sand taken in the Hupselse Beek, (HB) experimental watershed. The structure of soil samples was visualized by means of computer tomography, and in addition the infiltration outflow experiment for KH soil was performed in a magnetic resonance (MR) scanner. From MR images it was found that for the KH soil sample the character of flow changes gradually during the ponded infiltration. The volume flowing through the sample by gravity at the beginning of the experiment forming early outflow later decreases as the portion of the matrix flow increases. In the recurrent infiltration run into the "saturated" sample this decrease may be abrupt, accompanied by a significant decrease in the "steady state" infiltration rate. Both effects are unexpected from the point of view of Richards' theory.

1. INTRODUCTION

Since the time John Philip published his papers on infiltration theory the world has changed. Almost everyone can now use basic and/or advanced modeling techniques of unsaturated water flow and solute transport in the vadose zone having bought efficient user-friendly software packages. But a question remains, whether our understanding of the mechanism of the flow processes has developed adequately as well? In fact this question is similar to John Philip's questions or better his worries [*Philip*, 1991; *Nash at al.*, 1990].

The principles of continuum mechanics and the energy concept, expressed through the total matrix potential, allow us to write a number of continuity and motion equations for an elementary volume. For soil water flow, the generally accepted approach results in the classic Richards' equation (RE) and the advection-dispersion equation for the solute transport. In this principle, the pore geometry and the driving force mechanism are considered implicitly, defining media hydraulic properties averaged over the representative elementary volume (REV). In reality, all the processes are born at the molecular scale, where the field of interfacial forces forms.

In the classical flow theory, natural soil materials are described by soil hydraulic properties. Microscopic (pore) flow is generally related to macroscopic (Darcian) flow using capillary models [e.g. *Childs and Collis-George*, 1950; *Brooks and Corey*, 1964; *Burdine*, 1953; *Mualem*, 1976; *van Genuchten*, 1980]. The soil hydraulic characteristics are derived on a hypothetical bundle of capillaries related to the pore space geometry. The first derivative of the retention curve is mostly considered as a simple transformation of the pore size distribution curve. In the case of multi-structured media this distribution is treated as multi-modal [*Durner*, 1994]. Numerous assumptions have to be adopted, some of which are difficult to fulfill in reality. There are severe problems with the application of this con-

Environmental Mechanics: Water, Mass and Energy Transfer in the Biosphere
Geophysical Monograph 129
Copyright 2002 by the American Geophysical Union
10.1029/129GM22

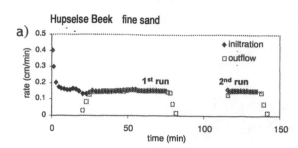

a)

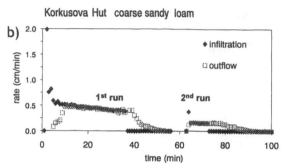

b)

Figure 1. Infiltration and outflow rates vs. time plotted for two recurrent ponded infiltration experiments on a 20 cm diameter and 20 cm high undisturbed soil sample a) for Hupselse Beek (HB) fine sand (a "regular" ponded infiltration); b) for Korkusova Hut (KH) coarse sandy loam (a typical example of the infiltration in this soil).

cept, in addition to the possible inadequacy of a particular capillary model itself [*Vogel and Císlerová*, 1988].

Alternatively, the unsaturated flow can be treated as multiphase flow. The multiphase approach makes it possible to reproduce soil hydraulic characteristics by pore-scale models based on microscale data. The pore-scale physics of fluid-fluid interfaces can be incorporated serving as internal system boundaries in case of preferential flow formation [e.g., *Tuller and Or*, 2001].

Regardless of the method used, there is the problem of convenient experiments to produce the accurate data necessary to interpret the measured macroscopic quantities. In standard observations, a more or less arbitrary size and shape of the soil sample is taken in the field to be subjected to flow measurements. The averaged data are measured over the sample, and thus the sample imposes in the real world the size and the shape of the REV. That means the *scale threshold* of the observation of involved flow and transport processes is given and some information from below this threshold is missed. One of the unresolved problems born below this threshold is the preferential flow often accompanied by flow instability. Non-invasive visualization of the pore geometry may be very helpful in studying this problem. We describe infiltration-outflow experiments that elucidate the observed flow effects.

2. THE PROBLEM DEMONSTRATION

Conventionally, when vertical infiltration is considered it is assumed that water propagates evenly through the soil sample forming a "piston like" wetting front. The fine sand from the Hupselse Beek (HB) experimental watershed [*Císlerová et al.*, 1990] exhibits this type of flow. Infiltration and outflow rates, the result of the ponded infiltration outflow experiment described below, measured on a sample of HB soil are presented in Figure 1a). The beginning of the outflow was reasonably delayed, the steady state infiltration rate is equal to the outflow rate and therefore to the saturated hydraulic conductivity of the sample. The drainage at the end of the infiltration run is negligible. In the subsequent infiltration performed on the "fully" saturated sample, steady state infiltration of equal rate continues.

In the country where the crystalline rocks prevail, in all scales we have to tackle the preferential flow accompanied by flow instabilities, that are widely observed on Dystric Cambisols [*Císlerová et al.*, 1988; *Pražák et al.*, 1994; *Tesař et al.*, 2001]. Typical shape of consecutive ponded infiltration runs for a sample of coarse sandy soil from Korkusova Hut (KH) is given in Figure 1b). The outflow commences quickly after infiltration begins because of preferential flow. In addition, both the infiltration and outflow rates continue to decrease and, at the end of infiltration, a certain volume of gravity driven water leaves the sample. In a second run performed on the "fully" saturated sample, the infiltration rate is substantially less from the beginning and continues decreasing. The volume of gravity driven outflow draining from the sample at the end of the experiment is smaller than after the initial infiltration run, demonstrating a change in the contributing flow domain.

Other examples of irregularities in the course of ponded infiltration and outflow also for other soil types, using HB fine sand soil as a reference, were discussed earlier [*Císlerová et al.*, 1990].

In natural porous media, heterogeneity appears at all scales, ranging from nanometer to kilometer scales. In the case of fast flow it is the very fine scale heterogeneity (100 μm) where the preferential flow starts to form. The KH coarse sandy loam belongs among highly heterogeneous soils, representing a typical "randomly structured medium" [*Nimmo*, 1997]. When we compare microphotographs of KH and HB soils (Figure 2), we see differences in the soil structure at all scales. There is an obvious difference between idealized bundles of capillaries and the natural pore spaces, that are multiscale and have a convoluted character of pore structure systems [*Or and Tuller*, 1999].

The data obtained from the infiltration outflow experiment may be related to the pore geometry non-invasively visualized by the X-ray CT (Computer Tomography) and

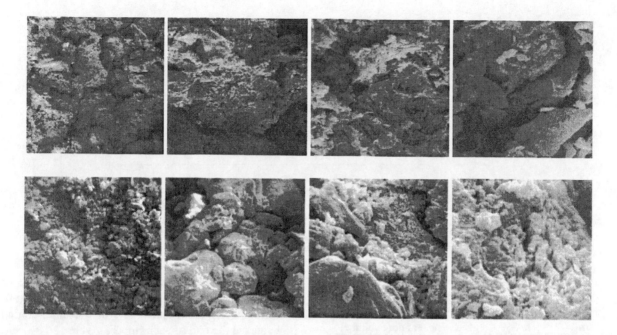

Figure 2. Microphotos of KH coarse sandy loam in the upper row, and HB fine sand in the lower row. The scale enlargement ratios from left to right are 19x, 99x, 990x and 4900x. In this sequence the central part of the photograph is always enlarged.

magnetic resonance imaging (MRI). Thus, more information can be obtained regarding a) the geometry of pore volumes contributing to the fast flow, b) the boundaries between the fast and slow flow domains. The images consist of voxels with the size given by the image resolution and the image thickness. While 3-dimensional CT scans of a dry sample describe the solid matter density distribution, in MR images provide complementary information on the „free" water distribution. Compared to CT, however, the character of the MRI information is more complicated. Treating the pore-size distribution and tortuosity in heterogeneous porous media, *Latour et al.*, [1995] classifies porous media samples as "regular", "well behaved" and "bad". The behavior of the regular group is similar to that of glass-sphere packs. The majority of rocks and soils however are distinctly irregular, due to varying shape and mineralogical composition of particular grains, complicated chemistry and structural and textural instability. Because of the presence of paramagnetic substances, soils are considered an inconvenient medium for MR quantitative analysis, which are at the edge of MRI possibilities [*Amin et al.*, 1997; *Hall et al.*, 1997].

3. INFILTRATION OUTFLOW EXPERIMENT

The laboratory infiltration outflow experiments mentioned above are performed on 1-6 liter undisturbed soil samples. A sample is placed on a digital balance in order to record changes of the weight throughout the experiment. A wide mesh supports the bottom of the sample allowing water to drain freely. The infiltrated volume and the draining outflow are recorded continuously. Tensiometers with pressure transducers are installed along the height of the sample and continuously read.

The measurements are done for a sequence of suction heads imposed by means of the tension infiltrometer [*Clothier and White*, 1981] placed on top of the sample. Between infiltration runs, fan-driven evaporation dries the sample to obtain comparable initial conditions for the each subsequent run. Each run lasts long enough to reach "steady state" flow conditions. In addition, consecutive ponded infiltration experiments are executed, separated by a period of drainage (shown in Figure 1). Details of the experimental set-up are presented by *Sněhota at al.*, [2001].

The collected data supply the information about changes of mass balance components during the flow process. For each imposed pressure head the values of hydraulic conductivities and the moisture contents are measured directly [*Sněhota at al.*, 2001]. The parameters of the retention curve and the relative hydraulic conductivities are later evaluated from transient infiltration outflow data by inverse modeling [*Vogel et al.*, 1999]. At the end of the complete set of experimental runs, the flow paths are visualized by adding a pulse of Brilliant Blue color into the ponded water of the last ponded infiltration run. Later when drained, the sample is sliced and photographed to reveal the distribution

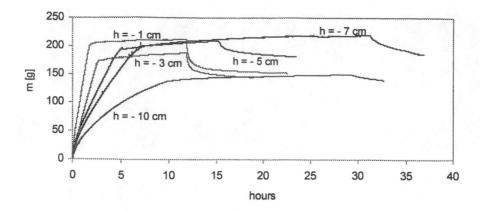

Figure 3. Changes of the sample mass (weight) in time for particular infiltration runs. Points where the curves break correspond to times when the outflow from the sample began.

Table 1. Overview of the inflow-outflow experiment settings and results

i	h_i [cm]	$q(h_i)$ [cm/h]	volume [g]	[% of pore volume]
1	-10	0.049	0	0
2	-7	0.255	10	1.1
3	-5	0.498	17	1.9
4	-3	1.066	21	2.4
5	-1	0.579	30	3.4
6	+0.6	61.56	54	6.2

of the dye. In this article we will pay attention to the sample mass (weight) developments in time (Figure 3) and to the volumes of water moving through the sample by gravity (Table 1). Data shown in Figure 3 were obtained for the undisturbed sample of Korkusova Hut coarse sandy loam 12 cm in diameter and 20 cm high. A sequence of pressure heads (h) of -15, -10, -7, -5, -3, -1 and +0.6 cm was applied as upper boundary conditions for particular infiltration runs starting at the same initial condition. The sample mass development illustrates the changes of the sample water content during the flow process. For all runs the slope of the curve breaks when the outflow begins. Even for visually equal outflow and inflow rates, similar to those given in Figure 1, a slow filling of the sample continues as recorded in the weight increase. It still continues after more then 30 hours. In subsidiary experiments on 100 cm³ samples to get the saturated moisture content the weight of the samples saturated for 48 hours was 8% greater than that achieved by a standard 24 hours saturation. When air-dried the volume of the samples decreased by approximately 0.5% in diameter. We note, however, that all infiltration outflow experiments were performed at the initial moisture content sufficiently great to avoid signs of shrinking. In Table 1 the volumes leaving the sample at the end of infiltration run are

given in grams and as a percentage of the total pore volume (in the case of ponded infiltration after the water on the top of the sample had disappeared). In the third column the flow rates at the end of each run are given.

4. COMPUTER TOMOGRAPHY

Based on measurements of the attenuation of X-ray beams and the numerical reconstruction of an image CT scanning produces the matrix of voxel numbers in Hounsfield units (HU), which can be correlated with various porous media properties. The relationship between HU and dry bulk density is reported to be linear, as is the relationship for voxel porosities. Dry soil is composed of both air and mineral solid. The HU measured for a voxel is some composite of air and solid; that means that the information contained in each voxel is at the averaged macroscopic level.

Figure 4 illustrates a three-dimensional reconstruction of CT data acquired for the air-dried undisturbed soil core of the KH soil from the infiltration-outflow experiment. Blanking out selected categories of voxel HUs in the image illuminates some interior features. In Figure 4, dense voxels, mostly stones, are shown. The dimensions of the sample are given in Table 2 (KH large).

In Figure 5 the normalized cumulative frequency of voxel HU values for three samples of KH and for one sample of HB are given. In the image, the relative HU value for air corresponds to zero, while for mineral solid we obtain 4095. When related to the voxel porosity distributions, the empty, air filled voxels scale to unity and the mineral solid of HU=4095 to zero. In Figure 5 the different range of values for both soils are evident. While the HU values of the images of KH samples are distributed over the whole range 0-4095, the HB matrix image consists of a narrow range of

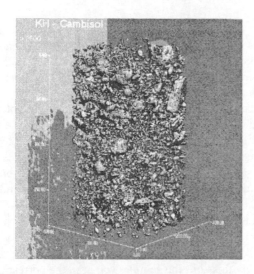

Figure 4. Three-dimensional reconstruction of CT data acquired for the air-dried undisturbed soil core of the KH soil. It shows dense voxels of HU greater than 2500 [mostly stones]. A quite complicated spatial structure of the porous medium is apparent. The 3D image consists of 23 819 481 voxels of 0.3 x 0.3 x 1mm.

HU values only, between 1250 – 2450. Figure 6 shows the particle size distribution of both soils. It can be seen that the differences in the ranges of particle diameter are very similar to differences in relative HU distributions. The mutual comparison of the frequency distributions of the three KH samples is also instructive. While in practice the particle size curves for one soil usually show little difference, the HU distributions of the three analyzed samples differ significantly, showing large variations in the pore space structure. By blanking out selected classes of HU values we may obtain "skeletons" of voxels visualizing empty or less dense regions and their connectivity throughout the sample.

It is straightforward to relate the part of these less dense regions to the pore volume of gravity driven water shown in Table 1. The classes of voxels representing volumes of different density give a basic idea of where the flow can take place. For the large KH sample this volume represents the voxels of HU < 170. While the network of empty voxels for the large KH sample is sparse, the network of voxels corresponding to the volume of gravity water of HU < 170 is continuous [*Císlerová et al.*, 2001]. This information is essential for the evaluation of flow domains within the sample, and is also very useful for the correct interpretation of MR images.

5. MAGNETIC RESONANCE IMAGING

For the sake of understanding the results, a simplified description of the MRI method follows (fundamentals are given e.g. by *Kean and Smith*, [1986]). When excited by a radio frequency pulse in a homogeneous magnetic field, each particular hydrogen proton of the water present in a soil sample produces a signal. The MR signal that we detect is the sum of all these. The parameters of the MR signal needed to quantify the entities of interest are the signal intensity, M_0, and the relaxation times, T_1, and T_2, which represent the rate of decay of the MR signal in longitudinal and transversal directions. Not all of hydrogen protons in the water in porous media produce an equal signal however, and both relaxation rates are dominated by relaxation at interfacial surfaces and are sensitive to pore size. In smaller pores the decay of the signal may be so fast that the MR parameters cannot be evaluated. In larger pores filled by reasonable numbers of water protons, the decay of the water-proton signal is close to the decay of bulk water. This is used to advantage when looking for the fast moving water during the flow, since in the case of KH, the "MR free"

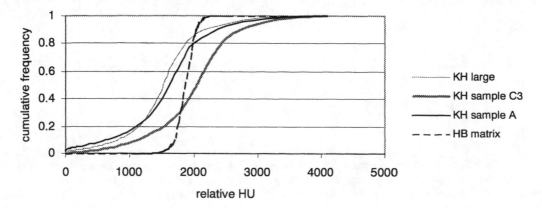

Figure 5. Cumulative frequency of voxel density for Korkusova Hut coarse sandy loam (KH) and Hupselse Beek fine sand (HB).

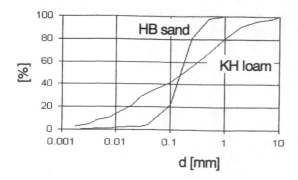

Figure 6. The particle size distribution of the both (KH and HB) soils.

In the experiment performed in the MR scanner, temporal changes of MR signal are traced during the continuous infiltration through a partially undisturbed soil core 4.5 cm in diameter and 7 cm high. The porosity of the sample was 0.49, the initial moisture content 0.23 and the bulk density 1.19 g/cm^3. To illustrate the development of the volume of "visible" water moving within the sample during the whole infiltration experiment, the sums of the signal intensities along the sample height were calculated from a series of 1D images as shown in Figure 8. The fast breakthrough of gravity driven water during the first stage of the ponded infiltration experiment is evident (it lasted for about 3 minutes). Later, when the flow is continuous the changes of MR signal are slower and the signal intensity decreases. As discussed above, in all infiltration outflow experiments performed on this soil, the sample weight in the corresponding period of time however steadily increases due to the increase of the soil moisture content (an example is given in Figure 3). The gradual decrease of the signal thus implies changes in the spatial distribution of water molecules during the flow from "MR visible" fraction (large pores) to "MR less visible" fraction (small pores). The sharp decrease of the signal intensity at the end of infiltration experiment shown in Figure 8 reflects the decrease due to drainage, when all gravity driven water drained from the sample. In Figure 9 two images of the central horizontal slice of the sample are shown. The image marked (Flow) reflects the distribution of the signal intensity during steady-state infiltration, in the image marked (Drained) the distribution of signal intensity at the end of the experiment is given when the gravity flowing water has drained out. In

water is the *only* water, which is *visible* (gives a signal). We may thus trace water contributing to preferential flow in the KH sample. Figure 7 shows that from the two soils, the HB fine sand gives significantly better bulk signal than the KH coarse sandy soil.

Based on the ability of MRI to visualize primarily gravity driven water in large pores, a method of quantitative evaluation of MRI in heterogeneous soils exhibiting preferential flow using the combination of several measuring categories has been suggested earlier [*Císlerová et al.*, 1999]. 1D imaging of the ponded infiltration experiment performed in the MR scanner is applied in combination with 2D imaging and T1 mapping. During the ponded infiltration experiment the fast breakthrough of gravity driven water (preferential flow) together with the temporal changes of the MRI visible water volume can be monitored.

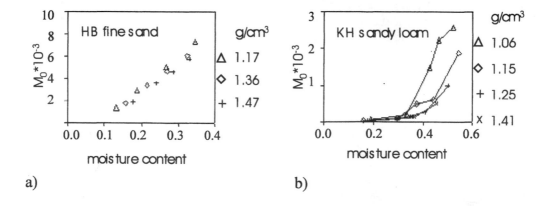

a)   b)

Figure 7. Bulk MR signal intensities for packings of varying bulk density and moisture content (a) for fine sand (Hupselse Beek), considered being a homogeneous soil material suitable for MRI, and (b) for coarse sandy loam (Korkusova Hut), a heterogeneous material with paramagnetic substances [*Votrubová et al.*, 1999]. While the plotted relationship is nearly linear in case (a), in heterogeneous soil (b) the signal is very weak and increases only when large pores are filling. Note the differences in the signal intensity scale. For *bulk* imaging this soil is not acceptable and was considered as unsuitable for MRI. However, in case of tracing water contributing to preferential flow this fact turns to advantage.

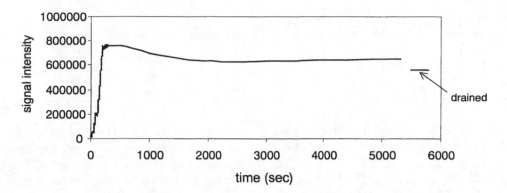

Figure 8. Temporal changes of the signal intensity summarized along the vertical axis of the whole sample acquired during the ponded infiltration and when the sample was drained. The infiltration rate was approximately 0.014 cm/sec.

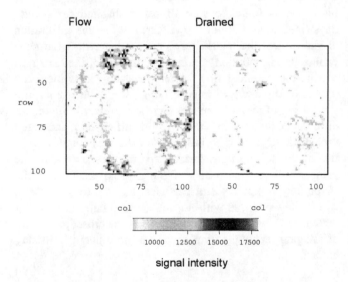

Figure 9. Two images of the central horizontal slice: the image marked (Flow) is taken during steady-state flow, and (Drained) at the end of the experiment when the gravity flowing water drained out. In both images, only 30% of all pixels are shown, those, which contain "MRI visible" water.

both images, only 30% of all pixels are shown, these which contain "MRI visible" water.

In an earlier experiment it was found that in the repeated infiltration the cumulative signal intensity stays much lower [Amin et al., 1994]. The detailed "visibility", related to the mobility of water occupying the pore space has to be gained from the T1 distribution maps [Císlerová et al., 1999].

The whole series of MR images including T1 maps was obtained during the ponded infiltration performed on the KH sample C3 (Table 2). The initial water breakthrough at the beginning of ponding was imaged, then the later "steady state" stage when the sample was fully saturated, followed by imaging of the sample when drained and then when the second infiltration run took place. There was an overnight period between the runs. The sequence of 2D images of the central slice for the saturated sample at the end of the first infiltration run (saturated 1 (S1)), drained sample (drained (D)) and for the saturated sample at the end of the second infiltration run (saturated 2 (S2)) were taken. Although there is not much difference between the images visible to the naked eye, (and therefore not shown here), they differ significantly both as a whole and in detail. By means of T1 mapping it was found that only 30 percent of the voxels in the central slice contain a reliable MR signal. The MR signal intensities integrated over the "MR visible" 30% portion of images S1, D and S2 are given in Table 3. The highest signal of the three is the one for the S1 run; the signal for S2 is much lower (90%).

To illuminate changes in water distribution in detail, a fixed line of 28 pixels of higher MR signal was selected arbitrarily. The MR signal intensity distribution acquired at these pixels during the three experiment stages S1, D and S2 is illustrated in Figure 10 in the form of diagram. For S2 the highest values are much lower than for S1 and the position of extremes is shifted during the drainage period. In some pixels with a high signal in case of the run S1 the signal vanished in run S2. This suggests that these pixels had to be almost empty (air filled) during the second infiltration run. It may be explained by the redistribution of water present in the sample during the interruption of infiltration when gravity water drains out of the sample. This detailed picture supports the hypothesis that water in large pores moves in the second run only near the walls, continuing to flow in smaller pores, in both cases being bound by the interfacial forces and not free enough to produce a signal anymore. The fast-flow region and driving force fields established during repeated infiltration events are evidently

Table 2. Description of the samples scanned by CT and MR

sample	diameter [mm]	height [mm]	porosity [-]	bulk density [g/cm^3]	dry weight [g]	volume [cm^3]	pore volume [cm^3]
KH large	120	200	0.39	1.26	2 848.6	2 260.8	872.7
KH C3	54.5	99	0.43	1.52	319.6	211.7	87.4
KH A	44.0	100	0.53	1.24	188.0	152.1	81.1

Table 3. Totals of the "reliable" MR signal intensities from the central slice of the sample C3 at different stages during the experiment

saturated 1	drained	saturated 2
7.61 E+08	6.20 E+08	6.83 E+08

different from those in the first run. This is consistent with observed drop of the outflow rate in the repeated infiltration runs, which is accompanied by the increase of the sample weight.

6. DISCUSSION

Using the concept of momentum balance, *German and Di Pietro*, [1999] concluded that pores within the range from 10 μm to 10 mm can, under certain specific conditions, participate in preferential flow. From the MR images of the flow it is evident that the amount of water participating in the preferential flow in the coarse sandy loam changes during the flow process and depends on the distribution of the soil moisture in smaller pores. Because the pore distribution is smooth, not bimodal but multimodal, as has been confirmed by the CT voxel porosity distribution, the size of the pores changes gradually and despite the larger open spaces there is not a firm boundary between different groups of pores. Due to the soil/pore heterogeneity the rates of the flow in the smallest pores and of the fast flow in the large pores differ by several orders of magnitude. Fast flow is initiated within seconds, the slow flow during many hours to days. The driving force fields are therefore not in equilibrium, and even during "steady state" flow they gradually change. For this soil, the final equilibrium may never be reached. In addition, the boundary between fast and slow flow domains for this soil is unstable, and moves with the gradual filling of the pore system.

The structure of the large pores network can be well evaluated from CT images, however no further conclusions about the parameters and contribution of the fast flow domain can be derived. For the infiltration experiments with the upper suction set close to zero, at the beginning the gravity flow dominates and the main portion of infiltrating water thus percolates through the sample via interconnected empty larger pores when available. Due to extremely contrasting flow rates only a small portion of infiltrating volume participates in the filling of the soil matrix obeying the rule that the smallest pores fill the first. However with gradual filling of the slow domain (soil matrix), a gradually increasing fraction of water volume moving through the sample participates in the flow in this domain being attracted by the matrix potential. Although the velocity field within the sample becomes less contrasting, the preferential flow continues and the "steady state" infiltration rate stays high. It was found to stay almost unchanged for several days (measured, but not shown here). When the infiltration is interrupted, the matrix flow, or in other words the slow capillary redistribution continues. This part of the flow represents the classical flow. After restarting infiltration, the impact of the newly formed driving force field results in a sharp drop of the volume that can flow preferentially and a different character of flow is identified by decreased "steady state" rate. The influence of the shape and properties of the soil particle surfaces, of the interactions of the present soil air phase and of the gradual changes in the shape and volume of soil particles, e.g., biotites or clays, which may swell, is with no doubts very important during the entire flow process. Because of the macroscopic nature of imaging technique employed, the more details still stay hidden.

7. CONCLUSION

The unstable hydraulic behavior was described during infiltration experiments performed on the undistorted soil sample. Similar effects were observed also in the field, playing an important role in forming of floods. The hydraulic interpretation of these phenomena is under way, together with further experiments. Important information is yet to be revealed from MRI data. The generalization to another soil types would not be justified at all, however for materials with wide ranges of pore size distribution similar effects may be expected.

Acknowledgements. The MRI was performed in cooperation with HSLMC Cambridge, for generous and friendly cooperation we are very grateful to Prof. Laurie Hall and Dr. Gao Amin. The authors are thankful to the editors and the reviewers for all helpful comments and patience. The work has been supported by research projects MSM 3402143, MSM 216200031, VAV/510/1/99 and AVCR A306 0001.

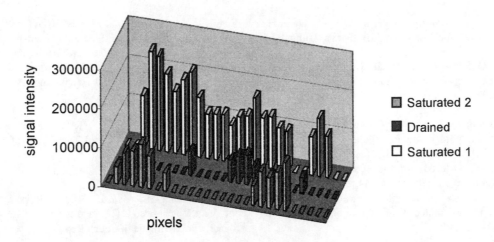

Figure 10. MR signal values in a fixed line of 28 pixels from the three runs S1, D and S2. Higher signal intensities during the run S1 give evidence that a larger volume was available as "free" water. The decrease of the signal during the second infiltration run S2 and the spatial shift of pixels with the highest signal may be a sign of changed driving force field.

REFERENCES

Amin, M.H.G., L.D.Hall, R.J.Chorley, T.A.Carpenter, K.S.Richards, and B.W. Bache, Magnetic resonance Imaging of Soil Water Phenomena, *Magn. Reson. Imaging*, 12, 319-321, 1994

Amin, M.H.G., R.J.Chorley, K.S.Richards, L.D.Hall, T.A.Carpenter, M.Cislerova, and T.Vogel, Study of infiltration into a heterogeneous soil using magnetic resonance imaging, *Hydrological Processes*, 11, 471-483, 1997.

Brooks, R.H., and A.T. Corey, Hydraulic properties of porous media, *Hydrology Papers No.3*, Colorado State University, Fort Collins, USA, 1964.

Burdine, N.T., Relative permeability calculations from pore size distribution data, *Petroleum Transactions, AIME*, 198, 71-78, 1953.

Childs, E.C., and Collis-George, The permeability of porous materials, *Proc. Royal Soc. London A*, 210, 392-405, 1950.

Císlerová, M., J. Šimůnek, and T. Vogel, Changes of steady-state infiltration rates in recurrent ponding infiltration experiments, *Journal of Hydrology*, 104,1-16, 1988.

Císlerová, M., T. Vogel, and J. Šimůnek, The Infiltration-Outflow Experiment Used to Detect Flow Deviations, in *Field-Scale Solute and Water Transport Through Soils*, edited by K.Roth, H.Flühler, W.A.Jury, and J.C.Parker, Birkhauser Verlag, Basel, pp.109-117, 1990.

Císlerová, M., J. Votrubová, T.Vogel, M.H.G. Amin, and L.D. Hall, Magnetic Resonance Imaging and Preferential Flow in Soils, in *Characterization and Measurement of the Hydraulic Properties of Unsaturated Porous Media*, edited by R.van Genuchten, and F.Leij, University of Riverside, pp. 397-412, 1999.

Císlerová, M. and J. Votrubová, Visualisation of porosity distribution by CAT, *a special issue of J.of Hydrology*, 2001, submitted

Clothier, B. E., and I.White, Measurement of Sorptivity and Soil Water Diffusivity in the Field, *SSSAJ*, 45, 241-245, 1981

Durner, W., Hydraulic conductivity estimation for soils with heterogeneous pore structure, *WRR* 30, 2, 211-223, 1994.

Germann, P. F. and L. Di Pietro, Scales and dimensions of momentum dissipation during preferential flow in soils, *WRR* 35, 5, 1443-1454, 1999.

Hall, L.D., M.H.G. Amin, M. Šanda, J.Votrubová, E. Dougherty, K.S. Richards, R.J. Chorley, and M. Císlerová, Detectable ranges of saturated soil water content by magnetic resonance imaging and MR properties of water in soils, *Geoderma*, 80, 431-448, 1997.

Kean, D.M., and M.A. Smith, Magnetic Resonance Imaging, Principles and Applications, W.Heinemann Medic. Books, London, UK, 1986.

Latour, L.L., R.L. Kleinberg, P.P. Mitra & C.H. Sotak, Pore-size distribution and tortuosity in heterogeneous porous media, *J. Magn. Reson. A*, 112, 83-91, 1995.

Mualem, Y., A new model for predicting the hydraulic conductivity of unsaturated porous media, *WRR*, 12, 513-522, 1976.

Nash, J.E., P.S. Eagelson, J.R. Philip, and W.H.van der Molen, The education of hydrologists, *Hydrol.Sci.J.*, 35, 597-607, 1990.

Nimmo, J., Modeling structural influences on soil water retention, *SSSAJ* 61, 712-719, 1997.

Or D., and M. Tuller, Liquid retention and interfacial area in variably saturated porous media: Upscaling from single-pore to sample-scale model, *WRR*, 35, 12, 3591-3605, 1999.

Pražák, J., M.Šír, F .Kubík, J. Tywoniak, and C. Zarcone, Oscillation Phenomena in Gravity-Driven Drainage in Coarse Porous-Media, *WRR*, 28, 7, 1849-1855, 1992.

Philip, J.R., Soil, Natural Science, and Models, *Soil Science*, 151, 1, 91-98, 1991.

Sněhota M, A. Robovská and M. Císlerová, Automated Set-Up Designed to Measure Hydraulic Parameters in Heterogeneous

Soil Close to Saturation. *Journal of Hydrology and Hydromechanics*, Slovak Academic Press Ltd., Bratislava, (submitted), 2001.

Tesař, M., M. Šír, O. Syrovátka, J. Pražák, L. Lichner, F. Kubík, Soil water regime in head water regions - observation, assessment and modelling, J. of Hydrology and Hydromechanics, SAV Bratislava, 2001, in print.

Tuller, M., D. Or, Hydraulic conductivity of variably saturated porous media: Film and corner flow in angular pore space, *WRR*, 37, 1257-1276, 2001.

van Genuchten, M.Th., A closed form for predicting the hydraulic conductivity of unsaturated soils, *SSSAJ*, 44, 892-898, 1980.

Vogel, T. and M. Císlerová, On the reliability of unsaturated hydraulic conductivity calculated from the moisture retention curve, *TIPM*, 3, 1-15, 1988

Vogel, T., M. Nakhaei and M. Císlerová, Description of Soil Hydraulic Properties Near Saturation from the Point of View of Inverse Modelling, in *Characterization and Measurement of the Hydraulic Properties of Unsaturated Porous Media*, edited by R.van Genuchten, and F.Leij, University of Riverside, pp. 693-703, 1999.

Votrubová, J., M.Šanda, M.Císlerová, M.G.Amin, and L.D.Hall, The Relationship between MR Parameters, and the Content of Water in Packed Samples of Two Types of Sandy Soil, *Geoderma*, 95, 267-282, 2000.

Dr. Milena Císlerová, Dr. Alice Robovská, Dr. Tomáš Vogel, and Jana Votrubová, CTU Faculty of Civil Engineering, Thákurova 7, 166 29 Prague 6, Czech Republic

Effect of Forced Convection on Soil Water Content Measurement With the Dual-Probe Heat-Pulse Method

Gerard J. Kluitenberg and Joshua L. Heitman

Department of Agronomy, Kansas State University, Manhattan, Kansas

The dual-probe heat pulse (DPHP) method is useful for measuring soil volumetric water content (θ) near heterogeneities such as the soil surface, but it does not consider convective heat transfer that may result from soil water movement (forced convection). In this study, we examined the effect of forced convection on estimates of soil water content using three different DPHP sensor orientations. Heat transfer theory that explicitly accounts for forced convection was used to test this effect. For Orientation I, the parallel heater and temperature probes were in a plane normal to the direction of steady water flow. The temperature probe was directly downstream from the heater probe for Orientations II and the temperature probe was upstream from the heater probe for Orientation III. A simple model based on instantaneous heating of the sensor gave excellent approximations of error in θ for Orientations II and III. Estimates of absolute error in θ (Δθ) for Orientation I required a model based on pulsed heating of the sensor. Forced convection causes θ to be underestimated for Orientation II and overestimated for Orientation I and III. The magnitude of these errors increased logarithmically with increasing water flux density, but the error for Orientation I was substantially smaller than that for Orientations II and III. We conclude that the effect of forced convection may be large enough to render the DPHP method useless for Orientations II and III. It does not, however, appear to limit the practical utility of DPHP sensors when placed in Orientation I.

INTRODUCTION

The dual-probe heat-pulse method introduced by *Campbell et al.* [1991] has proven useful for measuring soil volumetric water content [*Tarara and Ham*, 1997; *Bremer et al.*, 1998; *Bristow*, 1998; *Ham and Knapp*, 1998; *Song et al.*, 1998, 1999; *Basinger*, 1999; *Bremer and Ham*, 1999; *Bristow et al.*, 2001]. Because of their small size, dual-probe heat-pulse (DPHP) sensors may be particularly use-

ful for measuring water content near heterogeneities. *Philip and Kluitenberg* [1999] tested this expectation and found that heterogeneity errors are small provided that the heterogeneity is no closer than the probe separation (typically 0.006 m). Estimates of heterogeneity errors were later sharpened by *Kluitenberg and Philip* [1999].

The soil surface was one form of heterogeneity examined by *Philip and Kluitenberg* [1999] and *Kluitenberg and Philip* [1999]. The performance of DPHP sensors near the soil surface is of considerable interest and practical importance. Spatial and temporal observations of near-surface water content are needed to improve our understanding of near-surface hydrology [*Nielsen et al.*, 1996] and have proven useful in verifying remotely-sensed soil moisture estimates [*Famiglietti et al.*, 1999; *Georgakakos and Bau-*

Environmental Mechanics: Water, Mass and Energy Transfer in the Biosphere
Geophysical Monograph 129
Copyright 2002 by the American Geophysical Union
10.1029/129GM23

mer, 1996]. Wind erosion model development has also benefited from near-surface water content measurements [*Durar et al.*, 1995]. Many other examples could be cited.

Heterogeneity imposed by the presence of the soil surface is only one of several issues that must be addressed to evaluate the expected success in measuring near-surface water content with DPHP sensors. One issue of considerable importance is the possibility of heat convection resulting from soil water movement (forced convection). The heat transfer theory underlying the DPHP method of *Campbell et al.* [1991] considers only conductive heat transfer and implies that forced convection is negligible. Inasmuch as water flux densities are often greatest near the soil surface, it is important to determine whether forced convection due to infiltrating water will cause error in water content measurements with DPHP sensors.

In this study, we examine the effect of forced convection on estimates of soil volumetric water content for three sensor orientations. We first review the theory that forms the basis for the DPHP method and then introduce alternative heat transfer models that explicitly account for forced convection. The alternative models are then used to determine expected error in volumetric heat capacity estimates due to the effect of forced convection. These error estimates are then transformed to determine expected error in volumetric water content estimates.

DUAL-PROBE HEAT-PULSE METHOD

Campbell et al. [1991] proposed a sensor with two parallel probes extending from a plastic block. One probe contained a temperature sensor and the other probe contained a heater element that was used to introduce a heat pulse. By assuming that the heater probe approximates instantaneous heating of an infinite line source embedded in an infinite medium, *Campbell et al.* [1991] developed an inverse relationship between the maximum temperature rise at the temperature probe and the volumetric heat capacity of the medium. This relationship is

$$C = q/\left(e\pi r^2 T_m\right) \tag{1}$$

where C is the volumetric heat capacity (J m^{-3} K^{-1}) of the bulk soil (soil, water, and air), q is the quantity of heat liberated per unit length of heater (J m^{-1}), and T_m is the maximum temperature (K) rise measured at a distance r (m) from the heater. A value of r is determined for each sensor by making measurements with the sensor immersed in a medium of known heat capacity. Thus, r is treated as *apparent* rather than actual probe separation. Thereafter, C can be determined simply by obtaining measurements of q and T_m.

Neglecting the contribution from air, the heat capacity C is a weighted sum of the heat capacities of soil water and soil solid constituents

$$C = C_w\theta + \rho_b c_s \tag{2}$$

where C_w is the volumetric heat capacity of water (J m^{-3} K^{-1}), θ is the volumetric water content (m^3 m^{-3}), ρ_b is the bulk density (Mg m^{-3}), and c_s is the specific heat of the soil solid (mineral and organic) constituents (kJ kg^{-1} K^{-1}). Rearranging (2) yields the expression

$$\theta = \left(C - \rho_b c_s\right)/C_w \tag{3}$$

which shows that θ can be determined from measurements of C and ρ_b. The value of C_w is known and c_s can be estimated with sufficient accuracy for mineral soils.

It is evident from (3) that the effect of forced convection will be manifested in θ via error in determining C. Thus, it follows that fractional errors in θ will be directly proportional to fractional errors in C. It is for this reason that we first establish the effect of forced convection on C and later show how these errors are transformed into errors in θ.

HEAT TRANSFER MODELS

Governing Equation

Consider homogeneous, isotropic soil through which water moves at a constant rate in the x direction. Coupled conductive and convective heat transfer can be described by [*Marshall*, 1958; *Stallman*, 1965; *Ren et al.*, 2000]

$$\frac{\partial T}{\partial t} = \kappa\left(\frac{\partial^2 T}{\partial x^2} + \frac{\partial^2 T}{\partial y^2}\right) - V\frac{\partial T}{\partial x} \tag{4}$$

where T is temperature (K), t is time (s), κ is the thermal diffusivity (m^2 s^{-1}) of the bulk soil, and x and y are space coordinates (m). The heat-pulse velocity V (m s^{-1}), taken to be a positive quantity for flow in the positive x direction, is related to the water flux density J (m s^{-1}) by the expression

$$V = J\left(C_w/C\right) \tag{5}$$

This approach is based on the assumption that thermal homogeneity exists between solid, liquid, and gas phases in the soil. This assumption has not been tested and may require further evaluation.

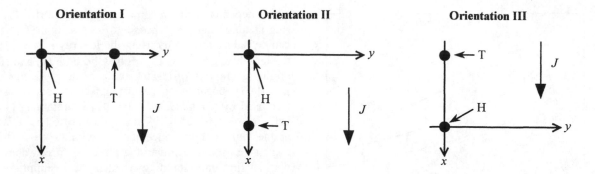

Figure 1. Three DPHP sensor orientations for measuring soil water content in the presence of a steady water flux density J in the positive x direction. The heater probe H is located at $(x = 0, y = 0)$ for all orientations. Temperature probe T is located at $(x = 0, y > 0)$ for Orientation I, $(x > 0, y = 0)$ for Orientation II, and $(x < 0, y = 0)$ for Orientation III.

Solutions for Infinite Line Source

We first consider an instantaneous release of heat from an infinite line source, which is normal to the x-y plane and passes through the point $(x, y) = (0, 0)$. The solution of (4) for this case is [*Marshall*, 1958]

$$T(x,y,t) = \frac{q}{4\pi\lambda t} \exp\left[-\frac{(x-Vt)^2 + y^2}{4\kappa t}\right] \quad (6)$$

where q is the quantity of heat liberated per unit length of heater (J m^{-1}) and λ is the thermal conductivity of the bulk soil (W m^{-1} K^{-1}), defined as the product of κ and C. Second, we consider an infinite line source heated at the rate q' (W m^{-1}) during the time interval $0 < t \leq t_0$. The solution of (4) for this case is [*Ren et al.*, 2000]

$$T(x,y,t) = \frac{q'}{4\pi\lambda} \int_0^t s^{-1} \exp\left[-\frac{(x-Vs)^2 + y^2}{4\kappa s}\right] ds \quad (7a)$$

for $0 < t \leq t_0$, and

$$T(x,y,t) = \frac{q'}{4\pi\lambda} \int_{t-t_0}^t s^{-1} \exp\left[-\frac{(x-Vs)^2 + y^2}{4\kappa s}\right] ds \quad (7b)$$

for $t > t_0$.

Character of the Solutions

Consider the three sensor orientations illustrated in Figure 1. Heater probe H is located at $(x = 0, y = 0)$ for all orientations, but the location of temperature probe T varies for each orientation. Probe T is located at $(x = 0, y > 0)$ for Orientation I, $(x > 0, y = 0)$ for Orientation II, and $(x < 0, y$

$= 0)$ for Orientation III. We now explore the behavior of (6) and (7) in the context of these sensor orientations. Equation (6) gives the transient temperature response for an instantaneous release of heat from an infinite line source in the presence of forced convection (Figure 2a). Equation (7) gives the transient temperature response for pulsed heating of an infinite line source in the presence of forced convection (Figure 2b). The behavior of (7) is similar to that of (6) except that pulsed heating causes a delay in the transient temperature response and a slightly smaller temperature rise. These differences in temperature response are illustrated for Orientation I (Figure 2b). Pulsed heating results for Orientations II and III (not shown) give nearly identical delays in transient temperature response and decreases in temperature rise. Thus, for all orientations, the maximum temperature rises predicted by (6) and (7) are similar in magnitude but occur at distinctly different times.

The results presented in Figure 2a show that forced convection causes asymmetry in the spatial temperature field. This can be seen by noting that each sensor orientation provides a measurement at a different location within the spatial temperature field near the heater probe (Figure 1). Forced convection causes an increase in maximum temperature rise downstream from the heater (Orientation II) and a decrease in maximum temperature rise upstream from the heater (Orientation III). Although it is not evident from Figure 2a, forced convection also causes a slight reduction in maximum temperature rise ($\approx 2\%$) for Orientation I. It follows that measurements of T_m obtained for estimating C and θ with the DPHP method are influenced by forced convection.

Although forced convection causes asymmetry in the temperature field, it appears that the temperature maxima occurred at the same time for all sensor orientations (Figure 2a). We show later that forced convection decreases the time t_m at which the maximum temperature rise is achieved, but that t_m is the same for all sensor orientations.

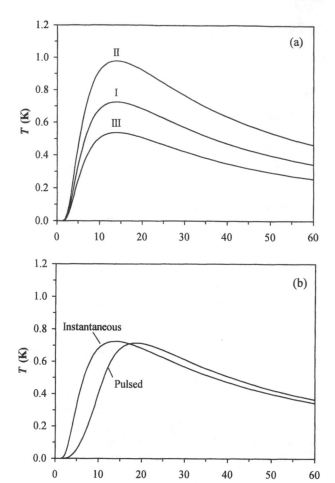

Figure 2. (a) Transient temperature response for an infinite line source with instantaneous heat input. Results were obtained by using (6) with the constants $q = 700$ J m^{-1}, $\kappa = 6.33 \times 10^{-7}$ m^2 s^{-1}, $C = 3.07$ MJ m^{-3} K^{-1}, and $V = 6.33 \times 10^{-5}$ m s^{-1} ($\upsilon = 0.3$). Separation between heater and temperature probes was 0.006 m for all orientations. (b) Transient temperature response for an infinite line source in Orientation I. Curve for instantaneous heat input is identical to that in upper panel. Curve for pulsed heat input was obtained by using (7) with $q' = 87.5$ W m^{-1} and $t_0 = 8$ s, which yields the same total heat input used for instantaneous heating (700 J m^{-1}). Values for κ, C, and V were the same as those used for the instantaneous heat input calculations.

ERROR ANALYSIS METHODOLOGY

Expressions for Volumetric Heat Capacity

Rearranging (6) yields an expression for the volumetric heat capacity

$$C = \frac{q}{4\pi\kappa T_m t_m} \exp\left[-\frac{(x - Vt_m)^2 + y^2}{4\kappa t_m}\right] \qquad (8)$$

which depends on the maximum temperature rise T_m occurring at time t_m. In the absence of convection ($V = 0$), (8) can be reduced to (1) by using the identity $r^2 = x^2 + y^2$ and the fact that $t_m = r^2/4\kappa$. Unfortunately, both (1) and (8) rely on the assumption of instantaneous heating. *Kluitenberg et al.* [1993] showed that the instantaneous heating assumption used by *Campbell et al.* [1991] in obtaining (1) typically causes error of $< 1\%$ in estimates of C, but this result applies only in the absence of convection ($V = 0$). Thus, (8) cannot be used to evaluate the effect of forced convection without first understanding whether the assumption of instantaneous heating is reasonable.

An alternative approach is to use an expression for C that explicitly accounts for pulsed heating. Such an expression can be obtained by rearranging (7b) to give

$$C = \frac{q'}{4\pi\kappa T_m} \int_{t_m - t_0}^{t_m} s^{-1} \exp\left[-\frac{(x - Vs)^2 + y^2}{4\kappa s}\right] ds \qquad (9)$$

which, in the absence of velocity ($V = 0$), reduces to [*Bristow et al.*, 1994]

$$C = \frac{q'}{4\pi\kappa T_m}\left[\text{Ei}\left(\frac{-r^2}{4\kappa(t_m - t_0)}\right) - \text{Ei}\left(\frac{-r^2}{4\kappa t_m}\right)\right] \qquad (10)$$

where $-\text{Ei}(-x)$ is the exponential integral.

Consistent with the use of (1) in the method of *Campbell et al.* [1991], we assume that measurements of T_m are available for use in (8) - (10), but that κ, V, and t_m must be specified or calculated. We next explore the relationship between κ, V, and t_m.

Relationships for Time to Temperature Maxima

Differentiating (6) with respect to time and setting the result to zero gives

$$4\kappa t_m = x^2 + y^2 - (Vt_m)^2 \qquad (11)$$

A useful dimensionless form of (11) is

$$\tau_m = \frac{2}{1 + \sqrt{1 + \upsilon^2}} \qquad (12)$$

where τ_m, dimensionless time to the temperature maximum, is defined as

$$\tau_m = \frac{4\kappa t_m}{x^2 + y^2} \qquad (13)$$

and υ, dimensionless velocity, is defined as

$$\upsilon = \frac{V\sqrt{x^2+y^2}}{2\kappa} \qquad (14)$$

From (14) we see that υ becomes $Vy/2\kappa$ for Orientation I, $Vx/2\kappa$ for Orientation II, and $|Vx/2\kappa|$ for Orientation III. Figure 3 shows that $\tau_m = 1$ in the absence of convection and decreases as υ increases.

Differentiating (7) with respect to time and setting the result to zero yields

$$\ln\!\left(1-\frac{t_0}{t_m}\right) = \frac{V^2 t_0}{4\kappa} - \frac{x^2+y^2}{4\kappa}\,\frac{t_0}{t_m(t_m-t_0)} \qquad (15)$$

A dimensionless form of (15) is

$$\ln\!\left(1-\frac{\tau_0}{\tau_m}\right) = \frac{\upsilon^2\tau_0}{4} - \frac{\tau_0}{\tau_m(\tau_m-\tau_0)} \qquad (16)$$

where τ_0, dimensionless heating duration, is defined as

$$\tau_0 = \frac{4\kappa t_0}{x^2+y^2} \qquad (17)$$

Values of τ_m satisfying (16) for different values of υ and τ_0 were obtained using a Van Wijngaarden-Dekker-Brent iterative technique [*Press et al.*, 1989]. Results for select values of τ_0 in Figure 3 show that τ_m increases as heating duration τ_0 increases. But notice that the shapes of the τ_m-υ curves remain largely unaffected by the heating duration τ_0.

Examination of (12) and (16) reveals that they are independent of sensor orientation. Thus, the results shown in Figure 3 hold for all orientations. In other words, the effect of convection in reducing τ_m (or t_m) is manifested similarly for all orientations. This confirms our earlier observation (Figure 2a) that t_m is decreased as a result of forced convection, but that it is decreased by the same amount for all sensor orientations.

Heat Capacity Error Analysis

For the instantaneous heating model, fractional error in C, denoted $\Delta C/C$, is calculated by computing the error introduced when (1) is used instead of (8) to estimate C. In this case $\Delta C/C$ is

$$\frac{\Delta C}{C} = \frac{\dfrac{q}{e\pi T_m r^2}}{\dfrac{q}{4\pi\kappa T_m t_m}\exp\!\left[-\dfrac{(x-Vt_m)^2+y^2}{4\kappa t_m}\right]} - 1 \qquad (18)$$

which reduces to

$$\frac{\Delta C}{C} = \tau_m \exp\!\left[\frac{\upsilon^2\tau_m}{4}+\frac{1}{\tau_m}-\frac{Vx}{2\kappa}-1\right] - 1 \qquad (19)$$

from which π, q, and T_m have been eliminated. Evaluation of (19) is accomplished by first specifying values for $Vx/2\kappa$ and υ. Substituting υ into (12) provides a corresponding value for τ_m. Using the values for $Vx/2\kappa$, υ, and τ_m in (19) give a corresponding value for $\Delta C/C$.

For the pulsed heating model, $\Delta C/C$ is calculated by computing the error introduced when (10) is used instead of (9) to estimate C. In this case $\Delta C/C$ is

$$\frac{\Delta C}{C} = \frac{\dfrac{q'}{4\pi\kappa T_m}\left[\mathrm{Ei}\!\left(\dfrac{-r^2}{4\kappa(t_m-t_0)}\right)-\mathrm{Ei}\!\left(\dfrac{-r^2}{4\kappa t_m}\right)\right]}{\dfrac{q'}{4\pi\kappa T_m}\displaystyle\int_{t_m-t_0}^{t_m} s^{-1}\exp\!\left[-\dfrac{(x-Vs)^2+y^2}{4\kappa s}\right]ds} - 1 \qquad (20)$$

Following *Kluitenberg and Warrick* [2001], (20) becomes

$$\frac{\Delta C}{C} = \frac{W\!\left(\dfrac{1}{\tau_m},0\right)-W\!\left(\dfrac{1}{\tau_m-\tau_0},0\right)}{\exp\!\left(\dfrac{Vx}{2\kappa}\right)\left[W\!\left(\dfrac{1}{\tau_m},\upsilon\right)-W\!\left(\dfrac{1}{\tau_m-\tau_0},\upsilon\right)\right]} - 1 \qquad (21)$$

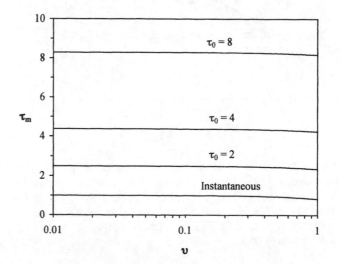

Figure 3. Time to temperature maximum (τ_m) as a function of velocity (υ) and heating duration (τ_0). Results for instantaneous heat input are from (12). Results for pulsed heat input are from (16). The dimensionless variables τ_m, υ, and τ_0 are defined in (13), (14), and (17), respectively. Note that $\tau_m = 1$ for instantaneous heating when $\upsilon = 0$.

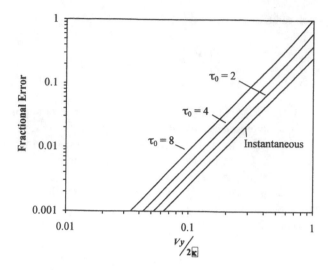

Figure 4. Fractional error in volumetric heat capacity ($\Delta C/C$) for Orientation I as a function of velocity (υ) and heating duration (τ_0). As noted in the text, $\upsilon = Vy/2\kappa$ for Orientation I. Results for instantaneous heating were obtained from (19); results for pulsed heating were obtained from (21). Dimensionless heating duration is defined in (17).

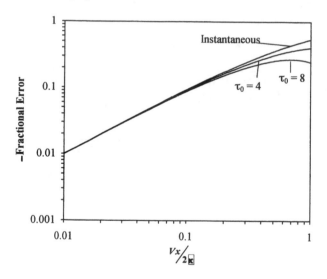

Figure 5. Fractional error in volumetric heat capacity ($\Delta C/C$) for Orientation II as a function of velocity (υ) and heating duration (τ_0). As noted in the text, $\upsilon = Vx/2\kappa$ for Orientation II. Results for instantaneous heating were obtained from (19); results for pulsed heating were obtained from (21). Dimensionless heating duration is defined in (17).

from which π, q, and T_m have been eliminated. The function W in (21) is the *well function for leaky aquifers*, defined as [*Hantush*, 1964, p. 321]

$$W(u,\beta) = \int_u^\infty z^{-1} \exp\left(-z - \beta^2/4z\right) dz \qquad (22)$$

The procedure for evaluating (21) is similar to that for evaluating (19) except that a value for τ_0 is also required and (16) is used to calculate τ_m. The method of *Kluitenberg and Warrick* [2001] was used to evaluate W with error $< 1 \times 10^{-5}$.

Water Content Error Analysis

Examination of (3) yields the expression

$$\Delta\theta = \frac{C}{C_\mathrm{w}} \frac{\Delta C}{C} \qquad (23)$$

for absolute error in water content ($\Delta\theta$), which is linearly related to fractional error in heat capacity. We used $C_\mathrm{w} = 4.18$ MJ m^{-3} K^{-1} in subsequent calculations.

ERROR ANALYSIS RESULTS

Error in Heat Capacity

Fractional error in heat capacity ($\Delta C/C$) is plotted as a function of υ for Orientations I, II, and III in Figures 4, 5, and 6, respectively. Results for the instantaneous heating model in Figures 4-6 were obtained from (19); results for the pulsed heating model were obtained from (21). As indicated previously, $\upsilon = Vy/2\kappa$ for Orientation I, $\upsilon = Vx/2\kappa$ for Orientation II, and $\upsilon = |Vx/2\kappa|$ for Orientation III. The magnitude of $\Delta C/C$ increases with increasing velocity for

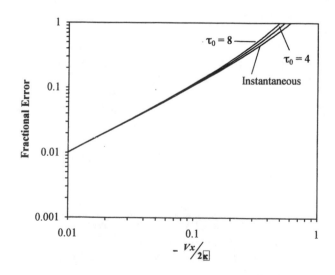

Figure 6. Fractional error in volumetric heat capacity ($\Delta C/C$) for Orientation III as a function of velocity (υ) and heating duration (τ_0). As noted in the text, $\upsilon = |Vx/2\kappa|$ for Orientation III. Results for instantaneous heating were obtained from (19); results for pulsed heating were obtained from (21). Dimensionless heating duration is defined in (17).

Table 1. Soil physical properties reported by *Ren et al.*, [2000]. Volumetric heat capacity (C) and thermal diffusivity (κ) were measured with disturbed soil material (ground, sieved, and repacked) at saturation.

Soil	Textural Class	Bulk Density	Organic Matter	C	κ
		Mg m^{-3}	%	MJ m^{-3} K^{-1}	m^2 s^{-1}
Hanlon	Sand	1.53	0.8	3.07	6.33×10^{-7}
Clarion	Sandy loam	1.38	2.3	3.19	5.54×10^{-7}
Harps	Clay loam	1.21	2.3	3.24	4.23×10^{-7}

all sensor orientations, but is positive (C overestimated) for Orientations I and III and negative (C underestimated) for Orientation II. It is also evident that $\Delta C/C$ is substantially smaller for Orientation I than for Orientations II and III over most of the dimensionless velocity range shown in Figures 4-6. Although it cannot be concluded from the figures, $\Delta C/C$ is zero in the absence of forced convection ($V = 0$) for both instantaneous and pulsed heat inputs.

Campbell et al. [1991] used DPHP sensors with an apparent probe spacing of $r \approx 0.006$ m and a heating duration of $t_0 = 8$ s. Others have used similar values for r and t_0. Typical values of thermal diffusivity for a mineral soil range from $\kappa \approx 2 \times 10^{-7}$ to $\kappa \approx 9 \times 10^{-7}$ m^2 s^{-1}. Thus, we find from (17) that τ_0 can be expected to fall in the range $0.2 < \tau_0 < 0.8$ under most circumstances. If indeed τ_0 is restricted to this range, (19) and (21) yield similar $\Delta C/C$ values for Orientations II and III in the velocity range $0.01 < \upsilon < 1.0$ (Figures 5 and 6). For Orientation II, values of $\Delta C/C$ obtained from (19) are within 1.5% of the values obtained from (21). For Orientation III, values of $\Delta C/C$ obtained from (19) are within 2.4% of those obtained from (21). We conclude that (19), which is simpler to evaluate than (21), provides excellent estimates of $\Delta C/C$ for Orientations II and III.

Unfortunately, (19) is not a good substitute for (21) when estimating $\Delta C/C$ for Orientation I with τ_0 restricted to the

range $0.2 < \tau_0 < 0.8$ (Figure 4). Values of $\Delta C/C$ obtained from (19) and (21) differ by as much as 10% over the velocity range $0.01 < \upsilon < 1.0$. Therefore, accurate estimates of $\Delta C/C$ for Orientation I require the use of (21).

Error in Water Content

Figures 4-6 provide general $\Delta C/C$ results that are applicable for a wide range of soil thermal properties. Similar generalization is not possible in presenting results for $\Delta\theta$. Instead, we present $\Delta\theta$ results for several examples based on the soils studied by *Ren et al.*, [2000]. Physical properties of these soils are reported in Table 1. Note that the thermal properties (Table 1) were obtained with disturbed soil material (ground, sieved, and repacked) at saturation. Inasmuch as we expect J to be largest for saturated soil, these soil materials, at saturation, yield worst-case estimates of error in $\Delta\theta$. For all example calculations, (21) was used to obtain $\Delta C/C$, heating duration was fixed at $t_0 = 8$ s, and (23) was used to calculate $\Delta\theta$ from $\Delta C/C$ and C/C_w.

Results for the Hanlon sand (Table 2) show that the magnitude of $\Delta\theta$ increases with increasing velocity for all sensor orientations, but is positive (θ overestimated) for Orientations I and III and negative (θ underestimated) for Orientation II. It is also evident that $\Delta\theta$ is substantially smaller for Orientation I than for Orientations II and III for the range of fluxes shown in Table 2. A flux density of $J =$

Table 2. Fractional error in volumetric heat capacity ($\Delta C/C$) and absolute error in volumetric water content ($\Delta\theta$) as a function of water flux density (J) for saturated Hanlon sand (see Table 1). As noted in the text, $\upsilon = Vy/2\kappa$ for Orientation I, $\upsilon = Vx/2\kappa$ for Orientation II, and $\upsilon = |Vx/2\kappa|$ for Orientation III.

J	υ	Orientation I		Orientation II		Orientation III	
		$\Delta C/C$	$\Delta\theta$[a]	$\Delta C/C$	$\Delta\theta$[b]	$\Delta C/C$	$\Delta\theta$[c]
m s^{-1}			m^3 m^{-3}		m^3 m^{-3}		m^3 m^{-3}
1×10^{-6}	6.45×10^{-3}	1.09×10^{-5}	< 0.001	-6.42×10^{-3}	-0.005	6.49×10^{-3}	0.005
3×10^{-6}	1.94×10^{-2}	9.84×10^{-5}	< 0.001	-1.91×10^{-2}	-0.014	1.97×10^{-2}	0.014
1×10^{-5}	6.45×10^{-2}	1.09×10^{-3}	0.001	-6.15×10^{-2}	-0.045	6.78×10^{-2}	0.050
3×10^{-5}	1.94×10^{-1}	9.81×10^{-3}	0.007	-1.68×10^{-1}	-0.123	2.26×10^{-1}	0.166
1×10^{-4}	6.45×10^{-1}	1.06×10^{-1}	0.078	-4.20×10^{-1}	-0.308	1.11×10^{0}	0.814

[a]Error of $\Delta\theta = 0.01$ occurs at $J = 3.538 \times 10^{-5}$ m s^{-1}.
[b]Error of $\Delta\theta = -0.01$ occurs at $J = 2.133 \times 10^{-6}$ m s^{-1}.
[c]Error of $\Delta\theta = 0.01$ occurs at $J = 2.089 \times 10^{-6}$ m s^{-1}.

Table 3. Absolute error in volumetric water content ($\Delta\theta$) as a function of flux density (J) for the water-saturated soils listed in Table 1. Results are for Orientation I.

	$\Delta\theta^a$		
J	Hanlon	Clarion	Harps
m s^{-1}	------------------ m^3 m^{-3} ------------------		
1×10^{-5}	0.001	0.001	0.002
2×10^{-5}	0.003	0.004	0.007
3×10^{-5}	0.007	0.009	0.015
5×10^{-5}	0.020	0.025	0.041
7×10^{-5}	0.039	0.048	0.078
1×10^{-4}	0.078	0.096	0.156

[a]Error of $\Delta\theta = 0.01$ occurs at $J = 3.538 \times 10^{-5}$ m s^{-1} for Hanlon, $J = 3.174 \times 10^{-5}$ m s^{-1} for Clarion, and $J = 2.462 \times 10^{-5}$ m s^{-1} for Harps.

1×10^{-5} m s^{-1} (3.6 cm h^{-1}) causes errors of $\Delta\theta = 0.001$ for Orientation I, $\Delta\theta = -0.045$ for Orientation II, and $\Delta\theta = 0.050$ for Orientation III. We anticipate that, for a coarse-textured soil such as the Hanlon (92.1% sand), the effect of forced convection on DPHP sensors in Orientation I will be negligible except for the most extreme flux densities. For DPHP sensors in Orientations II or III, the effect of forced convection may be large enough to render the method useless.

The Clarion and Harps soils exhibited larger water content errors than the Hanlon at a given flux density J (Table 3). These results are presented to illustrate the sensitivity of $\Delta\theta$ to changes in soil thermal properties. The larger θ errors were caused by the lower thermal diffusivities of the Clarion and Harps soils (Table 1). But this effect was offset slightly by the higher volumetric heat capacities of these soils. Decreases in C also cause $\Delta\theta$ to increase. Note, however, that we would generally anticipate smaller flux densities for the Clarion and Harps soils because of their higher clay content than the Hanlon. This is similar to what would be expected in the presence of unsaturated flow. Values for C and κ generally will be lower for unsaturated soil, but flux densities can be expected to be orders of magnitude smaller than those for saturated flow.

SUMMARY AND CONCLUSIONS

We evaluated the effect of forced convection on soil water content determination with DPHP sensors for three sensor orientations. Heat transfer models that explicitly account for forced convection were used to achieve this purpose. One of the models was based on the assumption that the heater probe of a DPHP sensor delivers its heat to the soil instantaneously. A second model was based on the

more realistic assumption that the heater probe imparts heat to the soil over a finite period of time (pulse input). Both models were used to develop expressions for fractional error in volumetric heat capacity ($\Delta C/C$) and absolute error in volumetric water content ($\Delta\theta$). We have shown that the simpler expression, based on the instantaneous heating model, gives excellent approximations of $\Delta C/C$ and $\Delta\theta$ for Orientations II and III. We have also shown that the expression based on the pulsed heating model must be used in order to obtain accurate error estimates for Orientation I.

We conclude from our results that, for all sensor orientations, error in C and θ increase in magnitude logarithmically as water flux density increases. But C and θ are underestimated for Orientation II and overestimated for Orientations I and III. We also note that lower soil thermal properties (C and κ) enhance the effect of forced convection and cause greater error in C and θ. Our results also show that errors in C and θ are substantially smaller for Orientation I than for Orientations II and III for all but the highest water flux densities. Hence, the effect of forced convection is predicted to be significantly smaller for DPHP sensors installed in Orientation I than for those installed in Orientations II and III.

For DPHP sensors in Orientations II or III, the effect of forced convection may be large enough to render the method useless. For DPHP sensors in Orientation I, the theory shows that relatively high water flux densities are required for forced convection to cause significant error in θ measurements. Forced convection, therefore, does not appear to limit the practical utility of DPHP sensors when placed in Orientation I.

Finally, we emphasize that our results were obtained using an assumption of uniform water flow in the vicinity of the DPHP sensor. Thus, our estimates of error in C or θ may not be useful if there is strong spatial heterogeneity in water flux density at the spatial scale of the sensor.

REFERENCES

Basinger, J. M., Laboratory and field evaluation of the dual-probe heat-pulse method for measuring soil water content, M.S. thesis, Kansas State University, Manhattan, KS, 1999.

Bremer, D. J., and J. M. Ham, Effect of spring burning on the energy balance in a tallgrass prairie, *Agric. For. Meteorol.*, 97, 43-54, 1999.

Bremer, D. J., J. M. Ham, C. E. Owensby, and A. K. Knapp, Responses of soil respiration to clipping and grazing in a tallgrass prairie, *J. Environ. Qual.*, 27, 1539-1548, 1998.

Bristow, K. L, Measurement of thermal properties and water content of unsaturated sandy soil using dual-probe heat-pulse probes, *Agric. For. Meteorol.*, 89, 75-84, 1998.

Bristow, K. L., G. J. Kluitenberg, C. J. Goding, and T. S. Fitzgerald, A small multi-needle probe for measuring soil thermal properties, water content and electrical conductivity, *Comput. Electron. Agric.*, 31, 265-280, 2001.

Bristow, K. L., G. J. Kluitenberg, and R. Horton, Measurement of soil thermal properties with a dual-probe heat-pulse technique, *Soil Sci. Soc. Am. J.*, 58, 1288-1294, 1994.

Campbell, G. S., C. Calissendorff, and J. H. Williams, Probe for measuring soil specific heat using a heat-pulse method, *Soil Sci. Soc. Am. J.*, 55, 291-293, 1991.

Durar, A. A., J. L. Steiner, S. R. Evett, and E. L. Skidmore, Measured and simulated surface soil drying, *Agron. J.*, 87, 235-244, 1995.

Famiglietti, J. S., J. A. Devereaux, C. A. Laymon, T. Tsegaye, P. R. Houser, T. J. Jackson, S. T. Graham, M. Rodell, and P. J. van Oevelen, Ground-based investigation of soil moisture variability within remote sensing footprints during the Southern Great Plains 1997 (SGP97) Hydrology Experiment, *Water Resour. Res.*, 35, 1839-1851, 1999.

Georgakakos, K. P., and O. W. Baumer, Measurement and utilization of on-site soil moisture data, *J. Hydrol.*, 184, 131-152, 1996.

Ham, J. M., and A. K. Knapp, Fluxes of CO2, water vapor, and energy from a prairie ecosystem during the seasonal transition from carbon sink to carbon source, *Agric. For. Meteorol.*, 89, 1-14, 1998.

Hantush, M. S., Hydraulics of wells, *Adv. Hydrosci.*, 1, 281-432, 1964.

Kluitenberg, G. J., J. M. Ham, and K. L. Bristow, Error analysis of the heat pulse method for measuring soil volumetric heat capacity, *Soil Sci. Soc. Am. J.*, 57: 1444-1451, 1993.

Kluitenberg, G. J., and J. R. Philip, Dual thermal probes near plane interfaces, *Soil Sci. Soc. Am. J.*, 63, 1585-1591, 1999.

Kluitenberg, G. J., and A. W. Warrick, Improved evaluation procedure for heat-pulse soil water flux density method, *Soil Sci. Soc. Am. J.*, 65, 320-323, 2001.

Marshall, D. C., Measurement of sap flow in conifers by heat transport, *Plant Physiol.*, 33, 385-396, 1958.

Nielsen, D. R., M. Kutilek, and M. B. Parlange, Surface soil water content regimes: Opportunities in soil science, *J. Hydrol.*, 184, 35-55, 1996.

Philip, J. R., and G. J. Kluitenberg, Errors of dual thermal probes due to soil heterogeneity across a plane interface, *Soil Sci. Soc. Am. J.*, 63, 1579-1585, 1999.

Press, W. H., B. P. Flannery, S. A. Teukolsky, and W. T. Vetterling, *Numerical recipes in Pascal. The art of scientific computing*, Cambridge Univ. Press, New York, 1989.

Ren, T., G. J. Kluitenberg, and R. Horton, Determining soil water flux and pore water velocity by a heat pulse technique, *Soil Sci. Soc. Am. J.*, 64, 552-560, 2000.

Song, Y., J. M. Ham, M. B. Kirkham, and G. J. Kluitenberg, Measuring soil water content under turfgrass using the dual-probe heat-pulse technique, *J. Am. Soc. Hort. Sci.*, 123, 937-941, 1998.

Song, Y., M. B. Kirkham, J. M. Ham, and G. J. Kluitenberg, Dual probe heat pulse technique for measuring soil water content and sunflower water uptake, *Soil & Tillage Res.*, 50, 345-348, 1999.

Stallman, R. W., Steady one-dimensional fluid flow in a semi-infinite porous medium with sinusoidal surface temperature, J. Geophys. Res., 70, 2821-2827, 1965.

Tarara, J. M., and J. M. Ham, Measuring soil water content in the laboratory and field with dual-probe heat-capacity sensors, *Agron. J.*, 89, 535-542, 1997.

G. J. Kluitenberg and J. L. Heitman, Department of Agronomy, Kansas State University, Manhattan, KS 66506. (gjk@ksu.edu, jheitman@ksu.edu).

Momentum Transfer to Complex Terrain

John Finnigan

CSIRO Atmospheric Research, Canberra, Australia

By definition the boundary layer is the layer of the atmosphere that responds directly to the character of the earth's surface. Usually, the land surface is heterogeneous on a variety of scales and this heterogeneity is reflected in the boundary layer. We have classified surface heterogeneity into three types: complex surfaces such as plant canopies and urban areas, where the horizontal scale of heterogeneity is small; changing surface cover such as that found in farmland or between major changes in land use; and topography. For each class of heterogeneity we compare the changes that occur in the boundary layer with the canonical boundary layer over homogeneous flat terrain. Our emphasis is on changes to the windfield and surface stress that occur in near-neutral stratification and we do not discuss scalar fields or the effects of changes in the surface energy balance. In discussing topography we deal only with low hills whose influence on the flow is confined within the boundary layer.

1. INTRODUCTION

Although John Philip is best known for his contributions to soil science, he also devoted a great deal of his attention to the turbulent atmospheric boundary layer. The building he designed for the CSIRO Division of Environmental Mechanics, which he founded and led for over twenty years, housed what was then the largest boundary layer wind tunnel in Australia. From its earliest days, the micrometeorological work that he encouraged combined simulations in this tunnel, field experiments and analytic theory to extend the compass of micrometeorology or natural aerodynamics from a focus on the 'flat earth' that is so convenient for theorists and so rarely encountered in practice, to the real world of heterogeneous surfaces, tall plant canopies and complex topography. Indeed, John's own most significant contributions to micrometeorology addressed the important question of scalar transfer from heterogeneous surfaces [*Philip, 1987; 1996a,b; 1997*].

In this paper, the present state of understanding of the way momentum is transferred to real surfaces is summarized. As will become apparent, a substantial degree of commonality can now be traced in transfer mechanisms in what are, at first sight, quite different physical configurations. I am sure that this emerging synthesis would have appealed to John's deep conviction that unity in the description of physical phenomena is to be found not far beneath surface dissimilarities.

The boundary layer is the layer of the atmosphere that is influenced directly by the roughness and energy balance of the surface. Atmospheric properties like windspeed, temperature and scalar concentrations vary rapidly through this layer, changing from their surface values to merge with the synoptic state above the boundary layer. The boundary layer is the only region of the atmosphere that is neutrally or unstably stratified for much of the time as the layers of air in contact with the ground respond to the friction and solar heating of the surface and become turbulent. Much of the character of the boundary layer, therefore, is impressed upon it by the particular nature of the underlying surface.

Environmental Mechanics: Water, Mass and Energy Transfer in the Biosphere
Geophysical Monograph 129

Copyright 2002 by the American Geophysical Union
10.1029/129GM24

At a sufficiently small scale all natural surfaces exhibit some spatial inhomogeneity. When this is of a scale large enough to cause sensible spatial variations in the mean and turbulent properties of the boundary layer, then we regard the underlying ground as a 'complex' surface. Although, more often than not, we find the elements of complexity in arbitrary combinations, it is useful to divide them into three main classes, each with its own characteristic features. These are: *Complex Surfaces*, including vegetation canopies and urban areas; *Changing Surface Cover*, including transitions between surfaces of different roughness such as farmland to forest or water to land; and *Topography* ranging from small hills to mountains.

Complex Surfaces

The distinguishing feature of complex surfaces is that, while the horizontal scale of inhomogeneity in the elements that make up the surface-plants or houses, for example-is relatively small, the elements are high enough that we are interested in the properties of the atmosphere between them, not just because we might walk amongst them as in a city or forest but because it is necessary to recognize that the exchange of quantities like heat and momentum between them and the atmosphere occurs over some height range rather than just at the ground surface.

Changing Surface Cover

Here we are concerned with inhomogeneity at much larger horizontal lengthscales, ranging from simple changes between one surface type and another to continual changes such as might be seen in farms with fields planted with different crops. It is characterized by the appearance of *internal boundary layers* over each new surface. If the new surface continues sufficiently far downstream without further change, the new internal boundary layer replaces the old boundary layer and eventually a new geostrophic balance is struck between the surface and the synoptic flow above the boundary layer. If the surface character changes continually, however, the impact of each internal layer only extends up to some *blending height*, above which the total boundary layer behaves as if it were flowing over a surface with properties that are some average of the different patches.

Topography

Hills and valleys affect boundary layer flow because the pressure field, that develops as the atmosphere flows over them, accelerates and decelerates the near-surface flow. In a relatively thin layer near the surface, analogous to an in-

ternal boundary layer, changes in turbulent stresses strongly affect the mean flow but at higher levels the changes in mean windspeed are essentially inviscid. The pressure field that develops about any given hill is strongly dependent upon the stratification of the atmosphere flowing over it, which can be characterized by a Froude number. Hence, the scale of the topography profoundly affects the resultant boundary layer flow patterns [*Carruthers and Hunt*, 1990]. Those over a very large hill, whose pressure field is largely determined by the displacement of the stratified synoptic flow above the boundary layer, are quite different to those over a smaller hill, where flow displacement is confined within the neutral or unstable boundary layer. Here we will confine our attention to smaller hills.

In this chapter we will be concerned with the boundary layers that develop over terrain with these different kinds of complexity and will concentrate especially on two aspects of their description: the windfields that we observe within them and the surface stresses beneath them. Our introduction of spatially averaged equations below suggests one motivation for this. Mathematical models used for climate or weather prediction have horizontal resolutions between 50km and 500km so that the windspeed averaged over grid cells between $50km^2$ and $500km^2$ in area has to be related to some average of the surface properties within the cell. At the same time we want to know how to relate measurements of windspeed and other variables at points in an evolving boundary layer to the surrounding landscape.

Although this paper will concentrate on the windfield, surface roughness and stress this is not meant to discount the equally important relationships between surface energy balance, heat and water vapour fluxes. Rather it is a response to space limitations and in recognition of this our attention will be primarily on boundary layers near neutral stratification, where the energy balance is less important. In particular, we have not discussed stably stratified nocturnal boundary layers as universal scaling laws are much less applicable in their case.

2. GOVERNING EQUATIONS

In discussing flow over individual roughness elements, surface patches or hills we adopt familiar Reynolds averaged flow equations. Conservation of momentum is expressed as,

$$\frac{\partial \overline{u_i}}{\partial t} + \overline{u_j}\frac{\partial \overline{u_i}}{\partial x_j} = -\frac{\partial \overline{p}}{\partial x_i} - 2\,\varepsilon_{ijk}\Omega_j\,\overline{u_k} + \frac{\partial \tau_{ij}}{\partial x_j} + \delta_{i3}\,g\,\frac{\overline{\theta}}{T_0} \qquad (1)$$

and conservation of mass by the continuity equation,

$$\frac{\partial u_i}{\partial x_i} = \frac{\partial \overline{u_i}}{\partial x_i} = \frac{\partial u_i{'}}{\partial x_i} = 0 \quad (2)$$

The time average operator is denoted by an overbar,

$$\bar{a}(t) = \frac{1}{T}\int_{t-T/2}^{t+T/2} a(t')dt' \quad (3)$$

and the velocity vector is split into mean and fluctuating parts,

$$u_i(t) = \bar{u}_i + u_i'(t) \quad \text{where,} \quad \overline{u_i'(t)} = 0 \quad (4)$$

We use a right-handed rectangular Cartesian coordinate system, x_i (x,y,z) with x_1 (x) aligned with the mean velocity at the surface and x_3 (z) normal to the ground surface. Velocity components aligned with x_i (x,y,z) are denoted by u_i (u,v,w) with u_1 (u) the streamwise and u_3 (w) the vertical component. ε_{ijk} is the alternating unit tensor and δ_{ij}, the Kronecker delta. The acceleration due to gravity is g, $\bar{\theta}$ is the averaged potential temperature, T_0 a reference temperature and we have made the Boussinesq assumption [eg. *Businger*, 1982]. Ω_i is the angular rotation vector of the earth, p is the kinematic pressure and ν the kinematic viscosity. The kinematic momentum flux tensor τ_{ij} includes both turbulent and viscous stresses,

$$\tau_{ij} = -\overline{u_i'u_j'} + \nu\frac{\partial\bar{u}_i}{\partial x_j} \quad (5)$$

We are particularly interested here in the average flow that develops over various scales of surface heterogeneity and so we now introduce momentum and continuity equations in which mean quantities are averaged spatially as well as in time.

The volume average of a scalar or vector function ϕ_j is defined as,

$$\langle\phi_j\rangle(\mathbf{x},t) = \frac{1}{V}\iiint_V \phi_j(\mathbf{x}+\mathbf{r},t)d^3\mathbf{r} \quad (6)$$

and $\phi_j = \langle\phi_j\rangle + \phi_j''$ where $\langle\phi_j''(\mathbf{x},t)\rangle = 0$

The averaging volume V, which excludes solid roughness elements, consists of a horizontal slab, extensive enough in the x-y plane to eliminate variations in flow structure on the scale of individual roughness elements but thin enough to preserve any important variation of properties in the vertical. When it is convenient, we will allow the thickness of the slab to be vanishingly small so that $\langle\ \rangle$ can also denote an areal average over the x-y plane. In canopies where a significant fraction of the total volume is occupied by solid elements, which may be the case in

dense urban areas, V should reflect the ratio of solid to open space and may be a function of height.

Where the surface roughness elements intersect the spatial averaging slab, differentiation and volume averaging do not commute. Instead it can be shown [*Raupach and Shaw*, 1982; *Finnigan*, 1985],

$$\left\langle\frac{\partial\phi_j}{\partial x_i}\right\rangle = \frac{\partial\langle\phi_j\rangle}{\partial x_i} - \frac{1}{V}\iint_{S_I}\phi_j n_i dS, \quad (7)$$

where the surface S_I is the sum of all the solid surfaces that intersect the averaging volume V, and n_i is the unit normal vector pointing away from S_I into V.

With these definitions, spatially averaged continuity and momentum equations become,

$$\frac{\partial\langle\bar{u}_i\rangle}{\partial x_i} = \frac{\partial\overline{u_i''}}{\partial x_i} = 0 \quad (8)$$

$$\frac{\partial\langle\bar{u}_i\rangle}{\partial t} + \langle\bar{u}_j\rangle\frac{\partial\langle\bar{u}_i\rangle}{\partial x_j} = -\frac{\partial\langle\bar{p}\rangle}{\partial x_i} - 2\varepsilon_{ijk}\Omega_j\langle\bar{u}_k\rangle$$
$$+ \frac{\partial\langle\tau_{ij}\rangle}{\partial x_j} + \delta_{i3}\,g\,\frac{\langle\bar{\theta}\rangle}{T_0} + f_{Pi} + f_{Vi} \quad (9)$$

$$\langle\tau_{ij}\rangle = -\langle\overline{u_i'u_j'}\rangle - \langle\overline{u_i''}\,\overline{u_j''}\rangle + \nu\frac{\partial\langle\bar{u}_i\rangle}{\partial x_j} \quad (10)$$

$$f_{Pi} = \frac{1}{V}\iint_{S_I}\bar{p}n_i dS\ ; \quad f_{Vi} = -\frac{\nu}{V}\iint_{S_I}\frac{\partial\bar{u}_i}{\partial n}dS \quad (11)$$

The kinematic momentum flux tensor, $\langle\tau_{ij}\rangle$ as well as the conventional turbulent and viscous stresses now includes the dispersive flux term, the second term on the right hand side of equation (10), which results from any spatial correlations in the time-averaged velocity field. The viscous flux is usually negligible in the high Reynolds Number flows we discuss in this Chapter. In defining $\langle\tau_{ij}\rangle$ we have ignored small terms that arise because the space and time averaging operators do not strictly obey Reynolds averaging rules.

f_{Pi} and f_{Vi} are (minus) the volume averaged sums of the pressure and viscous forces, respectively, exerted on every solid element that intersects the averaging volume V. Together they constitute the aerodynamic drag on unit mass of air within V. These two terms are identically zero when z is above the level of the highest roughness element.

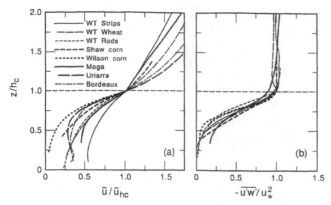

Figure 1. Normalized mean velocity and shear stress profiles measured in a variety of canopies ranging from wind tunnel models a few centimeters high (prefix WT), through natural cereal crops (Shaw and Wilson corn) to forests (Moga, Uriarra, Bordeaux). Full details may be found in Kaimal and Finnigan (1994) (see further reading).

3. COMPLEX SURFACES

The defining characteristic of a complex surface is that the region of interaction between the atmosphere and the surface is extended in the vertical rather than being confined to the ground plane. The most obvious and best studied examples are plant canopies but urban landscapes share many of the same characteristics and cityscapes are often referred to as 'Urban Canopies'. Canopies, whether natural or urban, have obvious structure in the vertical but are horizontally homogeneous on the scale of many plants or buildings. It is a practical impossibility to deal directly with the spatial complexity of the airflow between the canopy elements so below the level of the roughness elements we apply the equations defined in the last section and describe canopy flows in terms of areally averaged variables.

In figure (1) we have plotted mean velocity and shear stress profiles typical of natural and artificial canopies on flat ground [*Kaimal and Finnigan*, 1994; *Raupach et al*, 1996]. They illustrate some key features of flow close to and within complex surfaces. Looking first at Figure (1b), the shear stress profile, we observe above the canopy the constant stress layer expected in a steady, non-accelerating atmospheric surface layer (ASL) but within the canopy we see that momentum is absorbed steadily as aerodynamic drag on the canopy and that little stress is exerted on the underlying ground surface. An immediate consequence of this is seen in the velocity profiles plotted in Figure (1a). Again, above the canopy we see the expected boundary layer profiles but within the canopy the profiles are roughly exponential with a marked inflection point at the canopy top.

One effect of the absorption of momentum over the height of the canopy is that the origin of coordinates for the logarithmic velocity profile above the canopy is displaced from the ground a distance d, the *displacement height* so that in the neutral surface layer the velocity profile must be written,

$$\bar{u}(z) = \frac{u*}{\kappa} Log\left(\frac{z-d}{z_0}\right) \quad (12)$$

where κ is Von Karman's constant ($\square$ 0.4) and $u_* = \sqrt{\tau_0} = \sqrt{\langle\tau_{13}\rangle(h_C)}$ is the friction velocity, h_C being the height of the canopy. The displacement height d is equal to the mean level of momentum absorption on the solid elements [*Jackson*, 1981],

$$d = \int_0^{hc} z \frac{\partial \tau}{\partial z} dz \bigg/ \int_0^{hc} \frac{\partial \tau}{\partial z} dz \quad (13)$$

The inflection in the velocity profile at the top of the canopy has very important consequences for the turbulent structure and scaling laws. An inflected profile of this kind is inviscidly unstable and spontaneously generates energetic turbulent eddies. The size of the eddies that are generated is of order h_C and they dominate the turbulence structure in a layer known as the 'Roughness Sub-Layer', which extends from the ground up to two or three canopy heights [*Raupach et al*, 1996, *Finnigan*, 2000]. When the canopy elements are sufficiently sparse, a dynamically significant inflected profile may not be present although the area-averaged velocity will still display an inflexion point. In this case, the size of the dominant eddies will be linked to the size of the separated flow regions and wakes behind individual elements, which will still often be of order h_C, especially in urban canopies.

The Roughness Sub-Layer (RSL) forms the lowest distinct layer of the atmospheric boundary layer over complex surfaces. Between $3h_C$ and h_C turbulence statistics such as integral lengthscales depart from Monin-Obukhov scaling, where the lengthscale is $(z-d)$, and become constant with height, scaling on L_S, the natural lengthscale of the shear at the canopy top,

$$L_S = \langle \bar{u}(h_C) \rangle \big/ \left[\partial \langle \bar{u}(h_C) \rangle \big/ \partial z \right] \quad (14)$$

or more conveniently on $(h\text{-}d)$, which is approximately proportional to L_S [*Raupach et al*, 1996; *Kaimal and Finnigan*, 1994].

In this region also, vertical diffusion by the turbulence becomes more 'efficient' in the sense that eddy diffusivities for momentum, K_M and scalars, K_C,

$$K_M = \frac{\tau}{\partial \langle \bar{u} \rangle / \partial z} ; \qquad K_C = \frac{F_C}{\partial \langle \bar{c} \rangle / \partial z} \quad (15)$$

(where F_C is the kinematic flux density of an arbitrary scalar c) become up to twice as large as their Monin-Obukhov, ASL counterparts. Within the canopy layer itself ($h_C > z > 0$) turbulent transport is not a diffusive process because the large canopy eddies responsible for transport, are the same size as the scale of mean vertical gradients in momentum and scalars. Indeed, locally, the eddy-diffusivities may even become negative [*Denmead and Bradley*, 1985].

The effects on other turbulence moments of the vertical distribution of sources and sinks and the special character of RSL turbulence are equally profound. We have a particular interest in these features of complex surfaces for at least two reasons. First because plant canopies form a biologically active lower boundary to the atmosphere at which incoming solar radiation is partitioned into the sensible and latent heat fluxes which drive atmospheric mixing over land. Second because in cities, it is the turbulence generated though these mechanisms that is responsible for ventilating urban streets.

4. CHANGING SURFACE COVER.

Moving now to larger scale, we consider patches of surface cover that contain many individual roughness elements so that their characteristics can be described by averaged quantities like roughness lengths, z_0 or displacement heights, d (Equation 12). We will look first at simple changes of surface roughness such as those between bare soil and an irrigated crop. Once we have established the nature of simple transitions from one type of surface to another we will be in a position to describe the boundary layer over patchy surfaces.

4.1. Local Advection: Yhe Wind Field

Local advection refers to situations where the effects of surface changes do not propagate above δ_{ASL}, the depth of the surface layer. Imagine a situation where an equilibrium boundary layer, where the airflow in the surface layer is characterized by a logarithmic profile with roughness length z_{01} and displacement height d_1, encounters a new surface with roughness length z_{02} and displacement height d_2. We will assume that the boundary is perpendicular to the surface wind vector. As the airflow encounters the new surface it either slows down because of increased surface friction (smooth-rough, $z_{02} > z_{01}$) or speeds up because the surface friction falls (rough-smooth: $z_{02} < z_{01}$). The effect of this acceleration or deceleration, which is initially confined to the air layers in contact with the new surface, is diffused vertically by turbulence and the effect of the change is felt through a steadily growing *internal boundary layer* of depth $\delta_i(x)$ (see Figure 2a) [*Garratt*, 1990; *Kaimal and Finnigan*, 1994].

The effects of the change are also transmitted by pressure forces that are associated with any change in streamline height that follows if d_1 is not equal to d_2 and this pressure perturbation is not confined to the internal boundary layer. Its effect is negligible, however, except very close to the transition even when the change in displacement height is significant and for the rest of this section we will ignore it. We will also avoid writing $z-d$, assuming that the origin of the z coordinate is adjusted appropriately to include d.

The *strength* of the roughness change can be characterized by the ratio of the roughness lengths, M^* or its logarithm, M,

$$M^* = \frac{z_{01}}{z_{02}}; \qquad M = \ln\left(\frac{z_{01}}{z_{02}}\right) = \ln(z_{01}) - \ln(z_{02}) \quad (16)$$

Within the internal boundary layer the flow displays characteristics of the downstream surface. Outside it, apart from the small perturbation caused by the pressure pulse at the transition, the flow field is identical to that upwind (see Figure 2b). The internal boundary layer depth $\delta_i(x)$ is usually defined, therefore, as the height at which the downwind velocity $\bar{u}_2(z)$ or shearing stress $\tau_2(z)$ attain fixed fractions, eg. 99%, of their upwind values at the same height. Unless we state otherwise, henceforth we will assume that $\delta_i(x)$ is defined in terms of the velocity.

The growth of the internal boundary layer is caused by turbulent diffusion and, if we take the characteristic diffusion velocity as $u_{*2} = \sqrt{\tau_{02}}$, the downstream friction velocity, then we can write,

$$\frac{d\delta_i}{dx} = \frac{Bu_{*2}}{\bar{u}_2(z)} \quad (17)$$

To integrate equation (17) we need an expression for $\bar{u}_2(z)$ and, for $\delta_i < \delta_{ASL}$, the simplest assumption is that,

$$\bar{u}_2(z) = \frac{u_{*2}}{\kappa} \ln\left(\frac{z}{z_{02}}\right) \quad (18)$$

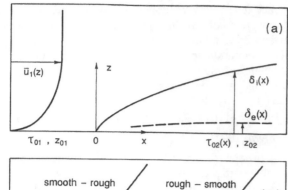

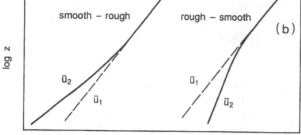

Figure 2. (a) Schematic diagram of internal boundary layer growth. The inner equilibrium region is marked by the dashed curve. This region is not expected to begin until some distance after the roughness change. (b) Logarithmic velocity profiles after a roughness change. The upwind equilibrium profile is denoted by a dashed line.

Then stipulating that $\delta_i(x) = 0$ at $x=0$, and locating the origin of coordinates at the roughness change, we obtain [*Panofsky and Dutton,* 1984],

$$\frac{\delta_i(x)}{x}\left[\ln\left(\frac{\delta_i(x)}{z_{02}}\right) - 1\right] = B\kappa \quad (19)$$

Equation (19) provides a qualitative description of the growth of the internal boundary layer and, with $B \square 1.25$ provides a good quantitative measure of $\delta_i(x)$ for smooth-rough changes and for moderate rough-smooth transitions ($M \lesssim 2$). When the rough–smooth change is larger ($M > 3$), equation (19) tends to overestimate the growth in $\delta_i(x)$ because then, diffusion downstream of the roughness change is controlled for some distance by the slowly decaying upstream turbulence [*Antonia and Luxton,* 1972].

To obtain equation (19) we assumed that the velocity profile within the internal boundary layer was logarithmic all the way up to $\delta_i(x)$. This is a gross oversimplification, however. In Figure (2b) we have identified an inner *equi-*

librium layer, $\delta_e(x)$ at the bottom of the internal boundary layer. Only in this layer has the flow attained local equilibrium with the new surface with the shearing stress $\tau_2(z)$ approximately constant with height and the velocity profile $\bar{u}_2(z)$ obeying equation (18). An estimate for $\delta_e(x)$ can be obtained by first writing an approximate equation for the streamwise momentum balance that ignores any pressure perturbation at the roughness change and also assumes that the changes in the flow field are small,

$$\bar{u}_1(z)\frac{\partial \Delta u}{\partial x} \square \frac{\partial \Delta \tau}{\partial z} \quad (20)$$

where $\Delta u = \bar{u}_2(z) - \bar{u}_1(z)$ and $\Delta \tau = \tau_2(z) - \tau_1(z)$. If we now insist that for local equilibrium to obtain below $\delta_e(x)$, the integral from $z=0$ to δ_e of the advection term on the left hand side of equation (20) must be negligible compared to the perturbation in surface stress, $\tau_{02} - \tau_{01} = u_{*2}^2 - u_{*1}^2$, we obtain [*Mason,* 1988],

$$\frac{\delta_e(x)}{x} \square 2\left(\frac{u_{*2}}{\bar{u}_2}\right)^2 \quad (21)$$

whence,

$$\frac{\delta_e(x)}{x}\ln^2\left(\frac{\delta_e(x)}{z_{02}}\right) \square 2\kappa^2 \quad (22)$$

For the kinds of roughness changes often studied in micrometeorology, the slope $\delta_i/x \approx 1/10$ while $\delta_e/x \approx 1/100$. Hence δ_e corresponds to the height-to-fetch requirements traditionally adopted as a rule of thumb by researchers who wish to apply one-dimensional formulae downwind of a change in surface cover.

For $\delta_i > z > \delta_e$ we have a blending region, where the velocity profile changes smoothly between $\bar{u}(z) = u_{*2}/\kappa \ln(z/z_{02})$ and $\bar{u}(z) = u_{*1}/\kappa \ln(z/z_{01})$. In this region and downwind of the immediate vicinity of the transition, the velocity and shear stress perturbations are *self preserving*, that is, they can be written as functions of a velocity scale u_0 and a dimensionless height $\eta(x) = z/\delta_i(x)$,

$$\Delta u(z) = \bar{u}_2(z) - \bar{u}_1(z) = \frac{u_0}{\kappa}g(\eta) \quad (23)$$

$$\Delta \tau = \tau_2(z) - u_{*1}^2 = \left[u_{*2}^2 - u_{*1}^2\right]h(\eta) \quad (24)$$

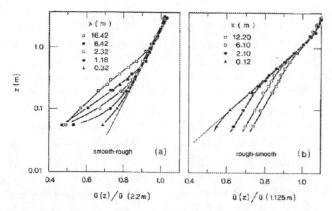

Figure 3. The development of logarithmic velocity profiles after a roughness change (Data taken from Bradley (1968) see Kaimal and Finnigan (1994) for details). (a) smooth-rough: z_{01}=0.02 mm, z_{02}=2.5 mm, M= -4.8 (b) rough-smooth: z_{01}=2.5 mm, z_{02}=0.02 mm, M= +4.8.

A good choice for the velocity scale is $u_0 = u_{*2} - u_{*1}$ and the functions $g(\eta)$ and $h(\eta)$ can be found by substituting equations (23) and (24) into the equations of motion and making a closure assumption to relate $\tau_2(z)$ to $\overline{u}_2(z)$. Several theories have been developed in this way [eg, *Mulhearn*, 1977] and we shall encounter one of them when we consider continually changing surfaces.

Typical examples of the velocity profiles that develop following smooth-rough (M= -4.8) and rough-smooth (M=+4.8) changes are illustrated in Figures (3a) and (3b) [*Bradley*,1968]. In each case we see the internal boundary layer deepening with downstream distance and the velocity profile slowing in the smooth-rough case and accelerating in the rough-smooth. In both cases, the lower part of the internal boundary layer is occupied by a logarithmic profile in equilibrium with the new surface although the true depth of the equilibrium region is exaggerated by the logarithmic height scale. Measured in terms of physical distance the equilibrium region appears to be established more slowly in the rough-smooth case but in terms of dimensionless distance, x/z_0 there is little difference between the two transitions in the rate at which equilibrium is reached.

4.2. Local Advection: Surface Stress

In Figures (4a) and (4b) we have plotted measurements of surface shearing stress from the experiment that furnished the velocity profiles of Figure (3a,b) [*Bradley*,1968]. These results are typical of those from experiments at a range of scales [*Kaimal and Finnigan*,1994]. Two features are noteworthy: the overshoot in stress at the transition and the rapid attainment of a new equilibrium.

The overshoot phenomenon is easily explained. In the case of a smooth-rough transition, the airstream, travelling relatively rapidly over the smooth surface, generates a high stress on first encountering the increased roughness. As the region of decelerated flow thickens into an internal boundary layer, the velocity of the air in contact with the surface slows and the surface stress falls. In a rough smooth transition we see a stress undershoot with a relatively slow airstream generating lower stress when the surface roughness falls but the stress then rising as the flow accelerates.

Although sophisticated models of the magnitude of the stress change have been developed, a simple expression can be derived by assuming that the velocity profile obeys equation (12) with $u_* = u_{*2}$ and $z_0 = z_{02}$ for the full depth of the inner region and then with $u_* = u_{*1}$ and $z_0 = z_{01}$ after a sharp discontinuity at $z = \delta_i$ [*Elliot*, 1958]. Matching the two layers leads directly to,

$$\frac{\tau_{02}}{\tau_{01}} = \left[1 - \frac{M}{\ln(\delta_i/z_{02})}\right]^2 \quad (25)$$

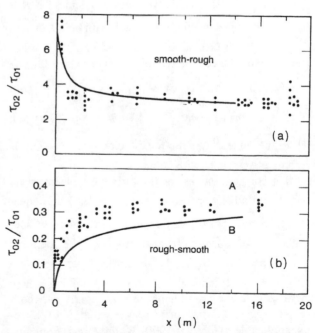

Figure 4. Surface shear stress development after roughness changes (Data taken from Bradley (1968) see Kaimal and Finnigan (1994) for details). (a) smooth-rough: z_{01}=0.02 mm, z_{02}=2.5 mm, M=-4.8 (b) rough-smooth: z_{01}=2.5 mm, z_{02}=0.02 mm, M=+4.8. The dotted line represents equation (21) with $\delta_i(x)$ calculated using equation (15).

The result of equation (25) is plotted on top of the data points in Figures (4a) and (4b) and it is clear that it performs quite well in the smooth-rough case but underestimates the stress change for the rough-smooth transition. Equation (25) relies on an accurate expression for δ_i and we have already noted that equation (19), which is used to generate the curves in Figure (4a,b), underestimates the growth rate of $\delta_i(x)$ in the rough-smooth case because it discounts the influence of the energetic upstream turbulence on the diffusion of the new internal boundary layer.

4.3   Advection on Larger Scales

The formulae we have derived above and the reasoning behind them strictly apply to internal boundary layers that are no deeper than δ_{ASL} because we have assumed that the mean velocity $\overline{u}_1(z)$ may be described by the logarithmic law. Above δ_{ASL}, both the characteristic velocity and length scales of the turbulence change. The length scale becomes $O(z_i)$, the depth of the whole boundary layer, while the velocity scale depends upon whether the boundary layer is neutrally or unstably stratified. In a neutral boundary layer, the turbulent velocity scale is u_* and, at higher levels, $\overline{u}(z)$ changes much more slowly with height than in the logarithmic surface layer. More usually, the surface layer is capped by a convective mixed layer, where the turbulent velocity scale is $w_* = \left[g/T_0 \left(\overline{w'\theta'} \right)_0 z_i \right]^{1/3}$ with $\left(\overline{w'\theta'} \right)_0$ the surface heat flux [Kaimal and Finnigan, 1994]. In the mixed layer the mean velocity $\overline{u}(z) = U_M$ is approximately constant with height.

Inserting constant values for the turbulent velocity scale (u_* or w_*) and advection velocity ($\overline{u}(z)$ or U_M) into equation (17), we see that we can expect $\delta_i(x)$ to grow linearly above the surface layer with a slope between $Bu_*/\overline{u}(z)$ and Bw_*/U_M as the boundary layer varies between neutral stratification and convective mixing. There are relatively few measurements in this regime but those that exist suggest that the surface layer value $B \approx 1.25$ remains applicable [Garratt, 1990, Kaimal and Finnigan, 1994].

The early attainment of a new equilibrium surface stress that is shown in Figure (4) belies the continual slow adjustment of this quantity as the internal boundary layer grows out of the surface layer. The new internal boundary layer replaces the old boundary layer when $\delta_i(x)$ equals the old boundary layer depth. This occurs at downstream distances of order $x/z_{02} = 10^6$ in neutral conditions but possibly much less in a convective boundary layer with a weak mean wind. Current understanding of the magnitude of the geostrophic drag coefficient u_*/G, where G is the geostrophic windspeed, suggests that in the smooth-rough case illustrated in Figure (4), the early equilibrium value of $\tau_{02}/\tau_{01} \square 3.5$ will fall to $\tau_{02}/\tau_{01} \square 2.0$ as the new boundary layer attains geostrophic balance [Taylor, 1969; Jensen, 1978]. For the neutral case, this occurs between the point $x/z_0 \approx 10^6$, at which the new boundary layer replaces the old and $x/z_0 \approx 10^8$. We can see by integrating equation (1) across the boundary layer that attaining a new balance between the surface drag and the geostrophic wind will also change the geostrophic departure, the angle between the surface and geostrophic wind direction. This angle will increase in a smooth-rough and decrease in a rough-smooth change [Taylor, 1969].

4.4   Patchwork Surfaces

Natural surfaces rarely consist of simple changes between two types, rather the surface cover changes continuously. To describe flow over these surfaces we generalize the concept of the internal boundary layer to define the blending height, h_B [Mason, 1988; Mahrt, 1996; Philip, 1996a, 1996b, 1997]. Figure (5) illustrates a hypothetical surface consisting of a set of N patches of different surface cover, each occupying a plan area a_i with streamwise extent L_i and having roughness lengths and displacement heights z_{0i} and d_i, respectively. Over each surface an internal boundary layer grows and reaches a depth $\delta_i(L_i)$ by the end of the patch. From the definition of the internal boundary layer we know that above δ_{iMAX}, the height of the deepest internal boundary layer, the velocity profile $\overline{u}(z)$ no longer varies horizontally but attains a spatially averaged value so we can identify the blending height with δ_{iMAX},

$$h_B = \delta_{iMAX} \quad (26)$$

If δ_{iMAX} is smaller than the depth of the surface layer, δ_{ASL}, then for

$\delta_{ASL} > z > h_B$ the velocity profile will be logarithmic with the form,

$$\left\langle \overline{u}(z) \right\rangle = \overline{u}(z) = \frac{\left\langle \tau_0 \right\rangle^{1/2}}{\kappa} \ln \left(\frac{z}{z_0^{eff}} \right) \quad (27)$$

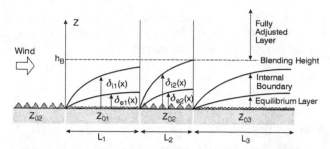

Figure 5. Schematic drawing of the flow structure over a series of surface patches with different roughness lengths z_{0i} and streamwise extents L_i.

where $\langle \ \rangle$ here denotes a volume average over a slab of vanishing thickness or, equivalently, an area average over the (x,y) plane.

A central problem over natural surfaces is to find an expression for the effective roughness length z_0^{eff} in terms of patch level roughness lengths z_{0i} and other accessible parameters such as the windspeed above δ_{iMAX} so that the area-averaged surface momentum flux $\langle \tau_0 \rangle$ can be inferred from windspeed measurements or parameterized in models that are unable to resolve the individual patches.

One approach to finding z_0^{eff} is to assume once again that the flow within each internal boundary layer, rather than being self preserving (equations (23), (24)), can be represented by logarithmic profiles with local roughness lengths and then to average the profiles across the x-y plane,

$$\langle \bar{u}(z) \rangle = \left\langle \frac{\tau_{0i}^{1/2}}{\kappa} \ln\left(\frac{z}{z_{0i}}\right) \right\rangle = \frac{\langle \tau_{0i} \rangle^{1/2}}{\kappa} \ln\left(\frac{z}{z_0^{eff}}\right) \quad (28)$$

whence,

$$z_0^{eff} = \frac{\langle u_{*i} \ln(z_{0i}) \rangle}{\langle \tau_{0i} \rangle^{1/2}}; \quad \text{and} \quad u_{*i} = \sqrt{\tau_{0i}} \quad (29)$$

Equation (29) is not a very useful formula because we do not, in general, know the stress τ_{0i} on each patch. The simplest recourse is to ignore the correlation between stress and roughness length and to write [*Taylor*, 1987]

$$z_0^{eff} \ \Box \ z_0^M = \langle \ln(z_{0i}) \rangle \quad (30)$$

Because, as we have seen, z_{0i} and τ_{0i} are positively correlated, z_0^M will always be an underestimate of z_0^{eff} but it

forms a useful reference value and provides a first estimate of z_0^{eff} when the variation in roughness length between patches is small.

A more accurate formula for z_0^{eff} has been derived by exploiting the fact that the flow in the internal boundary layers making up the blending region between δ_{ei} and h_B is self-preserving and assuming that in this region, a simple mixing-length expression is adequate to express the relationship between shear stress $\tau(z)$ and velocity shear $\partial \bar{u}/\partial z$ [*Goode and Belcher*, 1999]. These two assumptions allow the shear stress and velocity at the blending height to be related to the local values within the thin equilibrium layer over each patch so that the value of τ_{0i} required to weight the local roughness length z_{0i} in equation (29) can be inferred. The result is a formula for the effective roughness length that is most simply expressed as,

$$\left\langle \frac{\left[\ln\left(\frac{h_B}{z_0^{eff}}\right) + \frac{h_{ei}}{h_i - h_{ei}} \ln\left(\frac{h_i}{h_{ei}}\right) - 1 \right]^2}{\left[\ln\left(\frac{h_i}{z_{0i}}\right) + \frac{h_{ei}}{h_i - h_{ei}} \ln\left(\frac{h_i}{h_{ei}}\right) - 1 \right]^2} \right\rangle = 1 \ (31)$$

where $h_{ei} = \delta_{ei}(L_i)$ and $h_i = \delta_i(L_i)$. Values for h_B, h_i and h_{ei} are readily obtained using equations (26), (19) and (22).

Equation (31) is a much better estimate of z_0^{eff} than z_0^M but it also starts to underestimate the momentum-absorbing capacity of a heterogeneous surface when the streamwise lengthscale of the patches, L_i becomes small. This is because in deriving equation (31), it is assumed that the equilibrium value of stress, τ_{0i} applies over an entire patch a_i and the overshoots and undershoots in stress at the roughness transitions that we saw in Figure (4) have been ignored. Closer inspection of Figure (4) reveals that the smooth-rough and rough-smooth overshoot-undershoot in stress is not symmetrical. If we define the average of the equilibrium stress values over two adjacent patches as $\tau_{0A} = (\tau_{0R} + \tau_{0S})/2$, where subscript R refers to the rough and S the smooth surface then $\Delta\tau_0$, a non-dimensional deviation from τ_{0A} can be written as [*Schmid and Bunzli*, 1995],

$$\Delta\tau_0 = \frac{\tau_0 - \tau_{0A}}{\tau_{0R} - \tau_{0A}} \quad (32)$$

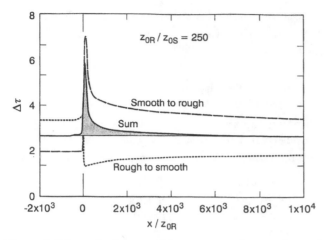

Figure 6. Schematic diagram of the asymmetry in smooth-rough and rough-smooth transitions using data taken from Figure (4).

and $\Delta\tau_0\left(x/z_{0R}\right)$ is plotted in Figure (6), which uses values of stress taken from the experiment shown in Figure (4). It is clear that, relative to τ_{0A}, there is an excess of stress associated with smooth-rough transitions within a region $10^3 > x/z_{0R} > 0$ so that as L_i, the streamwise extent of the rough patches of surface approaches $x/z_{0R} \,\square\, 10^3$, then z_0^{eff} will climb above the value predicted by equation (31). There is evidence from field studies, for example, that scattered small patches of woodland in a rural landscape can easily double z_0^{eff} relative to the value predicted by equation (31) [*Hopwood*, 1995].

This effect is most severe when the regions of high momentum absorption cannot be regarded as patches of surface roughness at all but instead are windbreaks, hedges, dykes, walls or other bluff structures. We can represent F_W the drag force on a unit cross-wind strip of an obstacle such as a windbreak, whose height is H_W by the expression,

$$F_W = \rho C_W U_H^2 H_W \quad (33)$$

where the drag coefficient C_W is of order 1 and U_H is the mean velocity at height H_W in the undisturbed upwind flow. Imagine now that we have an array of such obstacles aligned normally to the prevailing wind and spaced regularly a distance L_W apart on a surface that would have a roughness length z_0 in the absence of the windbreaks. Experiments have shown that the windbreak will effectively shelter a downwind region $x \sim 10H_W$, hence, using equa-

tion (12), we can readily calculate the spacing at which the momentum absorbed by the windbreaks and by the intervening surface is equal. This occurs at,

$$\frac{L_W}{H_W} \sim \frac{C_W \ln^2\left(z/z_0\right)}{\kappa^2} + 10 \quad (34)$$

With $z_0 \sim 0.002$m (typical of bare soil or snow) $L_W/H_W \sim$ 500 while for $z_0 \sim 0.01$m (crops) $L_W/H_W \sim 100$ so that in a typical rural landscape a large fraction of z_0^{eff} may be contributed by isolated upstanding obstacles and formulae like equation (27) must then be used with care.

4.5 Larger Scale Surface Variability

When the scale of individual surface patches L_i becomes much larger than a kilometre, then the blending height will be greater than the depth of the surface layer. At much larger scales ($L_i >> 10km$) then the new internal boundary layer will replace the entire planetary boundary layer and the regional surface stress can be calculated by averaging the contributions of essentially independent patches. At the intermediate scale, where $10km > L_i > 1km$, the blending height will be above the surface layer and formulae based on assumptions of logarithmic velocity profiles are inappropriate. Currently there are no simple descriptions of this scale of heterogeneity. Numerical models that can accommodate the diabatic influences that are usually important above the surface layer have been used in particular cases and the average surface stress can be considered to be bounded by the values appropriate to small and large scale heterogeneity.

5. TOPOGRAPHY

As we did in considering changing surfaces, we shall first describe the flow over an isolated hill and then go on to consider how the boundary layer adjusts to continuously hilly terrain. We will confine our attention to hills sufficiently small that the flow perturbations they cause are confined within the boundary layer. In practice this means that the hill height H and the hill horizontal lengthscale L satisfy $H<<z_i$ and $L<<h^*$, where h^*, the 'relaxation length' of the boundary layer is defined as $h^* = z_i U_0/u_*$ or $z_i U_0/w_*$ according to whether the flow is neutrally stratified or convectively unstable. The horizontal lengthscale L is defined as the distance from the hill crest to the half-height point. In continuously hilly terrain it can be more appropriate to use a characteristic wavelength λ as the horizontal lengthscale. For sinusoidal terrain, $L = \lambda/4$.

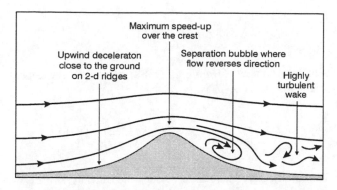

Figure 7. Schematic drawing of the flow over a 2D ridge showing the formation of a downstream separation region when the ridge is steep enough. On an axisymmetric hill, the upwind deceleration region is replaced by a region of lateral flow divergence.

In Figure (7) we have sketched the main features of the velocity field about an isolated hill. The figure could represent flow approaching an axisymmetric hill or a 2D ridge at right angles. Close to the surface, the flow decelerates slightly at the foot of the 2D ridge before accelerating to the summit. In the axisymmetric case the deceleration is replaced by a region of lateral flow divergence at the foot of the hill. The wind reaches its maximum speed above the hill top and then decelerates on the lee side. If the hill is steep enough downwind, a separation bubble forms in which the mean flow reverses direction. Whether the flow separates or not, a wake region forms behind the hill with a marked velocity deficit extending for at least $10H$ downwind.

The same information is made more concrete in Figure (8), where we plot velocity profiles well upwind, over the hill top and in the wake. The vertical coordinate z measures height above the local surface. In Figure (8) it is made dimensionless with the *inner layer height*, l, defined below. Upwind we have a standard logarithmic profile but on the hill top the profile is accelerated with the maximum relative speed-up occurring quite close to the surface at $z/l \sim 0.3$. In the wake we see a substantial velocity deficit extending to at least $z = H$.

Much of the understanding we now have about the dynamics of flow over hills derives from linear theory, which assumes that the mean flow perturbations caused by the hill are small in comparison to the upwind flow. Although strictly, linear theory is limited to hills of low slope, $H/L \ll 1$, its insights are applicable to much steeper hills. Linear theory supposes a division of the flow field into two main regions, an inner region of depth l and an outer region above, which are distinguished by essentially different dy-

namics (Figure 9). The balance between advection, streamwise pressure gradient and the vertical divergence of the shear stress can be expressed in an approximate linearized momentum equation,

$$U(z)\frac{\partial \Delta u}{\partial x} + \frac{\partial \Delta p}{\partial x} \sim \frac{\partial \Delta \tau}{\partial z} \quad (35)$$

where $\Delta u, \Delta p, \Delta \tau$ are the perturbations in streamwise velocity, kinematic pressure and shear stress that are induced by the hill and $U(z)$ denotes the undisturbed flow upwind of the hill. Well above the surface, perturbations in stress gradient are small and advection and pressure gradient are essentially in balance. Close to the surface an imbalance develops between these terms as the perturbation stress gradient grows. The inner layer height is defined as the level at which the left hand side of equation (35) equals the right hand side [*Hunt et al*, 1988].

A second interpretation of l is as the height at which the time taken for a turbulent eddy to be advected over the hill is equal to the *eddy turnover time*, that is the typical lifetime before the eddy, generated by interaction with the mean flow, is dissipated [*Kaimal and Finnigan*, 1994]. This interpretation tells us that for $z/l \ll 1$, the turbulence will be approximately in *local equilibrium*, that is, that production and dissipation of turbulent kinetic energy bal-

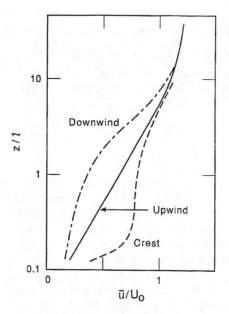

Figure 8. Profiles of mean velocity observed upwind, on the crest and in the wake region of a hill. The vertical scale is made dimensionless with the inner layer depth, l. Note the position of the maximum speed-up on the crest at $z \sim l/3$.

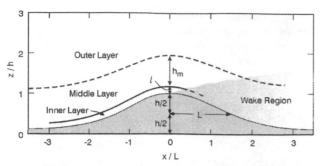

Figure 9. The different regions of the flow over a isolated hill, comprising: inner, middle, outer and wake layers and their associated lengthscales.

ance locally so that the relationship between the shear stress and the mean flow can be described by a *mixing length* or eddy-diffusivity . For $z/l \gg 1$, in contrast, the turbulence will experience *rapid distortion*, where changes to the turbulent stresses will depend on the cumulative straining of eddies by the mean flow as they are advected over the hill. In particular, for $z/l \gg 1$, the response of the mean flow is essentially inviscid because *changes* in turbulence moments have negligible effect on the mean flow. Hence, although the vertical structure of the undisturbed velocity profile $U(z)$ is entirely the result of turbulent stresses, perturbations to this profile are governed by inviscid dynamics except within the thin inner layer l (as long as $L \ll h^*$).

If the undisturbed upwind profile is taken as logarithmic, $U(z) = u_*/\kappa \ln(z/z_0)$, and we adopt a mixing length parameterization to relate the shear stress $\tau(x, y, z)$ to the velocity field over the hill, the first definition for the inner layer depth given above leads to an implicit expression for l [*Hunt et al,* 1988],

$$\frac{l}{L}\ln\left(\frac{l}{z_0}\right) = 2\kappa^2 \quad (36)$$

Equation (36) is very similar to the expression we found for the internal boundary layer height, equation (19). This is no accident as yet a third interpretation of l is as the height to which new vorticity, which is generated at the surface at a rate equal to the streamwise gradient of perturbation pressure, diffuses over the hill [*Finnigan et al,* 1990].

The pressure field that develops over the hill deflects the entire boundary flow over the obstacle. Its magnitude is determined, therefore, by the inertia of the faster flowing air in the outer region and is also related to the steepness of the hill so we expect,

$$\Delta p \sim \frac{H}{L}U_0^2 \quad (37)$$

Scaling arguments [*Hunt et al,* 1988] reveal that the appropriate definition of U_0 is,

$$U_0 = U(h_m); \qquad \frac{h_m}{L}\ln^{1/2}\left(\frac{h_m}{z_0}\right) \sim 1 \quad (38)$$

The middle layer height h_m divides the outer region into a *middle region* between l and h_m, where shear in the approach flow exerts an important influence on the flow dynamics, and an upper region, where the perturbations are described by potential flow. For a hill with $L = 200$m, $u_* = 0.3$m/s and $z_0 = 0.02$m, typical sizes of these scales are $l = 10$m, $h_m = 70$m, $U_0 = 6$m/s and $U(l) = 4.5$m/s. Note that in the linear theory, the vertical extent of the regions influenced by the hill depend only on the hill length, L. The hill height enters only through the influence of steepness H/L on the pressure perturbation that drives all other changes in the flow field.

The pressure perturbation falls to a minimum at the hill top and then rises again behind the hill and it propagates essentially unattenuated to the surface. Its scaling gives a strong clue as to why the relative speed-up peaks in the inner layer. Referring again to equation (35), except very close to the surface the momentum balance is dominated by the pressure gradient and the advection so that $U(z)\Delta u(x, z)/L \sim (H/L)U_0^2/L$. Within the inner layer, as the background flow $U(z)$ becomes much smaller than U_0, the velocity perturbation Δu must grow to compensate. Eventually, at the bottom of the inner layer the stress gradient dominates the momentum balance and reduces Δu so that the peak in speed-up is found at about $z \sim l/3$. The effects of this shifting balance can be clearly seen in the expression for the relative speed-up derived from linear theory [*Hunt et al,* 1988],

$$\frac{\Delta \bar{u}(x, z)}{U(z)} = \frac{H}{L}\left[\frac{U^2(h_m)}{U(l)U(z)}\right]\varsigma(x, z_0) \quad (39)$$

$\varsigma(x, z_0)$ is a function that factors the precise shape of the hill and the influence of surface roughness into the equation while both the dependence of the driving perturbation pressure gradient on hill slope H/L and the amplification of the speed-up by the ratio of background velocities across the middle layer, $U(h_m)/U(l)$ is evident.

5.1 Drag Force on Isolated Hills

We are particularly interested here in the drag force exerted by the hill upon the atmosphere. This is almost entirely a result of the asymmetry in the pressure field about the hill, which results in a net *form* or pressure drag on the obstacle as in equation (11). The hill also produces a negative perturbation in the surface shear stress, equal to the balance between the increase in stress as the wind accelerates to the hill crest and the extended region of reduced stress on the downslope and sheltered wake region. Even on low hills, however, this net reduction in τ_0 is an order of magnitude smaller than the increase in form drag. Over steep hills, flow separation ensures that the pressure on the lee side of the hill does not recover to its upstream value, leading to a net form drag but over lower hills without separation, the mechanism is more subtle.

The effect of the shear stress in the inner layer on the streamwise velocity perturbation Δu is to displace its peak value slightly upwind of the hill crest and to thicken the inner layer on the downslope side so that streamlines are not symmetrically disposed about the hill but are further from the surface on the downslope side. The asymmetric perturbation in vertical velocity Δw that accompanies Δu is amplified by the shear in the middle layer and acts to force an asymmetric component in the pressure perturbation Δp, which is determined primarily by the flow at $z = h_m$ and above. When integrated over the hill as in equation (11), this asymmetry in Δp results in form drag [*Belcher et al*, 1993]. Put more simply, the flow well above the hill acts like an inviscid flow over a surface defined by the streamlines at the top of the inner layer and it is this 'inviscid' flow that determines the pressure field acting on the surface. From this viewpoint it is easy to see that separation will set an upper limit on flow speed-up and form drag because, as the hill gets steeper, the upper level flow 'sees' the hill *plus* separation bubble as the lower boundary condition, effectively increasing L in equation (39) and limiting H/L.

Linear theory gives an exact expression for the drag on a 2D sinusoidal ridge of wavelength λ [*Belcher et al*, 1993],

$$|F_P| = 2\left[\frac{U_0}{U(l)}\right]^4 \left[\frac{\pi H}{\lambda}\right]^2 u_*^2 \lambda = \frac{\pi^2}{2}\left[\frac{U_0}{U(l)}\right]^4 \left[\frac{H}{L}\right]^2 u_*^2 L \quad (40)$$

where $F_P = \iint_{S_H} \overline{p}\, n_i\, dS$ is the streamwise pressure drag on a

ridge whose surface is S_H (see equation (11).

Comparing equations (40) and (39) we can see that the pressure asymmetry is proportional to the square of the velocity perturbation and so is proportional to the square of the hill slope H/L and to the 4th power of the shear amplification factor $U_0/U(l)$. This formula has been successfully extended to 3D hills by generalizing the hill slope H/L to A/S_h, where A is the frontal area of the hill and S_h its base area [*Wood and Mason*, 1993]. We then obtain,

$$|F_P| = 2\pi^2 \beta \left[\frac{U_0}{U(l)}\right]^4 \left[\frac{A}{S_h}\right] u_*^2 A \quad (41)$$

and β is an $O(1)$ shape factor for a particular hill. For 3D hills in the linear range, that is $A/S_h < 0.1$, β can be computed by treating the 3D hill as made up of 'slices' of 2D ridges with thickness dy and appropriate heights and wavelengths. For example, for an axisymmetric 3D hill of surface

height $Z_S(x,y) = H\cos\left(\pi\left\{(x/\lambda)^2 + (y/\lambda)^2\right\}^{1/2}\right)^2$, we

find that $\beta = \pi^2/8$ [*Wood and Mason*, 1993].

So far we have concentrated on low hills, H/L or $A/S_h \ll 1$. To extend these results to steeper hills we first rewrite the pressure drag formula as,

$$|F_P| = \frac{1}{2} C_d \langle\overline{u}\rangle^2 (Z_m) A \quad (42)$$

where Z_m is the height at which the area averaged windspeed $\langle\overline{u}\rangle(z)$ (or simply the upstream speed $U(z)$ for isolated hills) provides a proper scaling for the pressure force so that C_d is an $O(1)$ drag coefficient [*Wood and Mason*, 1993]. Equation (42) is commonly used in canopy studies, (see the earlier section on Complex Surfaces or *Kaimal and Finnigan*, 1994, Chapter 3) to parameterize the canopy drag force or generally to represent the drag on a bluff body (equation (33)). Over shallow hills $Z_m = h_m$ but with steep hills H may exceed h_m and we expect that the correct scaling velocity for steep hills that behave more like bluff bodies will be $\langle\overline{u}\rangle(Z_m)$; $H > Z_m > 0$. In practice, for steep hills we may take $U(H)$ or $\langle\overline{u}\rangle(H)$ as the scaling velocity when $H > h_m$ and $U(h_m)$ or $\langle\overline{u}\rangle(h_m) = U_0$ when $H \leq h_m$.

The second step is to assume that there is a smooth transition between the shallow hill result, equation (41) and the steep hill result equation (42) and to insist that they match when $H = h_m$. The formula that results is,

$$|F_P| = 2\pi^2 \beta \left[\frac{U_0}{U(l)} \right]^4 \left[\frac{A}{S_h} \right] u_*^2 A \left[\frac{\langle \overline{u} \rangle (Z_m)}{\langle \overline{u} \rangle (h_m)} \right]^2$$

$$\begin{cases} Z_m = h_m \text{ for } H < h_m \\ Z_m = H \text{ for } H \geq h_m \end{cases}$$

(43)

Despite the rather ad hoc derivation of this equation, which can be followed in more detail in *Wood and Mason, 1993*, it appears to describe the functional dependence of the form drag quite well up to values of hill slope $A/S_h \sim 0.2 - 0.3$ implying that the low slope, quasi-2D value for β can be used through this range. Care must be taken, however, because experimental determinations of hill drag are exceedingly rare and equations (41) and (43) have only been tested against numerical models so far. Nevertheless, they provide a point of departure for a consideration of the effect of hilly terrain on the whole boundary layer.

5.2 Effective Roughness Length of Hilly Terrain

Derivations of the logarithmic law in the ASL proceeds by an asymptotic matching argument that only applies in the height range $z_i \gg z \gg z_s$, where z_s is a characteristic size of the surface roughness elements. When most of the drag force on the surface is due to the form drag on hills, we expect that $z_s \sim H$ and, if H is a sensible fraction of z_i, we might not expect to observe a logarithmic region at all, the ASL being squeezed between the roughness sublayer that should exist above hills in analogy with flow over canopies, and the outer layer flow. Nevertheless, numerical models of flow over ranges of hills suggest that the area averaged velocity, $\langle \overline{u} \rangle (z)$ generally does have a logarithmic dependence through a layer that can occupy a much greater fraction of the PBL depth than the classic surface layer [*Wood and Mason, 1993*]. The experimental evidence is less convincing, partly because area averages are very difficult to measure, but they do not contradict the model results [*Kustas and Brutsaert*, 1986, *Grant and Mason*, 1990].

It seems reasonable then, in analogy with equation (27), to represent the flow over hilly terrain in the approximate range $z_i/2 > z > 2H$ as [*Wood and Mason, 1993, Xu and Taylor*, 1995],

$$\langle \overline{u} \rangle (z) = \frac{u_*^{eff}}{\kappa} \ln \left(\frac{z - d}{z_0^{eff}} \right) \quad (44)$$

so that,

$$u_*^{eff\,2} = \frac{\kappa^2 \langle \overline{u} \rangle (z)}{\left\{ \ln \left[(z - d)/z_0^{eff} \right] \right\}^2} = \left(-\langle F_P \rangle - \langle F_V \rangle \right) \quad (45)$$

where $\langle F_V \rangle$ is the counterpart of $\langle F_P \rangle$, resulting from surface friction. Now since the perturbations in $\langle F_V \rangle$ induced by a hill are much smaller than $|\langle F_V \rangle|$, we can write

$$u_*^{eff} \ \Box \ \left(-\langle F_P \rangle + u_*^2 \right)^{1/2} \quad (46)$$

The area averaged form drag is obtained by dividing equation (43) by S_D, the plan area of the domain under consideration so $\langle F_P \rangle = F_P/S_D$, and we approximate the undisturbed stress u_*^2 by assuming that it can be related to the area-averaged velocity at the pressure scale height, Z_m through the undisturbed roughness length z_0,

$$u_*^2 = \frac{\kappa^2 \langle \overline{u} \rangle^2 (Z_m)}{\left[\ln (Z_m/z_0) \right]} \quad (47)$$

This is not a bad assumption when $H \ll h_m$ and, when $Z_m \lesssim H$, the form drag term in equation (45) is dominant.

Finally, combining equations (39) to (41), we obtain an expression for the effective roughness length [*Wood and Mason*, 1993],

$$\frac{1}{\left\{ \ln \left[(Z_m - d)/z_0^{eff} \right] \right\}^2} = \frac{C_A}{\kappa^2} + \frac{1}{\left\{ \ln (Z_m/z_0) \right\}} \quad \text{and}$$

$$C_A = \frac{|F_P|}{S_D} = 2\pi^2 \beta \left[\frac{U_0}{U(l)} \right]^4 \left[\frac{A}{S_h} \right] \left[\frac{A}{S_D} \right] u_*^2 \left[\frac{\langle \overline{u} \rangle (Z_m)}{\langle \overline{u} \rangle (h_m)} \right]^2$$

with, $\begin{cases} Z_m = h_m \text{ for } H < h_m \\ Z_m = H \text{ for } H \geq h_m \end{cases}$

(48)

We can take the displacement height, d as the mean level of momentum absorption as in canopies (see earlier section on Complex surfaces or *Kaimal and Finnigan*, 1994, Chapter 3) or more simply as the average level of the terrain. Since we should not apply equation (44) or (45) with (48) too close to $z=H$, the precise specification of d is not critical.

Equation (48)(44) has been compared to mathematical models of the whole boundary layer and of the surface layer and it predicts z_0^{eff} to within better than 15% for slopes A/S_h <0.3. Comparison with field data is more difficult but equation (48) matches the available data reasonably well. We can infer from equation (48) that z_0^{eff} is an increasing function of slope A/S_h for low to moderate slopes and also that z_0^{eff} for a range of 2D ridges is substantially larger than for a range of *close-packed* 3D axisymmetric hills of the same wavelength, a result confirmed by numerical model studies that give values of $z_0^{eff\,2D}/z_0^{eff\,3D}$ between 3 and 6 [*Wood and Mason*, 1993]. An idea of the actual change that hills induce in the effective roughness of a natural surface can be gained by noting that, for 2D ridges with $z_0 \approx 0.1m$ and $A/S_h \approx 0.3$, equation (44) predicts that $z_0^{eff}/z_0 \approx 15$ while for close-packed 3D axisymmetric hills with the same z_0 and A/S_h, we find $z_0^{eff}/z_0 \approx 4$.

6. SUMMARY

We have seen both similarities and differences between boundary layers over three kinds of complex surface. Canopies absorb momentum from the boundary layer primarily by pressure drag on their roughness elements be they plants or buildings. In the close-packed canopy environment very high levels of turbulence ensure that the flow patterns around individual elements are quite different to those they would experience in isolation. Instead we observe a critical modification of the mean velocity profile and the consequent generation of large energetic turbulent eddies. Their effect is felt through a roughness sub-layer extending to two or three canopy heights, where details of turbulent transport as well as length and velocity scales depart substantially from those of the surface layer above.

Over continuously changing horizontal surfaces we also observe a blending region where turbulence properties and scaling depart from standard equilibrium ASL forms. The depth of this zone is related to the size of the surface patches and to the rate at which turbulence can diffuse information about the surface changes vertically. For sufficiently large patches the whole boundary layer readjusts to the new surface. Except very close to the edges of distinct patches, pressure effects are negligible but, if the surface is covered with scattered bluff objects like windbreaks or buildings, these can be responsible for a large fraction of the total drag of the landscape.

Over natural hilly landscapes we also find a region of altered mean flow extending up to $z\sim L$, where L is the horizontal lengthscale of the hills, but the depth of this region is determined by the pressure field that develops around the hill. Turbulent diffusion, in contrast, affects only a shallow surface layer. Increased momentum absorption by the hills is almost entirely the result of pressure drag and even for very shallow hills, this substantially exceeds the drag of a flat surface with the same surface texture. Unlike the roughness elements in canopies, the flow patterns around individual hills are not qualitatively affected by their neighbours, at least for hills with slopes commonly found in nature ($A/Sh \lesssim 0.3$).

REFERENCES

Antonia, R.A. and R.E. Luxton, The response of a turbulent boundary layer to a step change in surface roughness. Part 2: Rough to smooth, *J. Fluid Mech.*, 53, 737-757, 1972.

Belcher, S. E., T.M.J. Newley and J.C.R. Hunt, The drag on an undulating surface induced by the flow of a turbulent boundary layer, *J. Fluid Mech.*, 249, 557-596, 1993.

Bradley, E.F. A micrometeorological study of velocity profiles and surface drag in the region modified by a change in surface roughness, *Q. J. Roy. Meteorol. Soc.* 94, 361-379, 1968.

Businger, J.A., Equations and Concepts, in Nieuwstadt, F.T.M and H. van Dop (Eds.) Atmospheric Turbulence and Air Pollution Meteorology. *D. Reidel, Dordrecht, Holland.* 358p. 1982.

Carruthers, D. J. and J.C.R. Hunt, Fluid Mechanics of Airflow over Hills: Turbulence, Fluxes and Waves in the Boundary Layer, In W. Blumen (Ed.) Atmospheric Processes over Complex Terrain. *American Meteorological Society, Meteorological Monographs, Vol 23, No 45. Boston, USA.* Pp323. 1990.

Denmead, O.T. and E.F. Bradley, Flux-gradient relationships in a forest canopy. In *The Forest-Atmosphere Interaction*, B.A. Hutchison and B.B. Hicks (Eds.). *D. Reidel Publishing Co., The Netherlands*, pp 421-442, 1985.

Elliot, W.P., The growth of the atmospheric internal boundary layer, *Trans. Amer. Geophys. Union*, 39, 1048-1054, 1958.

Finnigan, J.J., Turbulent transport in flexible plant canopies. in *The forest-atmosphere interaction*. (eds. B.A. Hutchison and B.B. Hicks), *D. Reidel publishing Co., The Netherlands*, pp. 443-480, 1985.

Finnigan, J.J., Turbulence in Plant Canopies. *Annu. Rev. Fluid Mech. 32*: 519-572, 2000.

Finnigan, J.J., M.R. Raupach, E.F. Bradley and G.K. Aldiss, A wind tunnel study of turbulent flow over a two-dimensional ridge, *Boundary-Layer Meteorol.* 50, 277-317, 1990.

Garratt, J. R., The Internal Boundary layer- A Review, *Boundary-Layer Meteorol.* 50, 171-203, 1990.

Goode, K. and S.E. Belcher, On the parameterisation of the effective roughness length for momentum transfer over heterogeneous terrain. *Boundary Layer Meteorol.* 93, 133-154, 1999.

Grant A.L.M. and P.J. Mason, Observations of boundary layer structure over complex terrain, *Q. J. Roy. Meteorol. Soc*, 116, 159-186, 1990.

Hopwood, W.P., Surface transfer of heat and momentum over an inhomogeneous, vegetated land surface, *Q. J. Roy. Meteorol. Soc.* 121, 1549-1574, 1995.

Hunt, J.C.R., S. Leibovich, and K.J. Richards, Turbulent Shear Flow over Low Hills, *Q. J. Roy. Meteorol. Soc.* 114, 1435-1470, 1988.

Jackson, P.S., On the displacement height in the logarithmic velocity profile, *J. Fluid Mech.*, 111, 15-25, 1981.

Jensen, N.O. Change of surface roughness and the planetary boundary layer, *Q. J. Roy. Meteorol. Soc.*, 104, 351-356, 1978.

Kaimal, J. C. and J.J.Finnigan, Atmospheric Boundary Layer Flows: Their Structure and Measurement. *Oxford University Press, New York.* Pp 289, 1994.

Kustas, W.P. and W. Brutsaert, Windprofile constants in a neutral atmospheric boundary layer over complex terrain, *Boundary-Layer Meteorol.*, 34, 35-54, 1986.

Mahrt, L. The bulk aerodynamic formulation over heterogeneous surfaces, *Boundary-Layer Meteorol.*, 78, 87-119, 1996.

Mason, P.J., The formation of areally-averaged roughness lengths, *Q. J. Roy. Meteorol. Soc.*, 114, 399-420, 1988.

Mulhearn, P.J. Relations between surface fluxes and mean profiles of velocity, temperature and concentration downwind of a change in surface roughness, *Q. J. Roy. Meteorol. Soc.* 103, 785-802, 1977.

Panofsky, H.A. and J.A. Dutton, Atmospheric turbulence: models and methods for engineering applications, *Wiley-Interscience, New York,* 397p, 1984.

Philip, J.R. Advection, evaporation and surface resistance, *Irrig. Sci.*, 8, 104-114, 1987.

Philip, J.R., One-dimensional checkerboards and blending heights, *Boundary-Layer Meteorol.*, 77, 135-151, 1996a.

Philip, J.R., Two-dimensional checkerboards and blending heights, *Boundary-Layer Meteorol.*, 80, 1-18, 1996b.

Philip, J.R., Blending heights for winds oblique to checkerboards, , *Boundary-Layer Meteorol.*, 82, 263-281, 1997.

Raupach, M.R. and R.H. Shaw, Averaging procedures for flow within vegetation canopies. *Boundary-Layer Meteorol.* 22, 79-90, 1982.

Raupach, M.R., J.J. Finnigan, Y. Brunet, Coherent eddies and turbulence in vegetation canopies: the mixing layer analogy. *Boundary-Layer Meteorol.*, 78, 351-382, 1996.

Schmid, H. P. and B. Bunzli, The influence of surface texture on the effective roughness length. *Q. J. Roy. Meteorol. Soc. 121*: 1-22, 1995.

Taylor, P.A., The planetary boundary layer above a change in surface roughness, *J. Atmos. Sci.*, 26, 432-440, 1969.

Taylor, P.A., Comments and further analysis on effective roughness lengths for use in numerical three-dimensional models, *Boundary-Layer Meteorol.*, 39, 403-418, 1987.

Wood, N. and P. Mason, The pressure force induced by neutral turbulent flow over hills. *Q. J. Roy. Meteorol. Soc. 119*: 1233-1267, 1993.

Xu, D. and P.A.Taylor, Boundary-layer parameterization of drag over small-scale topography, *Q. J. Roy. Meteorol. Soc.* 121, 433-443, 1995.

Dr J J Finnigan, CSIRO Atmospheric Research, GPO Box 1666, ACT 2601, AUSTRALIA, john.finnigan@csiro.au

Diffusion of Heavy Particles in a Turbulent Flow

CSIRO Land and Water, Canberra, ACT 2601, Australia

Simple expressions are derived for the far-field eddy diffusivity and particle velocity statistics of a cloud of heavy particles in isotropic turbulence, in terms of statistics of the fluid flow. The analysis leads to the following outcomes:
(1) Using a linearised solution of the equation of motion for a heavy particle in a fluid and a kinematic analysis of dispersion, the particle diffusivity K_P is expressed in terms of the time scale T_P of the fluid (not particle) velocity along the particle path. This yields the simple result $K_P/K_F = T_P/T_F$, where K_F is the diffusivity and T_F the Lagrangian time scale for passive fluid elements.
(2) The time scale T_P is evaluated in terms of the particle-fluid relative speed V_R, using two alternative hypotheses ("elliptic" and "triangular") about the mixed space-time velocity covariance function in isotropic turbulence. Under both hypotheses, T_P depends explicitly on the mean of V_R (the drift velocity) and implicitly on the variance of V_R.
(3) The theory is compared with two sets of observations, providing reasonable overall agreement and experimental support for the "elliptic" hypothesis.
(4) The theory and observations together provide an estimate for the ratio of the fluid Lagrangian time scale to the Eulerian turbulent time scale.

1. INTRODUCTION

The motion of suspended heavy particles in a turbulent flow is not only a basic physical problem, but also has relevance in many geophysical and engineering disciplines. In the atmospheric environment, for example, the transport of heavy particles is a crucial process in sand and dust transport by wind; the drift of agricultural sprays to their target destinations or to other receptors where they may have undesired effects; the transport and eventual deposition of natural and anthropogenic aerosols to vegetation, soil, water and biotic receptors; and many related phenomena. Because of its significance both as a fundamental problem and in applications, the dispersion of heavy particles has been studied for many years, using theoretical analysis, numerical methods and experiments. The following brief survey

Environmental Mechanics: Water, Mass and Energy Transfer in the Biosphere
Geophysical Monograph 129
Copyright 2002 by the American Geophysical Union
10.1029/129GM25

examines in turn the contributions of each of these three approaches.

Theoretical analysis of heavy-particle dispersion began with the work of *Tchen* [1947], also described by *Hinze* [1975], which determined the variance and covariance function for the heavy-particle velocity in terms of fluid velocity statistics. These are fundamental results. However, Tchen also assumed that particles move with fluid elements, which led him to conclude - incorrectly in general - that the far-field eddy diffusivities for particles and passive scalars such as heat are identical. The essence of the problem is that the dispersion of heavy particles in a turbulent flow is different from the dispersion of a passive scalar, or an ensemble of marked but otherwise passive fluid elements. Two main factors contribute [*Yudine*, 1959]: drift due to body forces such as gravity, and particle inertia. First, gravity (or any other body force) gives heavy particles a drift or settling velocity relative to the fluid and causes them to continually change their fluid environment. This drift effect was called "trajectory crossing" by *Yudine* [1959] and *Csanady* [1963]. Second, inertia causes heavy

301

particles to accelerate less rapidly than the fluid in response to fluctuating stress fields, so that the fluid swirls about them. This effect also involves heavy particles crossing the trajectories of fluid elements, but it was labelled separately as the "inertia" effect by *Csanady* [1963]. Arguing that the inertia effect is small and analysing the drift effect alone, *Csanady* [1963] deduced the following expression for the ratio of the diffusivities of heavy particles and fluid elements in homogeneous turbulence:

$$K_P / K_F = \left[1 + (\beta V_D / \sigma)^2 \right]^{-1/2} \qquad (1)$$

where K_P and K_F are respectively the eddy diffusivities of heavy particles and fluid elements, V_D is the drift velocity of the particles, σ is the velocity standard deviation in the direction of the drift, and β is an O(1) constant. This result has been widely used. It predicts that K_P/K_F is always less than 1 for particles with a finite drift velocity.

The combined effects of drift and inertia were studied theoretically by *Reeks* [1977; 1983; 1991], *Pismen and Nir* [1978], *Nir and Pismen* [1979], and *Gouesbet et al.* [1984], using relationships between Lagrangian and Eulerian fluid velocity statistics [*Corrsin*, 1963; *Lundgren and Pointin*, 1976]. The resulting theories involve integral or integro-differential equations and are too complex to be suitable for most applications. A rather simpler, algebraic theory, similar in some respects to the present work, was developed by *Wang and Stock* [1993]; see also *Shao* [2000]. A feature of all these theories incorporating both drift and inertia is that they predict that in some circumstances K_P/K_F can exceed 1, so that particles disperse faster than fluid elements. This is a consequence of the inertial effect, and is predicted to occur when drift is small and when the autocorrelation function for the fluid velocity along the particle path has a negative loop [*Gouesbet et al.*, 1984].

Theoretical analysis has also been used to examine more subtle features of particle motion in a fluid, including the consequences of the added-mass and time-history (Basset) terms in the equation of motion [*Tchen*, 1947; *Hinze*, 1975; *Maxey and Riley*, 1983; *Gouesbet et al.*, 1984; *Mei et al.*, 1991]. Provided the particle is much more dense than the fluid (by a factor of 100 or more), these effects are not quantitatively significant. Also, *Maxey* [1987] has analysed an interaction between inertia and gravity which increases the drift velocity of a heavy particle in homogeneous turbulence by up to a few percent.

Turning now to *numerical methods*, random-flight simulations of heavy-particle motions in a turbulent flow have been carried out by *Hunt and Nalpanis* [1985], *Walklate* [1987], *Sawford and Guest* [1991], and *Kaplan and Dinar* [1992]. In this approach, Langevin-style stochastic equations are used to mimic the equations of motion for fluid elements and heavy particles. The equations are

solved by numerically constructing thousands of realisations of individual trajectories. In formulating the stochastic equation, it is necessary to confront the same central problem tackled by all the theoretical approaches mentioned above, that is, establishing a relationship between the Eulerian velocity, the fluid Lagrangian velocity and the fluid velocity along a heavy-particle path. Other numerical approaches include higher-order closure modelling of a cloud of heavy particles as a continuum [*Shih and Lumley*, 1986] and direct numerical simulation of the turbulent velocity field and the resulting particle trajectories [*Squires and Eaton*, 1991].

Finally (in this brief introductory review), there have been few laboratory *experiments* on the dispersion of heavy particles. Investigations in wind-tunnel (grid) turbulence have been carried out by *Snyder and Lumley* [1971], *Wells and Stock* [1983] and *Ferguson* [1986]. The first two of these are described in some detail later, as they are used for experimental comparisons.

Turning to the present paper, this work seeks to undertake a minimalist treatment of a complex problem. Such an attempt is natural in a volume commemorating the scientific life of John Philip, as he held parsimony to be one of the highest scientific virtues and was a master in its practice. The subject matter is also appropriate: JRP wrote a number of papers on problems in turbulent dispersion, concentrating especially on arguments based on the classic kinematic analysis of *Taylor* [1921], for example, *Philip* [1968].

Let me indicate something of the Philip influence. In the early 1990s, my colleague Yaping Shao and I began to grapple with the physical processes involved in wind erosion and the transport of sand and dust by wind; see *Shao* [2000] for a recent thorough review. One of our approaches at that time was the random-flight method described above. The JRP attitude to numerical methods in general, and stochastic methods in particular, was unequivocal: not parsimonious, not good science. After a seminar by Yaping describing that line of work, he handed us, the following day, several closely handwritten pages headed *Particle motions in turbulence - an essay by A SIMPLE ENGINEER*, with heavy underlining. The note was similar in approach to the work of *Tchen* [1947] mentioned above, with the trademark JRP creativity in the use of special cases. Though the essay by a simple engineer was never published, I acknowledge the profound influence of its writer - encompassing, for me, a striving for economy in scientific description but not a blanket rejection of numerical methods.

This paper is minimalist in several senses. First, it is partly based on simple kinematics in the spirit of *Taylor* [1921] (as are many of the papers reviewed above). Second, its dynamical analysis uses a linearised version of the equation of motion which provides a simple solution for part of the problem. Third, the treatment ignores the more subtle aspects (noted above) of particle motion in a turbulent fluid. The fourth and most severe respect in which this is a minimal treatment is that it concentrates only on the case of far-field diffusion in homogeneous turbulence, where a gradient-diffusion approach to scalar dispersion is valid. The main aim is to relate the fluid and particle diffusivities in the far field. This approach does not address the fact that in many real-world turbulent flows, a gradient diffusion approach is invalid because the length scales of the active, transporting turbulent eddies are comparable with the length scales in the mean flow, leading to such things as countergradient fluxes or negative diffusivities in flows in vegetation canopies [Corrsin, 1974; Denmead and Bradley, 1987]. While many authors including myself [Finnigan and Raupach, 1987] have criticised gradient-diffusion theory on such grounds and have sought alternatives, two factors cause gradient-diffusion theory to remain important: first, in many simple flows it works surprisingly well, key examples being atmospheric and laboratory surface layers. The main reason is that the relationships between fluxes and gradients are dimensionally constrained in these simple flows [Tennekes and Lumley, 1972]. Second, even in more complicated situations such as flows in vegetation canopies, a gradient-diffusion description can provide a base upon which non-diffusive aspects of the transport can be added as perturbations [Raupach, 1989].

The aim here is to develop and test simple, approximate, algebraic expressions for determining the far-field eddy diffusivity and particle velocity statistics for a cloud of heavy particles in homogeneous turbulence, in terms of statistics of the fluid flow. The argument proceeds in three main steps, respectively described in Sections 2 to 4. In Section 2, the equation of motion for a heavy particle in a fluid flow is presented and solved in linearised form. This well-known solution [Tchen, 1947] depends on the fluid velocity along the particle path, which differ from either conventional Lagrangian (fluid-following) or Eulerian (fixed-location) velocity statistics. In Section 3, the solution of the equation of motion is combined with basic kinematics to produce an expression for the particle diffusivity in terms of the covariance function for the fluid velocity along the particle path. This expression is also well known, dating from Pismen and Nir [1978]. In Section 4, the required covariance function is characterised by a time scale T_P, which is evaluated using two alternative hypotheses about the mixed space-time velocity covariance function. The expression for T_P depends on the particle-fluid relative velocity V_R and its variance, which in turn is a function of T_P. A closed solution for the whole problem is thus obtained. With the development complete, the last two sections of the paper present results and comparisons with observations (Section 5) and conclusions (Section 6).

The main notation is as follows: the fluid velocity vector is $U_i(\mathbf{x},t)$. The trajectories of a fluid element and a heavy particle (both starting from position vector $\mathbf{x} = 0$ at time $t = 0$) are $X_i(t)$ and $Y_i(t)$, respectively. The heavy particle velocity is $V_i(t) = dY_i/dt$. Overbars denote ensemble averages and small letters fluctuating quantities: $U_i = \bar{U}_i + u_i$. The Lagrangian fluid velocity (along a fluid element trajectory) is written as $U_{Fi}(t)$, while the fluid velocity along the trajectory of a heavy particle is $U_{Pi}(t)$, so that

$$\left.\begin{array}{l} U_{Fi}(t) = \bar{U}_{Fi}(t) + u_{Fi}(t) = U_i\big(\mathbf{X}(t),t\big) \\ U_{Pi}(t) = \bar{U}_{Pi}(t) + u_{Pi}(t) = U_i\big(\mathbf{Y}(t),t\big) \end{array}\right\} \quad (2)$$

The relative velocity of the particle with respect to the fluid is $V_{Ri} = V_i - U_{Pi}$, and the relative speed is $V_R = |V_{Ri}|$.

2. DYNAMICS OF A HEAVY PARTICLE IN A FLUID

2.1. Equation of Motion

The equation of motion for a spherical particle in a fluid is taken to be

$$\frac{dV_i}{dt} = -\frac{3C_D(\mathrm{Re})}{4\,r_{PF}\,d} V_{Ri}|V_{Ri}| + g_i \quad (3)$$

where d is the particle diameter, r_{PF} the particle-to-fluid density ratio, $g_i = (0,0,-g)$ the gravitational acceleration vector or body force vector per unit particle mass, C_D the particle drag coefficient and $\mathrm{Re} = V_R d/\kappa$ the particle Reynolds number, with κ the kinematic viscosity. Equation (3) involves an assumption:

• Compared with the full equation of particle motion in a fluid [Hinze, 1975; Maxey and Riley, 1983; Gouesbet et al., 1984; Mei et al., 1991], Equation (3) neglects several terms including the added-mass term and the Basset or history term. However, these terms are negligible when the density ratio r_{PF} is large, as for liquid droplets or solid particles in air. Consideration is accordingly restricted to large r_{PF}. (Here and elsewhere, assumptions are highlighted by dots).

The next simplifying assumption is:

• Equation (3) can be linearised [Owen, 1964] and split into mean and fluctuating parts, giving

$$\frac{d\bar{V}_i}{dt} = \frac{\bar{U}_{Pi} - \bar{V}_i}{\tau} + g_i \;; \qquad \frac{dv_i}{dt} = \frac{u_{Pi} - v_i}{\tau} \quad (4)$$

where τ is a particle relaxation time scale defined by

$$\tau = \frac{4\,r_{PF}\,d}{3C_D(\bar{V}_R d/\kappa)\,\bar{V}_R} \quad (5)$$

Here $\bar{V}_R$ is an average value rather than instantaneous value as in Equation (3), and defines the point about which linearisation occurs. A linearised equation of motion is exact only in the small Reynolds number limit, but Equation (4) is a tenable approximation at larger Re, provided that C_D and τ are evaluated at a mean $\bar{V}_R$.

Formally, the solutions of the mean and fluctuating parts of Equation (4) are:

$$
\left.
\begin{aligned}
\bar{V}_i(t) &= \bar{V}_i(0)\,e^{-t/\tau} \;+\; \tau^{-1}\int_0^t \bar{U}_{Pi}(s)\,e^{(s-t)/\tau}\,ds \\
&\qquad\qquad + \; g_i\tau\!\left(1-e^{-t/\tau}\right) \\
v_i(t) &= v_i(0)\,e^{-t/\tau} \;+\; \tau^{-1}\int_0^t u_{Pi}(s)\,e^{(s-t)/\tau}\,ds
\end{aligned}
\right\} \quad (6)
$$

The second term on the right hand side of each solution shows that the particle velocity is given by low-pass filtering the fluid velocity along the particle path, with time constant τ. The other terms are a decaying contribution from initial conditions and a gravitational drift term in the case of the mean velocity. The long-time limit of the mean solution in steady conditions (constant $\bar{U}_{Pi}$) yields the drift velocity, equal to the mean relative velocity vector:

$$
\bar{V}_{Ri} = \bar{V}_i - \bar{U}_{Pi} = g_i\tau \qquad (7)
$$

The magnitude of the drift velocity is written as $V_D = g\tau$. The mean relative speed in the steady state is therefore

$$
\bar{V}_R = \left(\bar{V}_{Ri}\bar{V}_{Ri} + \overline{v_{Ri}v_{Ri}}\right)^{1/2} = \left(V_D^2 + \overline{v_{Ri}v_{Ri}}\right)^{1/2} \qquad (8)
$$

and includes contributions from both the drift velocity and the fluctuating relative velocity induced by particle inertia. Summation over repeated Roman indices (i) is understood, so that $\overline{v_{Ri}v_{Ri}} = \overline{v_{R1}v_{R1}} + \overline{v_{R2}v_{R2}} + \overline{v_{R3}v_{R3}}$.

2.2. Particle Velocity Spectra and Covariance Functions

Using a statistical solution of the linearised equation of motion, Equation (4), the spectra, covariance functions and variances of the fluctuating particle velocity (v_i) and relative velocity $(v_{Ri} = v_i - u_{Pi})$ can be found in terms of statistics of the forcing velocity u_{Pi}, the fluctuating fluid velocity along the particle path. It is convenient to work with Fourier transforms, denoted by a tilde:

$$
\left.
\begin{aligned}
v_i(t) &= \int_{-\infty}^{\infty} e^{i\omega t}\,\tilde{v}_i(\omega)\,d\omega \\
\tilde{v}_i(\omega) &= \frac{1}{2\pi}\int_{-\infty}^{\infty} e^{-i\omega t}\,v_i(t)\,d\omega
\end{aligned}
\right\} \quad (9)
$$

where $i = \sqrt{(-1)}$ and ω is natural frequency. The Fourier transform of the fluctuating part of Equation (4) is

$$
i\omega\tilde{v}_i = \left(\tilde{u}_{Pi} - \tilde{v}_i\right)/\tau \qquad (10)
$$

giving immediately the Fourier-transformed solutions for v_i and v_{Ri} in response to forcing by u_{Pi}, the fluctuating fluid velocity along the particle trajectory:

$$
\left.
\begin{aligned}
\tilde{v}_i &= \left(\frac{1}{1+i\omega\tau}\right)\tilde{u}_{Pi} \\
\tilde{v}_{Ri} &= \tilde{v}_i - \tilde{u}_{Pi} = \left(\frac{-i\omega\tau}{1+i\omega\tau}\right)\tilde{u}_{Pi}
\end{aligned}
\right\} \quad (11)
$$

Hence, the (tensor) spectra of v_i and v_{Ri} are

$$
\left.
\begin{aligned}
\Phi_{(v)ij}(\omega) &= \frac{\tilde{v}_i\tilde{v}_j^*}{T_{total}} = \left(\frac{1}{1+\omega^2\tau^2}\right)\Phi_{(uP)ij}(\omega) \\
\Phi_{(vR)ij}(\omega) &= \frac{\tilde{v}_{Ri}\tilde{v}_{Rj}^*}{T_{total}} = \left(\frac{\omega^2\tau^2}{1+\omega^2\tau^2}\right)\Phi_{(uP)ij}(\omega)
\end{aligned}
\right\} \quad (12)
$$

where $\Phi_{(uP)ij}(\omega)$ is the spectrum of u_{Pi}, T_{total} is the (large) total record length subject to Fourier transformation, and asterisks denote complex conjugates.

The covariance functions for u_{Fi} (Lagrangian fluid velocity), u_{Pi} (fluid velocity along a particle path), v_i (particle velocity) and v_{Ri} (particle velocity relative to fluid) are

$$
\left.
\begin{aligned}
R_{(uF)ij}(t) &= \overline{u_{Fi}(s)u_{Fj}(s+t)} = \int_{-\infty}^{\infty} e^{i\omega t}\,\Phi_{(uF)ij}(\omega)\,d\omega \\
R_{(uP)ij}(t) &= \overline{u_{Pi}(s)u_{Pj}(s+t)} = \int_{-\infty}^{\infty} e^{i\omega t}\,\Phi_{(uP)ij}(\omega)\,d\omega \\
R_{(v)ij}(t) &= \overline{v_i(s)v_j(s+t)} = \int_{-\infty}^{\infty} e^{i\omega t}\,\Phi_{(v)ij}(\omega)\,d\omega \\
R_{(vR)ij}(t) &= \overline{v_{Ri}(s)v_{Rj}(s+t)} = \int_{-\infty}^{\infty} e^{i\omega t}\,\Phi_{(vR)ij}(\omega)\,d\omega
\end{aligned}
\right\} \quad (13)
$$

where each covariance function is also the Fourier transform of the corresponding spectrum. Using Equation (12), the covariance functions for v_i and v_{Ri} are

$$
\left.
\begin{aligned}
R_{(v)ij} &= \int_{-\infty}^{\infty} e^{i\omega t}\left(\frac{1}{1+\omega^2\tau^2}\right)\Phi_{(uP)ij}(\omega)\,d\omega \\
R_{(vR)ij} &= \int_{-\infty}^{\infty} e^{i\omega t}\left(\frac{\omega^2\tau^2}{1+\omega^2\tau^2}\right)\Phi_{(uP)ij}(\omega)\,d\omega
\end{aligned}
\right\} \quad (14)
$$

Equations (12) and (14) relate the particle velocity statistics to statistical properties of u_{Pi}. These are characterised as follows. First, consideration is restricted to one coordinate direction α, denoted by a Greek letter (using the convention that a repeated Greek index is not subject to summation, in contrast with a Roman index as in Equation (8)). Next, time scales $T_{F\alpha}$ and $T_{P\alpha}$ are defined for the fluid velocities $u_{F\alpha}$

(along a fluid-element trajectory) and $u_{P\alpha}$ (along a particle trajectory), such that

$$\overline{u_{F\alpha}^2}\,T_{F\alpha} = \int_0^\infty R_{(uF)\alpha\alpha}(t)\,dt \;; \quad \overline{u_{P\alpha}^2}\,T_{P\alpha} = \int_0^\infty R_{(uP)\alpha\alpha}(t)\,dt \quad (15)$$

Two assumptions are now made:

• In homogeneous turbulence, the variances of $u_{F\alpha}$ and $u_{P\alpha}$ are identical and are both equal to the Eulerian velocity variance σ_α^2. In a homogeneous, incompressible turbulent flow, the fluid Lagrangian variance $\overline{u_{F\alpha}^2} = \overline{u_\alpha^2(\mathbf{X}(t),t)}$ can be proved equal to σ_α^2 [*Tennekes and Lumley*, 1972]. The further assumption $\overline{u_{P\alpha}^2} = \overline{u_\alpha^2(\mathbf{Y}(t),t)} = \sigma_\alpha^2$ is an approximate extension of this result. It is exact in the limits $\tau/T_{F\alpha} \to 0$ (vanishing inertia, when the heavy particle behaves like a Lagrangian fluid element) and $\tau/T_{F\alpha} \to \infty$ (large inertia, when the particle motion becomes independent of the turbulence and the particle trajectory represents a purely Eulerian sample of the fluid velocity field). For intermediate values of $\tau/T_{F\alpha}$, *Maxey* [1987] has suggested that inertia produces a bias in particle trajectories toward regions of high strain rate or low vorticity, causing the time-averaged drift velocity to be slightly larger than its value in still fluid ($g\tau$), by up to a few percent near $\tau/T_{F\alpha} = 1$. A similar process can potentially cause $\overline{u_{P\alpha}^2}/\sigma_\alpha^2$ to differ slightly from 1 near $\tau/T_{F\alpha} = 1$. However, since this effect is neglected for the mean drift velocity, it is consistent to approximate $\overline{u_{P\alpha}^2}/\sigma_\alpha^2$ as 1 for all $\tau/T_{F\alpha}$.

• It is assumed that $u_{F\alpha}$ and $u_{P\alpha}$ have exponential autocorrelation functions and associated spectra:

$$\left. \begin{aligned} R_{(uP)\alpha\alpha}(t) &= \sigma_\alpha^2 \exp(-t/T_{P\alpha}) \\[6pt] \Phi_{(uP)\alpha\alpha}(\omega) &= \frac{\sigma_\alpha^2 T_{P\alpha}}{\pi\left(1+\omega^2 T_{P\alpha}^2\right)} \end{aligned} \right\} \quad (16)$$

and similarly for $u_{F\alpha}$. This is consistent with the assumption that both $u_{F\alpha}$ and $u_{P\alpha}$ behave as first-order Markov processes [*Sawford*, 1984, 1991].

Now, putting the expression for $\Phi_{(uP)ij}(\omega)$ from Equation (16) into Equation (14) and doing the integrals with contour integration or by other means, it is found that the covariance functions for v_i and v_{Ri} are

$$\left. \begin{aligned} R_{(v)\alpha\alpha}(t) &= \frac{\sigma_\alpha^2 T_{P\alpha}\left[\tau\,e^{-t/\tau} - T_{P\alpha}e^{-t/T_{P\alpha}}\right]}{\tau^2 - T_{P\alpha}^2} \\[10pt] R_{(vR)\alpha\alpha}(t) &= \frac{\sigma_\alpha^2 \tau\left[\tau\,e^{-t/T_{P\alpha}} - T_{P\alpha}e^{-t/\tau}\right]}{\tau^2 - T_{P\alpha}^2} \end{aligned} \right\} \quad (17)$$

Thus, even when the fluid velocities $u_{F\alpha}$ and $u_{P\alpha}$ have exponential autocorrelation functions, the particle velocity v_α and relative velocity $v_{R\alpha}$ do not. Taking the limit $t \to 0$ gives the variances of v_α and $v_{R\alpha}$:

$$\left. \begin{aligned} \overline{v_\alpha^2} &= \frac{\sigma_\alpha^2}{1+\tau/T_{P\alpha}} \\[10pt] \overline{v_{R\alpha}^2} &= \frac{\sigma_\alpha^2}{1+T_{P\alpha}/\tau} \end{aligned} \right\} \quad (18)$$

These simple results show that for very small particles (normalised relaxation time $\tau/T_{P\alpha} \to 0$), the particle velocity variance approaches the fluid velocity variance σ_α^2, and the relative velocity variance approaches zero. For very large particles which do not respond to the turbulence ($\tau/T_{P\alpha} \to \infty$), the situation is reversed.

Most of the above solutions for the spectrum, covariance function and variance of the particle velocity v_α were given by *Tchen* [1947], *Csanady* [1963] and others since, for instance *Hinze* [1975] and *Walklate* [1987]. The corollary results for the relative velocity $v_{R\alpha}$ are not usually given explicitly, though they are important for the determination of the particle diffusivity.

3. DIFFUSIVITIES FOR FLUID ELEMENTS AND HEAVY PARTICLES

3.1. Fluid Elements

Taylor [1921] used kinematic arguments to analyse the dispersion of an instantaneous point release of a passive scalar, or a cloud of passive marked fluid elements, in homogeneous turbulence. His essential result was expressed by *Batchelor* [1949] in terms of a time-dependent eddy diffusivity tensor K_{Fij} for marked fluid elements, such that

$$K_{Fij}(t) = \frac{1}{2}\left(\frac{d\,\overline{X_i(t)X_j(t)}}{dt}\right) \quad (19)$$

where $X_i(t)$ is the trajectory of a fluid element passing through the origin at time $t = 0$, and the overbar denotes an average over an ensemble of independent elements. The kinematic argument then yields

$$K_{Fij}(t) = \frac{1}{2}\int_0^t \left[R_{(uF)ij}(s) + R_{(uF)ji}(s)\right]ds \quad (20)$$

expressing K_{Fij} in terms of the Lagrangian fluid velocity covariance function $R_{(uF)ij}$ defined in Equation (13).

3.2. Heavy Particles

A similar analysis can be applied to the eddy diffusivity of heavy particles, by defining a time-dependent particle diffusivity tensor

$$K_{Pij}(t) = \frac{1}{2}\left(\frac{d\overline{Y_i(t)Y_j(t)}}{dt}\right) \qquad (21)$$

where $Y_i(t)$ is the particle trajectory. In this case the kinematic argument yields

$$K_{Pij}(t) = \frac{1}{2}\int_0^t \left[R_{(v)ij}(s) + R_{(v)ji}(s)\right]ds \qquad (22)$$

where $R_{(v)ij}$ is the particle velocity covariance function defined in Equation (13).

At this point the analysis for heavy-particle dispersion introduces dynamical information from the equation of particle motion, thus departing from the analysis for fluid element dispersion which is purely kinematic. By using Equation (6) (the explicit solution of the linearised equation of motion) in Equation (22), $R_{(v)ij}$ can be found in terms of statistics of the fluid velocity along the particle path:

$$R_{(v)ij}(t) = \frac{1}{2\tau}\int_{-\infty}^{\infty} R_{(uP)ij}(s)\exp\left(\frac{-|s-t|}{\tau}\right)ds \qquad (23)$$

Hence, $R_{(v)ij}$ is a convolution of $R_{(uP)ij}$ with an exponential filtering function. The particle eddy diffusivity tensor can now be expressed in terms of statistics of the fluid velocity along the particle path, using Equations (22) and (23) and integrating by parts once again. This leads to

$$\left.\begin{aligned} K_{Pij}(t) &= \frac{1}{2}\int_0^t \left[R_{(uP)ij}(s) + R_{(uP)ji}(s)\right]ds \\ &+ \frac{1}{2}\int_t^{\infty}\left[R_{(uP)ij}(s) + R_{(uP)ji}(s)\right]\exp\left(\frac{t-s}{\tau}\right)ds \\ &- \frac{\tau}{2}\left[R_{(uP)ij}(t) + R_{(uP)ji}(t)\right] \end{aligned}\right\} \quad (24)$$

which has the far-field limit

$$K_{Pij} \xrightarrow[t\to\infty]{} \frac{1}{2}\int_0^{\infty}\left[R_{(uP)ij}(s) + R_{(uP)ji}(s)\right]ds \qquad (25)$$

Hence, in the far field, K_{Pij} has the same form as the fluid eddy diffusivity K_{Fij} (Equation (20)), except that the Lagrangian fluid velocity covariance is replaced by the covariance of the fluid velocity along a particle path. These results for K_{Pij} were first given by *Pismen and Nir* [1978].

For dispersion in a single coordinate direction α, Equation (20) shows that the time dependent diffusivity $K_{F\alpha\alpha}(t)$

for passive marked fluid elements approaches the constant value $\overline{u_{F\alpha}^2}T_{F\alpha} = \sigma_\alpha^2 T_{F\alpha}$ in the far field limit $t \gg T_{F\alpha}$. This is the conventional eddy diffusivity in the α-direction for a homogeneous turbulent flow. Accordingly, in the far-field limit, the dispersion of the cloud of passive scalar or marked fluid elements is diffusive, satisfying the diffusion equation

$$\frac{\partial C}{\partial t} = K\frac{\partial^2 C}{\partial x_\alpha^2} \qquad (26)$$

where C is the scalar concentration and the diffusivity $K = K_{F\alpha\alpha} = \sigma_\alpha^2 T_{F\alpha}$. On the other hand, in the near field where t is of the order of or smaller than $T_{F\alpha}$, the diffusion equation (26) is not satisfied and the dispersion is non-diffusive. Applying the same principles to the dispersion of heavy particles in one coordinate direction α, Equation (25) shows that the one-dimensional particle diffusivity in the far field approaches $\overline{u_{P\alpha}^2}T_{P\alpha}$, so that a cloud of dispersing heavy particles obeys a diffusion equation similar to Equation (26) with C being the particle concentration and with diffusivity $K = K_{P\alpha\alpha} = \overline{u_{P\alpha}^2}T_{P\alpha}$. By combining these expressions for the fluid-element and particle diffusivities in the far field and using the approximation $\overline{u_{F\alpha}^2} = \overline{u_{P\alpha}^2} = \sigma_\alpha^2$, a very simple result is obtained:

$$\frac{K_{P\alpha\alpha}}{K_{F\alpha\alpha}} = \frac{\overline{u_{P\alpha}^2}T_{P\alpha}}{\overline{u_{F\alpha}^2}T_{F\alpha}} \approx \frac{T_{P\alpha}}{T_{F\alpha}} \qquad \text{(far field)} \qquad (27)$$

Hence, the problem of finding the far-field diffusivity for a cloud of heavy particles in homogenous turbulence reduces to that of finding the fluid eddy diffusivity and the ratio of the time scales $T_{P\alpha}/T_{F\alpha}$ (recalling that $T_{P\alpha}$ is the time scale for the *fluid* velocity along the particle trajectory, not the particle velocity itself, or the time scale for $U_\alpha(\mathbf{Y}(t),t)$ rather than $V_\alpha(t)$).

4. THE COVARIANCE FUNCTION AND TIME SCALE FOR THE FLUID VELOCITY ALONG A HEAVY-PARTICLE TRAJECTORY

The difference between $T_{P\alpha}$ and $T_{F\alpha}$ is caused by the fact that heavy particles continually change their fluid environment, that is, move from one fluid element trajectory to another, because of both inertia and drift due to external body forces on the heavy particles (typically gravity). This motion of the particle relative to the fluid occurs at the relative velocity $V_{R\alpha} = V_\alpha - U_{P\alpha}$, which has a mean part $\overline{V}_{R\alpha}$ (equal to the drift velocity) and a fluctuating part $v_{R\alpha}$

associated with particle inertia. Both parts contribute to the mean relative speed $\overline{V}_R = \left(V_D^2 + \overline{v_{Ri}v_{Ri}}\right)^{1/2}$.

The determination of $T_{P\alpha}$ proceeds in two steps: first, the covariance function $R_{(uP)\alpha\alpha}(t)$ is expressed in terms of Lagrangian and Eulerian covariance functions. This yields an expression for the time scale ratio $T_{P\alpha}/T_{F\alpha}$, which depends on fluid velocity statistics (σ_α and $T_{F\alpha}$) and the mean relative speed $\overline{V}_R$. Second, the problem is closed by using the dynamical solutions given in Section 2 to express $\overline{v_{Ri}v_{Ri}}$ and thence $\overline{V}_R$ as a function of $T_{P\alpha}$. The two steps yield a pair of algebraic equations which together determine $T_{P\alpha}$ and $\overline{V}_R$ fully, with given flow properties σ_α and $T_{F\alpha}$ and particle properties τ (the relaxation time) and $V_D = g\tau$ (the drift velocity). The first of these steps largely follows previous analyses cited in Section 1, while the second is novel.

4.1. The Covariance Function $R_{(uP)\alpha\alpha}(t)$

The first problem is to express the covariance function for the fluid velocity along the particle path, $R_{(uP)\alpha\alpha}(t) = \overline{u_{P\alpha}(s)\,u_{P\alpha}(s+t)} = \overline{u_\alpha(0,0)\,u_\alpha(\mathbf{Y}(t),t)}$, in terms of measurable properties of the fluid turbulence. The starting point, following *Csanady* [1963] and many others since, is to note that when the (scalar) relative speed $\overline{V}_R$ is very small compared with a turbulent velocity scale (say σ), particle trajectories and fluid-element trajectories are nearly identical so that $R_{(uP)\alpha\alpha}(t)$ approaches the fluid Lagrangian covariance function:

$$R_{(uP)\alpha\alpha}(t) \rightarrow R_{(uF)\alpha\alpha}(t) \quad \text{as} \quad \overline{V}_R/\sigma \rightarrow 0 \qquad (28)$$

On the other hand, when $\overline{V}_R$ is very large compared with σ, a heavy-particle trajectory is a slice through a turbulence field which is effectively static compared with the particle motion, so that $R_{(uP)\alpha\alpha}(t)$ is determined by the Eulerian flow properties. In this case

$$R_{(uP)\alpha\alpha}(t) \rightarrow R_{(uE)\alpha\alpha}\left(\overline{V}_R t\right) \quad \text{as} \quad \overline{V}_R/\sigma \rightarrow \infty \qquad (29)$$

where $R_{(uE)\alpha\alpha}$ is the two-point fluid Eulerian velocity covariance function defined by

$$R_{(uE)\alpha\alpha}(\mathbf{r}) = \overline{u_i(\mathbf{x},t)\,u_j(\mathbf{x}+\mathbf{r},t)} \qquad (30)$$

This is a function of the separation vector $\mathbf{r}$ alone because the turbulence is assumed to be homogeneous and stationary. In Equation (29), $\mathbf{r}$ is estimated as the product $\overline{\mathbf{V}}_R t$ where $\overline{\mathbf{V}}_R$ is the mean relative velocity vector.

For intermediate values of $\overline{V}_R/\sigma$, $R_{(uP)\alpha\alpha}(t)$ is determined with the following three assumptions.

- Guided by Equations (28) and (29), it is assumed that

$$R_{(uP)\alpha\alpha}(t) = f\left(\overline{\mathbf{V}}_R t, t\right) \qquad (31)$$

where the function f characterises the two-point, two-time covariance function $\overline{u_\alpha(0,0)\,u_\alpha(\mathbf{Y}(t),t)}$.

- The dependences of f on t and $\overline{\mathbf{V}}_R t$ are replaced by dependences on the normalised scalar variables $t/T_{F\alpha}$ and $\overline{V}_R t/L_\delta$, where $T_{F\alpha}$ is the fluid Lagrangian time scale and L_δ is the Eulerian length scale in the direction of the mean drift or body force (cartesian index δ). This direction may or may not be the same as the direction of dispersion (cartesian index α). As discussed below, the two important cases are when the drift is parallel to the direction of dispersion or normal to it.

- The function f in Equation (31) is determined by contours of constant covariance $R_{(uP)\alpha\alpha}(t)$ on a plane with axes $t/T_{F\alpha}$ and $\overline{V}_R t/L_\delta$. Two possible models are:

$$\left.\begin{array}{ll} \left(\dfrac{t}{T_{F\alpha}}\right)^2 + \left(\dfrac{\overline{V}_R t}{L_\delta}\right)^2 = \text{constant} & \text{(elliptic)} \\[3mm] \dfrac{t}{T_{F\alpha}} + \dfrac{\overline{V}_R t}{L_\delta} = \text{constant} & \text{(triangular)} \end{array}\right\} \quad (32)$$

where the shapes describe the appearance of the contours on the $\left(t/T_{F\alpha},\ \overline{V}_R t/L_\delta\right)$ plane. The elliptic model was proposed by *Csanady* [1963] and the triangular model by *Walklate* [1987], in both cases on intuitive grounds. Here these models are taken as broadly spanning a range of possibilities, and both are tested against data.

Combining Equation (32) with the requirements of Equations (28) and (29), it follows that

$$\left.\begin{array}{ll} R_{(uP)\alpha\alpha}(t) = \exp\left[\dfrac{-t/T_{F\alpha}}{\left(1 + \left(\overline{V}_R T_{F\alpha}/L_\delta\right)^2\right)^{1/2}}\right] & \text{(elliptic)} \\[5mm] R_{(uP)\alpha\alpha}(t) = \exp\left[\dfrac{-t/T_{F\alpha}}{1 + \left(\overline{V}_R T_{F\alpha}/L_\delta\right)}\right] & \text{(triangular)} \end{array}\right\} \quad (33)$$

Expressions for the fluid time scale along a particle trajectory, $T_{P\alpha}$, are now readily found by matching Equation (33) with the second of Equation (13). It follows that

$$\left.\begin{array}{ll} T_{P\alpha} = T_{F\alpha}\left(1 + \left(\overline{V}_R T_{F\alpha}/L_\delta\right)^2\right)^{-1/2} & \text{(elliptic)} \\[3mm] T_{P\alpha} = T_{F\alpha}\left(1 + \left(\overline{V}_R T_{F\alpha}/L_\delta\right)\right)^{-1} & \text{(triangular)} \end{array}\right\} \quad (34)$$

In the fluid-Lagrangian limit $\bar{V}_R T_{F\alpha}/L_\delta \to 0$, Equation (33) relaxes to the exponential form for $R_{(uF)\alpha\alpha}(t)$ in Equation (13). In the Eulerian limit $\bar{V}_R T_{F\alpha}/L_\delta \to \infty$, a consequence of the above assumptions is that the Eulerian covariance function $R_{(uE)\alpha\alpha}(r)$ in the drift direction δ also takes an exponential form:

$$R_{(uE)\alpha\alpha}(r) = \sigma_\alpha^2 \exp(-r/L_\delta) \qquad (35)$$

Exponential covariance functions are not fully consistent with known properties of the fine structure of turbulence. In the Lagrangian case, departures occur at times of the order of the temporal Taylor microscale and smaller [*Sawford*, 1991]. In the Eulerian case, the actual slope of the spectrum in the inertial subrange is −5/3 according to Kolmogorov inertial subrange similarity theory [*Tennekes and Lumley*, 1972; *Hunt and Nalpanis*, 1985], rather than −2 as required by an exponential covariance function (see Equation (16)). However, as argued by *Wang and Stock* [1993], the use of exponential covariance functions is justifiable if the main properties of the dispersion are determined by the integral scales $T_{F\alpha}$ and L_δ, rather than the details of the shapes of the covariance functions. Since the particle diffusivity is an integral property of $R_{(uP)\alpha\alpha}(t)$ (Equation (25)) and the particle dispersion $\overline{Y_\alpha^2}$ is a double integral, this is a reasonable position to take. The benefit is that exponential covariance functions make the analysis mathematically straightforward.

4.2. Closing the Problem

Equation (34) specifies $T_{P\alpha}$ in terms of the fluid turbulence properties σ_α, $T_{F\alpha}$ and L_δ and the mean relative speed $\bar{V}_R$, which depends on both the drift velocity V_D and the relative velocity variance $\overline{v_{Ri}v_{Ri}}$. Of these two, V_D is an external parameter but $\overline{v_{Ri}v_{Ri}}$ depends on the particle-turbulence interaction. However, Equation (18) already provides a specification of $\overline{v_{Ri}v_{Ri}}$ in terms of $T_{P\alpha}$ and the particle relaxation time τ. From Equations (8) and (18), $\bar{V}_R$ is given by

$$\bar{V}_R = \left(V_D^2 + \frac{\sigma_1^2}{1+T_{P1}/\tau} + \frac{\sigma_2^2}{1+T_{P2}/\tau} + \frac{\sigma_3^2}{1+T_{P3}/\tau}\right)^{1/2} \qquad (36)$$

Hence, $\bar{V}_R$ involves all three components of the relative velocity variance, even in a treatment of one-dimensional particle dispersion. For isotropic turbulence the fluid statistics are independent of direction ($\sigma_1 = \sigma_2 = \sigma_3$ and $T_{F1} = T_{F2} = T_{F3}$), but it does not follow that the fluid velocity time scale along particle trajectories ($T_{P\alpha}$) is independent

of direction, because there is a preferred direction for particle dispersion induced by the drift velocity. Thus, even in isotropic turbulence, the components of $T_{P\alpha}$ cannot be assumed equal. In anisotropic turbulence the fluid statistics σ_α and $T_{F\alpha}$ also vary with direction, introducing additional sources of inequality among the components of $T_{P\alpha}$. In principle, this requires that, even in isotropic turbulence, four variables ($\bar{V}_R$, T_{P1}, T_{P2} and T_{P3}) must be treated as unknowns in the four equations formed from the three components of Equation (34) and the scalar Equation (36). However, this complication can be avoided by a modest additional approximation:

• For the purpose of calculating $\bar{V}_R$ with Equation (36) (but not elsewhere), the three time scales T_{P1}, T_{P2} and T_{P3} may be related to the time scale $T_{P\alpha}$ in the direction of dispersion by assuming that the ratios among T_{P1}, T_{P2} and T_{P3} are the same as those among the fluid time scales T_{F1}, T_{F2} and T_{F3}. Equation (36) then becomes

$$\bar{V}_R = \left(V_D^2 + \sum_{\gamma=1}^{3} \frac{\sigma_\gamma^2}{1 + \left(\dfrac{T_{P\alpha}}{\tau}\right)\left(\dfrac{3T_{F\gamma}}{T_{F1}+T_{F2}+T_{F3}}\right)}\right)^{1/2} \qquad (37)$$

The solution for $T_{P\alpha}$ is now complete, since Equations (34) and (37) form a pair of coupled nonlinear algebraic equations which uniquely specify $\bar{V}_R$ and a single value of $T_{P\alpha}$ in the direction of dispersion. These can be reduced to just one closed equation for $T_{P\alpha}$ (a cubic in $T_{P\alpha}$ in for both the triangular and elliptic models for $R_{(uP)\alpha\alpha}$) but in practice the simplest method of solution is a coupled iteration of Equations (34) and (37) starting from the initial estimates $\bar{V}_R = V_D = g\tau$, $T_{P\alpha} = T_{F\alpha}$. This converges rapidly.

4.3. The Time Scale $T_{F\alpha}$ and the Length Scale L_δ

To use the solution formed by Equations (34) and (37), it is necessary to determine the scales $T_{F\alpha}$ and L_δ. The fluid Lagrangian time scale $T_{F\alpha}$ is estimated using the relationship proposed by *Corrsin* [1963]:

$$T_{F\alpha} = \beta L_\alpha/\sigma_\alpha \qquad (38)$$

where β is a constant of order 1 and L_α is the length scale in the direction of dispersion (α). This may differ from L_δ, the length scale in the direction of drift (δ).

It remains to identify the relationship between L_α and L_δ. As foreshadowed above, the two important cases are when the drift is parallel to the direction of dispersion or normal to it. Several authors [*Csanady*, 1963; *Wang and Stock*, 1993] have dealt with this issue by appealing to analytic results for isotropic turbulence. The same approach is followed here, on the grounds that the experiments used below for tests of the predictions took place in grid (decaying, approximately isotropic) turbulence.

A standard result [see *Hinze*, 1975] enables the Eulerian (two-point, one-time) velocity covariance function for isotropic turbulence to be expressed as

$$R_{(uE)ij}(\mathbf{r}) = \overline{u_i(\mathbf{x},t)\,u_j(\mathbf{x}+\mathbf{r},t)}$$
$$= \sigma^2\left[\frac{f(r)-g(r)}{r^2}r_ir_j + g(r)\delta_{ij}\right] \quad (39)$$

where $\sigma = \sigma_\alpha$ (the subscript α being unnecessary for isotropic turbulence), $r^2 = r_ir_i = \mathbf{r}\cdot\mathbf{r}$, and f and g are the longitudinal and transverse velocity correlation functions. These are the instantaneous velocity correlations (normalised with σ^2 so that $f(0) = g(0) = 1$) between velocity components at two points separated by a distance r, such that the velocity components are parallel to the separation vector $\mathbf{r}$ in the case of $f(r)$, and normal to $\mathbf{r}$ in the case of $g(r)$. Equation (38), which is a consequence only of the requirements of isotropy, implies that the full two-point covariance function $R_{(uE)ij}$ (a second-rank tensor) is determined by the two scalar functions $f(r)$ and $g(r)$. If the additional requirement of fluid incompressibility is imposed, then $\partial u_j/\partial x_j = 0$ and $\partial R_{(uE)ij}/\partial r_j = 0$, which implies that

$$g(r) = f(r) + \frac{r}{2}\frac{\partial f}{\partial r} \quad (40)$$

If longitudinal and transverse length scales L_f and L_g are defined for $f(r)$ and $g(r)$ by

$$L_f = \int_0^\infty f(r)\,dr; \qquad L_g = \int_0^\infty g(r)\,dr \quad (41)$$

then Equation (40) shows that

$$L_g = L_f/2 \quad (42)$$

irrespective of the shapes of $f(r)$ and $g(r)$.

The length scale L_α in Equation (38) is always a longitudinal length scale, since the velocity fluctuations responsible for dispersion in the direction α are parallel with that direction. The length scale L_δ is a longitudinal length scale when the drift is parallel to the dispersion and a transverse length scale when the drift is normal to the dispersion. Hence, for isotropic turbulence, it follows that

$$L_\alpha/L_\delta = \gamma, \quad \text{with} \begin{cases} \gamma = 1 \ (\text{drift parallel to dispersion}) \\ \gamma = 2 \ (\text{drift normal to dispersion}) \end{cases} \quad (43)$$

Note that the term "longitudinal" in this section refers to the direction of the dispersion, not the direction of the flow, and also identifies a spatial length scale for velocity components parallel (rather than normal) to the direction of separation. Also, the proportionality constant β in Equation (38) is independent of direction. For any direction, β relates the Lagrangian time scale $T_{F\alpha}$ to the longitudinal (rather than transverse) length scale L_α.

4.4. Character of the Solution

In summary, by using Equations (27), (34), (38) and (43), the solution for the far-field eddy diffusivity for a cloud of heavy particles in homogenous turbulence can be written as

$$\left.\begin{aligned} \frac{K_{P\alpha\alpha}}{K_{F\alpha\alpha}} &= \frac{T_{P\alpha}}{T_{F\alpha}} = \left(1 + \left(\frac{\beta\gamma\bar{V}_R}{\sigma_\alpha}\right)^2\right)^{-1/2} \quad (\text{elliptic}) \\ \frac{K_{P\alpha\alpha}}{K_{F\alpha\alpha}} &= \frac{T_{P\alpha}}{T_{F\alpha}} = \left(1 + \left(\frac{\beta\gamma\bar{V}_R}{\sigma_\alpha}\right)\right)^{-1} \quad (\text{triangular}) \end{aligned}\right\} \quad (44)$$

where $\bar{V}_R$ and γ are respectively given by Equations (37) and (43). The coupled Equations (44) and (37) are solved iteratively, or (with more algebra and little gain in numeric efficiency) by the formation of a cubic equation in $T_{P\alpha}$.

The *Csanady* [1963] result, Equation (1), is a special case of this theory (for the elliptic choice of the covariance function $R_{(uP)\alpha\alpha}(t)$ when the second term in the right-hand bracket of Equation (37) is omitted, so that $\bar{V}_R = V_D = g\tau$ in Equation (44). This corresponds to estimating $\bar{V}_R$ from drift only and neglecting inertia. By retaining the full Equation (37), the inertia effect is included.

Other statistics of the particle motion may also be computed from $T_{P\alpha}$, including the covariance functions and variances for the particle velocity and relative velocity (Equations (17) and (18), respectively).

The nature of the solution is shown by plotting the particle-to-fluid diffusivity ratio ($K_{P\alpha\alpha}/K_{F\alpha\alpha} = T_{P\alpha}/T_{F\alpha}$) in Figure 1, and the particle velocity variance normalised by the fluid velocity variance ($\overline{v_\alpha^2}/\sigma_\alpha^2$) in Figure 2, both as functions of $V_D/\sigma_\alpha = g\tau/\sigma_\alpha$, the ratio of the drift velocity to the turbulent velocity scale. The drift velocity V_D increases with increasing particle diameter (d), body force per unit particle mass (g), and particle-to-fluid density ratio (r_{PF}). In both figures, the three panels show the effect of varying (a) the fluid Lagrangian time scale $T_{F\alpha}$, (b) the fluid velocity scale σ_α, and (c) the formulation for the covariance function $R_{(uP)\alpha\alpha}(t)$, among the elliptic and triangular choices defined in Equation (32). In both Figures 1 and 2, panels (a) and (b) use the elliptic formulation for $R_{(uP)\alpha\alpha}(t)$. The main features of these results are:

(1) The particle diffusivity is always less than the fluid diffusivity and the particle velocity variance less than the

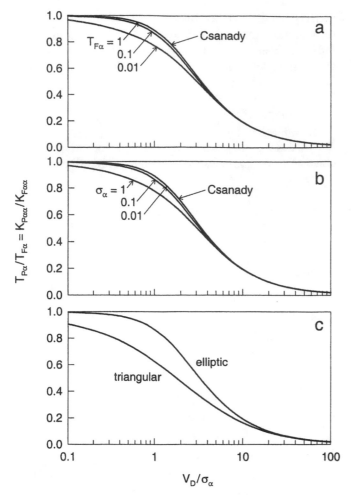

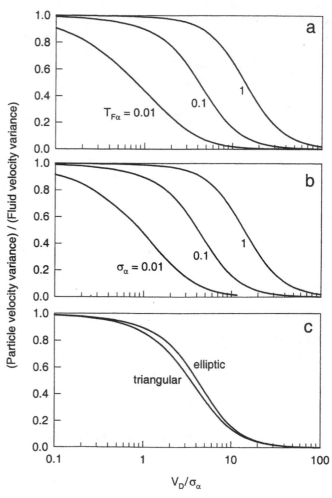

weaker (smaller σ_α) turbulence (Figures 1a, 1b). The same trends appear in $\overline{v_\alpha^2}/\sigma_\alpha^2$ (Figures 2a, 2b).

(4) The inertia effect causes $K_{P\alpha\alpha}/K_{F\alpha\alpha}$ and $\overline{v_\alpha^2}/\sigma_\alpha^2$ to decrease below their values without inertia. The inertia-free values (the same as the predictions from the *Csanady* [1963] theory) coincide with the upper curves in Figures 1a, 1b, 2a and 2b. The effect of inertia becomes progressively more significant in faster (smaller $T_{F\alpha}$) and stronger (larger σ_α) turbulence. For most atmospheric flows the ranges of $T_{F\alpha}$ and σ_α are such that neglect of inertia causes an error of no more than a few percent, but this may not apply in other flows.

(5) The choice between the elliptic and triangular formulations for $R_{(uP)\alpha\alpha}(t)$ has a significant influence on

Figure 1. The diffusivity ratio $K_{P\alpha\alpha}/K_{F\alpha\alpha}$ as a function of $V_D/\sigma_\alpha = g\tau/\sigma_\alpha$, the ratio of the drift velocity to the turbulent velocity scale. Conditions held constant unless otherwise specified: particle-to-fluid density ratio $r_{PF} = 2200$ (an approximate value for quartz grains in air); fluid Lagrangian time scale $T_{F\alpha} = 0.1$ s; fluid velocity scale $\sigma_\alpha^2 = 0.1$ m s^{-1}; constants $\beta = 0.5$ and $\gamma = 1$; elliptic formulation for $R_{(uP)\alpha\alpha}(t)$, the fluid velocity covariance function along a particle trajectory, $R_{(uP)\alpha\alpha}(t)$. (a) Effect of varying $T_{F\alpha}$; (b) effect of varying σ_α; (c) effect of varying the formulation for $R_{(uP)\alpha\alpha}(t)$ among the elliptic and triangular choices defined in Equation (32). The *Csanady* [1963] prediction is obtained as $T_{F\alpha} \to \infty$ and as $\sigma_\alpha \to 0$, and coincides with the uppermost curve in panels (a) and (b).

fluid velocity variance. The two converge ($K_{P\alpha\alpha}/K_{F\alpha\alpha} \to 1$, $\overline{v_\alpha^2}/\sigma_\alpha^2 \to 1$) as $V_D/\sigma_\alpha \to 0$.

(2) With increasing V_D/σ_α, $K_{P\alpha\alpha}/K_{F\alpha\alpha}$ falls off more rapidly than $\overline{v_\alpha^2}/\sigma_\alpha^2$.

(3) For a given V_D/σ_α (or particle diameter), $K_{P\alpha\alpha}/K_{F\alpha\alpha}$ becomes progressively closer to 1 in slower (larger $T_{F\alpha}$) and

Figure 2. The particle velocity variance normalised by the fluid velocity variance, $\overline{v_\alpha^2}/\sigma_\alpha^2$. Conditions and panel descriptions as for Figure 1.

Table 1. Properties of the spherical particles used in the dispersion experiments of *Snyder and Lumley* [1971] (SL71) and *Wells and Stock* [1983] (WS83): particle density (ρ_P), particle diameter (d), drift velocity ($V_D = g\tau$) and relaxation time (τ). For WS83 particles, values of V_D are with no applied electric field.

	ρ_P (kg m^{-3})	d (μm)	$V_D = g\tau$ (m s^{-1})	τ (s)
SL71 particles				
Hollow glass	260	46.5	0.0167	0.0017
Corn pollen	1000	87	0.198	0.020
Solid glass	2500	87	0.442	0.045
Copper	8900	46.5	0.483	0.049
WS83 particles				
5 μm glass	2475	5	0.000188	0.192
57 μm glass	2420	57	0.2316	0.0245

$K_{P\alpha\alpha}/K_{F\alpha\alpha}$, and to a lesser extent on $\overline{v_\alpha^2}/\sigma_\alpha^2$ (Figures 1c, 2c). These quantities are lower for the triangular than the elliptic formulation. The difference in $K_{P\alpha\alpha}/K_{F\alpha\alpha}$ is large enough to be experimentally testable.

5. COMPARISONS WITH OBSERVATIONS

5.1. The Experiments

Two significant sources of data for testing theories of particle dispersion in a turbulent flow are provided by the experiments of *Snyder and Lumley* [1971] and *Wells and Stock* [1983], hereafter SL71 and WS83 respectively. Both experiments used homogeneous, decaying grid turbulence in a wind tunnel, this being perhaps the closest approximation to isotropic turbulence that is easily achieved in a laboratory flow. WS83 arranged their flow field to be as close as possible to that of SL71, facilitating comparisons between the flow measurements in the two experiments.

There were also some major differences between the experiments, particularly in the choices of dispersing particles and in the orientation of the wind tunnel, which was vertical in the SL71 experiment and horizontal in WS83. This has the important implication that the primary direction of dispersion (transverse to the flow) was parallel with the body force (gravity) in the WS83 experiment, and perpendicular with the body force in SL71.

In both experiments, grid turbulence was generated in a square-duct working section of breadth about 0.4 m, using a square grid at the upstream end of the working section (streamwise coordinate $x = 0$). Both grids consisted of square bars of size $b = 4.76$ mm, made into a square-lattice mesh with spacing $M = 25.4$ mm. In both cases the mean flow speed was $U = 6.55$ m s^{-1}. The only essential difference between the wind tunnels (other than the orientation, which does not affect the flow) was in the effective length

of the working section: this was about $x/M = 180$ or $x = 4.6$ m for SL71, and $x/M = 90$ or $x = 2.3$ m for WS83.

SL71 used four kinds of spherical particle (copper, solid glass, corn pollen, hollow glass) with properties given in Table 1. WS83 used just two kinds of particle, glass beads with diameters 5 μm and 57 μm (Table 1), but modified their drift velocities by giving the particles controlled electric charges and setting up an electric field across the wind tunnel to either oppose (cancel) the gravitational force or augment it. In both experiments the particles were introduced into the flow from an effectively point source near the tunnel centreline, at $x/M = 20$ for SL71 and at $x/M = 15$ for WS83. Particle positions and velocities were measured photographically in SL71 and by laser doppler anemometry in WS83.

5.2. Flow Fields

Because of the significance of the turbulence properties in any comparison of a dispersion theory with experiment, it is important to make a careful assessment of the velocity measurements in both SL71 and WS83. Given the nearly identical flow configurations, it is expected that the flow fields should be practically identical, providing a useful cross-check.

Figure 3 shows a comparison of several flow properties as functions of a normalised distance $(x-x_0)/M$. The effective origin x_0 is chosen to optimise the fit of the data to the basic decay law for grid turbulence,

$$\frac{U^2}{\sigma_u^2} = \frac{U^2}{\sigma_w^2} = D\left(\frac{x-x_0}{M}\right) \qquad (45)$$

where σ_u^2 and σ_w^2 are the velocity variances in the streamwise and cross-stream directions, respectively, and D is a dimensionless decay constant. The fitted parameters were $D = 41$, $x_0/M = 14.0$ for SL71, and $D = 55$, $x_0/M = 8.0$ for WS83. In both experiments, the turbulence was measured to be very nearly isotropic ($\sigma_u = \sigma_w$) and the velocity standard deviation for both components is henceforth denoted σ_α. This obeys $\sigma_\alpha/U = D^{-1/2}((x-x_0)/M)^{-1/2}$, shown together with the measurements from both experiments in Figure 3a. Here, and for other power-law fits to the data described below, the WS83 parameters ($D = 55$, $x_0/M = 8.0$) are used. The power law for σ_α/U from Equation (45) was well satisfied by the velocity variance data from both experiments, but the measured turbulent velocities throughout the SL71 data were about 15% higher than those in the WS83 data under otherwise identical conditions.

Using Equation (45), power laws can be deduced for other turbulence properties. The dissipation rate ε is of interest because it provides one way to estimate the time scale

$T_{F\alpha}$. From the turbulent kinetic energy equation for homogeneous, decaying grid turbulence, ε obeys

$$\frac{3U}{2}\frac{\partial\sigma_\alpha^2}{\partial x} = -\varepsilon \qquad (46)$$

which yields the power law

$$\left(\frac{M}{U^3}\right)\varepsilon = \frac{3}{2D}\left(\frac{x-x_0}{M}\right)^2 \qquad (47)$$

This is compared with the measurements in Figure 3b. Not surprisingly, the agreement is very good, because the quoted measured values were partly derived using Equation (46). However, independent dissipation measurements in both experiments were also made by integrating the area beneath a dissipation spectrum (the spectrum of $\partial u/\partial t$), with good agreement in both cases.

A power law for the Lagrangian fluid time scale $T_{F\alpha}$ can be found by using the inertial-subrange law for the Lagrangian structure function [*Monin and Yaglom*, 1975; *Thomson*, 1987; *Sawford*, 1991]:

$$B_{(uF)\alpha\alpha}(t) \underset{\text{definition}}{=} \overline{u_{F\alpha}(s)u_{F\alpha}(s+t)} \underset{\substack{\text{inertial}\\\text{subrange}\\\text{law}}}{=} C_0\varepsilon t \qquad (48)$$

where C_0 is the Kolmogorov Lagrangian structure function constant. The structure function is also related to the covariance function $R_{(uF)\alpha\alpha}(t)$ defined in Equation (13), by $B_{(uF)\alpha\alpha}(t) = 2\sigma_\alpha^2 - R_{(uF)\alpha\alpha}(t)$. If the covariance function is exponential with time scale $T_{F\alpha}$, this relationship implies that $B_{(uF)\alpha\alpha}(t) = 2\sigma_\alpha^2 t/T_{F\alpha}$ in the inertial subrange ($t \ll T_{F\alpha}$). Equating with Equation (48), it follows that

$$T_{F\alpha} = \frac{2\sigma_\alpha^2}{C_0\varepsilon} \qquad (49)$$

Then, using Equations (45) and (47) to specify power laws for σ_α and ε, the power law for $T_{F\alpha}$ is obtained:

$$\left(\frac{U}{M}\right)T_{F\alpha} = \frac{4}{3C_0}\left(\frac{x-x_0}{M}\right) \qquad (50)$$

In Figure 3c, this prediction is compared with two values inferred from the measurements: a "dissipation" value from Equation (49) with ε inferred from Equation (46) with $C_0 = 4.5$ (see Section 6), and a "length scale" value inferred from measured values of the streamwise Eulerian length scale L_u, using Equation (38) with $\beta = 0.4$ (Section 6). The L_u values were tabulated directly by SL71, and in the case of WS83 were inferred from tabulated values of the fixed-point Eulerian time scale for u (T_{Eu}) as $L_u = UT_{Eu}$. In both experiments the agreement between Equation (50) and the

"dissipation" values is excellent, again not surprisingly considering the way these values were inferred. The agreement between the "length scale" values and Equation (50) is very good for WS83 except at small x/M, but for SL71 the length scale values of $T_{F\alpha}$ exceed the other three estimates and Equation (50) by a factor between 1.5 and 2, with the agreement improving as x/M increases.

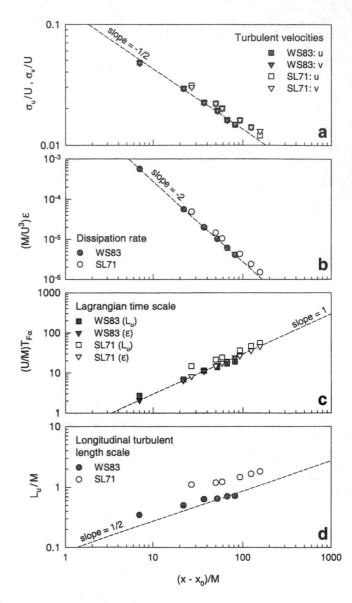

Figure 3. Normalised flow statistics as functions of normalised distance $(x-x_0)/M$, for the experiments of *Snyder and Lumley* [1971] (SL71) and *Wells and Stock* [1983] (WS83): (a) velocity standard deviation σ_α/U; (b) dissipation rate $(M/U^3)\varepsilon$; (c) fluid Lagrangian time scale $(U/M)T_{F\alpha}$; (d) Eulerian length scale L_u/M. Power-law fits are from Equations (45) to (51).

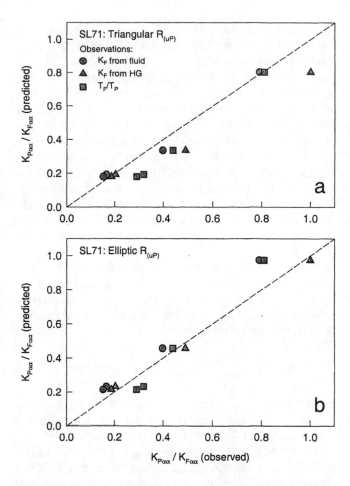

Figure 4. Predicted particle-to-fluid diffusivity ratio $K_{P\alpha\alpha}/K_{F\alpha\alpha}$ compared with the observations of *Snyder and Lumley* [1971], using (a) the triangular formulation and (b) the elliptic formulation for $R_{(uP)\alpha\alpha}(t)$, from Equation (32). Different observation methods are described in the text.

Finally, a power law for the streamwise Eulerian length scale L_u can be obtained by using Equation (38) and the power laws already established for σ_α and $T_{F\alpha}$:

$$\frac{L_u}{M} = \frac{4}{3\beta C_0 D^{1/2}} \left(\frac{x - x_0}{M} \right)^{1/2} \qquad (51)$$

Figure 3d compares this prediction with direct measurements. The SL71 results exceed those from WS83 by a factor between 1.5 and 2, consistent with the pattern for the "length scale" estimates of $T_{F\alpha}$.

WS83 also compared their flow field with several earlier experiments on grid turbulence, including SL71. They found generally good agreement with earlier work, except for a few discrepancies concerning SL71 (in σ_α and L_u, as noted above, and in the effective origin x_0). This result mo-

tivated the present choice of WS83 rather than SL71 values of D and x_0/M in setting the power-law fits to the data.

From the point of view of the dispersion predictions, the most important outcome of the flow-field analysis is a set of estimates for σ_α and $T_{F\alpha}$ at a central point in the working section (in the x direction) at which the mean particle dispersion is to be predicted. That point is chosen to be $x/M = 60$ for WS83 and $x/M = 107$ for SL71, the fourth of six velocity measurement stations in each case. Because the tunnels were practically identical in the setup of the flow, the predictions are made with values of σ_α and $T_{F\alpha}$ respectively derived from Equations (45) and (50) with consistent (WS83) values of D and x_0/M. The resulting values are $\sigma_\alpha = 0.089$ m s^{-1}, $T_{F\alpha} = 0.114$ s (at $x/M = 107$, for SL71); and $\sigma_\alpha = 0.123$ m s^{-1}, $T_{F\alpha} = 0.060$ s (at $x/M = 60$, for WS83).

5.3. Particle Dispersion Measurements

In both experiments the observed dispersion rates $d\overline{Y^2}/dt = U\,d\overline{Y^2}/dx$ were nearly constant with x, despite the fact that the grid turbulence was slowly decaying. One interpretation of this constancy is that the far-field diffusivity $K_{F\alpha\alpha} = \sigma_\alpha^2 T_{F\alpha}$ is independent of x, from the power laws for σ_α [proportional to $((x-x_0)/M)^{-1/2}$] and $T_{F\alpha}$ [proportional to $((x-x_0)/M)$]. This supports the use of decaying grid turbulence for testing a dispersion theory, despite the fact that both σ_α and $T_{F\alpha}$ are changing with x.

The primary quantity used here for comparison between the theory and the experiments is $K_{P\alpha\alpha}/K_{F\alpha\alpha}$, for a transverse direction of dispersion α. From the SL71 data, three different observed values of $K_{P\alpha\alpha}/K_{F\alpha\alpha}$ are available:

(1) $K_{P\alpha\alpha}$ from the measured dispersion rate $U\,d\overline{Y^2}/dx$ and Equation (21), and $K_{F\alpha\alpha}$ from measurements of $\sigma_\alpha^2 T_{F\alpha}$;

(2) $K_{P\alpha\alpha}$ from the measured $U\,d\overline{Y^2}/dx$ and $K_{F\alpha\alpha}$ from measurements of dispersion for the lightest particles (hollow glass);

(3) $K_{P\alpha\alpha}/K_{F\alpha\alpha}$ from the measured time scale ratio $T_{P\alpha}/T_{F\alpha}$, using Equation (27).

From the WS83 experiment, two different observed values are available:

(4) $K_{P\alpha\alpha}$ from the measured $U\,d\overline{Y^2}/dx$ and $K_{F\alpha\alpha}$ from measurements of $\sigma_\alpha^2 T_{F\alpha}$ (as in (1));

(5) $K_{P\alpha\alpha}$ from particle velocity correlation measurements and Equation (22), and $K_{F\alpha\alpha}$ from $\sigma_\alpha^2 T_{F\alpha}$.

Figures 4 and 5 compare the observed and predicted values of $K_{P\alpha\alpha}/K_{F\alpha\alpha}$ for the SL71 and WS83 data, respectively, using both the triangular and the elliptic formulation for $R_{(uP)\alpha\alpha}(t)$. The only adjustable parameter available in the theory is β, which was set at $\beta = 0.4$ to provide best overall

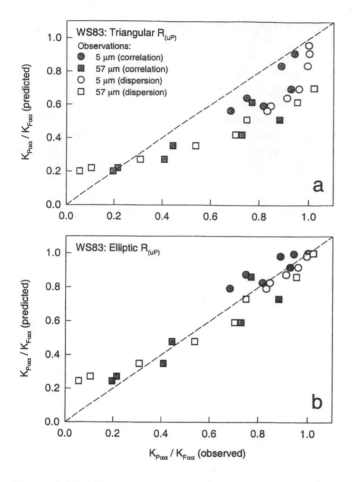

Figure 5. Predicted particle-to-fluid diffusivity ratio $K_{P\alpha\alpha}/K_{F\alpha\alpha}$ compared with the observations of *Wells and Stock* [1983]. Details as for Figure 4.

agreement between model and both sets of measurements (see Section 6). It is important to note that γ in Equations (43) and (44) is 2 for SL71 and 1 for WS83, because the direction of drift was normal to the direction of dispersion in SL71 (with a vertical tunnel) and parallel to the dispersion WS83 (with a horizontal tunnel).

The agreement between measurements and predictions in both experiments is good, provided the elliptic formulation for $R_{(uP)\alpha\alpha}(t)$ is used. The triangular formulation does not provide satisfactory agreement. The quality of the agreement is broadly similar for all methods of observing $K_{P\alpha\alpha}/K_{F\alpha\alpha}$ across both experiments, indicating that the observations are essentially in agreement as noted for methods (4) and (5) by WS83.

6. DISCUSSION AND CONCLUSIONS

6.1. Parameter Values

The only free parameter in the theory is β, which relates the fluid Lagrangian and Eulerian time scales ($T_{F\alpha}$ and

L_α/σ_α) through Equation (38). This has been chosen to optimise the agreement in Figures 4 and 5 between the measured and predicted values of $K_{P\alpha\alpha}/K_{F\alpha\alpha}$. The resulting value is $\beta = 0.40 \pm 0.1$, where the uncertainty takes account of both the scatter in the measurements and the uncertainty in other measured parameters, especially σ_α (recalling that there is some difference between SL71 and WS83; see Figure 3a). This procedure is effectively a determination of β, based on the model with the elliptic formulation for $R_{(uP)\alpha\alpha}(t)$. If β is optimised with the triangular formulation, a smaller value (around $\beta = 0.2$) is obtained, but the resulting fit to both data sets, especially WS83, is substantially worse that with the elliptic formulation. Hence, at least between these two choices, the triangular option can be rejected.

It is interesting to compare the value $\beta = 0.4$ with an inferred value of β for the adiabatic atmospheric surface layer, obtained by equating two estimates for the vertical eddy diffusivity ($K_{F\alpha\alpha}$ with $\alpha = 3$) for a passive scalar. The first (Lagrangian) estimate is

$$K_{F33} \underset{(a)}{=} \sigma_3^2 T_{F3} \underset{(b)}{=} \beta\sigma_3 L_3 \underset{(c)}{=} \beta\sigma_3 U T_{E3} \underset{(d)}{=} \frac{\beta\sigma_3 z}{2\pi n_{w(peak)}} \quad (52)$$

where the successive equalities come from (a) Equation (20) in the far-field limit; (b) Equation (38); (c) Taylor's hypothesis $L_3 = U T_{E3}$, where T_{E3} is the fixed-point time scale for the vertical velocity; and (d) the assumption $T_{E3} = 1/(2\pi f_{w(peak)}) = z/(2\pi U n_{w(peak)})$. Here z is height above ground and $n_{w(peak)}$ is the peak dimensionless frequency of the vertical velocity spectrum, plotted as $f\Phi_{33}(f)/\sigma_3^2$ against the dimensionless frequency $n = fz/U$ where f is frequency [s^{-1}]. Many observations, reviewed for example by *Kaimal and Finnigan* [1994], give $n_{w(peak)}$ in the range 0.3 to 0.5. The second (Eulerian) estimate of K_{F33} comes direct from surface-layer similarity theory:

$$K_{F33} = k_S u_* z \quad (53)$$

where k_S is the von Karman constant for a scalar and u_* is the friction velocity. Equating these two estimates gives

$$\beta = 2\pi n_{w(peak)} k_S (u_*/\sigma_3) \quad (54)$$

Taking $k_S = 0.4$, $n_{w(peak)} = 0.3$ to 0.5 and $\sigma_3/u_* = 1.25$ (a typical value for an atmospheric surface layer), it follows that β is 0.6 to 1.0.

This would seem to conflict with the estimate $\beta = 0.4$ from the present work, until it is recalled that the length scale L_3 from Equation (52) applies to vertical velocities along a streamwise axis of separation, because of the use of Taylor's hypothesis to transform the time axis of spectral measurement into a streamwise axis $x = Ut$. Hence, $L_3 = U T_{E3}$ is a transverse length scale rather than a longitudinal length scale as required by a dispersion theory. If the turbulence were isotropic, a factor $\gamma = 2$ should then appear in equality (b) of Equation (52). This therefore becomes $K_{F33} = \beta\gamma\sigma_3 L_3$, and Equation (54) becomes

$$\beta = 2\pi n_{w(peak)} k_S \left(u_* / \sigma_3 \right) / \gamma \qquad (55)$$

which (with $\gamma = 2$) implies $\beta = 0.3$ to 0.5, bracketing the result from the heavy-particle analysis. Of course surface-layer turbulence is far from isotropic, so γ cannot be assumed to be exactly 2. However, the comparison indicates the importance of the distinction between longitudinal and transverse length scales when using the hypothesis $T_{F\alpha} = \beta L_\alpha / \sigma_\alpha$.

The other parameter introduced in Section 5 is C_0, the Kolmogorov Lagrangian structure function constant defined by Equation (48). This appears only in the power laws for $T_{F\alpha}$ and L_u in grid turbulence (Equations (50) and (51)), not in the particle dispersion theory. The value used here, $C_0 = 4.5$, was chosen to provide good agreement between the "dissipation" and "length scale" values of $T_{F\alpha}$ in Figure 3c, for the WS83 measurements. Had the SL71 data been used instead, the resulting value of C_0 would have been somewhat lower (3.0 to 3.5). The effect of such a change on other predictions is insignificant.

Sawford [1991] showed that Reynolds number effects appear in Equation (49), making C_0 (defined to satisfy a relationship involving the integral Lagrangian time scale) a function of the Reynolds number $Re_\lambda = \sigma_\alpha \lambda / \kappa$ based on the Taylor microscale λ (κ is kinematic viscosity). He predicted that C_0 approaches a value of 7 as $Re_\lambda \to \infty$, and that when $Re_\lambda = 50$ (about the value for the SL71 and WS83 grid turbulence), the effective C_0 in Equation (49) (which he denoted $\tilde{C}_0$) is about 0.6 of the limiting value at large Re_λ, that is, about 4. This is close to the value of 4.5 used here.

6.2. Concluding Summary

Through the simple (though approximate) relationship $K_{P\alpha\alpha} / K_{F\alpha\alpha} = T_{P\alpha} / T_{F\alpha}$ (Equation (27)), it has been possible to develop an analytic theory for the eddy diffusivity of heavy particles in a turbulent flow. The focus is on the time scale $T_{P\alpha}$ for the fluid velocity ($u_{P\alpha}$) along a particle trajectory, rather than the covariance function for the particle velocity (v_α). Hence the present theory is simpler than alternative theories based on integral [*Reeks*, 1977; *Pismen and Nir*, 1978] or algebraic [*Wang and Stock*, 1993] equations which solve for the covariance function.

The theory has been compared with experiments, leading to:

(1) good overall agreement;

(2) ability to discriminate on the basis of the experimental evidence between "elliptic" and "triangular" formulations for the covariance function $R_{(uP)\alpha\alpha}(t)$ for $u_{P\alpha}$; and

(3) the estimate $\beta = 0.40 \pm 0.1$ for the ratio of the fluid Lagrangian time scale $T_{P\alpha}$ to the Eulerian turbulence time scale L_α / σ_α (Equation (38)).

This theory predicts that the heavy-particle diffusivity $K_{P\alpha\alpha}$ is always less than the fluid diffusivity $K_{F\alpha\alpha}$. Some more complex theories [*Reeks*, 1977; *Pismen and Nir*, 1978] predict that $K_{P\alpha\alpha}$ can exceed $K_{F\alpha\alpha}$ as a consequence of the inertial effect. However, this only occurs when the covariance function for $R_{(uP)\alpha\alpha}(t)$ has negative loops [*Gouesbet et al.*, 1984]. The present assumption that $R_{(uP)\alpha\alpha}(t)$ is exponential, without negative loops, accordingly rules out the possibility that $K_{P\alpha\alpha} > K_{F\alpha\alpha}$. In an experiment designed specifically to isolate the effects of inertia [*Wells and Stock*, 1983], no significant evidence that $K_{P\alpha\alpha} > K_{F\alpha\alpha}$ was found.

The predictions from these cases provide a description of the diffusivity of particles in homogeneous turbulence which is suitable for many environmental applications. It is reasonable to expect that this description is applicable in inhomogeneous environmental turbulent flows, to the same extent that a gradient-diffusion relationship describes scalar transfer in such flows. One reason for expecting this to be the case is that the time scale $T_{P\alpha}$ governing heavy-particle diffusion is less than the time scale $T_{F\alpha}$ for the diffusion of fluid elements, so the criteria for a diffusion-equation description of heavy-particle dispersion should at least be no more severe than equivalent criteria for fluid-element dispersion.

Acknowledgments. In addition to the influence of Dr John Philip recorded in the introduction, I wish to acknowledge conversations and influence from Dr Brian Sawford, and assistance with figures provided by Mr Peter Briggs.

REFERENCES

Batchelor, G. K., Diffusion in a field of homogeneous turbulence, I. Eulerian analysis, *Aust.J.Sci.Res.*, 2, 437-450, 1949.

Corrsin, S., Estimates of the relations between Eulerian and Lagrangian scales in large Reynolds number turbulence, *J.Atmos.Sci.*, 20, 115-119, 1963.

Corrsin, S., Limitations of gradient transport models in random walks and in turbulence, *Adv.Geophys.*, 18A, 25-60, 1974.

Csanady, G. T., Turbulent diffusion of heavy particles in the atmosphere, *J.Atmos.Sci.*, 20, 201-208, 1963.

Denmead, O. T. and E. F. Bradley, On scalar transport in plant canopies, *Irrig.Sci.*, 8, 131-149, 1987.

Ferguson, J. R., The effects of fluid continuity on the turbulent dispersion of particles, Washington State University, Pullman, 1986.

Finnigan, J. J. and M. R. Raupach, Transfer processes in plant canopies in relation to stomatal characteristics, in *Stomatal Function*, edited by E. Zeiger, G. D. Farquhar, and I. R. Cowan, pp. 385-429, Stanford University press, Stanford, CA, 1987.

Gouesbet, G., A. Berlemont, and A. Picart, Dispersion of discrete particles by continuous turbulent motions. Extensive discussion of the Tchen's theory, using a two-parameter family of Lagrangian correlation functions, *Phys.Fluids*, 27, 827-837, 1984.

Hinze, J. O., Turbulence, p. 790, McGraw-Hill, New York, 1975.

Hunt, J. C. R. and P. Nalpanis, Saltating and suspended particles over flat and sloping surfaces. I. Modelling concepts, in *Proceedings of the International Workshop on the Physics of Blown Sand*, edited by O. E. Barndorff-Nielsen, J. T. Moller, K. Romer Rasmussen, and B. B. Willetts, pp. 9-36, Institute of Mathematics, University of Aarhus, Aarhus, 1985.

Kaimal, J. C. and J. J. Finnigan, Atmospheric Boundary Layer Flows: their Structure and Measurement, p. 289, Oxford University Press, Oxford, 1994.

Kaplan, H. and N. Dinar, A stochastic model for the dispersion of a nonpassive scalar in a turbulent field, *Atmos.Environ.*, 26, 2413-2423, 1992.

Lundgren, T. S. and Y. B. Pointin, Turbulent self-diffusion, *Phys.Fluids*, 19, 355-358, 1976.

Maxey, M. R., The gravitational settling of aerosol particles in homogeneous turbulence and random flow fields, *J.Fluid Mech.*, 174, 441-465, 1987.

Maxey, M. R. and J. J. Riley, Equation of motion for a small rigid sphere in a nonuniform flow, *Phys.Fluids*, 26, 883-889, 1983.

Mei, R., R. J. Adrian, and T. J. Hanratty, Particle dispersion in isotropic turbulence under stokes drag and Basset force with gravitational settling, *J.Fluid Mech.*, 225, 481-495, 1991.

Monin, A. S. and A. M. Yaglom, Statistical Fluid Mechanics: Mechanics of Turbulence. Volume 2, p. 874, MIT Press, Cambridge, 1975.

Nir, A. and L. M. Pismen, The effect of a steady drift on the dispersion of a particle in turbulent fluid, *J.Fluid Mech.*, 94, 369-381, 1979.

Owen, P. R., Saltation of uniform grains in air, *J.Fluid Mech.*, 20, 225-242, 1964.

Philip, J. R., Diffusion by continuous movements, *Phys.Fluids*, 11, 38-42, 1968.

Pismen, L. M. and A. Nir, On the motion of suspended particles in stationary homogeneous turbulence, *J.Fluid Mech.*, 84, 193-206, 1978.

Raupach, M. R., A practical Lagrangian method for relating scalar concentrations to source distributions in vegetation canopies, *Quart.J.Roy.Meteorol.Soc.*, 115, 609-632, 1989.

Reeks, M. W., On the dispersion of small particles suspended in an isotropic turbulent fluid, *J.Fluid Mech.*, 83, 529-546, 1977.

Reeks, M. W., The transport of discrete particles in inhomogeneous turbulence, *J.Aerosol Sci.*, 14, 729-739, 1983.

Reeks, M. W., On a kinetic-equation for the transport of particles in turbulent flows, *Phys.Fluids A*, 3, 446-456, 1991.

Sawford, B. L., The basis for, and some limitations of, the langevin equation in atmospheric relative dispersion modelling, *Atmos.Environ.*, 18, 2405-2411, 1984.

Sawford, B. L., Reynolds number effects in Lagrangian stochastic models of turbulent dispersion, *Phys.Fluids A*, 3(6), 1577-1586, 1991.

Sawford, B. L. and F. M. Guest, Lagrangian statistical simulation of the turbulent motion of heavy particles, *Boundary-Layer Meteorol.*, 54, 147-166, 1991.

Shao, Y. P., Physics and Modelling of Wind Erosion, p. 393, Kluwer Academic Publishers, Dordrecht, 2000.

Shih, T. H. and J. L. Lumley, Second-order modelling of particle dispersion in a turbulent-flow, *J.Fluid Mech.*, 163, 349-363, 1986.

Snyder, W. H. and J. L. Lumley, Some measurements of particle velocity autocorrelation functions in a turbulent flow, *J.Fluid Mech.*, 48, 41-71, 1971.

Squires, K. D. and J. K. Eaton, Measurements of particle dispersion obtained from direct numerical simulations of isotropic turbulence, *J.Fluid Mech.*, 226, 1-35, 1991.

Taylor, G. I., Diffusion by continuous movements, *Proc.London Math.Soc.Series 2*, 20, 196-212, 1921.

Tchen, C. M., Mean Value and Correlation Problems Connected with the Motion of Small Particles Suspended in a Turbulent Flow (PhD Thesis, Delft), Martinus Nijhoff, The Hague, 1947.

Tennekes, H. and J. L. Lumley, A First Course in Turbulence, p. 300, MIT Press, Cambridge, Massachusetts, 1972.

Thomson, D. J., Criteria for the selection of stochastic models of particle trajectories in turbulent flows, *J.Fluid Mech.*, 180, 529-556, 1987.

Walklate, P. J., A random-walk model for dispersion of heavy particles in turbulent air flow, *Boundary-Layer Meteorol.*, 39, 175-190, 1987.

Wang, L. P. and D. E. Stock, Dispersion of heavy particles by turbulent motion, *J.Atmos.Sci.*, 50, 1897-1913, 1993.

Wells, M. R. and D. E. Stock, The effects of crossing trajectories on the dispersion of particles in a turbulent flow, *J.Fluid Mech.*, 136, 31-62, 1983.

Yudine, M. I., Physical considerations on heavy-particle diffusion, *Adv.Geophys.*, 6, 185-191, 1959.

Dr. Michael Raupach, CSIRO Land and Water, P.O. Box 1666, Canberra ACT 2601, Australia

A Simple One-dimensional Model of Coherent Turbulent Transfer in Canopies

Michael D. Novak

Faculty of Agricultural Sciences, University of British Columbia, Vancouver, British Columbia, Canada

Vertical turbulent heat flow in a vegetation canopy is described using a one-dimensional transient heat transfer equation with steady-state source density distribution. Fluxes are assumed proportional to local gradients at all levels, with a time-varying diffusivity that accounts for the intermittant coherent eddies that mediate most of the upward transfer of heat in canopies. Diffusivities also vary with height in a realistic way. Solution is achieved numerically using a finite element technique and calculated air temperatures are compared with values measured in a Douglas-fir forest. Good agreement with a measured daytime profile is achieved by adjusting input diffusivity parameters, but in retrospect these are shown to have some physical basis. The theory is able to describe the ramp-like behaviour often seen in temperature time series and the counter-gradient heat flow that occurred within the Douglas-fir canopy. Raupach's localized near-field theory based on Lagrangian methods is also tested and is shown to underestimate the near-field effect in the forest.

INTRODUCTION

The flow of air within and above vegetation and other surface canopies, such as plant-residue mulches, occurs at very high Reynolds number and is therefore highly turbulent. While the randomness of turbulence has often been emphasized, it is now recognised that vertical transfer of momentum and scalar quantities (e.g., heat and water vapour) within vegetation canopies and between canopies and the atmosphere is largely mediated by intermittant canopy-scale, coherent eddies [*Gao et al.*, 1989; *Raupach et al.*, 1996]. One signature of these coherent flow structures are the so-called "ramps" that occur in the time series of vertical wind speed and temperature, vapour density, CO_2 concentration, and other trace gases. For example, when heat is flowing from the canopy into the air near midday, air temperature traces often consist of repeated patterns in which temperature increases fairly steadily with time at all heights within and just above the canopy until a rapid drop occurs nearly simultaneously in all traces. The drop occurs first near the top of the canopy and is progressively delayed with decreasing height within the canopy. Both the delay and the drop occur quickly compared to the time during which the temperature is steadily increasing. They are associated with a canopy-scale coherent flow event that injects cooler air from above the canopy down into it and which, as a consequence, sweeps relatively warm canopy air up and out. On nights when the canopy is a sink for heat (with the ultimate loss of heat occurring by upward longwave radiation flow) and wind speeds exceed the threshold required for forced convection, only the directions change in the ramp patterns, i.e., the temperature steadily decreases before exhibiting a sudden increase. The period of steady tempera-

Environmental Mechanics: Water, Mass and Energy Transfer in the Biosphere
Geophysical Monograph 129
Copyright 2002 by the American Geophysical Union
10.1029/129GM26

ture change and the sudden drop or increase are referred to as "quiescent" and "gust" periods, respectively.

Researchers have exploited this ramp behaviour and developed "surface renewal" models that describe the fluxes of various scalar quantities exchanged between the surface and the atmosphere [Paw U et al., 1995; Chen et al., 1997a]. Most often, these models have been applied above the canopy to the total heat flux exchanged, although Chen et al. [1997b] showed that they could be used to predict the flux profile within a canopy. The greatest use for renewal models appears to be as a measurement tool, either in primary or backup mode to determine fluxes more cheaply and easily than eddy correlation, because high frequency measurements of wind speed are not needed and the monitoring frequencies for the scalar variables can be reduced. At present, however, renewal models are semi-empirical and a proper physical theory is lacking. Such a theory should allow the development of more complete predictive models of canopy exchange.

"K-theory" assumes that the vertical flux of any entity at any height is linearly proportional to the local vertical gradient in the concentration of that entity, with turbulent diffusivity, K, as proportionality factor. However, field measurements in canopies have demonstrated that K-theory is invalid within canopies. A common observation, in canopies ranging in scale from plant-residue mulches up to forests, is that the flux density is counter the sign of the local gradient, so that K is negative, which is both physically illogical and leads to numerical instability when used in predictive models. This behaviour occurs because the vertical size of the turbulent eddies that transport most of the flux (the coherent structures discussed above) is roughly given by canopy height, h, so that the exchange depends upon differences in concentration over vertical distances larger than implied by the local gradient. Raupach [1989] presented an analytical Lagrangian theory of vertical turbulent dispersion that explains the counter-gradient fluxes as the effect of the "near-field" on the profile of temperature. He showed that his theory was in good agreement with a more general random-flight model of turbulent dispersion. However, Baldocchi [1992] found that a similar random-flight model was not in good agreement with his measured scalar profiles in a soybean canopy, which he speculated might be due to its neglect of intermittancy. Lee [1992] showed that a random-flight model applied to a Douglas-fir forest predicted counter-gradient flow as observed but underestimated the magnitude of the effect. El-kilani [1996] suggested that the disagreement with random-flight and Lagrangian-type models is related to the universal assumption that released particles move independently throughout time, which clearly is violated due to the intermittent canopy-scale coherent eddies.

This paper presents numerical solutions of the transient heat transfer equation that describe the ramping of air temperature and the counter-gradient flow observed in canopies. The quiescent and gust periods are simulated by a time-varying turbulent diffusivity, so that despite the remarks above, K-theory is assumed. Although it is recognized that the exchange of air during the gust is primarily advective and two-dimensional [Brunet and Raupach, 1987], it is assumed to be one-dimensional and diffusive in the theory. El-kilani and Van Pul [1996] developed a complete canopy energy balance model based on the same concepts and validated it with measurements made in a maize canopy. Herein, the numerical solutions are compared with field measurements only for heat transfer in a Douglas-fir forest, with simplified boundary conditions, so as to better illustrate the fundamental elements of the theory and its sensitivity to basic input parameters. Comparison of the Raupach [1989] Lagrangian analytical theory with the same forest canopy data set is also presented.

METHODS

Theory for Time-Varying Diffusivity

The equation governing one-dimensional vertical heat transfer within and above a canopy, assuming K-theory to be valid, is given by

$$C\frac{\partial \theta}{\partial t} = \frac{\partial}{\partial z}\left(CK\frac{\partial \theta}{\partial z}\right) + S, \qquad (1)$$

where $\theta = \theta(z,t)$ is the potential temperature of the air (which in practice differs negligibly from actual air temperature in the domain of interest), t is the time, z is the height above the soil surface, $K = K(z,t)$ is the turbulent diffusivity, $S = S(z,t)$ is the sensible heat flux source density, and C is the volumetric heat capacity (at constant pressure) of the air. According to K-theory, the sensible heat flux density, H, is given by

$$H = -CK\frac{\partial \theta}{\partial z}. \qquad (2)$$

For both analytical and numerical efficacy, it is assumed that $K(z,t) = K_z(z)K_t(t)$, so that changing the time variable to

$$\tau = \int_0^t K_t(\delta)\,d\delta \qquad (3)$$

yields

$$\frac{\partial \theta}{\partial \tau} = \frac{\partial}{\partial z}\left(K_z \frac{\partial \theta}{\partial z}\right) + \frac{S}{CK_t}, \qquad (4)$$

which is the standard heat equation but now with the time-varying factor K_t in the source term. Although this assumption for K is somewhat restrictive, more complex expressions are probably not warranted. Note that K_t is dimensionless, so that K_z carries the dimensions of K.

To simulate the effects of the turbulent exchange, $K_t(t)$ is assumed to be a series of repeating step functions such that $K_t = K_1$ for $nT < t < t_1 + nT$ and $K_t = K_2$ for $nT + t_1 < t < (n+1)T$, where $n = 0, 1, 2, 3\ldots$ and the period $T = t_1 + t_2$, with t_1 and t_2 the time periods for which $K_t = K_1$ and K_2, respectively. K_1 applies to the quiescent period between gusts and K_2 to the gust period.

To simulate a real forest canopy (described later), $K_z(z)$ is chosen as follows, based on Lee [1992]:

$$K_z = \sigma_w^2(z)\, T_L(z), \qquad (5)$$

where σ_w, the standard deviation of the vertical wind speed, is given by

$$\sigma_w = \begin{cases} \sigma_w(z_r), & z > h \\[2mm] \sigma_w(z_r)\left(\dfrac{z + 0.3}{h}\right)^{1/2}, & 0 \le z \le h, \end{cases} \qquad (6)$$

and T_L, the Lagrangian time scale, is given by

$$T_L = \begin{cases} 0.43\,\dfrac{z}{\sigma_w(z_r)}, & z > h \\[3mm] 0.43\,\dfrac{h}{\sigma_w(z_r)}, & 0 \le z \le h. \end{cases} \qquad (7)$$

In these equations, the units of z are m, $h = 16.7$ m is the height of the trees, and $z_r = 23$ m is the reference height, which was the highest measurement level. The $\sigma_w(z)$ profile in (6) is based directly on measurements made with a 3-dimensional sonic anemometer during the field study. The form of the $T_L(z)$ profile in (7) is based on other studies from the literature, with the factor 0.43 determined by fitting (2), with $K = K_z$ given by (5)–(7), to daytime values of H measured by eddy correlation with the 3-dimensional sonic anemometer at $z = 2$ m, the lowest measurement height for H during the study (near-field effects were shown by Lee to be small at this height). The 0.43 is slightly greater than the expected value of 0.38 based on the usual assumptions that within the canopy $T_L = 0.3h/u_*$, with u_* the

friction velocity, and $\sigma_w = 1.25\,u_*$ above the canopy under neutral conditions [Raupach, 1989]. The offset for z of 0.3 m in (6) differs from the value of 0 in Lee [1992], but avoids a potentially troublesome singularity at $z = 0$ and has little effect on the solution except very near $z = 0$. The small discontinuity in K_z at $z = h$ that arises from this offset has little effect on the solution. It was assumed that $\sigma_w(z_r) = 0.6$ m s^{-1}, which was the daytime average over the three measurement days of interest. Although these expressions are specific to Lee's forest canopy, their form is typical of most canopies [Raupach, 1989].

In general, S, which represents the part of the net radiation absorbed by the trees that is available for sensible heat, changes with both z and t throughout the day. Herein though, comparison with field measurements is on a daytime average basis and so S is assumed to be independent of t. This is not considered to be a drastic assumption since the time scale of interest, determined by the quiescent and gust periods, is relatively short compared to the time scale with which S changes diurnally (except perhaps for partly cloudy conditions during which solar irradiance is highly variable). Following Lee [1992], S was assumed to be proportional to the measured vertical leaf area density distribution, i.e.,

$$S(z) = \begin{cases} 113.67\left(\dfrac{z - 5.5}{h - 5.5}\right)^{0.8}\left(1 - \dfrac{z - 5.5}{h - 5.5}\right)^{1.44} \\[2mm] \quad \text{W m}^{-3}, \quad 5.5 \le z \le h \\[3mm] 0, \quad 0 \le z < 5.5 \text{ m and } z > h. \end{cases} \qquad (8)$$

The value of H at $z = 0$, H_g, was assumed to be constant and equal to 45 W m^{-2}, which was the daytime average value over the three measurement days of interest at $z = 2$ m. The average value of H over any period of length T after the steady periodic state is reached, at any z, is therefore given by $H_g + \int_0^z S(z)\,dz$, with the value at $z = h$, H_t, equal to 230 W m^{-2}.

Analytical solution of (4) is possible using methods described in Carslaw and Jaeger [1959]. Novak [1997] presented such a solution for the case of $K_z(z) = 1$, on a semi-infinite domain. I also derived the solution to this case for a finite domain, with θ known and constant at some z_b, which yields similar results. For $K_z(z) \propto z$, as in (5)–(7), (4) can be transformed into a nonhomogeneous Bessel equation [Novak, 1986; Raupach, 1989] and then solved analytically.

However, herein solutions were determined numerically with a commercial partial differential equation solver, FlexPDE version 2.20c (available at www.pdesolutions.com), which uses a finite element technique. The steady-state solution ($K_t = 1$) and the time-varying solution as described above (but with K_z constant) were found to be in excellent agreement with analytical solutions to these problems, and time-averaged fluxes (once the solutions became periodic in time) for all numerical solutions presented were always in excellent agreement with the distribution expected from H_g and $S(z)$. I found that accurate numerical solution of (1) directly was very difficult when K_t changed dramatically but that (4), whose use is evidently critical for developing analytical solutions, was easily solved for all K_t. FlexPDE has the advantage that problems are formulated simply and naturally and numerical details are hidden from the user, including determination of the unstructured mesh, which is automatically adapted to the solution to meet specified accuracy criteria.

Lagrangian Theory of Raupach [1989]

Raupach [1989] developed an approximate Lagrangian theory based on the assumption that the mean vertical temperature profile in a canopy is the sum of near-field and "far-field" contributions, and that the near-field part can be determined assuming the turbulence to be locally homogeneous (hereafter referred to as the localized near-field theory, or LNF). The near-field contribution is from sources near enough upwind that their release temperature still persists at the measurement location while for the far-field the sources are far enough upwind that parcels of heat arriving there have undergone a random walk and so contribute diffusively. The near-field term, θ_n, is given by

$$\theta_n(z) = \int_0^\infty \frac{S(\delta)}{C\sigma_w(\delta)} \left\{ k_n \left[\frac{z-\delta}{\sigma_w(\delta)T_L(\delta)} \right] \right.$$
$$\left. + k_n \left[\frac{z+\delta}{\sigma_w(\delta)T_L(\delta)} \right] \right\} d\gamma, \qquad (9)$$

where $k_n(\zeta) = -k_n(-\zeta) = -1/\sqrt{2\pi}\ln(1-e^{-\zeta})+(1/2-\pi^2/\sqrt{72\pi})e^{-\zeta}$, and the far-field term, θ_f, is given by

$$\theta_f(z) = H_g \int_z^{z_r} \frac{1}{CK_z(\gamma)} d\gamma$$
$$+ \int_z^{z_r} \frac{1}{CK_z(\gamma)} \int_0^\gamma S(\delta) \, d\delta \, d\gamma + \theta_{fb}, \quad (10)$$

with

$$\theta_{fb} = \theta(z_b) - \theta_n(z_b). \qquad (11)$$

Here $\theta(z_b)$ is a known constant temperature specified at some $z = z_b$, usually chosen well above the canopy.

Field Experiments

All field measurements shown were extracted directly from *Lee* [1992] and his data set; see also *Lee and Black* [1993a,b]. Micrometeorological measurements were made in a Douglas-fir forest near Courtenay, on the east coast of Vancouver Island, British Columbia, during July 19–20 and July 26–August 1, 1990. The period from July 6–August 1 was rainless and the trees were moderately stressed with daytime overstory Bowen ratios between 1 and 3. The stand was planted in 1962 and thinned (to 575 stems ha^{-1}) and completely pruned (up to $z = 5.5$ m) in 1988. It had a (projected) leaf area index of 5.4, a height of 16.7 m, and was located on a 5° east-facing slope. The forest floor was littered with dead branches and trunks, with sparse understory vegetation less than 0.5 m tall.

Air temperature profiles were measured using fine-wire thermocouples installed at $z = 0.9, 2, 4.6, 7, 10, 12.7, 16.7,$ and 23 m throughout the experiment. Data were sampled at 10 Hz but averaged on-line, half-hourly, throughout each day. Wind speeds and H, including all high frequency turbulence statistics, were measured by two 3-dimensional sonic anemometers operating at 10 Hz (during daytime only, with averaging every half hour). One anemometer was installed permanently at $z = 23$ m, while the second was successively placed at one of $z = 2, 7, 10,$ and 16.7 m for two or three days at a time. For the most part herein, only data from July 19, 20, and 26 are used. On these days, the roving sonic anemometer was located at $z = 2$ m and H_g was assumed to be equal to H measured by it.

RESULTS

Figure 1 compares profiles of average θ from the numerical solution with a measured profile for the forest, which is an average of those daytime half-hour periods for which H_g/H_t was in the range 0.15–0.25 on July 19, 20, and 26. The measured temperatures were normalized as $[\theta(z) - \theta(z_r)]/\theta_* = \Delta\theta/\theta_*$, where $\theta_* = H_t/[C\sigma_w(z_r)]$ for each half hour. For the numerical solutions, $\theta = 20$ °C at $z_b = 50$ m was assumed (this was also the initial condition), and the profiles were normalized as were the measurements by subtracting the average $\theta(z_r)$ and dividing by $\theta_* = H_t/[C\sigma_w(z_r)] = 0.32$ °C with $H_t = 230$ W m^{-2}, $C = 1200$ J m^{-3} K^{-1}, and $\sigma_w(z_r) = 0.6$ m s^{-1}. Since these H_t and $\sigma_w(z_r)$, and $H_g = 45$ W m^{-2}, were daytime averages for the three measurement days, which includes periods for which H_g/H_t was outside the range 0.15–0.25, the parameters assumed may not be perfectly applicable to the subset

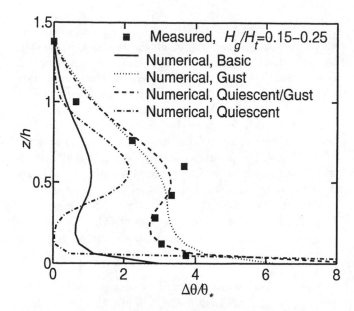

Figure 1. Numerically calculated profiles of normalized potential temperature in the Douglas-fir forest for the four K_t-parameter cases in Table 1. The numerical profiles are averages for the last computed period of length T. Also shown is the measured average profile for H_g/H_t in the 0.15–0.25 range.

of data with H_g/H_t in the 0.15–0.25 range, although because of the normalization this is expected to have a small effect. The $z_b = 50$ m was chosen because ramp behaviour was observed at $z = 23$ m. If the roughness sublayer is estimated to extend to $2h$ then ramps associated with coherent structures generated at $z = h$ by shear instability [*Raupach et al.*, 1996] should be discernible to that level; above $z = 2h$ the temperature is generally not constant but variations should not be correlated with those below this level. Averages for the numerical solutions were from the last calculated period of length T, which was chosen so that the solution had clearly achieved periodicity in time. The standard value of $C = 1200$ J m^{-3} K^{-1} was reduced by 5% within the canopy to roughly account for the volume of the canopy layer occupied by the trees.

The only differences between the four numerical runs shown in Figure 1 are in the K_t parameters, which are listed in Table 1. The $t_1 = 70$ s used in all the runs is the average time between ramps found by *Chen et al.* [1997a] for the $z = 23$ m data during all 9 measurement days. For the run labelled "basic", $K_1 = 0.3$ is a reasonable guess for the quiescent period based on *El-kilani and Van Boxel* [1996], and $t_2 = 5$ s and $K_2 = 300$ were chosen to guarantee that essentially all heat stored in the canopy volume is removed upwards during the gust

period, as suggested by *El-kilani and Van Pul* [1996]. For the run labelled "gust", gust parameters (only) have been modified by trial and error to achieve the best fit possible to the measured data. For the run labelled "quiescent/gust", both quiescent (only K_1) and gust parameters were modified to achieve a good fit to the measured data. For the run labelled "quiescent", the $K_1 = 0.02$ assumes diffusivity values that are probably unreasonably low from a physical standpoint.

Evidently, the theory with the basic input parameters is able to reproduce the general shape of the measured profile but yields normalized θ about 3 times too low within the canopy. Reducing K_1 but keeping the gust parameters such that virtually all heat within the canopy is removed during the gust periods does not increase canopy θ overall, only modifying profile shape (quiescent case). The only way to increase canopy θ is to adjust gust parameters so that heat removal during the gust is less effective, which allows a buildup of canopy heat. This was done by decreasing t_2 and K_2 as shown in Table 1 (gust case), although it could have been done by keeping t_2 constant and only adjusting K_2. A reasonably good fit to the measured profile is obtained by a combination of reducing K_1 and adjusting gust parameters (quiescent/gust case).

According to Figure 2, the average H from the last calculated period of length T is in excellent agreement with that expected from H_g and the profile of S, i.e., $H_g + \int_0^z S(z)\,dz$, for all four numerical cases. This demonstrates that the numerical calculations were accurate and that the (implicitly determined) time steps were adequate to capture all temporal variations. It is evident that average $H > 0$ at all z within and above the canopy, so that for both measurements and theory (quiescent/gust case) the average flux is counter the local average temperature gradient for $0.3 < z/h < 0.6$.

Time series of θ at selected z for the basic and quiescent/gust cases are shown in Figures 3 and 4, respectively. Ramping behaviour is evident in both cases as expected and both eventually reach steady periodicity, the basic case doing so almost immediately while the quiescent/gust case requires about 500 s, or 7 periods,

Table 1. K_t parameters (four cases) used in the numerical solutions, as described in the text.

	Quiescent Period		Gust Period	
	t_1 (s)	K_1	t_2 (s)	K_2
Basic	70	0.3	5	300
Gust	70	0.3	1	30
Quiescent/Gust	70	0.1	1	50
Quiescent	70	0.02	5	300

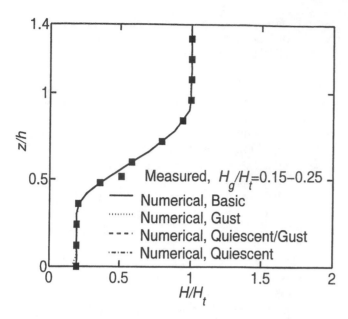

Figure 2. Numerically calculated profiles of normalized sensible heat flux density in the Douglas-fir forest for the four K_t-parameter cases in Table 1. The numerical profiles are averages for the last computed period of length T. Also shown is the average profile for H_g/H_t in the 0.15–0.25 range calculated by integrating (8), which was based on field measurements.

to do so. This is not considered significant because conditions in the field should always be near the steady periodic state. The amplitudes of the ramps (difference between maximum and minimum values of θ) are similar for both cases at all z. This is not surprising given that S is the same for both.

Measured values of ramp amplitude, determined using the ramp model of *Chen et al.* [1997a], are: 0.62 ± 0.31, 0.61 ± 0.28, 0.84 ± 0.40, 1.32 ± 0.43, and 1.23 ± 0.38 °C at $z = 2$, 7, 10, 16.7 (h), and 23 m, respectively, with the indicated range being 1 standard deviation. Since determination of ramp amplitude required data from the 3-dimensional sonic anemometer, simultaneous values were not measured at all z within and above the canopy. Therefore, it is not possible to generate a proper profile with these ramp data, but they give us a rough indication of it. While measured and calculated amplitudes match fairly well at and below $z = 10$ m, modelled values are apparently too low above this level.

Time series of H at selected z for the basic and quiescent/gust cases are shown in Figures 5 and 6, respectively. H varies strongly with t, with large spikes occurring during the gust periods. Detailed calculations show that about 90% of the vertical transfer occurs during gusts and only about 10% during the quiescent pe-

riods. The magnitudes of H during the gust periods are clearly not realistic, being much too large. Despite this, average values of H are correct, as shown in Figure 2. It may be that optimizing K_t parameters should have also included a good representation of the time series of H, but the focus herein was on average θ and H profiles.

Figure 7 tests whether the theory, with K_t parameters from the quiescent/gust case, can predict the effect of H_g/H_t on the profile of average θ. *Lee* [1992] generated the measured profiles from the half-hour average daytime data for July 19, 20, and 26 (the profile for H_g/H_t in the range 0.25–0.35 is not shown for clarity). Although the theory predicts the direction of the changes correctly, it underestimates their magnitudes, requiring that optimal values of K_t change somewhat with H_g/H_t.

Figures 8 and 9 repeat the above comparisons but for the *Raupach* [1989] LNF theory. While the LNF theory is able to reproduce the general shape of the average θ profile, it underestimates the magnitude of the "bump" in the profile associated with the peak value of S near $z/h = 0.55$. Calculated profiles are quite sensitive to the assumed value of $\sigma_w(z_r)$. Reducing it by only 10% to 0.54 m s^{-1} improved the agreement in the lower part of the canopy, while a 20% reduction to 0.48 m s^{-1} yielded good agreement for the mean θ within the canopy. Such reductions may not be excessive, given that $\sigma_w(z_r) = 0.6$ m s^{-1} is a daytime average for July 19, 20, and 26,

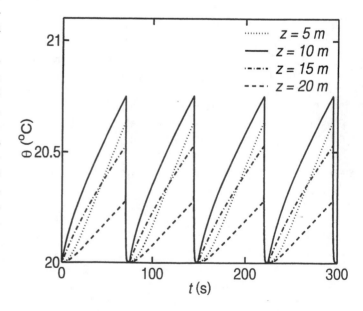

Figure 3. Time-series of numerically calculated potential temperature at the indicated heights in the Douglas-fir forest for the basic case (see Table 1).

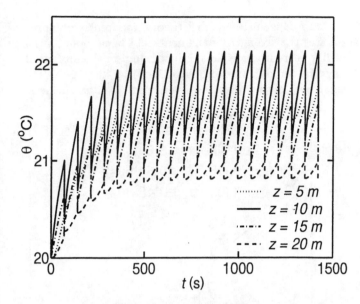

Figure 4. Time-series of numerically calculated potential temperature at the indicated heights in the Douglas-fir forest for the quiescent/gust case (see Table 1).

while the data for H_g/H_t in the range 0.15–0.25 is a subset of the daytime values. The LNF theory captures the trends, but not the magnitudes, of the effects of changing H_g/H_t, similar to the time-varying diffusivity theory. But changing H_g/H_t has a stronger effect on profile shape, with the bump eliminated for $H_g/H_t = 0.45$.

CONCLUDING DISCUSSIONS

The time-varying diffusivity theory can simulate the temperature profiles measured in the Douglas-fir forest studied by *Lee* [1992], but only after determining optimal difffusivity parameters by fitting to the desired profile. Both gust and quiescent period parameters were critical in achieving a good fit. Initially, it was hoped that the best solution would be insensitive to gust parameters and sensitive only to quiescent period parameters. During the quiescent period, turbulence is smaller in scale, K-theory is expected to be valid, and the possibility of developing a robust physically meaningful method to estimate K_1 is strong. But during the gust phase, turbulence is large-scale and vertical transfer is two-dimensional (at least) and advective, so that gust parameters for a one-dimensional model based on K-theory are artificial and difficult to estimate directly.

However, for the quiescent/gust case, the average K_t, given by $(K_1t_1 + K_2t_2)/T$, is 0.8. This is equivalent to using (5)–(7) with $\sigma_w(z_r) = 0.48$ m s^{-1}, which it was

seen yields a good fit on average for θ within the canopy with the LNF theory. If detailed future analysis shows that $\sigma_w(z_r)$ is appropriate for the subset of data with H_g/H_t in the range 0.15–0.25, then gust parameters can

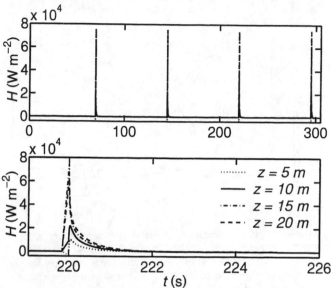

Figure 5. Time-series of numerically calculated sensible heat flux density at the indicated heights in the Douglas-fir forest for the basic case (see Table 1). The lower graph shows the gust phase during the second to last period in greater detail.

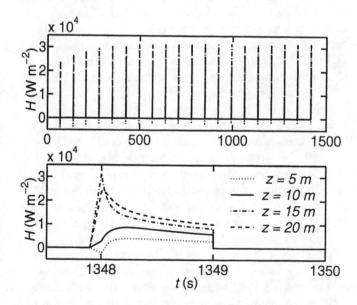

Figure 6. Time-series of numerically calculated sensible heat flux density at the indicated heights in the Douglas-fir forest for the quiescent/gust case (see Table 1). The lower graph shows the gust phase during the second to last period in greater detail.

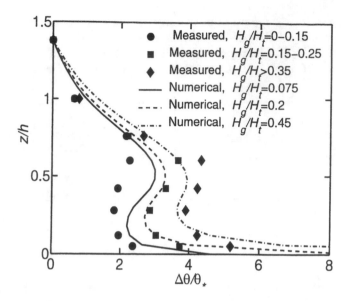

Figure 7. Numerically calculated profiles of normalized potential temperature in the Douglas-fir forest for different values of H_g/H_t. The numerical profiles are averages for the last computed period of length T. Also shown are measured average profiles for the indicated ranges of H_g/H_t.

be determined objectively in the theory by requiring that the average K be given by the far-field diffusivity profile for the period of interest. A second critical question is whether $K_1 = 0.1$ is physically realistic for the quiescent periods in the Douglas-fir forest. Answering this also requires further analysis, first to isolate quiescent periods in the measured temperature time series (*Chen et al.* [1997a] used a wavelet transform to identify the ramps) and then to assess turbulence during these.

With the proper diffusivity parameters, the model reproduces not only average θ profiles but also the small-scale ramping behaviour in time. Despite its K-theory basis, it simulates the strong counter-gradient vertical heat transfer that occurred near midday in the Douglas-fir forest. Mathematically, this is because when $K = K(t)$ the proportionality between averages of the flux and local gradient over some long time period can be very different from that occurring at each instant of time, to the extent that the sign of the proportionality can be reversed. This has been seen for soil heat transfer (which is dominated by conduction) when thermal properties vary with time [*Burn and Smith*, 1988]. The theory also reproduces the intermittant nature of turbulent heat transfer [*El-kilani and Van Boxel*, 1996; *Gao et al.*, 1989] as nearly all of the upward flux occurs during the short gust periods, although the magnitudes of the fluxes during the gusts is greatly overestimated.

The *Raupach* [1989] LNF theory apparently underestimates the strength of the near field in canopies, since it did not reproduce the distinct bump in the measured daytime temperature profiles. Based on his Figure 9, a small part of this might be due to neglect of w skewness,

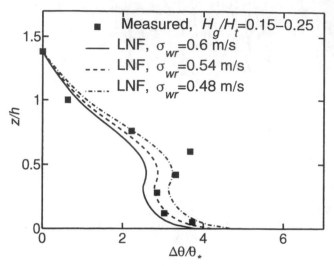

Figure 8. Profiles of normalized potential temperature in the Douglas-fir forest calculated with the *Raupach* [1989] localized near-field theory for different values of $\sigma_w(z_r)$. Also shown is the measured average profile for H_g/H_t in the 0.15–0.25 range.

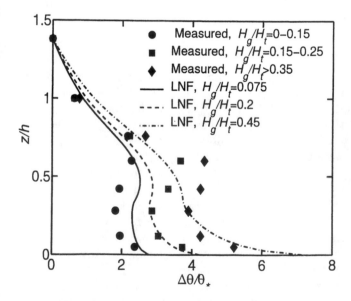

Figure 9. Profiles of normalized potential temperature in the Douglas-fir forest calculated with the *Raupach* [1989] localized near-field theory for different values of H_g/H_t. Also shown are measured average profiles for the indicated ranges of H_g/H_t.

which reached values down to -0.5 in the Douglas-fir forest. *El-kilani* [1996] makes some suggestions as to how to include the effects of the intermittant coherent structures in a Lagrangian analysis of turbulent transfer due to canopy sources, and it may be that when they are incorporated in such a model the near field will be correctly described. The one-dimensional time-varying diffusivity theory presented herein is another approach, which does account for the distinct shape of the daytime temperature profile, as well as the observed ramping behaviour and flux intermittancy in time.

REFERENCES

Baldocchi, D.D., A Lagrangian random-walk model for simulating water vapor, CO_2, and sensible heat flux densities and scalar profiles over and within a soybean canopy, *Boundary-Layer Meteorol., 61*, 113–144, 1992.

Brunet, Y., and M.R. Raupach, A simple renewal model for transfer in plant canopies, poster paper presented at the International Symposium of Flow and Transport in the Natural Environment: Advances and Application, Canberra, ACT, Aug. 31 to Sept. 4, preprinted abstract published by the Australian Academy of Science, Canberra, 2 pp., 1987.

Burn, C.R., and C.A.S. Smith, Observations of the "thermal offset" in near-surface mean annual ground temperatures at several sites near Mayo, Yukon Territory, Canada, *Arctic, 41*, 99–104, 1988.

Carslaw, H.S., and J.C. Jaeger, *Conduction of Heat in Solids*, 2nd ed., 510 pp., Clarendon, Oxford, 1959.

Chen, W., M.D. Novak, T.A. Black, and X. Lee, Coherent eddies and temperature structure functions for three contrasting surfaces. Part I: Ramp model with finite microfront time, *Boundary-Layer Meteorol., 84*, 99–123, 1997a.

Chen, W., M.D. Novak, T.A. Black, and X. Lee, Coherent eddies and temperature structure functions for three contrasting surfaces. Part II: Renewal model for sensible heat flux, *Boundary-Layer Meteorol., 84*, 125–147, 1997b.

El-kilani, R.M.M., Intermittant canopy turbulent transport and its effect on the use of random walk models to describe heat and mass transfer within plant canopies, paper presented at the 22nd Conference on Agricultural and Forest Meteorology, Atlanta, Georgia, Jan. 28 to Feb. 2, preprinted abstract published by the American Meteorological Society, Boston, p. J50–52, 1996.

El-kilani, R.M.M., and J. Van Boxel, A quantification of the turbulent transport coefficient during the quiescence period, paper presented at the 22nd Conference on Agricultural and Forest Meteorology, Atlanta, Georgia, Jan. 28 to Feb. 2, p. 210–213, preprinted abstract published by the American Meteorological Society, Boston, 1996.

El-kilani, R.M.M., and W.A.J. Van Pul, Validation of an intermittency model for describing heat and mass transfer within a plant canopy, paper presented at the 22nd Conference on Agricultural and Forest Meteorology, Atlanta, Georgia, Jan. 28 to Feb. 2, preprinted abstract published by the American Meteorological Society, Boston, p. 214–217, 1996.

Gao, W., R.H. Shaw, and K.T. Paw U, Observation of organized structure in turbulent flow within and above a forest canopy, *Boundary-Layer Meteorol., 47*, 349–377, 1989.

Lee, X., Atmospheric turbulence within and above a coniferous forest, Ph.D. Thesis, 179 pp., University of British Columbia, Vancouver, January 1992.

Lee, X., and T.A. Black, Atmospheric turbulence within and above a Douglas-fir stand. Part I: Statistical properties of the velocity field, *Boundary-Layer Meteorol., 64*, 149–174, 1993a.

Lee, X., and T.A. Black, Atmospheric turbulence within and above a Douglas-fir stand. Part II: Eddy fluxes of sensible heat and water vapour, *Boundary-Layer Meteorol., 64*, 369–389, 1993b.

Novak, M.D., A simple analytical model of coherent turbulent transfer in canopies, paper presented at the 12th Symposium on Boundary Layers and Turbulence, Vancouver, British Columbia, July 28 to Aug. 1, preprinted abstract published by the American Meteorological Society, Boston, p. 389–390, 1997.

Novak, M.D., Theoretical values of daily atmospheric and soil thermal admittances, *Boundary-Layer Meteorol., 34*, 17–34, 1986.

Paw U, K.T., J. Qiu, H.B. Su, T. Watanabe, and Y. Brunet, Surface renewal analysis: A new method to obtain scalar fluxes without velocity data, *Agric. For. Meteorol., 74*, 119–137, 1995.

Raupach, M.R., A practical Lagrangian method for relating scalar concentrations to source distributions in vegetation canopies, Q. J. R. Meteorol. Soc., 115, 609–632, 1989.

Raupach, M.R., J.J. Finnigan, and Y. Brunet, Coherent eddies and turbulence in vegetation canopies: The mixing-layer analogy, *Boundary-Layer Meteorol., 78*, 351–382, 1996.

M.D. Novak, Faculty of Agricultural Sciences, University of British Columbia, 266B–2357 Main Mall, Vancouver, B.C., V6T 1Z4, Canada. (e-mail: novk@interchange.ubc.ca)

The Concept of the Soil-Plant-Atmosphere Continuum and Applications

M.B. Kirkham

Department of Agronomy, Kansas State University, Manhattan, Kansas

J.R. Philip pioneered the concept of the soil, plant, atmosphere as a thermodynamic continuum (SPAC) for water transfer. A central, currently unresolved problem arises as a result of the SPAC concept. How can water move to the top of tall trees in a water continuum from soil to leaf surface without cavitation? A suction pump can lift water only to the height due to atmospheric pressure (1.0 atm = 10.33 m). However, trees are taller than 10.33 m. To get water to the top of skyscrapers, standing tanks are used. In plants, there are no standing tanks, pumps (hearts), or valves that can move water up trees. At present, the cohesion theory is the theory generally accepted as the one that explains the way that water ascends in plants. With the advent of the pressure probe, the necessary high tensions for the cohesion theory have not been found. Consequently, in 1995 a new theory was put forward which postulates that solutes in the parenchyma cells of the tissue around the tracheary elements cause an imbibing of water, and this creates a pressure on them that prevents water cavitation. However, the theory has been questioned both thermodynamically and anatomically. The challenging problem now is to use the SPAC concept on a fine scale and determine the water potential gradients in the xylem tissue to see how water can be at the top of tall trees.

INTRODUCTION

John R. Philip pioneered the concept of the soil, plant, atmosphere as a thermodynamic continuum for water transfer. He defined the SPAC as follows [*Philip*, 1966, p. 246]. "Because water is generally free to move across the plant-soil, soil-atmosphere, and plant-atmosphere interfaces it is necessary and desirable to view the water transfer system in the three domains of soil, plant, and atmosphere as a whole. Under some circumstances, and for some purposes, we can, of course, isolate certain parts of the total system and study only certain modes of water transfer; but a general appreciation of the plant water relations of the whole plant in

nature must involve the soil-plant-atmosphere continuum (SPAC)." In an earlier paper, Philip (1957) discusses the soil-plant-atmosphere continuum and diagrams it (Fig. 1), but he does not use the abbreviation SPAC. The 1966 review is more cited than the 1957 paper, because it is easily accessible in the literature.

Philip (1966) was the first to use the term SPAC. It now is part of the standard terminology of soil-plant water relationships and appears in textbooks and reviews, sometimes with accreditation to Philip [e.g., *Scott*, 2000, p. 329], but usually without [e.g., *Kramer* and *Boyer*, 1995, p. 201; *Steudle*, 2001]. Adaptations of the term SPAC are in common use. The computer-controlled environmental systems for studying whole-plant responses, developed by the United States Department of Agriculture, are called SPAR units. SPAR stands for a Soil-Plant-Atmosphere Research system [*Phene et al.*, 1978]. In addition, to SPAR, SVATS (soil-vegetation-atmosphere-transfer schemes) are in use and can be seen as the logical extension of the SPAC

Environmental Mechanics: Water, Mass and Energy Transfer in the Biosphere
Geophysical Monograph 129
Copyright 2002 by the American Geophysical Union
10.1029/129GM27

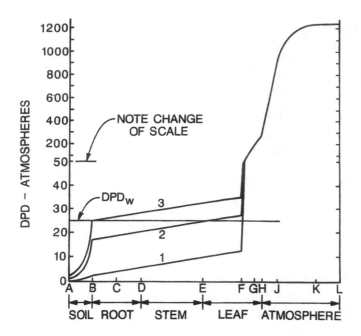

Figure 1. The soil-plant-atmosphere continuum, showing energy profiles, (1), during normal transpiration; (2), during temporary wilting; (3), at permanent wilting. DPD = diffusion pressure deficit. Points on the transpiration path: A. Soil (a definite distance from plant root); B. Surface of root hairs and of absorbing epidermal cells; C. Cortex; D. Endodermis; DE. Vessels and tracheids in xylem; E. Leaf veins; F. Mesophyll cells; FG. Intercellular space and substomatal cavity;   GH. Stomatal pore; HJ Laminar sub-layer; JK. Turbulent boundary layer; KL. Free atmosphere. (Redrawn from Philip, 1957).

concept from plant to vegetation and from local boundary layer to the larger scale of the convective boundary layer (*Raupach*, 1995; *Raupach and Finnegan*, 1995).

HISTORY

The concept of water moving through a soil-plant-atmosphere continuum was present in the literature, even before Philip's (1966) widely cited review. Stephen Rawlins and Paul Waggoner at the Connecticut Agricultural Experiment Station in the USA discussed the idea (*Rawlins*, 1963; *Waggoner*, 1965).   Philip (1966, p. 257) credits Gradmann (1928) for providing the initial steps toward the formulation of the SPAC. Gradmann recognized, for an isothermal system, the thermodynamic equivalence of water in the transpiration stream within the plant and as vapor in the atmosphere, and, in consequence, the existence of systematic gradients of potential in the plant and atmosphere, with continuity of potential at the interface. Gradmann's Figure 1 (1928, p. 3; reproduced here as Fig. 2) represents the "Schema des Saugkraftabfalles" (Diagram of the Decrease in Suction Force"), and it has three lines: AB, BC, and CD.

Line CD shows the fall in suction force between the air and the surface of the plant; line BC shows the fall from the outside surface of the plant to the plant; and line AB shows the fall from the plant to the soil. The total drop goes from 1000 atm (in the air) to 0 atm (in the soil). Gradmann's insight was neglected by other workers for nearly 20 years, when van den Honert (1948) drew attention to it in the English literature [*Philip*, 1966]. Philip (1966; 1996) also acknowledges that the basic ideas behind the SPAC were "largely implicit in the 1952 review of Richards and Wadleigh". Richards and Wadleigh (1952) discuss the use of the free-energy function for expressing the energy status and driving forces of water in soils and plants. They use the term "soil-plant-water system" [*Richards* and *Wadleigh,* 1952, p. 157].

Van den Honert (1948) said, "It was Gradmann's idea to apply an analogue of Ohm's law to this water transport as a whole."   Richards and Wadleigh also noted, "Gradmann (1928) applied an analog of Ohm's law to this water transport as a whole, to the effect that the potential drop across a given part of the system is directly proportional to the resistance" [*Richards* and *Wadleigh,* 1952, p. 175]. However, Gradmann does not mention Ohm in his paper. But Gradmann does

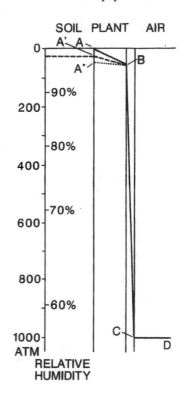

Figure 2. Diagram of the decrease in suction force. (Redrawn from Gradmann, 1928, and translated into English; original German words at the top were Boden, Pflanze, Luft, and the "relative humidity" at the bottom was abbreviated R.F. by Gradmann).

state (1928, p. 3), "Das Gefälle wird zwar innerhalb der Pflanze von A bis B nicht so gleichmäßig sein, sondern jeweils proportional den Teilwiderständen..." (the A and B refer to line AB in his figure described above). So he recognized that the fall in suction force was proportional to the resistance of each part of the system. Gradmann realized that he was working with a linear flow law, like Ohm's law, without mentioning Ohm specifically.

Despite the uncertainty about who originated the idea that water flow through the soil-plant-atmosphere system is similar to the flow of electricity, the paper published by van den Honert (1948) is the most cited paper on the topic and has been used to model movement of water in the soil-plant-atmosphere continuum [e.g., *Cowan*, 1965; *Scott Russell*, 1977, p. 94; *Wind* and *Mazee*, 1979]. Now hydraulic systems are modelled as an analogue to a simple electrical circuit with no acknowledgement to van den Honert [*Williams* et al., 1996].

Ohm's law states that the potential difference, or voltage drop, V, measured in volts, across any part of a conductor is equal to the current in the conductor, I, measured in amps, multiplied by the resistance, R, of that part, measured in ohms, or [*Schaum*, 1961, p. 147]:

$$V = IR \qquad (1)$$

When Ohm's law is applied to the soil-plant-atmosphere continuum, the following analogies are made: V is the potential difference between any two parts in the system. The potential in each part of the system is the (total) water potential (ψ_w), which is measured, for example, with a thermocouple hygrometer or pressure chamber and is usually expressed using the unit of MPa. I is the flow of water or the transpiration rate. This is what Nobel (1974, e.g., p. 142) calls J_v or volume flow measured in units such as m s^{-1}. R is the hydraulic resistance. Its units depend upon how V (or ψ_w) and I have been defined. Combined resistance/capacitance networks also have been developed, and addition of capacitance to the SPAC in trees resulted in a number of papers in the 1970s and 1980s, such as the one by Landsberg et al. [1976]. Capacitance in trees is important, because they can store large quantities of water in the wood. Nobel [1983, p. 514] calculates the capacitance of a tree truck to be 1.5 x 10^{-2} m^3 (MPa)$^{-1}$ and that of much smaller stems, such as a young tomato or sunflower, to be 1.5 x 10^{-5} m^3 (MPa)$^{-1}$. Whitehead and Jarvis (1981, p. 107) report that coniferous species store up to 70% of the total water of the aerial parts in the stems. The change in stem diameter of trees can be used as an estimate of transpiration, because more shrinkage means a higher transpiration rate [*Herzog et al.*, 1995].

Ohm's law is for steady-state conditions. The question arises, "Does one have steady state in the plant?" Richards and Wadleigh state, "[T]he plant is a dynamic organism"

(1952, p. 178), which seems to refute steady state. Van den Honert said his considerations were "confined to steady-state conditions" [1948, p. 147]. Zhang and Kirkham [1999] calculated hydraulic resistance in two ways: (1) using an Ohm's law analogue, which assumes that the relation between flux (transpiration) and difference in water potentials of the soil and plant is linear; and (2) using an equation that considers diurnal changes in leaf water content along with transpiration and difference in water potentials. Because change in leaf-water content during a day was small, hydraulic resistances calculated by the two methods resulted in similar values.

In developing his Ohm's law analogue, van den Honert [1948] uses the now obsolete term "diffusion pressure deficit" (D.P.D.), which is equal to the osmotic pressure (O.P.) minus the turgor pressure (T.P.) of classical plant physiology [*Kirkham* et al., 1969]. The D.P.D. is Gradmann's "suction force" in the plant. Today the D.P.D. is the sum of the osmotic potential (ψ_s) and turgor potential (ψ_p), and the matric potential (ψ_m) and gravitational potential (ψ_g) energies of the plant are neglected. Van den Honert (1948) states, "In the steady-state, the rate of water transport dm/dt is the same in all successive parts. If we call the resistances in root cells, xylem, leaf cells and in the gaseous part R_r, R_x, R_l and R_g respectively, the D.P.D. values on either side of each respective part P_o, P_1, P_2, P_3 and P_4, we have:

$$\mathrm{d}m/\mathrm{d}t =$$
$$(P_1 - P_0)/R_r = (P_2 - P_1)/R_x = (P_3 - P_2/)R_l = (P_4 - P_3)/R_g \qquad (2)$$

This equation now appears in textbooks on plant-water relations [*Slatyer*, 1967, p. 223; *Kramer*, 1983, p. 190; *Baker*, 1984].

CURRENT PROBLEM

A central, currently unresolved problem arises as a result of the SPAC theory. How can water move to the top of tall trees in a water continuum from soil to leaf surface without cavitation? The presence of water at the top of giant trees seems to defy the laws of physics. Let us consider why is it hard for water to get to the top of trees. A suction pump can lift water only to the barometric height, which is the height that is supported by atmospheric pressure from below (1.0 atm = 10.33 m of water) [*Salisbury* and *Ross*, 1978, p. 49]. If a hose or pipe is filled with water, sealed at one end, and then placed in an upright position with the open end down and in water, atmospheric pressure will support the water column to 10.33 meters theoretically. At this height the pressure equals zero, and above this height the water will turn to vapor. My father, Don Kirkham, and his students at Iowa State University tried to see how far they could climb the outside back stairs of the Agronomy Building with a hose which had its bottom in a water bucket on the ground. The column of water in the hose collapsed well before they climbed 10.33 meters.

So how does water get to the top of tall trees? The tallest tree in the world is 111.6 m, and in 1872 a tree estimated to be over 150 m tall was felled in Victoria, Australia [*Salisbury and Ross*, 1978, p. 49].

Let us first consider how water gets to the top of skyscrapers. Wooden tanks that hold water are used to raise water in cities. If one lives in a tall building and is not getting a good strong shower, the spigot is probably too close to the holding tank. In those buildings whose plumbing requires the help of gravity to create sufficient water pressure, a tank needs to be elevated at least 7.6 m above a building's highest standpipe. One gets 6896.3 Newtons/m^2 or 0.068963 bar of pressure for every 7.0 m in height [*Weber*, 1989]. Animals have hearts and valves to move blood to the head.

However, in plants, no standing tanks, pumps (hearts), or valves have been observed. If one looks through Katherine Esau's books on plant anatomy, one sees no such structures [*Esau*, 1965; 1977]. Gradmann himself said that there is no proof that pressure or suction pumps are in plants (1928, p. 3, footnote no. 1). So, again we ask, how does water get to the top of tall trees?

COHESION THEORY

At present, the <u>cohesion theory</u>, or sap-tension theory, is the theory generally accepted as the one that explains most satisfactorily the way that water ascends in plants. Here we shall use interchangeably the terms "sap" and "water in the xylem tissue." We recognize that the fluid in the xylem tissue is not pure water, but a dilute aqueous solution [*Nobel*, 1974, p. 393]. Even in mangroves, which grow in salt water, the sap in the xylem tissue is very nearly salt free and lowers the temperature at which water freezes by <0.1°C [*Hammel* and *Scholander*, 1976, p. 32].

The cohesion theory, as first described in English by Dixon and Joly (1895; see *Steudle*, 2001, for earlier references), assumes that diffusion of water from the non-collapsible xylem elements in contact with the leaf cells creates a state of tension within the water columns in the xylem vessels. This tension is possible because of the cohesion of water molecules and their adhesion to the hydrophilic walls of the xylem elements. Tension in the water columns is assumed to lift water from the roots to the leaves, in addition to reducing the free energy of the water in the root xylem tissue until water diffuses from the soil into the root during absorption of the water. The cohesion hypothesis assumes continuity of water columns, laterally and vertically, in the conducting elements of the xylem tissue. These water columns ultimately are placed under tensile strain.

The cohesion theory of the ascent of sap was foreshadowed by Stephen Hales (1677-1761), an English clergyman, physiologist, chemist, and inventor, famous for his pioneering studies in animal and plant physiology, Julius von Sachs (1832-1897), a German botanist and outstanding plant physiologist, and Eduard Strasburger (1844-1912), a German botanist and one of first to realize the importance of the nucleus and chromosomes in heredity. They all concluded that transpiration produces the pull causing the ascent of sap [*Kramer*, 1983, p. 282].

Even though most plant physiologists feel that the cohesion theory is probably the correct explanation for the rise of water in plants, it has limitations. The main difficulty is that it postulates a system of potentially great instability and vulnerability. It is clear, however, that the water-conducting system in plants must be both stable and invulnerable. The objections that have been raised in plant physiology textbooks concerning the theory include three major points [*Kramer*, 1983, p. 283; *Salisbury* and *Ross*, 1978, p. 58-60]:

1. The tensile strength of water is inadequate under the great tensions necessary to pull water to the top of plants, especially tall plants.
2. There is insufficient evidence for the existence of continuous water columns (that is, water columns under tension are not stable and cavitate).
3. It seems impossible to have tensive channels in the presence of free air bubbles, which can occur when trees in cold climates freeze and then thaw.

These are the classical objections, with which not all plant physiologists now agree. We examine each problem below.

FIRST PROBLEM OF COHESION THEORY

Is the tensile strength of water adequate to pull water to the top of plants? Nobel's (1974, p. 46-47, 52-53) theoretical considerations show that the calculated value for the tensile strength of water is large (1800 MPa) and would permit rise of water in plants even under great tensions. Tensions in higher plants probably never exceed 100 atm. Lower plants like fungi apparently can grow in soil with a tension (or absolute value of matric potential) of |400| bars [*Harris*, 1981, p. 26].

What are values of the tension of water that have been measured experimentally? Scholander et al. (1955), who centrifuged water in glass tubes, observed tensive values from 10 to 20 atm (10.13 to 20.26 bar) without producing cavitation of water. When the experiments were repeated using plant material, they observed much lower values (1-3 atm or 1.01 to 3.04 bar). Also, they were unable to fit hydrostatic pressures in transpiring grape vines into a pattern that followed the cohesion theory. Measured pressure, done with a pressure chamber [*Scholander* et al., 1965], did not indicate cohesion tension at any time, and hydrostatic pressure in transpiring tall vines were higher at the top rather than lower, as they should have been if the transpiring stream were under tension. Measurements taken on Douglas fir trees, however, did follow the pattern that one would expect if water were rising in the plants according to the cohesion theory [*Scholander* et al., 1965]. That is, the hydrostatic pressure at the top of the trees was more negative than at the bottom of the trees. However, later experiments by others with trees did show the expected gradients of water potential through the stems and branches of trees (e.g., see *Whitehead and Jarvis*, 1981, their figure on p. 90). Schill et al. [1996] also found that the osmotic potential of xylem sap decreased with height in maple.

Experiments demonstrate that water can withstand negative pressures (tensions) up to about 300 bars without breaking [*Nobel*, 1974, p. 52]. Recent work confirms the high tensile strength of water and demonstrates that xylem can support large negative pressures (*Holbrook* et al., 1995; *Pockman* et al., 1995). The observed tensile strength depends upon the wall material, the diameter of the xylem vessel, and any solutes present in the water. Local imperfections in the semicrystalline structures of water, such as those caused by H^+ and OH^-, which are always present, even in pure water, reduce the observed tensile strength from the maximum value predicted based on hydrogen bond strengths.

The ability to hear the water columns break is supporting evidence that the columns are under tension, and, when they cavitate, the sound can be picked up acoustically. Milburn and Johnson (1966) developed an acoustic detector, and subsequent experimenters have monitored cavitation using the technique [*Tyree* et al., 1986; *Jackson and Grace*, 1996; *Hacke and Sauter*, 1995].

SECOND PROBLEM OF COHESION THEORY

Let us now consider the second problem with the cohesion theory. Are water columns in the xylem tissue stable under tension? Much has been written about the instability of water columns under tension and the ease with which they break by cavitation in glass capillary tubing [*Kramer*, 1969, p. 275]. It has been suggested that if they break as easily in the xylem of trees, they would soon become inoperative because of shocks such as those caused by swaying in the wind. There is evidence of widespread fracture of stretched water columns and a high percentage of gas-filled, nonfunctional elements under field conditions [*Greenidge*, 1957; *Scholander*, 1958]. However, it seems probable that the nature of the walls of the dead xylem tissue, which is filled with imbibed water, makes the water columns in the stems of plants more stable than

those in glass tubes. If cavitation caused by air entry should occur in the conducting tubes of the xylem tissue, the matric potential component attributed to the hydrophilic nature of the surfaces involved can be expected to maintain surface films of water capable of transporting water up the stem [*Gardner*, 1965].

THIRD PROBLEM OF COHESION THEORY

Let us now consider the third problem. Microscopic observations have shown that air blockage occurs when some trees in cold climates are frozen [*Johnson*, 1977]. Inability to restore the water columns in the spring may well be the factor that excludes certain trees and especially vines with large vessels from these regions [*Salisbury and Ross*, 1978, p. 60]. But how do trees grow in such regions? Imagine a northern tree thawing in the spring. As the ice melts, the tracheids become filled with liquid containing the many bubbles of air that had been forced out by freezing. As melting continues and transpiration begins, tension begins to develop in the xylem tissue. Because of the small dimensions of the tracheids involved, the pressure difference across the curved air-water interface bounding the bubbles would be considerable, resulting in much higher pressure in an air bubble than would exist in the water. Any bubbles which form should dissolve fairly readily restoring the integrity of the water column [*Gardner*, 1965]. Hammel (1967) found no evidence for cavitation of the xylem sap of twigs of hemlock after freezing. He interpreted his results to mean that the bordered pits on the tracheids of gymnosperms function to isolate the freezing sap in each tracheid so that the expansion of water upon freezing not only eliminates any existing tension but also develops positive pressure in the sap. Dissolved gases frozen out of solution may then be redissolved under this positive pressure as melting occurs. However, he found that freezing stem sections of angiosperms invariably increased the resistance to sap flow leading to wilting and death in a few hours. Studies of wood in the spring indicate that about 10% of the tracheids are filled with vapor, but the remaining 90% appear ample to handle sap movement [*Salisbury and Ross*, 1978, p. 60; *Kramer*, 1983].

Also new wood that forms in the spring carries water with it. Jaquish and Ewers [2001] found that stems of two ring-porous trees, *Sassafras albidum* and *Rhus typhina* were 100% embolised in the early spring and became conductive by late June following leaf expansion and maturation of new early wood vessels. Dyes indicated that the stem conduction was restricted almost exclusively to the current year's growth ring. Stems became totally embolized again by early October, before the first freezing temperatures. In contrast, woody roots of both species maintained low embolism values, many conductive growth rings, and high conductivity values regardless of the season. No positive root pressures were detected in either species. Embolism results not only from freezing, but also from drought [*Nardini et al.*, 2001]. Conifers vary in their vulnerability to drought. Scots pine is one of the most vulnerable ones, with a threshold water potential for cavitation between −2.5 and −0.55 MPa [*Jackson et al.*, 1995]. Problems remain in explaining how refilling occurs in trees in the absence of freeze-thaw and in the absence of root pressure.

Gymnosperms with their tracheids are especially well adapted to cold climates. Trees and vines with large, long vessels, are practically absent from cold climates, but are abundant in the tropics. In a study of moisture relations in tall lianas, Scholander et al. (1957) found that allowing vessels of a cut vine to become plugged with air caused a lowered hydrostatic pressure in the plant, but did not reduce the rate of water uptake, indicating that water movement was shifted to the numerous tracheids of the stem. Again, they found no direct evidence of the cohesion theory.

PRESSURE PROBE MEASUREMENTS

For several decades, the cohesion theory was accepted, and essentially no experiments on the topic between about 1960 and 1995 were published. A probe had long been available to measure pressure in large cells of algae [*Steudle and Zimmermann*, 1971]. In recent years, this probe has been miniaturized, so small cells of multicellular plants can be measured [*Boyer*, 1995]. However, early measurements made with it contradicted the cohesion theory. They showed the following [*Canny*, 1995a]: 1) The necessary high tensions in the xylem are not present and the tension in the xylem is around 2 bars; 2) The necessary gradient of tension with height is not present; 3) The measurements of tension with the pressure chamber, believed to verify the cohesion theory, conflict with those made with the pressure probe.

Canny (1995a), therefore, put forward a theory called the "compensating-pressure theory" to account for rise of water in plants. He noted that the xylem tissue has ray cells throughout it. These are living parenchyma cells. He said, "I now propose an entirely different resolution: that the compensating pressure is provided by the tissue pressure of xylem parenchyma and ray cells, pressing onto the closed fluid spaces of the tracheary elements and squeezing them. The driving force is provided, as in the Cohesion Theory, by evaporation and the tensions generated in curved menisci in the wet cell walls of the leaf. The force is transmitted, as in the Cohesion Theory, by tension in the water in the tracheary elements. But this tension is kept within the operating range by the compression from tissue pressure around the tracheary elements. The gravitational gradient of tension up a tall tree would then be compensated by increasing tissue pressure of the xylem parenchyma with height, and the need for a tension gradient to sustain the standing columns would vanish." Key

to his theory is the endodermis, the inner layer of cortical cells in roots which contains the Casparian strips. They are bands of lignin and suberin in the walls of the endodermal cells [*Esau*, 1977, p. 504] and force water and solutes moving through the apoplast of the root into living cells. A differentially permeable membrane, such as that in a living endodermal cell, appears essential for root pressure [*Kramer*, 1983, p. 223]. Canny (1995a) postulates that root pressure pumps water up the stem. He discusses his theory in other papers [*Canny*, 1995b; 1997; 1998; 2001]. The theory of Canny (1995a) has support from the work by Kargol et al. (1995), who say that water is transported along the xylem vessels by "graviosomotic mechanisms." They also show the importance of a root pump in getting water up a plant.

Canny's theory has been challenged by Comstock (1999). He points out that Canny's theory "suffers from fatal flaws". Tissue pressure is likely to be ubiquitous but small. Extreme reinforcement would be needed to sustain the tissue pressures postulated in his model, not just one or two cell layers with thickened walls. Canny postulates a pump-and-valve system, which is essential to the working of his model, but no viable mechanism has been identified. If the water potential gradient is such that the parenchyma cells have a lower water potential than the cells of the xylem tissue, water would move from the tracheary elements to the parenchyma. Comstock (1999) says, "[W]ater would be pouring out of the xylem into the surrounding parenchyma to equilize the water potentials of the two compartments, exactly the opposite of what Canny claims is happening in his putative refilling mechanism."

Tyree (1999) takes issue with Canny's concept of "tissue pressure." Tissue pressure arises when the volume change of some living cells exerts a pressure on adjacent living or dead cells. Tyree (1999) says, "Contrary to previous assertions, tissue pressure cannot cause a permanent change in pressure potential or water potential of adjacent cells. Tissue pressure induces only a transitory increase of pressure and water potential. After equilibrium is reestablished, the same or a more negative pressure or water potential results. The idea that tissue pressure can prevent or repair xylem embolism is without merit."

Canny (2001), in responding to Comstock (1999), said, "It is not helpful to isolate parts of the system mentally and try to assign them water potentials." But we do need to do this. Only by knowing the water potential gradient can we determine the direction of water movement in the SPAC.

Others defend the cohesion theory [*Pockman et al.*, 1995; *Tyree*, 1997; *Stiller and Sperry*, 1999; *Wei et al.*, 1999a, 1999b, 2000; *Steudle*, 2001]. Most of the negation of the cohesion theory comes from measurements made with the xylem pressure probe. Zimmermann et al. (1994, 1995) say that the cohesion theory requires a reappraisal, because direct measurements of the xylem pressure in single vessels of higher plants and tall trees, by means of the xylem pressure probe technique, indicate that the xylem tension in the leaves of intact, transpiring plants is often much smaller than that predicted for transpiration-driven water ascent through continuous water columns. The probe may be measuring inaccurate values [*Tomos and Leigh*, 1999]. The ultimate tension limit of the probe is somewhere between 1.6 and 1.8 MPa [*Wei et al.*, 2000]. The probe is incapable of measuring higher tensions either because of an imperfect seal between the probe and the xylem wall or the creation of micro-fissures in the xylem cell wall when the probe is inserted. In both cases, cavitation via "air-seeding" is proposed to occur at pressures less negative than those normally sustained by the xylem [*Tomos and Leigh*, 1999]. Wei et al. [1999a] report that their new, direct measurements of xylem pressure support the cohesion-tension theory. Xylem pressure probes used by Wei et al. [1999a] differed from those of other researchers, because they used a cell pressure probe filled with silicone oil instead of with water. They put the probe directly in the tracheary elements. If the pressure probe can be used to monitor pressure (or tension) in the tracheary elements, then it seems to me that the pressure probe could be inserted into a ray cell (parenchyma cell) to see what pressure exists there that is pushing against the tracheary elements. This would confirm or refute Canny's theory.

EXPERIMENTAL MEASUREMENTS NEEDED

Experiments to study tensive values of water in plants have been done with plants that have been punched with manometers [*Scholander et al.*, 1955], cut [*Scholander et al.*, 1957], sawed [*Greenidge*, 1955], punctured with a pressure probe [*Tomos* and *Leigh*, 1999], frozen [*Cochard et al.*, 2000], or otherwise disturbed. If it would be possible to study plants under natural conditions when they are intact, one might come to a better understanding as to what tension water is under in plants, and, if tensions are built up, if they are sufficient to account for the rise of water.

New equipment is being developed that can be used to measure non-destructively the characteristics of water transport in the SPAC, such as nuclear magnetic resonance imaging (NMR) [*Scheenen et al.*, 2000]. Equipment can be gotten to the top of giant trees. Tall platforms have been constructed to access the top of forest canopies (12 m high) [*Ellsworth*, 1999]. See the cover of Plant, Cell Environment, May, 1999, for a photo of a platform with a man standing on it at the top of a forest canopy. However, the problem would be using the equipment. For example, the pressure probe requires a totally stable environment for its use [*Boyer*, 1995] and could not provide useful data in a gondola suspended from a canopy crane. Maybe Scholander's measurements were correct [1955, 1957] and the tension does not increase with height. (I was told that Scholander got his measurements of tension in the top of tall vines and trees by

shooting branches down with a gun and then putting them in his pressure chamber!) The unanswered question is how much transpiration does actually occur from the top of tall trees.

CONCLUSIONS

In spite of difficulties of demonstrating in some experiments appreciable values of tension in water columns of plants, most plant physiologists continue to assume that high tension values are readily obtainable and that the cohesion theory is correct [e.g., see *Wei et al.*, 1999a]. On the assumption that water moves through the SPAC according to potential gradients, one has to assume that the cohesion theory works. Van den Honert (1948) assumed that it was valid and said "[T]he correctness of the cohesion theory will be taken for granted".

Feelings get heated when scientists are either defending or refuting the cohesion theory. A plant physiologist told me when I was a graduate student that the cohesion theory was valid, and I should not waste my time thinking about it. Almost 80 years ago, when a physicist published a book questioning the validity of the cohesion theory [*Bose*, 1923], plant physiologists who reviewed the book used strong language to show that he was wrong. For example, MacDougal and Overton (1927) said, "Every page of Bose's book on the ascent of sap … is utterly lacking in scientific significance. Such books appearing on the lists of scientific publications constitute a menace and danger to sound science." The Bose questioning the cohesion theory was Sir Jagadis Chunder Bose, who was the teacher of Satyendra Nath Bose [*Ghosh*, 1992]. S.N. Bose was the Bose of the Bose-Einstein condensation (BEC) [*Wyatt*, 1998]. The cohesion theory has been vigorously defended by plant physiologists for many years, but more experiments based on sound physical theory are needed to accept its assumptions. The challenging problem now is to use the SPAC theory on a fine scale and determine the water potential gradients in the xylem tissue to see the direction of movement of water. Philip's concluding words (1966, p. 265) are as appropriate now as they were 35 years ago: "The few years since the SPAC concept has had wide acceptance have seen an increasing stream of experiments and observations pertinent to the problems of plant-water relations in the field. The SPAC has been a fruitful stimulus to meaningful research, and has thus proved its worth as a scientific idea. My only concern is that this heartening upsurge of effort should not be vitiated by a failure of self-criticism."

REFERENCES

Baker, D.A., Water relations, In *Advanced Plant Physiology*, edited by M.B. Wilkins, pp. 297-318, Pitman Pub. Ltd., London, 1984.

Bose, J.C., *The Physiology of the Ascent of Sap.*, Longmans, Green and Co., London, 277 pp., 1923.

Boyer, J.S., *Measuring the Water Status of Plants and Soils,* Academic Press, San Diego, 178 pp., 1995. (See Chap. 4, "Pressure Probe.")

Canny, M.J., A new theory for the ascent of sap - cohesion supported by tissue pressure, *Ann. Bot.,* 75, 343-357, 1995a.

Canny, M.J., Apoplastic water and solute movement: New rules for an old space, *Annu. Rev. Plant Physiol. Plant Mol. Biol.,*46, 215-236, 1995b. (see p. 229 and following).

Canny, M.J., Vessel contents of leaves after excising-a test of Scholander's assumption, *Amer. J.Bot.,* 84, 1217-1222, 1997.

Canny, M.J., Applications of the compensating pressure theory of water transport, *Amer. J. Bot.,* 85, 897-909, 1998.

Canny, M.J., Contributions to the debate on water transport, *Am. J. Bot.,* 88, 43-46, 2001.

Cochard, Hervé, Christian Bodet, Thierry Améglio, and Pierre Cruiziat, Cryo-scanning electron microscopy observations of vessel content during transpiration in walnut petioles, Facts or artifacts? *Plant Physiol.,* 124, 1191-1202, 2000.

Comstock, Jonathan P., Why Canny's theory doesn't hold water, *Amer. J. Bot.,* 86, 1077-1081, 1999.

Cowan, I.R., Transport of water in the soil-plant-atmosphere system, *J. Appl. Ecol.,* 2, 221-239, 1965.

Dixon, H.H., and J. Joly, On the ascent of sap, *Phil. Trans. Roy. Soc.,* London, B186, 563-576, 1895.

Ellsworth, D.S., CO_2 enrichment in a maturing pine forest: Are CO_2 exchange and water status in the canopy affected?, *Plant, Cell Environ.,* 22, 461-472, 1999.

Esau, K., *Plant Anatomy,* Second ed. John Wiley and Sons, New York, 767 pp., 1965.

Esau, K., *Anatomy of Seed Plants.,* Second edition, John Wiley and Sons, New York, 550 pp., 1977.

Gardner, W.R., Dynamic aspects of soil-water availability to plants, *Annu. Rev., Plant Physiol.,* 16, 323-342, 1965.

Ghosh, Amitav, Vignette: Looking Toward Calcutta, *Nature,* 257, 1775, 1992. (one page only)

Gradmann, H., Untersuchungen über die Wasserverhältnisse des Bodens als Grundlage des Pflanzenwachstums, *I. Jahrbücher für wissenschaftliche Botanik,* 69, 1-100, 1928.

Greenidge, K.N.H., Studies in physiology of forest trees, III, The effect of drastic interruption of conducting tissues in moisture movement, *Amer. J. Bot.,* 42, 582-587, 1955.

Greenidge, K.N.H., Ascent of sap, *Annu. Rev. Plant Physiol.,* 8, 237-256, 1957.

Hacke, U., and J.J. Sauter, Vulnerability of xylem to embolism in relation to leaf water potential and stomatal conductance in *Fagus sylvatica* f. *purpurea* and *Populus balsamifera. J. Exp.Bot.,* 46, 1177-1183, 1995.

Hammel, H.T., Freezing of xylem sap without cavitation, *Plant Physiol.* 42, 55-66, 1967.

Hammel, H.T., and P.F. Scholander., *Osmosis and tensile solvent, Springer-Verlag,* Berlin, 133 p., 1976.

Harris, R.F., Effect of water potential on microbial growth and activity, in *Water Potential Relations in Soil Microbiology,* edited by J.F. Parr, W.R. Gardner, and L.F. Elliott, p. 23-95, SSSA Special Pub. No. 9. Soil Sci. Soc. Amer., Madison, Wis., 1981.

Herzog, K.M., R. Häsler, and R. Thum, Diurnal changes in the radius of a subalpine Norway spruce stem: their relation to the sap flow and their use to estimate transpiration, *Trees,* 10, 94-101.

Holbrook, N.M., M.J. Burns, and C.B. Field, Negative xylem pressures in plants: A test of the balancing pressure technique, *Science,* 270, 1193-1194.

Jackson, G.E., and J. Grace, Field measurements of xylem cavitation: Are acoustic emissions useful?, *J. Exp. Bot.,* 47, 1643-1650, 1996.

Jackson, G.E., J. Irvine, and J. Grace, Xylem cavitation in Scots pine and Sitka spruce saplings during water stress, *Tree Physiol.,* 15, 783-790, 1995.

Jaquish, L.L., and F.W. Ewers, Seasonal conductivity and embolism in the roots and stems of two clonal ring-porous trees, *Sassafras albidum* (Lauraceae) and *Rhus typhina* (Anacardiaceae), *Amer. J. Bot.,* 88, 206-212, 2001.

Johnson, R.P.C., Can cell walls bending round xylem vessels control water flow?, *Planta,* 136, 187-194, 1977.

Kargol, M.N., T. Kosztolowicz, and S. Przestalski, About the biophysical mechanisms of the long-distance water translocation in plants, *Int. Agrophysics,* 9, 243-255, 1995.

Kirkham, M.B., W.R. Gardner, and G.C. Gerloff., Leaf water potential of differentially salinized plants, *Plant Physiol.,* 44, 1378-1382, 1969.

Kramer, P.J., *Plant and Soil Water Relationships. A Modern Synthesis,* McGraw-Hill,, New York, 482 p., 1969.

Kramer, P.J., *Water Relations of Plants,* Academic Press, New York, 489 p., 1983.

Kramer, P.J. and J.S. Boyer, *Water Relations of Plants and Soils,* Academic Press, San Diego, 495 pp., 1995.

Landsberg, J.J., T.W. Blanchard, and B. Warrit, Studies on the movement of water through apple trees, *J. Exp. Bot.,* 27, 579-596, 1976.

MacDougal, D.T., and J.B. Overton, Sap flow and pressure in trees, *Science,* 65, 189-190, 1927.

Milburn, J.A., and R.P.C. Johnson, The conduction of sap. II. Detection of vibrations produced by sap cavitation in *Ricinus* xylem, *Planta,* 69, 43-52, 1966.

Nardini, A., M.T. Tyree, and S. Salleo, 2001, Xylem cavitation in the leaf of *Prunus laurocerasus* and its impact on leaf hydraulics, *Plant Physiol,* 125, 1700-1709, 2001.

Nobel, P.S., *Introduction to Biophysical Plant Physiology,* W.H. Freeman and Co., San Francisco, 488 pp., 1974.

Nobel, P.S., *Biophysical Plant Physiology and Ecology,* W.H. Freeman and Co., San Francisco, 608 pp, 1983.

Phene, C.J., D.N. Baker, J.R. Lambert, J.E. Parsons, and J.M. McKinion, SPAR - A Soil-Plant-Atmosphere Research System, *Trans. ASAE,* 21, 924-930, 1978.

Philip, J.R., The physical principles of soil water movement during the irrigation cycle. *Internationl Commission on Irrigation and Drainage, Third Congress,* Algiers, R. 7, Question 8, 8.125 – 8.154, 1957.

Philip, J.R., Plant water relations: Some physical aspects, *Annual Review of Plant Physiology,* 17, 245-268, 1966.

Philip, J.R. History of Hydrology Film Interviews, John R. Philip Interviewed by Stephen Burges, July 5, 1995, American Geophysical Union, Source Material VI-001-2812, Washington, D.C., 1996.

Pockman, William T., John S. Sperry, and James W. O'Leary, Sustained and significant negative water pressure in xylem, *Nature,* 378, 715-716, 1995.

Raupach, M.R., Vegetation-atmosphere interaction and surface conductance at leaf, canopy and regional scales, *Agric. Forest Meteorol.*, 73, 151-179, 1995.

Raupach, M.R., and J.J. Finnigan, Scale issues in boundary-layer meteorology: Surface energy balances in heterogeneous terrain, *Hydrol. Processes*, 9, 589-612, 1995.

Rawlins, S.L., Resistance to water flow in the transpiration stream, In *Stomata and Water Relations in Plants*, Bull. 664, The Connecticut Agricultural Experiment Station, New Haven (Ed. I. Zelitch), pp. 69-85, 1963.

Richards, L.A., and C.H. Wadleigh, Soil water and plant growth, in *Soil Physical Conditions and Plant Growth*, edited by Byron T. Shaw, p. 73-251, Volume II in the series of monographs AGRONOMY, Academic Press, New York, 1952.

Salisbury, F.B., and C.W. Ross, *Plant Physiology,* Second edition, Wadsworth Pub. Co., Inc., Belmont, California, 436 pp., 1978.

Schaum, Daniel, *Schaum's Outline of Theory and Problems of College Physics,* Sixth ed., Schaum Pub. Co., New York, 270 pp., 1961.

Scheenen, T.W.J., D. van Dusschoten, P.A. de Jager, and H. Van As., Quantification of water transport in plants with NMR imaging, *J. Exp. Bot.*, 51, 1751-1759, 2000.

Schill, V., W. Hartung, B. Orthen, and M.H. Weisenseel, The xylem sap of maple (*Acer platanoides*) trees – sap obtained by a novel method shows changes with season and height, *J. Exp. Bot.*, 47, 123-133, 1996.

Scholander, P.F., The rise of sap in lianas, *In The Physiology of Forest Trees,* A symposium held at the Harvard Forest, April, 1957, The Ronald Press Co., New York, 1958.

Scholander, P.F., W.E. Love, and J.W. Kanwisher, The rise of sap in tall grapevines, *Plant Physiol.*, 30, 93-104, 1955.

Scholander, P.F., B. Rund, and H. Leivestad, The rise of sap in tropical liana, *Plant Physiol.*, 32, 1-6, 1957.

Scholander, P.F., H.T. Hammel, E.D. Bradstreet, and E.A. Hemmingsen, Sap pressure in vascular plants, *Science,* 148, 339-346, 1965.

Scott, H. Don, *Soil Physics,* Agricultural and Environmental Applications, Iowa State University Press, Ames, 421 pp., 2000.

Scott Russell, R., *Plant Root Systems: Their Function and Interaction with the Soil,* McGraw-Hill Book Co., UK, Ltd., London, 298 pp., 1977.

Slatyer, R.O., *Plant-Water Relationships*, Academic Press, London, 366 pp., 1967.

Steudle, E., The cohesion-tension mechanism and the acquisition of water by plant roots, *Annu. Rev. Plant Physiol. Plant Mol. Biol.*, 52, 847-875, 2001.

Steudle, E., and U. Zimmermann, Hydraulic conductivity of *Valonia utricularis*, *Z. Naturforsch.*, 26b, 1302-1311, 1971.

Stiller, V., and J.S. Sperry, Canny's compensating pressure theory fails a test, *Amer. J. Bot.*, 86, 1082-1086, 1999.

Tomos, A. D., and R. A. Leigh, The pressure probe: A versatile tool in plant cell physiology, *Annu. Rev. Plant Physiol. Plant Mol. Biol.,* 50, 447-472, 1999.

Tyree, M.T., The cohesion-tension theory of sap ascent: current controversies, *J. Exp. Bot.*, 48, 1753-1765, 1997.

Tyree, M.T., The forgotten component of plant water potential: A reply – Tissue pressures are not additive in the way M.J. Canny suggests, *Plant Biol.*, 1, 598-601, 1999.

Tyree, M.T., E.L. Fiscus, S.D. Wullschleger, and M.A. Dixon., Detection of xylem cavitation in corn under field conditions, *Plant Physiol.*, 82, 597-599, 1986.

Van den Honert, T.H., Water transport in plants as a catenary process, *Disc. Faraday Soc.*, 3, 146-153, 1948.

Waggoner, P.E., Decreasing transpiration and the effect upon growth, In *Plant Environment and Efficient Water Use*, American Society of Agronomy and Soil Science Society of America, Madison, Wisconsin (Eds. W.H. Pierre, D. Kirkham, J. Pesek, and R. Shaw), pp. 49-72, 1965.

Weber, Bruce, Keeping up the pressure, *The New York Times Magazine*, 3 September 1989, p. 62, 1989. (one page only)

Wei, Chunfang, E. Steudle, and M.T. Tyree, Water ascent in plants: do ongoing controversies have a sound basis?, *Trends Plant Sci.*, 4, 372-375, 1999a.

Wei, Chunfang, Melvin T. Tyree, and Ernst Steudle, Direct measurement of xylem pressure in leaves of intact maize plants, A test of the cohesion-tension theory taking hydraulic architecture into consideration, *Plant Physiol.*, 121, 1191-1205, 1999b.

Wei, Chunfang, E. Steudle, and M.T. Tyree, Reply…Water ascent in plants, *Trends Plant Sci.*, 5, 146-147, 2000.

Whitehead, D., and P.G. Jarvis, Coniferous forests and plantations, In *Water Deficits and Plant Growth. Vol. VI. Woody Plant Communities*, Academic Press, New York (Ed. T.T. Kozlowski), pp. 49-152, 1981.

Williams, M., E.B. Rastetter, D.N. Fernandes, M.L. Goulden, S.C. Wofsy, G.R. Shaver, J.M. Melillo, J.W. Munger, S.-M. Fan, and K.J. Nadelhoffer, Modelling the soil-plant-atmosphere continuum in a *Quercus-Acer* stand at Harvard Forest: The regulation of stomatal conductance by light, nitrogen and soil/plant hydraulic properties, *Plant, Cell Environment,* 19, 911-927, 1996.

Wind, G.P., and A.N. Mazee, An electronic analog for unsaturated flow and accumulation of moisture in soils, *J. Hydrol.,* 41, 69-83, 1979.

Wyatt, Adrian F.G., Evidence for a Bose-Einstein condensate in liquid ^{4}He from quantum evaporation, *Nature,* 391, 56-59, 1998.

Zhang, Jingxian, and M.B. Kirkham, Hydraulic resistance of sorghum (C_4) and sunflower (C_3), *Journal of Crop Production*, 2, 287-298, 1999.

Zimmermann, U., F.C. Meinzer, R. Benkert, J.J. Zhu, H. Schneider, G. Goldstein, E. Kuchenbrod, and A. Haase, Xylem water transport: Is the available evidence consistent with the cohesion theory? *Plant, Cell Environ.*, 17, 1169-1181, 1994.

Zimmermann, U., F. Meinzer, and F.-W. Bentrup, How does water ascend in tall trees and other vascular plants?, *Ann. Bot.*, 76, 545-551, 1995.

Department of Agronomy, 2004 Throckmorton Hall, Kansas State University, Manhattan, Kansas, 66506-5501, USA

Rootzone Processes, Tree Water-Use, and the Equitable Allocation of Irrigation Water to Olives

Steve Green and Brent Clothier

Environment & Risk Management Group, HortResearch, Palmerston North, New Zealand

Horst Caspari [1] and Sue Neal

Environment & Risk Management Group, HortResearch, Blenheim, New Zealand

John Philip's first job in 1947, at Griffith in Australia's Murrumbidgee Irrigation Area, was to develop means by which irrigation practices could become sustainable. Subsequently, through his analytical endeavors he created revolutionary new understanding of mass and energy transfers in the soil-plant-atmosphere continuum. Here we describe applications and modeling that have directly benefited from John Philip's insights and perspicacity. We have used a new means for determining the radiation interception by an isolated olive tree, and we have employed these results to interpret and model the measured rates of tree water-use from heat-pulse measures of sapflow. These parameters are used in a risk assessment framework, along with measures of the soil's hydraulic character to provide a basis for establishing guidelines for the equitable and sustainable allocation of water for the irrigation of olive trees in Marlborough, New Zealand. We find that small 2-year old olive trees use about 25 litres a week, whereas mature 8-year old trees transpire about 525 L/wk. Our model developed to establish irrigation allocations, SPASMO, used a 28-year sequence of local weather records. For the Fairhall stony silt loam, we find that an irrigation allocation of 230 mm will meet the needs of olives 9 years in 10. Average requirements would be met with just 140 mm. Only 35 mm would be required to meet the needs of olives 90% of the time on the Woodbourne deep silt loam. Apposite measurements and apt modeling are shown capable of guiding regulatory authorities in managing the complexity of allocating water to olive irrigationists.

---

[1] *Present address:* Colorado State University, Grand Junction, Colorado, U.S.A

Environmental Mechanics: Water, Mass and Energy Transfer in the Biosphere
Geophysical Monograph 129
Copyright 2002 by the American Geophysical Union
10.1029/129GM28

1. INTRODUCTION

Today, irrigation is responsible for some two thirds of the worldwide usage of fresh water. In developing countries, this allocation to irrigation rises to 90 percent (Postel, 2001). Yet, as much as half of all water diverted for agriculture never yields any food ... [so] the challenges we face are to use the water we have more efficiently, [and] to re-think our priorities for water use (Gleick, 2001).

Table A1. Water holding properties of Marlborough soils. TAW is the total available water in the top 1.0 m of soil between field capacity (-0.01 MPa) and wilting point (-1.5 MPa). The readily available water (RAW), we take for olives to 65% of TAW.

Soil Type	% stones	TAW [mm/m]	RAW [mm/m]
Fairhall	44-77	41	24
Renwick	25-50	73	46
Wairau	-	157	66
Woodbourne	-	176	110

Back in 1947, a callow engineer named John Robert Philip was seconded to the CSIR's Irrigation Research Station at Griffith in NSW's Murrumbidgee Irrigation Area. There, in his words (Philip, 1992), he found "… agricultural scientists struggling to understand the hydraulics of furrow irrigation". He noted that "… the question of just how water is held and moves in soil are central to the scientific study of the land sector. Surprisingly, a proper understanding of these processes came very late … until the mid-1950's, they were treated more or less at the folklore level" (Philip, 1977). At Griffith, Philip discovered that "… my own modest stock of mathematics and physics seemed to shed light where no light had been (Philip, 1992). As we show here, Philip's mathematical musings have lead to knowledge that has improved our ability to model the fate of irrigation water applied to soil. Once efficiently in the soil, irrigation water then needs to be effectively available to the plant.

"Water moves in all parts of the complicated soil-plant-atmosphere system down the gradient of potential (Philip, 1957) … this concept gives a useful first picture, but it must be emphasized that for real plants the three-dimensional disposition of roots, leaves and other plant parts makes for a more complicated problem (Philip, 1977). In developing equitable and sustainable irrigation strategies for olive trees, we have had to determine what controls water use from an olive tree in a grove containing trees of various ages. This has demanded that we respond to Philip's challenge by determining the impact of canopy architecture on water use. We have used a Whirligig (McNaughton et al., 1992), in conjunction with heat-pulse measurements of sapflow (Green and Clothier, 1988), to model the transpiration of an olive tree, following Green and McNaughton (1997).

Our mechanistic models of soil-water movement and tree-scale transpiration we then employ in a risk assessment framework, using a long sequence of weather data, to determine the probability that any given amount of water will be required for irrigation on various soil types. A decade ago, John Philip expressed fear that simulation modelling will mean "… that by 2066 we shall be deep into the electronic Dark Ages" (Philip, 1991). He asserted that "… modelling is rather … a pleasurable and harmless pastime". We are not blinded by this assertion. Rather we consider that by its ability to accommodate rationally temporal variability, apposite modelling provides a robust risk-assessment framework.

This year in Marlborough, New Zealand's prime viticultural region, rainfall for the three months of summer has been the lowest on record with just a trace of precipitation falling. Furthermore, as irrigated vineyards expand, and as owners of new olive plantings seek irrigation consents, pressure is mounting on the ground and surface water resources. The District Council is thus seeking guidelines for the equitable and sustainable allocation of irrigation water to grapes and olives. Nuttle (2000) pondered whether ecosystem managers could rely on mechanistic models to guide their decisions. He concluded that a whole system approach was the key, and that "… observation, experiment and modelling together are the essential components of the whole-system approach". Here we present the results of both an experimental study and modelling analysis of allocation of irrigation water to olives.

2. EXPERIMENTS

2.1. Field Site

An experiment on olive trees (cv Barnea) of ages 2, 4 and 8 years old was carried out on the Ponder Estate (41°30`S, 173°52`E) in Marlborough between October 1999 and April 2000. The site is flat and the trees are planted in rows 6 m apart, at a tree-spacing of 5 m. The ground is covered in grass except for a 2 m wide herbicide strip along each row. The Renwick stony silt loam at the site is deep, yet because it comprises 25-50% stones, its total available water holding capacity (TAW) is just 73 mm of water per m depth of soil (Table 1). We take TAW as being that water held in the soil between potentials of -0.1 and -15 bar. The readily available water (RAW), we take for olives to be 65% of TAW (Allen et al., 1998), and so it is just 46 mm/m. The orchard trees are irrigated using low pressure drippers (2 L/h) connected to a single drip line. There is one dripper per tree on the 2 and 4 yr-old trees, and three drippers per tree on the 8 yr-old trees. During the summer of our experiment the irrigation was turned on for 4 hours, once every three days.

2.2. Measurements

Six olive trees were instrumented with heat-pulse equipment to monitor instantaneous rates of tree transpira-

Figure 1. Measurement of energy and light interception by a medium-sized olive tree. The 'Whirli-gig' rotates at 3 rpm and measures the amount of visible and all-wave solar energy intercepted by the isolated tree.

tion (Green and Clothier, 1988). Each tree had two sets of heat-pulse probes installed into the trunk. Sap flow was monitored for a period of about 8 weeks during the middle of the summer. A data logger (Model CR10, Campbell Sci., Utah, USA) was used to activate the heat-pulse equipment and to record sap flow every half hour. The measured sap flow was then summed up over 24 hours to calculate the daily water use of the trees. Details of the heat-pulse system can be found in Green (1998).

A weather station in the middle of the orchard recorded half-hourly averages of net all-wave radiation, global PAR radiation, air temperature, relative humidity, wind speed and rainfall. These data were used to calculate instantaneous rates of whole-tree transpiration, E [kg s^{-1}], as described below.

The 'Whirligig' radiometer of McNaughton et al., 1992 was set up around one of the 4 yr-old olive trees and used to measure the total amount of solar radiant energy intercepted by the tree canopy (Fig. 1). Light interception by the other olive trees was estimated using an array of 20 PAR sensors (Biggs et al., 1971) distributed uniformly over a square grid located close to the ground surface.

An array of 16 rain gauges, made from 185 mm dia. plastic funnels and 4 L plastic bottles, was distributed over a regular grid located close to the ground under each of the instrumented trees. The catch from each rain gauge was recorded manually, shortly after each large rainfall event. Readings from each rain gauge were subsequently compared against the accumulated rainfall recorded by the weather station in order to determine what fraction of rainfall enters the tree's root-zone.

Next to each rain gauge, a set of 2-wire Time-Domain Reflectometry (TDR) probes (5 mm dia. stainless steel rods) was installed vertically into the soil to record changes in soil moisture at depths of 0.15, 0.30 and 0.45 m. The soil's volumetric water content, θ [m^3 m^{-3}], was measured manually, once every week, using a digital TDR (Tektronix Model 1502B, Oregon, USA) under computer control. Analysis of the waveform followed a procedure similar to that of Baker and Allmaras (1990), using the general equation of Topp et al. (1980) to calculate θ.

The measurements of light interception, tree water use, soil water content and throughfall of rain were used to parameterise a risk-assessment model of the daily water balance and the irrigation requirements of olives in Marlborough. A continuous record of 28 years of daily weather data (global short-wave radiation, air temperature, wind speed, rainfall) was also downloaded from the national climate

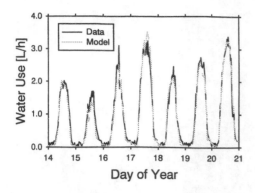

Figure 2. Measured water use of a 4 year old olive tree (data) during summer, compared to model calculations based on local weather. The measurements are the average sap flow recorded in two trees.

database (NIWA Ltd., New Zealand) for the purpose of making these calculations.

3. MODELLING

3.1. Tree-Scale Modelling

Total transpiration from the olive tree was modelled using a modified form of the Penman-Monteith equation. For this calculation, the total leaf area of the tree, A_T [m^2], was divided into a fraction of sunlit leaves (a_1) and a complementary fraction of shaded leaves (a_2). Uniform leaf properties were assumed for each class of leaves. Since olive leaves are hypostomatous we used the equation

$$\lambda E = \sum_i a_i \left[\frac{sR_{n,i}\,r_{b,i} + \rho\,c_p D_a}{(s + 2\gamma)\,r_{b,i} + \gamma\,r_{s,i}} \right] A_T \qquad (1)$$

following Jarvis and McNaughton (1986). In this equation we have assumed that each class of leaves has an associated leaf stomatal and boundary-layer resistance equal to $r_{s,i}$ and $r_{b,i}$ [s m^{-1}], respectively. We model these resistances using empirical relationships based on local microclimate and leaf dimension (Green, 1993). Here, E is the total transpiration [kg s^{-1}] from all the leaves, $R_{n,i}$ is the net radiation flux density [W m^{-2}] of the i-th set of leaves, D_a is the ambient vapour pressure deficit of the air [Pa], λ is the latent heat of vapourization of water [J kg^{-1}], γ is the psychrometric constant [Pa], ρ is the density of moist air [kg m^{-3}], and c_P is the specific heat capacity of air [J kg^{-1} K^{-1}].

Data from the 'Whirligig' are used to calculate both $R_{n,i}$ and a_i from a direct measure of the total amount of all-wave radiation absorbed by the tree (Green and McNaughton, 1997). Implicit in this calculation is a knowledge of the

tree's total leaf area. This area was estimated by counting all leaves and removing a fraction (1 in 50 or 2%) whose area was measured using a leaf area meter (Licor Model 3100, Nebraska, USA). The leaf area of the tree in Fig. 1 was calculated to be about 13.5 m^2. A simple allometric relationship between branch circumference and leaf area was then derived so that the total leaf area of the other trees could be estimated.

A time sequence of tree water use, as calculated by the big-leaf model of Eq. [1], is shown in Fig 2. The same graph also shows our heat-pulse measurements of sap flow in the trunk of the olive tree. There is a very good correspondence between our sap flow measurements and the tree-scale model results. This gives us added confidence in our calculation procedures.

Between 14th and 21st of January the days were mostly sunny with little cloud cover. Yet there were large differences in tree water use. Transpiration rates peaked at about 3.5 L h^{-1} on a bright, sunny day when the air temperature exceeded 32 $^{\circ}$C and the relative humidity fell below 20%. However, on other days we observed much lower rates of water use, because of changes to the local weather. For example, the 15th January was quite cool (15 $^{\circ}$C) and humid (RH > 60%), and transpiration rates were less than 2 L h^{-1}.

Tree water use is strongly influenced by changes in the temperature and humidity of the air. The data of Fig. 2 reflect a change in the predominant wind direction, from an easterly (a moist off-shore wind) to a north-westerly (a dry foehn wind) on 17th and 20th. The north-westerlies brought a much warmer, drier air mass into the valley and resulted in much higher transpiration rates on the 17th and the 20th, compared to the other days when the wind blew from the east (off shore) and the ambient air was cooler and more humid. Our tree-scale model is able to account for those weather-induced changes in tree water use.

When the sap flow is summed up over a week, we calculate the 2 yr-old trees used about 25 L of water, the 4 yr-old trees used about 175 L of water, and the 8 yr-old trees used about 525 L of water during the warmest week of the summer. We take these measured rates of water use to represent the maximum transpiration expected during one week, for olive trees of this size. These results are useful to develop a very simple 'rule of thumb' that can be used by growers to estimate the maximum water use of their olive trees each week in summer. Tree age is something that most growers can relate to.

3.2. Grove-Scale Modelling and Crop Factors

The more progressive farmers in the region, along with the resource managers at the District Council, are very keen to use advances in irrigation science to improve irrigation scheduling and to minimise any wastage of the regions precious water resources. Any savings in water also means a savings in the electricity costs associated with running the pumps for irrigation. A understanding of how crop water use responds to changes in the daily and seasonal weather is seen as a vital first step to improving irrigation management. The next step is to develop irrigation management that replaces just that amount of water used by the crop, and applies water only when the crop is in need of it. We will use grove-scale modelling to guide the water allocation process.

Weather watching has become a major interest for all farmers and growers in Marlborough because summer rainfall is unreliable, and because their crops are reliant on adequate soil moisture to achieve optimum productivity. The local newspaper regularly reports values for the potential evapotranspiration, ET_O [mm d^{-1}] as well as daily rainfall totals that farmers can use to guide their irrigation management. An appropriate 'crop factor' is needed to relate the actual water use of the olive trees to potential ET_O.

Our measurements of trunk sap-flow provide a direct means of determining the daily water-use of the trees (Fig. 1). Our weather data can also be used to calculate values of the potential ET_O. The procedure is based on guidelines given by the Food and Agriculture Administration (FAO) of the United Nations (Allen et al, 1999). From the modified Penman-Monteith equation, we obtain

$$ET_0 = \frac{\frac{s}{\lambda}(R_N - G) + \gamma \frac{900}{(T+273)} u_2 (e_s - e_a)}{s + \gamma (1 + 0.34 u_2)} \qquad (2)$$

where R_N [MJ m^{-2} d^{-1}] is the net radiation, G [MJ m^{-2} d^{-1}] is the ground heat flux, e_s [kPa] is the saturation vapour pressure at the mean air temperature T [$^{\circ}$C], e_a [kPa] is the mean actual vapour pressure of the air, u_2 [m s^{-1}] is the mean wind speed at 2 m height, and the remaining terms have been described above. We use daily weather records from the national climate database (NIWA Ltd., New Zealand) to calculate values of the ET_O. A continuous record spanning the last 28 years of climate data, between the years 1972 and 2000, was extracted from the database.

The link between a reference evaporation rate, ET_O, and the actual crop water use, ET_C, is made using the crop factor approach. A crop factor, K_C, is used to approximate the influence of canopy architecture and plant-physiological characteristics. For routine calculations of crop water use, the following equation is used:

$$ET_C = K_C . ET_O \qquad (3)$$

where K_C is a dimensionless number that varies between about 0.2 and 1.1 for most crops (Allen et al., 1998).

By integrating the sap flow measurements over a whole day (midnight to midnight) we are able to calculate the daily water use of the olive trees. For the 8 yr-old trees, this ranged between about 30-75 L per day during the middle of summer (Fig 3). A tree water use of 75 L d^{-1} corresponds to an effective transpiration rate of just 2.5 mm d^{-1} when the results are expressed on a 'per unit ground area' basis. Similar rates of daily water use have been reported for mature olive trees growing in the south of Spain (Fernandez and Moreno, 1999).

A very good correspondence was found between the daily water use and the potential ET_O (Fig. 3). On days when the reference ET_O equaled 5 mm d^{-1}, the corresponding transpiration rate from the trees, ET_C, was equal to 2.5 mm d^{-1}. It follows from Eq. [3] that the ratio ET_C/ET_O gives a direct measure of the crop factor, K_C. Here, we calculate $K_C = 0.50$ for the 8 yr-old olive trees in the Marlborough.

Trees at the experimental site were all planted at the same spacing, i.e. 6 m by 5 m, but their leaf canopies were at different stages of development. As expected, the appropriate crop-factor for olives depends on the size, or age, of the olive tree. In terms of water use, the 2 yr-old trees used just 25 L (= 1 mm) and the 4 yr-old trees used just 175 L (= 6 mm) of water per week, respectively, for the same evaporative demand. During the middle of summer the average weekly value of ET_O in Marlborough is about 35 mm per week. Thus, we calculate the crop factor for 2 and 4 yr-old olive trees to be 0.05 and 0.17, respectively.

It is interesting to note that our value of K_C is very close, numerically, to the corresponding value of % light interception as determined from the array of PAR sensors placed on the ground under the tree canopy. The fractional crop cover of the 8 yr-old trees was found to be 0.53, and the corresponding value of the crop factor was $K_C = 0.50$.

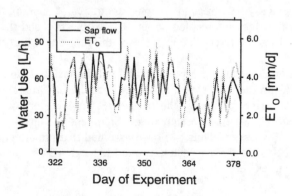

Figure 3. Measured sap flow of an 8 yr-old olive tree compared to the potential evapotranspiration, ET_O, calculated from local weather data. The measurements are the average sap flow in two trees. The trees were on a 6 m by 5 m spacing. The ratio of tree water use per unit ground area (L m^{-2} d^{-1}) to ET_O is a measure of the crop factor, K_C (=0.5).

Similar findings have been reported in the literature. For example, Fernandez and Moreno (1999) found an 'effective' crop factor of between 0.35 and 0.50 for a mature olive orchard with mature trees planted at a 7 x 5 m spacing, and covering 34% of the ground surface. Allen et al. (1998) suggest a mid-season value of $K_C = 0.65$ as being typical for olive trees that provide 60% ground coverage by the canopy. It appears that % light interception is a very good substitute for the crop factor of olive trees. Variations in K_C between orchards are expected because of varietal factors that effect tree shape and stomatal response, and also other orchard factors such as soil type, planting density and irrigation management.

Olives are often recognized as being drought tolerant, and they have an ability to reduce water loss when they experience a shortage of water in their root zones (Dichio et al., 1993). In our experiment, the root zone soil of the 8 yr-old trees did become quite dry over the summer. TDR measurements of soil moisture in the root-zone of the 8 yr-old olives indicated the tree roots had depleted more than 70% of the total available soil water from the top 0.5 m of the root zone. It is possible that the 8 yr-old trees could have been under a mild degree of water stress during our experiment. The trees were growing on very stony soils and receiving only a small amount of irrigation (8 L d^{-1}) relative to their actual water use (30-75 L d^{-1}).

The remaining factor needed to complete the parameterisation of our simple water balance model is an estimate of the fraction of rainfall that enters the root zone. This fraction was determined by simply taking the amount of rainfall collected in the array of rain gauges under the olive trees and comparing it to the amount of rain fall in the open area of the orchard. A very good linear relationship was found between throughfall and incident rain, when taken over periods of a week or more (data not shown). On average, the 2-yr old trees intercepted about 5% of the rain, the 4 yr-old trees intercepted about 10% of the rain and the 8 yr-old trees intercept about 15% of the rain that fell at the site.

Some of the intercepted rainfall will be lost as evaporation from the wet leaves, and the remainder will run down the tree stem and enter the root zone. For the purpose of modeling, we have assumed any intercepted rain is lost from the system. Thus, we have assumed that the effective rainfall is just 85% of actual rainfall for a grove of 8 yr-old olive trees.

4. TEMPORAL RISK ASSESSMENT

In this section, we describe our risk assessment model, SPASMO (Soil-Plant-Atmosphere-System-Model) that was used to determine the soil water balance and calculate the irrigation requirements of an olive grove in Marlborough.

The calculations run on a daily basis and assume model parameters derived from field experiments using the 8 yr-old olive trees.

The SPASMO model considers water movement through a 1-dimensional soil profile that extends from the soil surface to a depth of 6.0 m. The model calculates the water balance of a cropped soil by considering the inputs (rainfall and irrigation) and losses (plant uptake, evaporation, runoff and drainage) of water from the soil profile. The soil's physical and hydraulic properties are defined using data from the NZ Soils database (Landcare Research Institute, New Zealand) to describe the water retention properties, the stone fraction and the bulk density of the soil. We use a continuous record of average daily weather, from the national climate data base (NIWA Institute, New Zealand).

Water transport through the soil profile is modelled using a water capacity approach (Hutson and Wagenet, 1993) that considers the soil to have both mobile and immobile pathways for water transport. The mobile domain is used to represent the soil's macropores (e.g. old root channels, worm holes and cracks) and the immobile domain represents the soil matrix. After any rainfall or irrigation events, water is allowed to percolate through the soil profile whenever the soil is above field capacity. The infiltrating water first fills up first the immobile domain and, once this domain is full, it then refills the mobile domain as the water travels progressively downward through the soil profile. Subsequently, on days when there is no significant rainfall, there is a slow approach to equilibrium between the mobile and immobile phases, driven by a difference in water content between the two domains.

Crop water use is determined from meteorological data using the FAO Penman-Monteith method (Allen et al., 1998) to get a standard reference value for potential evapotranspiration, ET_O. We use a dual crop-factor approach to describe the combined water loss from the crop and the soil surface. The depth-wise pattern of the crop's fine roots and the local soil-water potential are used to determine the pattern of water uptake in the root zone (Green and Clothier, 1998). We have assumed the tree roots extend to a depth of 2.0 m and that root-length density decreases exponentially with increasing soil depth (Moreno et al., 1996).

Crop water use is assumed to proceed at the maximum rate when soil water is non-limiting. The crop can tolerate a certain 'water deficit' in the root zone, and will grow and transpire at the optimum rate while there is enough water in the root zone. However, once the root-zone water deficit increases above a given threshold, then the crop will begin to exhibit symptoms of water stress that impact on crop water use and productivity. Irrigation is applied automatically to the crops, on basis of need, whenever the water deficit in the root-zone declines below the threshold value. For olives, which can tolerate a dry root-zone, we have as-

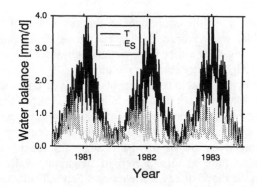

Figure 4. The water balance of an olive grove in Marlborough, as represented by transpiration (black) and soil evaporation (grey).

sumed irrigation is required whenever the depletion level exceeds 65% of the available soil water (Allen et al., 1998). The calculations have allowed for a daily irrigation rate of 2.8 mm d^{-1} when the soil is dry and the crops are in need of water (Fig. 5). An irrigation of 2.8 mm is the maximum daily value that is allowed for by the District Council.

Figure 4 shows the daily water loss from the olive grove due to crop transpiration and soil evaporation, as calculated by the SPASMO model. The corresponding temporal pattern of soil water content and the daily irrigation required to meet the tree's water demands is shown in Fig. 5. Large olive trees are expected to have a peak transpiration rate of 3.0-3.5 mm d^{-1} during the middle of summer, but they use less than 1.0 mm d^{-1} during the winter (Fig. 4). The evaporative loss of water from the bare-soil strip under the trees is predicted to reach a maximum of just 1.5 mm per day in the springtime, whenever the soil is quite wet. A reduced radiant-energy load (> 45% of incident) and a small fractional ground area (~33% of the land area) are the main reasons for the relatively low evaporation rates from the bare soil.

Over the summer time, evaporation from the bare soil declines further, below about 0.5 mm d^{-1}, as the surface soil becomes drier (Fig. 4). Soil evaporation does not decline all the way to zero because there is sufficient rainfall and frequent irrigation to maintain soil water content above the air-dry value. Water loss from the grassed inter row will peak at about 2.0 – 2.5 mm d^{-1} in the spring and thereafter decline over the summer as the readily available soil water is progressively depleted (calculations not shown).

Figure 4 suggests that an irrigation of 2.8 mm d^{-1} is reasonable for large olive trees, since it matches approximately the peak water use of trees planted at a spacing of 5 m by 6 m. However, a daily irrigation of 2.8 mm would exceed the actual tree water-use by a factor of about 3 for the 4 yr-old trees, and by a factor of about 20 for the 2 yr-old trees. As

expected, smaller trees require more frequent irrigation of a smaller amount.

Irrigation will not be required every day through the summer, and it is never required in the wintertime. This is because there is often enough rainfall to top-up the amount of readily-available soil water that is stored in the root-zone (Fig. 5). In the wet summers of 1980 and 1981, irrigation was not necessary until late December, and then it was only needed every couple of days through until the end of March. In contrast, in the drier summer of 1982 irrigation was required almost every day, from November through until the end of March (Fig. 5). The variability and uncertainty of rainfall has a large influence on the timing and the total amount of irrigation required.

With regard to the depth-wise pattern of water uptake, we know from our measurements that the surface roots are the most active. In our model we make sure that these surface roots are the first to deplete the soil water when it becomes available. Meanwhile, as soon as the surface soil dries, we know from our measurements that the trees will switch their uptake activity to become more reliant on water that is deeper in their root zone (Clothier and Green, 1996). Once most of the readily-available soil-water gets depleted from the root zone, then additional irrigation is needed to maintain the tree water status.

If regular irrigation is supplied to the soil surface at a rate that approximately matches tree water use, then most of the tree's water uptake will be via the surface roots and so water extraction by the deeper roots will cease. Between January and March, we predict that the soil-water content will hardly change near the bottom of the root zone (Fig. 5). This is because of a declining water uptake by the deeper roots, combined with a reduced drainage over the summer time. Thereafter, the root zone water content will be recharged slowly by winter rainfall.

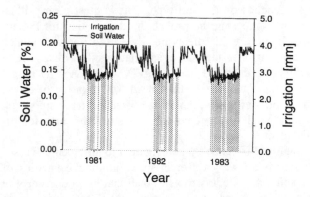

Figure 5. Predicted changes in the soil's volumetric water content and automatic irrigation of 2.8 mm whenever the threshold soil water content is reached.

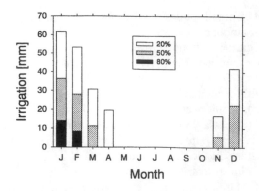

Figure 6. Risk assessment for the irrigation requirements of olives on a Renwick stony silt loam in Marlborough.

As expected, the irrigation requirements for olive are greatest in the middle of summer when the root-zone soil water is at its highest level of depletion. We have carried out a risk assessment for the monthly irrigation requirements of olives and the results are shown in Fig 6. For olives on a Renwick silt loam soil, we calculate an average of 36 mm of irrigation is required in the month of January. Half of the time the olives will need more irrigation, and half the time they will need less.

During January, there is a 20% probability that more than 62 mm of irrigation will be required. Yet, there is also a 20% probability that they will need less that 14 mm of irrigation. Rainfall variability is the main reason for such a wide range in irrigation needs. According to our calculations, an allocation of 62 mm of water in January should be sufficient to meet the water demands of olives in four out of five years.

5. SPATIAL RISK ASSESSMENT

Crop irrigation depends not only on the time of year, but it also depends on the soil type. This is most clearly illustrated by comparing the probability distribution for the annual irrigation requirements of olives on a Renwick stony silt loam against the same calculations for olives on a Woodbourne deep silt loam (Fig 7). The SPASMO model predicts there to be very little need for irrigation of olives on a Woodbourne soil, other than when the trees are young and they need frequent watering of small amounts for establishment.

The big difference in the irrigation requirements is associated with the very different water holding capacities of these soils. A Woodbourne silt loam that has a deep profile and very few stones, can store more that 170 mm of water in the top 1.0 m of the soil profile (Table 1). Contrast this with a Renwick silt loam that has 25 – 50% stones and a much lower storage capacity of just 72 mm in the top 1.0 m. During the dry summer months, olives on the Woodbourne

soil would have the equivalent of about 4 to 6 weeks more 'available' water stored in their root-zone. This additional store of root-zone water is almost enough water to carry the trees through the dry summer months.

The annual water requirements of olives on other soil types in Marlborough is shown in Figure 7. Half of the time we estimate that mature olive trees will not need any irrigation if they grown on the Woodbourne deep silt loam. The same trees grown on the Wairau silt loam will, on average, require about 75 mm of irrigation, while those trees on the more stoney Renwick and Fairhall soils will, on average, need more than 130 mm of irrigation.

For the purpose of irrigation allocation, we suggest a less conservative figure that would satisfy the crop water demands in four out of five years. Olives should be allocated the following amount of irrigation water: 193 mm per year on a Fairhall stoney silt loam, 185 mm per year on a Renwick silt loam, 154 mm per year on a Wairau silt loam, and 78 mm per year on a Woodbourne deep silt loam. It is clear from Fig. 7 that soil type plays a key role in determining what is a reasonable water allocation for olives.

The annual irrigation requirements of olives (Fig. 7) appear to be much less than would be calculated from adding up the corresponding monthly totals at the same probability of exceedence (Fig. 6). This is because the driest month does not always coincide with the driest year.

The District Council have decided to make water allocations on an annual basis, and they are using Fig. 7 to set their guideline values. These are based on local soil data and take into account 28 years of local weather data that includes at least two periods of extended drought in Marlborough region. If the District Council were to allocate irrigation water on the basis of a maximum allowable water take each month, then they could use results similar to those of Fig. 6 for guidance. That information would also be useful in determining design factors for any storage dam or water augmentation schemes that might, in the future, be proposed for this water short region.

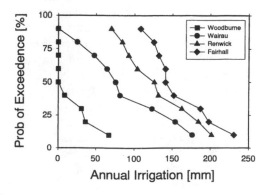

Figure 7. A risk assessment of the annual irrigation requirements of olives for a range of soils in Marlborough.

Apposite measurements and apt modeling are shown capable of guiding regulatory authorities in managing the complexity of allocating water to olive irrigationists. It is only through the endeavors, insights and perspicacity of scientists such as John Robert Philip that we have the tools to provide regulators and growers with the means to allocate and manage irrigation sustainably.

Acknowledgments. This research was conducted via co-investment under FRST Contract C06X0004 "Knowledge Tools for Environmental Action", and through Marlborough District Council Contract 9307 "Determination of the Irrigation Requirements for Olives and Grapes Growing in Marlborough". The enthusiastic support and advice of Mike Ponder and Tony Harvey of Ponder Estate is greatly appreciated.

REFERENCES

Allen, R.G., Pereira, L.S., Raes, D., and Smith, M., Crop Evapotranspiration: Guidelines for computing crop water requirements. FAO Irrigation and Drainage Paper No: 56, FAO, Rome, 301 pp., 1998.

Baker, J.M., and Allmaras, R.R., System for automating and multiplexing soil moisture measurement by time-domain reflectometry, *Soil Science Society of America Journal*, 54:1-6, 1990.

Biggs., W.W., Edison, A.R., Eastin, J.W., Brown, J.W., Maranville, J.W., and Clegg, M.D., Photosynthesis light sensor and meter. *Ecology*, 52: 126-131, 1971.

Clothier, B.E., and Green, S.R., Roots: the big movers of water and chemical in soil. *Soil Science*, 162: 534-543, 1997.

Dichio, B., Xiloyannis, C., Celano, G., and Angelopoulis, K., Response of olive trees subjected to various levels of water stress, *Acta Horticultrae*, 356:211-214, 1993.

Fernandez, J.E., and Moreno, F., Water use by the olive tree. In: *Water use in crop production*, editied by M.B. Kirkham pp. 101-162, Haworth Press Inc., New York, 1999.

Gleick, P.H., Making every drop count, *Scientific American,* 284, 2: 29-33, 2001.

Green, S.R., Radiation balance, transpiration and photosynthesis of an isolated tree, *Agricultural and Forest Meteorology*, 64: 201-221, 1993.

Green, S.R., Measurement of sap flow by the heat-pulse method. An instruction manual for the HPV system. HortResearch Internal Report No. IR98/44, 49pp, 1998.

Green, S.R., and B.E. Clothier, Water use by kiwifruit vines and apple trees by the heat-pulse technique, *Journal Experimental Botany* 39:115-123, 1988

Green, S.R. and K.J. McNaughton, Modelling effective stomatal resistance for calculating transpiration from an apple tree, *Agricultural and Forest Meteorology*, 39:115-123, 1997.

Hutson, R.J., and Wagenet, J.L., A pragmatic approach for modeling pesticides. *Journal of Environmental Quality*, 22: 494-499, 1993

Jarvis, P.W.G., and McNaughton, K.G., Stomatal control of transpiration: scaing up from leaf to region. *Advances in Ecological Research*, 15: 1-49., 1986.

McNaughton, K.G., S.R. Green, T.A. Black, B.R. Tynan, W.R.N. Edwards, Direct measurement of net radiation and photosynthetically active radiation absorbed by a single tree, *Agricultural and Forest Meteorology*, 62:87-107, 1992.

Moreno, F., Fernandez, J.E., Clothier, B.E., and Green, S.R., Transpiration and root water uptake by olive trees. *Plant and Soil*, 184: 85-96, 1996.

Nuttle, W.K., Ecosystem managers can learn from past successes, in *EOS*, 81, 25: 278 and 284, 2000.

Philip, J.R., The physical principles of soil water movement during the irrigation cycle, in *Proceedings of the Third International Congress of Irrigation and Drainage*, San Francisco, pp 8.125-8.154, 1957.

Philip, J.R., Water on Earth, in *WATER Planets, Plants and People,* edited by A.K. McIntyre, pp. 35-59, Australian Academy of Science, Canberra, Australia, 1977.

Philip, J.R., Soils, natural science and models, *Soil Science* 151:91-98, 1991

Philip, J.R. (*pers comm*) Text of "Speech at a Dinner Marking John Philip's Formal Retirement from CSIRO, The Lobby Restaurant, Canberra, 26th February, 1992".

Postel, S., Growing more food with less water, *Scientific American*, 284, 2: 34-37, 2001.

Topp, G.C., Davis, J.L., and Annan, A.P., Electromagnetic determination of soil water content: measurements in coaxial transmission lines, *Water Resources Research*, 16:574-582, 1980.

Horst Caspari, State Viticulturalist, Colorado State University, 3168 B½ Road, Grand Junction, CO 81503, U.S.A. (hcaspari@coop.ext.colostate.edu)

Brent Clothier, HortResearch, PB 11-030, Palmerston North, New Zealand 5301 (bclothier@hortresearch.co.nz)

Steve Green, HortResearch, PB 11-030, Palmerston North, New Zealand 5301 (sgreen@hortresearch.co.nz)

Sue Neal, HortResearch, HortResearch, PB 1007, Blenheim, New Zealand (sneal@hortresearch.co.nz)